Society & Technological Change

Ninth Edition

Rudi Volti
Professor Emeritus, Pitzer College

Jennifer L. Croissant
University of Arizona

Long Grove, Illinois

For information about this book, contact:
Waveland Press, Inc.
4180 IL Route 83, Suite 101
Long Grove, IL 60047-9580
(847) 634-0081
info@waveland.com
www.waveland.com

10-digit ISBN 1-4786-5104-0
13-digit ISBN 978-1-4786-5104-8

Printed in the United States of America

7 6 5 4 3 2

Content

CHAPTER 12

CHAPTER 13

CHAPTER 14

About the Authors

Rudi Volti is Emeritus Professor of Sociology at Pitzer College, where he was a founding member of the program in Science, Technology and Society of the Claremont Colleges. His books and articles have covered a variety of topics on the interaction of technology and society, including technology transfer to East Asia, the history of the engineering profession, the origin of frozen foods, commercial air travel, and the history of automobile engines. His personal encounters with modern technology centered on cars, motorcycles, and model railroading.

Jennifer Croissant is an Associate Professor in the Department of Gender & Women's Studies at the University of Arizona, in Tucson, AZ. With degrees in Engineering Science and Mechanics (BS, Penn State 1987), Technology & Policy (SM, MIT 1989), and Science and Technology Studies (PhD, RPI, 1994), she's been interested in the ways that science and technology shape and are shaped by social organizations and practices. Recent projects include studies of interdisciplinarity and collaboration, agnotology (studies of ignorance) and absence, and case studies of abandoned technologies. Jen spends much too much time in front of screens, and not nearly enough time outdoors with her hobby of coaching track and field (athletics), especially hurdles, javelin, and pole vault—trying to explain the complicated technical stuff.

Preface

When the first edition of *Society and Technological Change* came out in 1988, Microsoft's initial public offering had occurred only two years earlier, tweets were something birds did, and Mark Zuckerberg had not yet entered kindergarten. Since that time, ongoing technological changes and new ways of interpreting the interaction of technology and society have provided new opportunities to revise and expand succeeding editions. Even so, the animating spirit of the book remains the same. This ninth edition continues to explore the many ways in which various technologies have influenced our lives. At the same time, it shows how these technologies have themselves been shaped by social, economic, cultural, and political forces, and that the study of technology is important not just for its own sake but also for what it tells us about the kinds of societies we make for ourselves.

This book is oriented to the growing number of courses on technology and society. It will also be useful in other courses that explore technology's role in human affairs. It presents perspectives, theories, and facts that should help the reader to understand the consequences of technological changes, as well as the forces that have produced these changes. Many specific examples of the interaction between technological change and other changes will be introduced, for general processes are often best understood through references to particular instances.

The rapid pace of technological change during the opening years of the twenty-first century may have led to an overuse of the word "revolutionary," but it also provides the basis for significant new discussions of the reciprocal interactions of technology and society. In particular, this book now devotes a considerable amount of space to issues that scarcely existed when the book was first published: the internet, social media, genetic engineering, and challenges to intellectual property and personal privacy. In similar fashion, the book covers instruments of warfare that have emerged in recent years such as cruise missiles, smart bombs, drones, and cyber attacks.

One of modern technology's strongest influences has been on the development of the cluster of political, cultural, social, and economic changes that are subsumed in the term "globalization." New material in this edition covers offshoring and technology transfer, appropriate technologies in poor countries, new media and social movements in authoritarian societies, and the extent to which the world's cultures are converging toward a common pattern.

Economic globalization has been blamed for growing inequalities in the distribution of wealth and income. No less culpable has been technological change. This book looks into the connection between technological advance and rising social and economic inequality. Considered here are technological unemployment, the changing rewards for particular skills, and the ongoing evolution of twenty-first century economies.

Some of the most important issues involving technology and society center on health, both the health of humans and the health of Earth. In regard to the latter, the broad issue of sustainability is addressed by expanded coverage of climate change and the use of sources of energy other than fossil fuels. As far as human health is concerned, advances in genetics research are giving rise to new healing technologies. At the same time, however, technologies that alter an organisms genetic makeup also pose many practical and ethical problems discussed in a chapter devoted to these matters.

The preparation of this edition has also provided an opportunity to update and extend many pertinent facts and statistics. These include new data on climate change, the costs of medical care, unemployment, the distribution of income, sales of recorded music, fracking, alternative energy sources, the use of various media (including email, mobile phones, and social media), future employment prospects, and government support of research and development.

Although this edition has quite a lot of new material, no pretense is made that it presents an all-encompassing view of technology and society. We have added sections on technological diffusion, commodity chains, and the social construction of technology as frameworks for organizing analysis of technology. We've also added chapters on expertise and professions, accidents and disasters, and additional materials on information and computing technologies, as they are rapidly changing. Much has been left out because of space limitations, and our own limitations of time, energy, and expertise. At the same time, the systematic study of the interactions between technology and society is a relatively recent endeavor, and many gaps remain to be filled. It can only be hoped that this book will provide a foundation for thought and future study. If annoyance at the inadequacy of coverage leads the reader to undertake more extensive explorations of some of the topics presented, then this book will have served its purpose.

Acknowledgements

Both Rudi and Jen would like to thank colleagues who have adopted the prior editions of this textbook and provided feedback that has shaped this nineth edition. We are also very grateful for the editorial staff at Waveland Press, particularly Diane Evans and Dakota West, who shepherded us through the complex process of changing publishers and then through the process of getting the manuscript into production.

Rudi wishes to acknowledge Pitzer College colleagues and scholars whom he has met through the Society for the History of Technology (SHOT); across decades, they have become friends who have helped to shape his interests and thinking. Very special gratitude goes to Jennifer Croissant who, with grace, expertise, energy, and the perfect touch, joined Rudi as co-author when his health declined. Without her this edition would not have come to fruition. Finally, Rudi recognizes and thanks his wife, Ann Helton Stromberg, who played a key role in this book's preparation with her careful reading of every chapter and unfailing encouragement.

Jen would like to thank Jacob Licker, a graduate assistant who stepped in at the eleventh hour to bring the book into shape with final manuscript preparations. She would also like to recognize November Papaleo Prentiss, a former teaching assistant who made a significant time commitment for making suggestions improving the book. And of course, Jen would like to recognize her spouse, R. D. Willits, Jr., for his forbearance in having to listen to the challenges of revising this textbook.

The authors would finally like to thank the generations of instructors and students who have engaged this textbook in its many iterations, rising to the challenge of thinking about society and technological change, and challenging us to be more clear and attentive to new technologies and social issues.

Be curious.

PART 1

Defining Technology

The ability to create and use a great variety of technologies is one of the distinguishing characteristics of humans, but what exactly is meant by "technology"? The term is a familiar one, but like many words in current circulation it carries with it a multitude of meanings and assumptions. For a start, it encompasses much more than the array of digital devices that have been integral parts of our lives for only a few decades. Our folk definitions of technology tend toward "presentism,"[1] or using contemporary technology as the benchmark for defining technologies of the past. At one point in time (the late nineteenth century) the ordinary typewriter was a high-tech marvel, as was the basic stethoscope (the 1860s), or the development of early irrigation systems nearly 8,000 years ago was a civilization-transforming set of techniques allowing for the expansion of agriculture and human social complexity.

Technology also goes well beyond the "wave of gadgets," famously recounted by a student to historian T. S. Ashton, as a major impetus for the industrial revolution.[2] So then what is it? Chapter 1 offers a definition of technology that is meant to be precise yet elastic enough to cover the many connotations of the word. Although technology is often associated with particular items of hardware, the ultimate basis of technology is knowledge, and Chapter 1 delineates the ways of thinking that are associated with technological advance. Chapter 1 also includes an effort to disentangle technological advance from an even more slippery concept: "progress."

In Chapter 2 we continue the discussion by noting that many technological changes do not necessarily make things better for everyone, as is implied by the word "progress." To the contrary, technologies may affect individuals and groups in different ways, leaving some better off, while others are left in a worse position. This aspect of technological change is often ignored or dismissed as a "side effect," making it hard to resist the temptation to seek "technological fixes" for problems that require more than just the introduction of new devices or processes. In Chapter 2 we consider the kinds of situations where technological fixes are likely to be successful and others where they are doomed to failure.

Notes

1. Alexandra Walsham, *Introduction: Past and … Presentism, Past & Present*, Volume 234, Issue 1, February 2017, Pages 213–217, https://doi-org.ezproxy3.library.arizona.edu/10.1093/pastj/gtw054
2. Asthon, T. S. (1998). *The Industrial Revolution: 1760–1830.* Cambridge, UK: Oxford University Press.

Chapter 1

The Nature of Technology: "More than Google"

Today's technology leaves us both exhilarated and terrified. Recent technological developments have presented us with such marvels as spacecraft leaving the solar system, instant access to quintillions of bytes of data[1] of information on every conceivable topic via the internet, and diseases cured through gene therapy. At the same time, however, the seemingly inexorable march of technology has produced global climate change, overpopulation, and the threat of nuclear annihilation. On many occasions, technological change has also produced social disruptions, as when automation destroys jobs in a particular industry, a new weapon upsets the balance of power between nations, or hacking produces data breaches that disrupt security and lead to financial losses. And when technologies fail, some of them do so in a big way, as exemplified by the loss of the *Challenger* and *Columbia* space shuttles, the massive Deepwater Horizon oil spill in the Gulf of Mexico, the catastrophic failure of the Fukushima nuclear plant in Japan, and the disastrous breaching of the levees in New Orleans in the wake of Hurricane Katrina.

Despite all the crises, disruptions, and disasters that have accompanied it, modern technology is still generally viewed in a favorable light according to public opinion surveys. Although significant minorities of respondents express their disapproval of specific technologies like nuclear power and genetically modified foods, the positive achievements of technology as a whole are seen to substantially outweigh the negative ones.[2] At the same time, however, certain firms employing advanced technologies do not enjoy the same degree of support, and specific technologies come under scrutiny. For example, according to one recent survey, 55 percent of respondents agreed that firms in the vague category of "technology companies" have too much power and influence.[3] Social media platforms receive criticism over their role in the spread of disinformation,[4] and social media has the potential to be hazardous to the mental health of adolescents.[5] So the general category of "technology" is held in high esteem while specific technologies sometimes generate significant criticism.

Affirmation of technology in general is based more on faith than on understanding. When confronting technology, most of us are poorly informed spectators, seemingly incapable of understanding esoteric realms of lasers, microprocessors, gene splicing, nanomaterials, and rapid technological change across diverse sectors. This inability to understand technology and perceive its effects on our society and on ourselves is one of the greatest, if most subtle, problems of an age that has been so heavily influenced by technological change. Yet ignorance need not be a permanent condition. One can only hope to comprehend the inner workings of just a small number of the most significant technologies, but it is still possible to come to a better understanding of the major causes and consequences of technological change. All technologies—be they high-definition televisions, reinforced concrete bridges, or farming practices—have some basic features in common. It will be the task of this chapter to show what they are.

Defining Technology

Gaining an understanding of the meaning of words is often the beginning of knowledge. We need a precise definition of what is meant when we use the term technology. The linguistic roots of the word "technology" can be traced to the Indo-European stem *tekhn-*, which seems to have referred to woodworking. It is the source of the Greek word *tekne*, which can be variously translated as "art," "craft," or "skill." It is also the root of the Latin word *texere*, "to weave," which eventually took on the larger meaning of fabrication or construction. By the early eighteenth century, the word had come close to its present meaning when an English dictionary defined it as "a Description of Arts, especially the Mechanical." In 1831, Jacob Bigelow published *Elements of Technology*, the first book in English with the word "technology" in its title. As he defined it, technology consisted of "the principles, processes, and nomenclatures of the more conspicuous arts, particularly those which involve applications of science."[6]

Tools and Techniques

Technologies are developed and applied so that we can do things not otherwise possible, or so that we can do them cheaper, faster, and more easily. The capacity of human beings to employ technologies sets us apart from other creatures. To be sure, beavers build dams, otters crack open shellfish with rocks, and chimpanzees use sticks to extract termites from their nests. But no other animal comes close to humans in the ability to create tools and techniques—the first two elements in our definition of technology—and no other creature is so dependent on them. The development of technology is in large measure responsible for the survival and expansion of a species that lacks many of the innate abilities of other animals. Left with only our basic physical capabilities, we humans cannot match the speed of a cheetah, the strength of an elephant (or even a chimpanzee), or the leaping ability of a kangaroo. We do not possess the eyesight of an eagle or the defensive armament of a porcupine, and we are among the 25 percent of all species that are incapable of flying. All in all, humankind is a physically unimposing bunch. But compensating for this physical weakness are the intelligence and sociality that are the ultimate source of technology. Humans stand apart from all other animals in their ability to gain and transmit knowledge, and to use this knowledge to develop tools and techniques. Without this capacity to invent and use a great variety of technologies, the human species would have never been able to establish itself on virtually every part of the globe.

Reliance on technology is as old as humanity itself. Whatever evils have accompanied the use of a particular technologies, it is pointless to indict technology as being somehow "unnatural." Our past as well as our future as a species is inextricably linked to our capacity to shape our existence through the invention and application of implements and techniques that allow us to transcend our meager physical endowments. It is certainly true, as Jacob Bronowski observed, that "to quarrel with technology is to quarrel with the nature of man—just as if we were to quarrel with his upright gait, his symbolic imagination, his faculty for speech, or his unusual sexual posture and appetite."[7]

Organizing Humanity

Tools and techniques have been of unquestioned importance in allowing the physical survival of the human species. Still, they are not the whole story. It is necessary to add some elements to our definition of technology that go beyond the usual identification of technology with pieces of hardware and ways of manipulating them. The first of these is *organization.* This follows from the fact that the development, production, and employment of technologies require a group effort. Even a relatively simple technology, such as one centering on the use of earthenware pots, requires a complex network of material suppliers, potters, toolmakers, marketing agents, and consumers capable of making good use of the pots. Of course, one person can learn all these skills adequately if not expertly, but the day is not long enough for him or her to do them all on a scale that produces a reasonable degree of efficiency. In the case of a complex technology like a computerized manufacturing system, there is no possibility of a single individual developing even a tiny fraction of the requisite skills.

For a technology to be developed and used, the energies and skills of many individuals must be combined and coordinated through some organizational structure. Organization may be likened to the software that controls and guides a computer; without an operating system and application programs, a computer is a useless arrangement of capacitors, transistors, resistors, and other bits of hardware. In similar fashion, an organizational structure allows the integration of diffuse human and material inputs for the attainment of a task. From this standpoint, there is considerable merit in historian of technology Lewis Mumford's assertion that the first "machine" was not a physical object, but the organizational structures that the Egyptian pharaohs employed to build the pyramids.[8]

When technology is seen as a combination of devices, skills, and organizational structures, it becomes natural to think of it as a *system*, the next element in our definition. For an individual technology to operate effectively, more is required than the invention of a particular piece of hardware; it must be supported by other elements that are systematically interconnected. When Thomas Edison began to work on electrical illumination, he realized that this technology would require the development of such a system. The invention of a practical, long-lasting lightbulb rested on the development of a serviceable filament and the use of an improved vacuum pump that evacuated the interior of the bulb, thereby preventing the combustion of the filament. But by itself, a lightbulb was useless. An effective electrical generator was needed to supply the current that produced the incandescence of the filament. A network of electrical lines had to be strung up between the generator and individual homes, shops, and factories. And of course metering devices were necessary so that users could be accurately billed for the electricity they used. Edison and his associates worked out these problems, and in so doing brought large-scale electrical illumination to the world.[9]

The development of all the elements of a technological system can be an uneven process, for technological advance often entails the resolution of tensions that are generated when one part of the technological system changes. This process is exemplified by the development of the modern airplane. Early biplanes with their drag-inducing wires and struts could not make effective use of more powerful engines. The availability of these engines became a strong inducement to the design of aerodynamically cleaner aircraft. The faster aircraft that resulted from the marriage of streamlined airframes and powerful engines produced a new problem: dangerously high landing speeds. This, in turn, stimulated the invention of wing flaps and slats. By the 1940s, aerodynamic theory made it apparent that improved airframes could achieve still higher speeds if provided with even more powerful engines; this possibility gave a strong stimulus to the development of the turbojet.[10]

For an example of the interplay of devices, skills, and organizational patterns, we can take note of Lewis Mumford's analysis of the technology of handwriting.[11] Two hundred years ago, the standard writing instrument was a goose-quill pen. Based on an organic product and sharpened by the user, it represented the handicraft technologies typical of its time. Cheap and crude, it called for a fair degree of skill if it was to be used effectively. In contrast, the steel-nib pen of the nineteenth century was a typical artifact of the industrial age, the product of a complex manufacturing process. Less adaptable than the quill, it was mass-produced in many different forms to meet specialized needs. Although Mumford's ideas were formulated before the invention of the ballpoint pen in the 1940s, his analysis fits this implement perfectly. Made from a variety of artificial materials and manufactured to close tolerances, the ballpoint pen could only be produced through sophisticated industrial processes. It is completely divorced from the organic world and requires very little skill from its user. Indeed, the technological artistry embodied in the pen itself stands in sharp contrast to the poor quality of the writing that so often comes from the hand that wields it.

A technological system does not emerge all at once with every one of its components neatly fitting together. In addition to changes in tools, techniques, and organizational structures, many social, psychological, economic, and political adjustments may be required for the support of a technological system. Technological change is not always a smooth process, and many of the necessary changes may entail considerable pain and disruption. Seeing technology as a system should help us to understand that technological change is closely connected with a variety of associated changes, and that the creation of a technological system may be fraught with tension and discomfort. Technological systems can also be fragile—disruptions in supply chains during COVID-19 simultaneously challenged organizational and technological systems.[12]

Much of what has just been said can be incorporated into a schematic definition of technology: **a system created by humans that uses knowledge and organization to produce objects and techniques for the attainment of specific goals**. Useful as it may be, this definition of technology is incomplete and possibly misleading in one important respect. The last part of the definition implies that technological change comes about as a response to existing needs: its purpose is "the attainment of specific goals." In the first place, one could legitimately ask *whose* goals are to be attained? This is an important issue, but it is best left for the next chapter. For now, we should note that although it is a human creation, technology does not always respond to existing needs; a new technology may in fact create its own needs. The development of technology on occasion exemplifies a phenomenon that has been dubbed "the law of the hammer": give a child a hammer, and to the child everything starts looking like a nail.

The history of technology is replete with examples of inventions looking for problems to solve. One example is found in almost every medicine chest: a bottle of aspirin. One of the most common uses of aspirin is to suppress fevers that accompany illnesses. But medical research (as well as some ancient practices) demonstrates that running a moderate fever is a therapeutic process that aids in a patient's recovery; it is the body's way of naturally combating infection. Yet since the introduction of aspirin in the early 1900s, fever has been seen as a problem requiring intervention. As one medical researcher has noted, "It's no surprise that society's deep worries about fever closely followed the synthesis of aspirin, the first drug that could safely reduce it."[13] Similarly, the drug sildenafil (generic for Viagra), started out as an experimental drug to reduce chest pain, but its primary side-effect was experienced very positively even though it did little in terms of pain relief. The drug goal was redefined and its availability facilitated the redefinition of erectile dysfunction from an unavoidable byproduct of aging into a treatable disorder.[14] In short, a new technology created its own need. Today, we are at a point where recent innovations like email and smartphones are now considered so essential that even a temporary loss of them produces symptoms akin to drug withdrawal.

It is also important to note that the goals achieved with a technology do not have to be "practical" ones. Some technologies have been developed so that we can grow more food or construct more comfortable buildings, but others have been developed simply for the challenge and enjoyment of solving technological problems, a proclivity that historian Robert Post has described as "technological enthusiasm."[15] Investors in epochal inventions ranging from nineteenth-century railroads to products from the latest high-tech startup have been motivated by the opportunity for financial gain, but the excitement occasioned by apparent technological wizardry should not be discounted.

Flight seems to be a particularly attractive object of technological enthusiasm. Immensely expensive technological endeavors such as the supersonic Concorde airliner and crewed space exploration programs are hard to justify on practical grounds, although their supporters have made valiant efforts to do so. Their primary purpose seems to be the elevation of national prestige by demonstrating a nation's collective ability to solve daunting technological problems. At the same time, many other technologies have a dual nature; they serve a practical purpose, but they are not valued only for this reason. An outstanding example is the automobile. It would be hard to justify the enormous resources employed for the building and operation of cars if transportation were the only goal. For many people (especially one of the authors), cars are objects of inherent fascination. Technological features like variable valve timing and active suspension systems have little to do with utilitarian transportation. The appeal is at least as much in the sophisticated technologies themselves as in the purposes that they serve.

Technological Advance and the Image of Progress

The development of technology is a dynamic and cumulative process. It is dynamic because a technology is never perfect; there is always room for improvement. As Henry Ford said of his firm, "If we have a tradition, it is this: Everything can always be done faster and better."[16] Technology is cumulative, for one advance paves the way for another. The lessons learned in working with an existing technology very often provide materials, tools, and, most importantly, a knowledge base for the next stage of development.

The dynamic and cumulative nature of technological change sets it apart from many other human endeavors. Ignoring for the moment the social consequences of technology, the process of technological change is usually one of continuous improvement in the internal workings of a particular technology: as they evolve, engines develop more power and are more efficient, integrated electronic circuits pack more components on a single chip, aircraft fly higher and faster.

The process of technological advance can be portrayed with the following diagram, in which the horizontal axis represents time, and the vertical axis represents just about any aspect of technological advance: the speed of commercial airliners, the production of synthetic materials, the scope of solar power implementation, or computer processing speed. Although there are inevitable fits and starts over time, the general trend can be depicted as a frontier moving forward.

Multiple factors need to align to keep a technological system moving forward. Figure 1.1a is an optimistic representation of technological change, while Figure 1.1b represents a "reverse salient"—a component or subsystem or element of the technological system which holds back further development. Figure 1.1c illustrates the specific case of electrification in the U.S. in the late nineteenth century. The problem of long-distance transmission held back advancement of electrification.

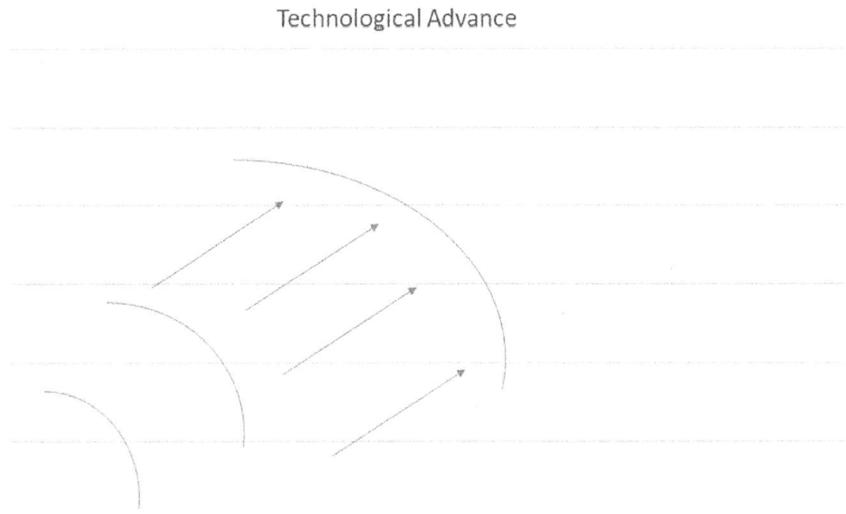

Figure 1.1a Technological advance. If technological advance were smooth and unimpeded, it would advance along a curve like this as multiple systems and parameters progress at similar rates, moving a technology.

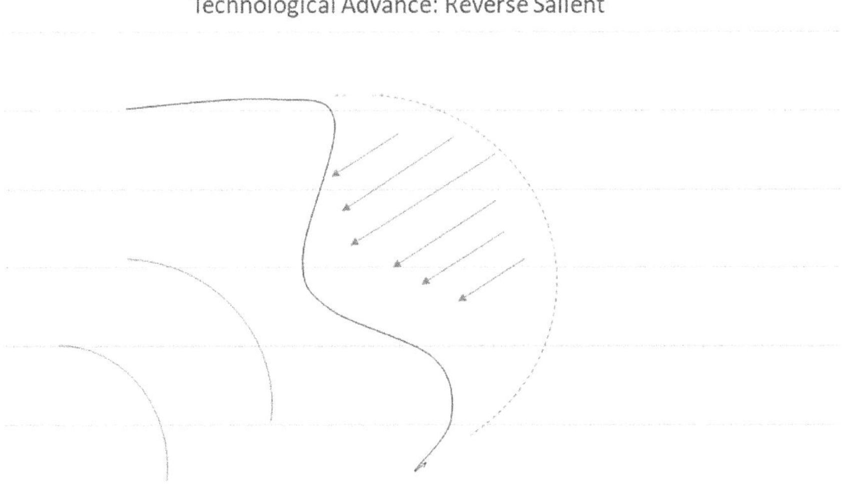

Figure 1.1b Technological advance with reverse salient. Technological advance can be impeded as components, infrastructure, culture, or other parameters of a technological system do not advance at the same rate. No one thing may be intentionally impeding progress, but subsystems and components or new problems that emerge can prevent a new technology from moving forward.

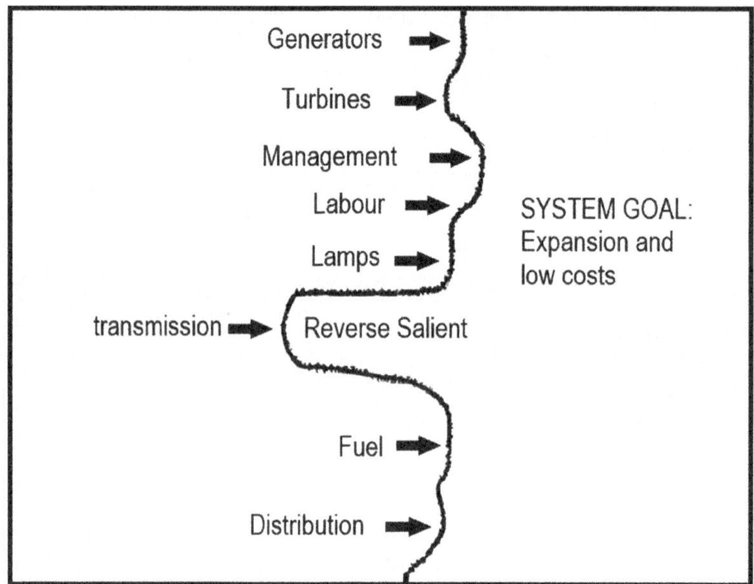

Figure 1.1c Technological advance vs. time in the nineteenth-century electrical system. Advancing electrification in the U.S. was held back by challenges with transmission. (Source: T. P. Hughes "The dynamics of technological change: salients, critical problems, and industrial revolutions," G. Desi (Ed.), *Technology and enterprise in a historical perspective*, Clarendon Press, Oxford (1992), p. 99)

Not all human endeavors can be fitted to this sort of curve. While technology tends to be dynamic and cumulative, the same cannot always be said of other manifestations of human creativity. Although there is ample room for debate, a good case can be made that succeeding generations of writers, composers, and painters have not produced works superior to the ones created by Shakespeare, Beethoven, and Vermeer. And while we continue to take great pleasure in the artistic creations of eras long past, few of us would be satisfied with the technologies that were prevalent in those times. We also see few indications that people are more humane than they were centuries ago. The present era certainly provides a multitude of horrifying examples of human cruelty, many of them augmented by enlisting technology in the service of destruction.

Still, when judged solely according to internal criteria, technology is one of the best examples of humankind's largely unrealized dream of continual progress. Technological progress, however, is not the same thing as progress in general. The fact that a society develops and uses advanced technologies does not guarantee that it will be equally advanced in other areas, such as equity, democracy, ethics, or moral reasoning.[17] Nazi Germany produced many technological triumphs, such as the all-conquering Mercedes and Auto Union grand prix racing cars of the late 1930s and the V-2 rocket used during World War II, but in its ideology and treatment of people it can only be described as barbaric. Conversely, many people from nonindustrialized societies have exhibited a high level of sophistication in their artistic creations, religious beliefs, and social relationships. The term "progress" can be used with some precision when applied to the development of technology per se, although even here problems can crop up because different standards of evaluation may lead to conflicting conclusions. Is it really "progress" when a new medical technology maintains an individual's life, but does so only at enormous expense while preserving nothing but the maintenance of organic functions? Does maintaining a "life" of this sort justify expenditures that otherwise might be used for expanded prenatal care or other preventative measures? Given the value judgments, ambiguities, and complexities surrounding the word "progress," its use is avoided here unless its meaning is clearly defined.

Technology as a Metaphor

Despite these qualifications, it is evident that beginning in the late eighteenth century and continuing today, technology's stunning advances have fueled a belief in generalized human progress. In this way, technology has operated as a metaphor—the transference of an idea from one area to another. Technology has provided many other metaphors that have affected our way of looking at ourselves and the world, as when human thought is made analogous to the operation of a digital computer.

A further example of the power of a technology to shape our way of thinking comes from the late eighteenth century. At that time, the designers of windmills and steam engines discovered the important principle of *feedback*, which the great twentieth-century mathematician Norbert Wiener defined as "a method of controlling a system by reinserting in it the results of its past performance."[18] When a steam engine begins to rotate too rapidly, a feedback device such as a flyball governor closes the valve that admits the steam, thereby

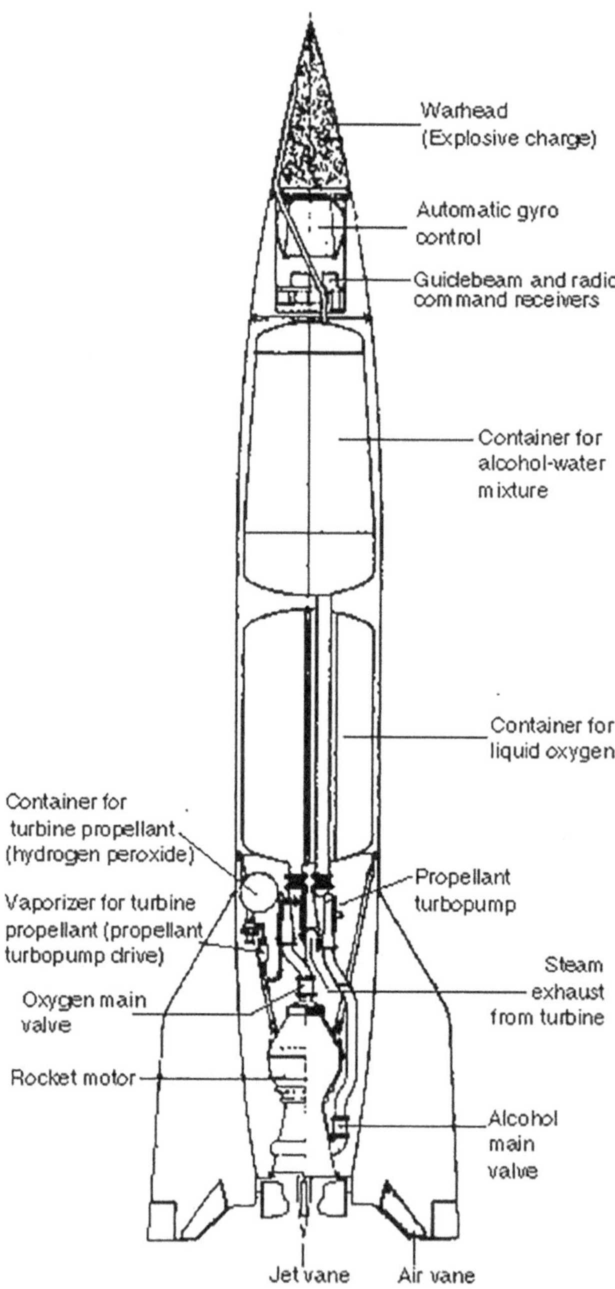

Figure 1.2a Built with slave labor, the V-2 rocket exemplified the technological prowess of Nazi Germany. (Source: NASA)

Figure 1.2b Damage from a V2 rocket landing in London, England, March 1945.

bringing the engine back into its proper operating range. When it slows down, the reverse happens, and the governor opens the valve to admit more steam.

During the late eighteenth century, the feedback principle offered a suggestive metaphor for the workings of the economic system: instead of being guided by a centralized authority, an economy might best be organized through the operation of a self-regulating market, with the actions of independent buyers and sellers providing the feedback. Thus, when buyers wanted a particular commodity, its price would be high, motivating sellers to produce more of it. If the price were low, less would be produced. In similar fashion, an increase in production would cause the price of a commodity to fall, so more of it would be purchased, while a drop in production would cause the price to rise, leading to a reduction of purchases. In this way, the actions of buyers and sellers in the market provide a feedback mechanism through which supply and demand are supposedly brought into equilibrium. It is probably no coincidence that the Scottish economist Adam Smith developed this basic concept

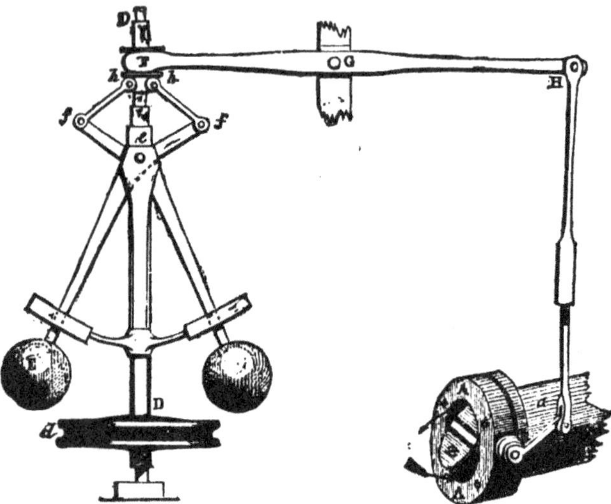

Figure 1.3 The flyball governor of a steam engine. Changes in the rotational speed of the vertical shaft at the top of the engine causes the two balls to move up or down, thereby controlling the linkage that opens and closes the throttle.

at the same time that the steam engine was being put into service.[19] Today, the widespread use of the feedback principle makes its apparent applicability to the economic system even more appealing, even though the real-world economy is hardly a neat closed system like a steam engine. Laws and regulations, as well as a host of other elements may strongly affect individual feedback loops, thereby preventing a complex economy from operating solely based on supply-and-demand signals. Technological development has supplied a useful metaphor in the feedback principle, but like all metaphors it cannot be taken as a literal depiction of reality.

Technology and Rationality

The development of technology has stimulated a belief that progress is a natural part of human life. At the same time, the progressive development of technology has itself been the product of a distinctive set of cultural values and mental processes that are characterized by a rational approach to the world and how it is to be controlled. Technological development is more than the random accumulation of tools, techniques, and organizational forms. Underlying the process is a set of attitudes and orientations that are collectively described as "rational."

What makes a technologically progressive society different from others is that its methods of problem solving are oriented toward a systematic scrutiny of the problem at hand, coupled with a systematic, empirically based examination of possible solutions and a logical selection of the most appropriate ones. Beyond this approach to the solution of problems lies another cultural attribute: the belief that solutions are *possible* and that constant changes are necessary to realize them. A society imbued with a rational ethos is dynamic and essentially optimistic, and it exhibits the confidence necessary to alter existing ways of doing things to gain benefits.

These abstract concepts may be illustrated through a simple example. All societies are faced with the problem of coping with the capriciousness of the weather. A great deal of human suffering has been the result of the vagaries of rainfall, and history provides many examples of the tragic consequences of drought and floods. A variety of responses are possible when people are confronted with this problem. The simplest is to succumb to despair, and perhaps try to find meaning in it by attributing the drought to fate or God's will. A more active approach might be to offer prayers, perform a special ceremony, or make a sacrifice. These latter activities only coincidently meet with success. There is no empirically verifiable connection between rituals and the circumstances that produce a drought, a fact that could be demonstrated by a systematic inquiry into the long-term connection between prayers and the incidence of rainfall.

Attitudes and behaviors of this sort stand in sharp contrast with rational ones. With logic and empirical observation it is possible to develop ways of dealing with problems like drought that are both more effective and more closely connected to the way the world actually works. A systematic and empirical observation of weather patterns might allow the prediction of a drought so that necessary steps can be taken to alter farming practices and conserve water. Other solutions could be the development of drought-resistant crops, improved methods of conserving water, and the distillation of seawater. It might also be possible to artificially stimulate rainfall through cloud seeding. In short, a rational approach to problem solving is continuously concerned with identifying and developing appropriate means for achieving specific ends.

Figure 1.4 Science cannot prevent natural disasters, such as tornados, but it can and has helped develop better ways of predicting when they will occur to reduce injuries and fatalities. On May 11, 1982, a wide, multiple vortex tornado with four or five vortices apparent to the left of the big one, hit Altus Air Force Base and produced F3 damage. This was one of several observed by NSSL storm intercept teams within a few miles of Altus, OK, that day. (Source: National Severe Storms Laboratory, NOAA)

The Limits of Rationality

These remarks are not meant to convey the ethnocentric belief that modern Western culture is superior to all others. The intention here is not to ridicule the beliefs and practices of people and societies that use nonrational approaches to problem solving. There is no reason to believe that rationality has been and always will be the special attribute of a particular group of people. Moreover, modern societies often manifest behaviors and patterns of thought that are anything but rational, as when large numbers of people continue to find value in astrology, numerology, and the predictions of supposed psychics.

It is also important to recognize that rational ways of thinking do not confer moral superiority. To the contrary, the rigorous development and use of rational procedures can be accompanied by major moral and ethical transgressions. The rational method of problem solving, with its overarching concern for devising appropriate means for attaining ends, makes no distinction concerning the ends being pursued. There is nothing in the rational approach to the world that prevents the use of logically and empirically derived means in the service of goals that are neither rational nor ethically justifiable. Similarly, efficiency and expediency, as technically rational goals, can be used in the pursuit of horrifying ends.[20] We can take note of the words of Captain Ahab, the main figure in Herman Melville's novel *Moby Dick*: "All my means are sane, my motive and subject mad." Nazi Germany provides many ghastly historical examples of human destruction ensuing from rational thinking and its resultant technologies. As Albert Speer, Hitler's Minister of Armaments, ruefully noted, "The criminal events of these years were not only an outgrowth of Hitler's personality. The extent of the crimes was also due to the fact that Hitler was the first to be able to employ the implements of technology to multiply crime."[21]

Even when rationality is not used for manifestly immoral purposes, it can still leave a dubious spiritual legacy. The very strength of rationality and the scientific and technological accomplishments that flow from it lie in their matter-of-fact approach to the world. A rational approach to things is often accompanied by a reluctance to admit there are any forces incapable of withstanding logical and empirical scrutiny. As the great German sociologist Max Weber put it, the world defined by rational thought processes became "disenchanted" due to the loss of the gods, genies, and spiritual forces that people, not imbued with the spirit of rationality, used to explain their world.[22] A reliance on rationality makes it difficult to engage ethics and values and nonmonetary modes of reasoning in decision-making about technology, and can substitute control for care as an orientation to others. "Disenchantment" is a two-edged sword, as the everyday meaning of the word makes clear. To be disenchanted is to lose the sense of awe, commitment, and loyalty that is a necessary part of a meaningful existence. Weber's melancholy analysis of a world that has lost its enchantment is summarized by the French sociologist Julian Freund:

> With the progress of science and technology, man has stopped believing in magic powers, in spirits and demons; he has lost his sense of prophecy and, above all, his sense of the sacred. Reality has become dreary, flat and utilitarian, leaving a great void in the souls of men which they seek to fill by furious activity and through various devices and substitutes.[23]

Political philosopher Edmund Burke Similar voiced similar misgivings in the eighteenth century. Burke's primary concern was the destruction of traditional authority by modern mass movements, as exemplified by the French Revolution. Burke attributed much of the demonic energy of that movement to the spread of rational modes of thought that left no room for the traditional attitudes, values, and political structures that had long sustained European civilization. Burke's comment on the downfall of the queen of France, Marie Antoinette, thus contains a sharp indictment of the bearers of rational values who, in his estimation, were leading Europe to its doom:

> Little did I dream that I should have lived to see such disasters fallen upon her in a nation of gallant men, in a nation of men of honor and of cavaliers. I thought ten thousand swords must have leaped from their scabbards to avenge even a look that threatened her with insult. But the age of chivalry is gone. That of sophisters, economists, and calculators, has succeeded; and the glory of Europe is extinguished forever.[24]

Rationality also implies objectivity; coolness and detachment are part of the rational approach to understanding and changing the world. Guided by a rational outlook, scientific inquiry and technological application are usually based on the abstraction or isolation of the part of the natural world that is being studied or manipulated. This is not always a good thing, for it can produce a sharp separation between the individual and the rest of the world. The scientist or technologist seems to stand apart from the system that is being studied and manipulated, resulting in a kind of tunnel vision that may ignore the larger consequences of gaining and applying knowledge.[25] For example, in discovering a genetic marker for a serious disease, a researcher might not consider potential abuses of that discovery, such as insurance companies refusing coverage of people with that marker.

A logical, detached, and dispassionate approach to the world is suffused with what can be identified as a "masculine" approach to understanding and interacting with the world. In some ways, the idea or definition of technology itself is associated with masculinity. Some technologies have largely been a male domain, and for a variety of reasons continue to be so today,[26] but throughout history women have also made significant contributions to technological advance.[27] But important artifacts, such as the baby bottle, are not imagined to be technologies, or not often analyzed as part of technological systems, even though they are part of complex manufacturing systems, and have been controversial in terms of infant health and development and women's empowerment.[28]

When technologies become associated with a gender, that impacts their acceptance in the marketplace and culture. For example, in the early 1900s in the early emergence of automobility, the electric car was deliberately marketed for stereotyped "feminine" driving habits—such as errands and social calls. The early electric cars had heating and cooling, an electric starter, as opposed to a crank start, and were quiet and clean. Gasoline-powered vehicles were associated with the masculine ideal of touring and off-road racing.[29] Remember that at this time neither fuel source was widely available, and the interstate highway system not yet built. The highly gendered distinction between the electric and the gasoline-powered internal combustion engine limited the electric car market, as middle-class households were buying cars for the male breadwinner. The Great Depression of the 1930s led to the extinction of most of the electric car manufacturers of that generation.

Figure 1.5a A 1913 Detroit Electric Car. Electric cars were available and attractive for in-town errands and assumed to be a good car for women. (Source: Corvair Owner/CC BY-SA 2.0)

Figure 1.5b A 1913 Ford Model T, Gasoline Powered. Gasoline owners of the early twentieth century also suffered from "range anxiety" as a gasoline powered car at the time had about 100-mile range. Fuel had to be shipped ahead. (Source: Mariordo/CC BY-SA 3.0)

The complex relationship of gender and technology is illustrated by the history of the technological artifact most strongly associated with the present era—the digital computer. Its development has generally been viewed as the product of hyperrational male engineers, mathematicians, scientists, and technicians. In reality, many of the programmers of first-generation computers were women whose accomplishments have often been passed over in standard histories. More recently, the development of computer technology has depended on thought processes that are framed as relentlessly rational, objective, and logical, and associated with masculinity.[30] But at the same time coding has required an intuitive, interactive, and generally less structured approach.[31] These gendered stereotypes about styles of thought have long cultural antecedents and consequences for whether people see themselves fitting into various occupations, but not much empirical validity.[32]

Technological Determinism vs. Social Constructivism

Nothing worthwhile in life comes without some costs. So it is with technology; while it has expanded human power and made our lives materially richer, the advance of technology has created many problems—environmental degradation, alienation, and the threat of nuclear war, to name only the most obvious ones. There looms the possibility that technology is out of control. If this is so, what began more than a million years ago as a human creation has taken on a life of its own, with technology advancing according to its own inner dynamic, unrestrained by social arrangements, systems of governance, culture, and thought.[33] The belief that technology acts as an independent force in our life, unaffected by social process, is known as "technological determinism," and if it is true, we have become the servant of technology instead of its master. Technological determinism is connected to rationality, and the inclination of rationality to control and domination, to succumbing to what Rubinoff has called "the law of the possible": "What is possible for science to know science must know. What is possible for technology to do technology is obliged to do. Whatever is possible is obligatory."[34]

There can be little question that technology exerts a great influence on social, political, and economic relationships. Everything from antibiotics to zippers has affected our lives to some degree; many of these influences will be explored in subsequent portions of this book. But that is not the end of the story. As will be explored in Part 2, scholars of technology have argued that technology is shaped by social arrangements and processes. According to the social construction model, the emergence of particular technologies, choices between competing technologies, and the way these technologies are actually used owe a great deal to socially grounded forces such as political power, social class, culture, gender, and organizational dynamics. As will become clear, it requires a multidisciplinary approach from economics, sociology, and anthropology and cultural studies, as well as history, to even begin to track the complexity of technological change.

Asserting the supremacy of either technological determinism or social constructivism is not very useful. Straightforward cause-and-effect relationships can be found in some realms—Newtonian physics, for example—but technological and social change is better understood in terms of probabilities, reciprocal interactions, and feedback loops. Even William F. Ogburn, a sociologist often characterized as a technological determinist, on occasion took a more nuanced view of the subject:

> "The more that one studies the relationships between mechanical and social invention, the more interrelated they seem.... The whole interconnected mass [i.e., social institutions, customs, technology, and science] is in motion. When each part is in motion and banging up against some other part, the question of origins seems artificial and unrealistic. If one pushes the question to the extreme, origins are lost in a maze of causative factors."[35]

The wondrously complicated interactions of technology and society often result in unimagined consequences when new technologies emerge. To take one example, when the first digital computers appeared in the mid-1940s, they elicited modest expectations about their future applications. Today, the world as we know it is almost unimaginable without computers, as everything from air travel to the mapping of genomes is dependent on the storage, retrieval, and manipulation of information performed by computers. In similar fashion, ARPANET, the ancestor of the internet, was created in the 1960s to facilitate communications between a small number of defense and university laboratories. No one saw it evolving into a universal medium for the exchange of texts, images, and music. The history of computers and the internet would seem to lend credence to technological determinism; computers apparently generated their own momentum to become universal and essential components of contemporary life.

This is the story from a technological determinist standpoint, but social constructivists would challenge it by noting that the technical development of the computer in the 1950s and 1960s was heavily supported by military expenditures. This was also the case for the internet; ARPANET is the abbreviation of the name of the network sponsored by the U.S. Department of Defense's Advanced Research Projects Agency. Someone taking a social constructivist approach might also point out that the expansion of the market for computers was also powerfully stimulated by commercial enterprises like banks and insurance companies, and that this huge market supported the research and development that rapidly advanced computer technology and the internet.

A similar story could be repeated for most successful technologies. New technologies bring changes to many aspects of society, while at the same time social forces do much to stimulate and shape these technologies. To try to assign primacy to one or the other is to ignore a crucial feature of technological and social change. Both are dynamic processes characterized by the reciprocal interaction of a host of factors, some of them narrowly technical in nature, others not. No reasonable person could deny that technology has been a major force in making the world we live in, but it is important to always keep in mind that technology has not operated as an agent independent of the society in which it is embedded. Technologies are *designed*, and the design process always involves tradeoffs and reflects the priorities and aspirations of designers and others involved in bringing a technology from idea to reality.

Social constructivism therefore offers the possibility for more human agency than technological determinism, but it is not likely that the ability to influence the course of technological change will be evenly distributed among the whole population. Specifically, constructivist analyses have often shown how differences in power and access to resources have shaped technological change. Technologies may be devised, selected, and disseminated because they serve the interests of a particular group, possibly in opposition to the interests of other groups. Technology confers power, but this power is not only wielded over only the nonhuman universe. As C. S. Lewis reminds us, "Man's power over nature is really the power of some men over others with nature as their instrument."[36]

Living in a Technological Society

The development of technologies that are suited to our needs requires the informed participation of a wide range of people. Unfortunately, contemporary technology places severe limits on popular understanding. The complexity of current technologies preclude direct involvement by all but those immediately concerned with them. The rest of us are passive consumers, content to reap the benefits of rationally derived knowledge but woefully ignorant of it. This creates the fundamental paradox of modern society: technology generates massive power for human society, while as individuals we exert very little of that power. We have access to a wide range of powerful technologies, yet our inability to understand them often leaves us with feelings of impotence and frustration, as anyone who has experienced a computer crash will attest.

As has been noted, the application of rationality for the solution of human problems is both a consequence and cause of optimism and a willingness to accept constant change. Yet one cannot help but wonder if these characteristics can be sustained in an environment that sharply limits participation and inculcates widespread feelings of having little or no power over the process of technological change.

Strange notions can emerge when feelings of powerlessness are coupled with unquestioned faith in technology. The consequences of this combination are sometimes exhibited by fervent believers in alien spacecraft or UFOs (unidentified flying objects). Although convincing evidence of UFOs is lacking, a belief in their existence does not necessarily make one a crackpot. In some cases, however, a strident belief in the existence of UFOs takes on the characteristics of membership in a religious cult where the deities are superior beings who have produced an advanced technology. Alien spaceships represent a level of technical sophistication not attained on Earth, and some UFO enthusiasts entertain the hope that the aliens that created them will take over this planet and solve its problems. Faith in a higher technology may be combined with a mistrust of the "establishment," as a fair number of UFO adherents claim that their government is engaged in a massive conspiracy to prevent the public from being aware of the existence of UFOs. There is no denying that on occasion governments lie to their citizens, but a cover-up of the required magnitude would be impossible for even the most well-organized government to pull off. Still, conspiracy theories strike a resonant chord with people who feel that they have been excluded from decision-making, both political and technological. A quasi-religious belief in UFOs therefore combines an excessive confidence in technology in general with a distrust of the people and organizations that control it in actual practice.

Distrust flourishes when people have no ability to participate in decisions that shape their lives, and the inability to affect the course of technological change can produce a mixture of naive hope and paranoid reaction. A realistic sense of control, including a sense of having some control over technology, is essential for an individual's mental health. No less important, widespread participation in shaping technology is essential for democracy. Technology's benefits cannot be separated from its costs, and thus it becomes necessary to determine if the former justify the latter. If a society is truly democratic, such decisions will be made with as much citizen participation as possible. Moreover, the benefits and costs of technology are not shared equally, and the distribution of costs and benefits should be discussed and mediated in collaborative processes. We will return to these themes in Part 7, but first, we will take a closer look at how technology can affect people and groups in different ways.

DISCUSSION QUESTIONS

1. In your opinion, which recent technology has produced the greatest benefit? Which has produced the most harm? Are there any harmful elements to the beneficial technology, and has anything good come from the harmful one?
2. You have probably heard the old saying that "necessity is the mother of invention." Are new technologies usually a response to an existing need of some sort? Can you think of any technologies that created a need before most people were aware of it?
3. Are technologies "gendered"? Are some technologies identified with women and others with men? On what bases do we make these distinctions? Will this situation necessarily continue in the years to come?
4. Can you think of any technologies that were developed simply because of the technical challenges involved? How can these "impractical" technologies be justified?
5. How do you feel when a technological device upon which you depend malfunctions? What do these feelings tell you about your attitude toward technology in general?
6. Technology has intensified resource use and the production of pollution, such as with the explosion of plastic waste in the environment or the release of CO_2 into the atmosphere. What kind of social and technological advancements would be necessary to ameliorate these negative consequences of our current technological systems?

Notes

1. Branka Vuleta, "How Much Data Is Created Every Day? +27 Staggering Stats," *SeedScientific* (October 28, 2021) Accessed from https://seedscientific .com/how-much-data-is-created-every-day/.
2. National Science Foundation, "Science and Engineering Indicators: 2014," accessed on from www.nsf.gov/statistics/seind14/index.cfm/chapter-7/c7h. htm.
3. Carroll Doherty and Jocelyn Kiley, "Americans Have Become Much Less Positive about Tech Companies' Impact on the U.S.," *Pew Research Center* (July 19, 2019). Accessed from www.pewresearch.org/fact-tank/2019/07/29/ americans-have-become-much-less-positive-about-tech-companies-impact-on-the-u-s/.
4. Joshua A. Tucker, et al., "Social Media, Political Polarization, and Political Disinformation: A Review of the Scientific Literature" *SSRN* (March 19, 2018). Accessed from http://dx.doi.org/10.2139/ssrn.3144139
5. Elia Abi-Jaoude, Karline Treurnicht Naylor, and Antonio Pignatiello, "Smartphones, Social Media Use, and Youth Mental Health," *CMAJ* 192 (2020): E136–E141.
6. This paragraph is derived from Carl Mitcham, *Thinking through Technology: The Path between Engineering and Technology* (Chicago: University of Chicago Press, 1994), pp. 117–34.
7. Jacob Bronowski, "Technology and Culture in Evolution," *Philosophy of the Social Sciences* 1, (1971): 199.
8. Lewis Mumford, "Technics and the Nature of Man,"Technology and Culture 7, (July 1966): 303–17.
9. The Thomas Edison Papers, "Electric Light and Power System" (March 16, 2015). Accessed from http://edison.rutgers.edu/power.htm.
10. John B. Rae, *Climb to Greatness: The American Aircraft Industry, 1920–1960* (Cambridge, MA: MIT Press, 1968), p. 74; Edward Constant, *Origins of the Turbojet Revolution* (Baltimore: Johns Hopkins University Press, 1980).
11. Lewis Mumford, *Technics and Civilization* (New York: Harcourt, Brace and World, 1934), p. 110.
12. Anvar Jay Varadaraj (2020). "Tackling Supply Chain Disruption During COVID-19. *Industry Week*. May 22, 2020. Accessed from www .industryweek.com/supply-chain/article/21132223/tackling-supply-chain-disruption-during-covid19.
13. Edwin Kiester, Jr., "A Little Fever Is Good for You," *Science* 84, (November 1984): 172.
14. Meika Loe, *The Rise of Viagra: How the Little Blue Pill Changed Sex in America* (New York: NYU Press, 2004). On Viagra's profitability see https://fortune .com/2018/03/27/viagra-anniversary-pfizer/.
15. Robert C. Post, "Technological Enthusiasm," in Rudi Volti (Ed.), *The Encyclopedia of Science, Technology, and Society*, vol. 3 (New York: Facts on File, 1999), pp. 999–1001.

16. Quoted in Jonathan Hughes, *The Vital Few: The Entrepreneur and American Economic Progress* (New York and Oxford: Oxford University Press, 1986), p. 327.
17. Michael Adas, *Machines as the Measure of Man: Science, Technology, and Ideologies of Western Domination* (Ithaca and London: Cornell University Press, 1989).
18. Norbert Wiener, *The Human Use of Human Beings: Cybernetics and Society* (New York: DaCapo, 1988; originally published in 1954), p. 61.
19. Otto Mayr, "Adam Smith and the Concept of the Feedback System," *Technology and Culture* 12 (January1971): 1–22.
20. Steven B. Katz, "The Ethic of Expediency: Classical Rhetoric, Technology, and the Holocaust," *College English* 54 (1992): 255–275.
21. Albert Speer, *Inside the Third Reich* (New York: Macmillan, 1970), p. 212
22. This concept is explored by Weber in "Science as a Vocation," in H. H. Gerth and C. Wright Mills (Eds.), *From Max Weber: Essays in Sociology* (New York: Oxford University Press, 1958), pp. 129–56.
23. Julian Freund, *The Sociology of Max Weber* (New York: Pantheon, 1968), p. 24.
24. Edmund Burke, *Reflections on the Revolution in France* (New York: Holt, Rinehart and Winston, 1959; originally published in 1790), p. 91.
25. Richard Schlegel, "Why Can Science Lead to a Malevolent Technology?" *Centennial Review* 21 (Winter 1977): 14.
26. Kayla Matthews, "Why Does Gender Equality Remain Stagnant in Technology?" *Innovation and Tech Today*, (December 24, 2019) at https://innotechtoday.com/gender-equality-technology/
27. For a narrative of the historical processes that have led to the perception that technology is "men's work," see Ruth Oldenziel, *Making Technology Masculine: Men, Women, and Machines in America, 1870–1945* (Amsterdam: University of Amsterdam Press, 1999).
28. Penny Van Esterik, *Beyond the Breast-Bottle Controversy* (New Brunswick: Rutgers University Press, 1989).
29. Virginia Sharff, *Taking the Wheel: Women and the Coming of the Motor Age* (University of New Mexico Press, 1992); Michael Brian Schiffer, *Taking Charge: The Electric Automobile in America* (Collingdale, PA: Diane Publishing, 1994). See also: www.wisconsinhistory.org/Records/Newspaper/BA3778 "Race of First Steam Buggies to Madison" *Capital Times*, May 26, 1921.
30. Jennifer Light, "Programming," in Nina E. Lehrman, Ruth Oldenziel, and Arwin Mohun (Eds.), *Gender and Technology: A Reader* (Baltimore and London: Johns Hopkins University Press, 2003).
31. Sherry Turkle, *The Second Self: Computers and the Human Spirit* (New York: Simon and Schuster, 1984); Paul N. Edwards, "Industrial Genders: Hard/Soft," in Nina E. Lerman, Ruth Oldenziel, and Arwen Mohun, (Eds.), *Gender and Technology: A Reader* (Baltimore: Johns Hopkins University Press, 2003).
32. Cordelia Fine, *Delusions of Gender: How our Minds, Society, and Neurosexism Create Difference* (New York: W.W. Norton, 2011).

33. One of the most influential explorations of this idea is Langdon Winner, *Autonomous Technology: Technics-Out-of-Control as a Theme in Political Thought* (Cambridge, MA: MIT Press, 1977).
34. Lionel Rubinoff (1977) "Technology and the Crisis of Rationality: Reflections on the Death and Rebirth of Dialogue." *Philosophy Forum* 15(3–4):273–278.
35. William F. Ogburn, "Technology and Governmental Change," in Otis Dudley Duncan (Ed.), *On Culture and Social Change: Selected Papers* (Chicago: University of Chicago Press, 1964), pp. 132–33. Quoted in Rudi Volti, "Classics Revisited: William F. Ogburn, Social Change with Respect to Culture and Original Nature," *Technology and Culture*, 45 (April 2004): 399.
36. C. S. Lewis, *The Complete C. S. Lewis Signature Classics* (San Francisco: Harper San Francisco, 2002), p. 719.

CHAPTER 2

Winners and Losers: The Differential Effects of Technological Change

The last chapter may have seemed a bit negative in its assessment of technology and the culture that supports it. In one regard, however, there is no denying technology's positive consequences: technological advance has been the greatest single source of economic growth. If our material lives are better than those of our grandparents, it is largely because technological development has boosted the production of goods and services. Equally important, it has created entirely new products while at the same time improving the quality of existing ones.

Curiously, economists were slow to grasp this seemingly obvious fact. Conventional economic analysis identifies three basic "factors of production": land (which includes natural resources), labor, and capital. Any increase in production was therefore taken to be the result of an increase of these factors. This view began to change in the 1950s when the historical course of economic development in the United States was analyzed through the use of sophisticated statistical techniques. It then became apparent that increases in the traditional factors of production did not adequately explain the actual record of economic growth. The amount of land had remained constant, and capital accumulation and increases in the labor force accounted for only 10 to 20 percent of economic growth during the first half of the twentieth century.[1] Accordingly, the major source of economic growth was a "residual" factor of overwhelming importance. Most economists agree that technological advance is the main element of this residual, although organizational development and improved worker skills, along with economies of scale, are also key components. Still, as we have already seen, organization and skill are integral parts of technology, so it is reasonable to view technological change as the major source of economic growth, a conclusion that is widely supported by economists today.[2]

While technological development has been the primary source of economic advance, it has not been cost-free. One of the most pleasant myths about technology is that it can work its wonders without altering existing social arrangements. Americans in particular have often seen technological progress as the surest basis for progress in general and have tended to believe that technological solutions to problems are less painful than solutions that require political or social changes.[3] These beliefs are not easily sustained after an examination of the actual pattern of technological advance.

It is a truism that a particular technology can be used for either good or evil purposes; a construction team employs explosives to build a road, while a terrorist uses them for roadside bombs. But there is less appreciation for a more subtle point: technological change is often a subversive process that results in the modification or destruction of established social roles, relationships, and values. Even a technology that is used exclusively for benign purposes will cause disruptions by altering existing social structures and relationships. There are many technological changes that are small in scope, the effects of which are felt by only a few. A few technological changes are massive, and they lead to vast social restructuring. In either case, technology does not yield its benefits without exacting a cost.

Technology as a Subversive Force

The disruptive effects of technological change can readily be seen in the economic realm, where new technologies can lead to the destruction of obsolete firms, as when the fabled Pony Express rapidly lost its customers after telegraph wires had been strung across the West. Of course, sometimes the disruption is less apparent when technological innovation results in the creation of entirely new industries that are not in direct competition with existing ones. Many new industries and individual firms owe their existence to the emergence of a new technology. Witness, for example, the rapid growth

Figure 2.1 The last Blockbuster, a video rental store in Bend, Oregon. In September 2020, it partnered with Airbnb to provide the location as a limited-time 90s-themed overnight rental. (Source: Another Believer/CC BY-SA 4.0)

of personal computer manufacturing, peripheral equipment production, software publishing, and app development that followed the invention of the integrated circuit. Even so, lurking behind these successes were a number of failures, most notably the manufacturers of vacuum tubes and individual transistors, who faced a diminished market for their products.

Concerns about the disruptive effects of technological change are not new, as can be seen in an English magazine editor's fulminations against the first railroads in 1835: "Railroads, if they succeed, will give an unnatural impetus to society, destroy all the relations that exist between man and man, overthrow all mercantile regulations, and create, at the peril of life, all sorts of confusion and distress."[4]

Anyone convinced of the virtues of technological change could easily criticize this reactionary view by noting how the railroad stimulated economic development and produced many social benefits. Even so, there is more than a grain of truth in the concerns expressed by the agitated magazine editor. Technological changes, both major and minor, often lead to a restructuring of power relations, the redistribution of wealth and income, and alterations to human relationships.

One recent and much-debated instance of a disruptive new technology is the rise of ride-sharing services like Uber and Lyft. The use of private automobiles for the paid transportation of individuals is nothing new; a century ago, privately owned cars known as "jitneys" regularly picked up and delivered paying passengers.[5] Although it can be argued that they performed a useful service, jitneys were quickly banned by municipal governments due to safety concerns, but also because they were deemed unfair competition for public trolley and bus lines. The resurgence of ride-sharing services today has come as a result of the ubiquity of smartphones and GPS technology. After downloading an appropriate app, individuals can easily request a car and driver to pick them up and take them to their destination.

Ride-sharing services are convenient for their users, and they allow drivers to work as much or as little as they prefer with no immediate supervision. The convenience and generally lower fares offered by these services has led to their spectacular growth. Uber, the largest, began in San Francisco in 2010; a mere five years later, it was operating in 58 countries and was valued at $50 billion.[6]

At the same time, however, ride-sharing services—and especially Uber—have generated a storm of protest. Drivers of established taxi firms and their supporters have mounted vociferous and sometimes violent protests against Uber in New York, France, Brazil, Australia, and elsewhere.[7] Critics of Uber and similar services point to the iniquities that result from treating drivers as contractors rather than employees. As such, they are not eligible for pension and health care benefits, and ride-sharing firms avoid making contributions to Social Security payroll taxes. Issues have also been raised concerning the screening of drivers, vehicle safety, and insurance coverage.

In addition to energizing a considerable amount of mass protest, ride-sharing firms have met with government opposition. In California, the birthplace of so many new technologies, government agencies have challenged some of the practices of ride-sharing firms. In 2015, the California Labor Commission ruled that Uber drivers were employees, and not contractors, which would make them eligible for the benefits noted above. In the same year, New York taxi-enforcement officials seized nearly 500 cars over a six-week period, citing illegal street pickups by Uber drivers.[8] But government agencies elsewhere have been less decisive. In 2015, the Florida state agency charged with resolving unemployment compensation issues overturned an earlier decision that had labeled Uber drivers employees and not contractors.[9] In both California and Florida, the cases centered on single individual firms and the overall applicability of either of these legal rulings is debatable. As of this writing (2023), the National Labor Relations Board has suggested that Uber drivers are contractors.[10] The ICTs (information and communication technologies) that make Uber and other ride share services possible have been disruptive to established taxi companies and produced both challenges and opportunities for drivers as members of the "gig" economy.

On occasion, technological advance has fatally disrupted entire communities and the people living in them. One such place was Caliente, Nevada.[11] Caliente was a small town with a variety of civic amenities—schools, churches, a hospital, a theater, a park, and many prosperous small retail businesses. Many of its inhabitants were proud members of civic organizations such as the Chamber of Commerce, Rotary, the Masons, and the American Legion. It was a typical American small town, with typical American small-town values.

The life of the town was supported by a single industry: the servicing of steam locomotives. Caliente was an important division point on a transcontinental railroad, and many of the town's people worked as machinists, boilermakers, and repairmen. Their incomes in turn supported Caliente's commercial and civic establishments. But in the late 1940s, a "disruptive technology," the diesel-electric locomotive, rapidly replaced the steam locomotive. Diesels had many advantages; they were more fuel-efficient, hauled longer trains, and did less damage to the rails and roadbed. They also required less frequent servicing. When servicing was required, it took place in large, centralized shops. As a result, service facilities were eliminated at many division points, and Caliente was one of them. The town lost its economic base, and within a few years, it had become a shell of its former self. People moved out, homes were abandoned, and shops were boarded up. The local newspaper sadly noted, "Employees who have given the best years of their lives to this railroad are cut off without anything to which they can turn, many of them with homes in which they have taken much pride; while others, similarly with nice homes, are told to move elsewhere."[12]

Figure 2.2 Steam Engines and Railroad Towns. An old steam engine, location and date not specified.

Figure 2.3 Steam engines required a great deal of servicing and supplies along their routes. Servicing engines at coal and sand chutes at Argentine yard, Santa Fe R.R., Kansas City, Kansas. (Source: Jack Delano, 1939; Library of Congress).

The tragedy of this small town has been repeated in many other communities affected by technological change. Many places of employment have closed down as new products and processes have replaced old ones, leaving communities and their inhabitants in desperate straits. The technological advances that produced these dislocations may have benefited society as a whole, but at great cost to the people who were immediately affected.

Technological changes do not always result in the disruption or modification of an existing social order; sometimes they may help to preserve it, as happened when pneumatic molding machines were adopted by the McCormick reaper manufacturing plant in the 1880s.[13] These machines were not installed, as conventional analysis would lead us to think, in order to reduce costs or to produce a better product; in fact, they were deficient on both counts. They were installed for the sole purpose of eliminating the skilled workers who formed the backbone of the National Union of Iron Molders, an organization that was challenging the entrenched authority of McCormick's management. The molding machines allowed the replacement of skilled workers by unskilled ones, and three years later, having served their purpose, they were discarded by McCormick's management.

Groups that are threatened by a technological innovation are not always as helpless as the iron molders apparently were. Many affected parties have been able to defend themselves against changes in the way of doing things. To take one example, prefabricated buildings were vigorously resisted by many local construction workers' unions because they threatened their members' jobs. One sad tale is narrated by Peter Blake.[14] Shortly after the end of World War II, an enterprising manufacturer decided to mass-produce a so-called service core: a complete "package" containing kitchen, bathroom, and utility room, with all fixtures, pipes, ducts, and wires in place, ready to be plonked down in any suburban plot. The first twenty of these beautifully designed and beautifully made "packages" arrived on a site near Detroit; local union plumbers and electricians promptly refused to install them. Finally, after nine months of heated debate (during which the units, parked on a sidewalk, were exposed to weather and vandalism), the local unions agreed to handle the "packages"—by disassembling them on the sidewalk and then reassembling them, piece by piece, in each of the houses. The manufacturer, needless to say, thereupon went out of business.

The Luddites

In the cases previously described, the successful resistance to a new technology entailed some disruptive, even violent, acts. A resort to violence was also a feature of the case of the most famous example of resistance to technological change, the outbreaks of machine-smashing that occurred in early nineteenth-century England.[15] These attacks were the work of different groups who were collectively known as Luddites, a name that was derived from one Ned Ludlum (or Ludd), an apprentice stocking maker who, as legend had it, answered his master's reprimand by smashing his stocking frames with a hammer. There was really nothing new about these attacks; the breaking of machines by disgruntled workers had a long history in England, the earliest recorded episode taking place in 1663. But the Luddite disturbances that began in 1811 did represent a substantial increase in the scale of these attacks; by the following year, the government had to deploy 12,000 troops to restore order to the parts of England affected by the movement.

Since these attacks coincided with an era of rapid technological change, it is easy to draw the conclusion that they were motivated by the fear of many workers that their jobs would be lost to new machinery. The actual story is a bit more complicated. Luddite attacks occurred in a number of separate branches of the textile industry, and each was characterized by a distinctive set of motivations and responses. The Luddite movement began in the hosiery trades, where there had long been opposition to the use of wider stocking frames, which allowed the employment of poorly paid unskilled labor for the manufacture of an inferior product. The situation might have been resolved in a peaceful manner had it not been for the dire conditions encountered by many of England's working people at the time. The Napoleonic wars had resulted in the closure of many export markets, leading to a general trade depression. To make matters worse, a series of bad harvests led to sharp increases in the cost of food, and many workers found that their wages were insufficient to meet their basic needs. These conditions produced a fertile ground for the spread of "collective bargaining by riot," and Luddite attacks were soon fomented by shearers in the textile industry. Another occupational group, the handloom weavers, viewed the advance of steam-powered weaving machinery with understandable apprehension, and, following the example of workers in the hosiery trade, some of them attacked the factories housing mechanized looms, as well as the houses of their owners. Only in a few instances was the machinery itself directly attacked.

Luddite disturbances were expressly oriented toward blocking technological change in the cropping trade. Wool cloth was traditionally finished by raising the nap and then leveling the surface through the use of a heavy set of shears. The growing use of the gig mill, a device for raising the nap, along with the employment of a crude device for the mechanized cropping of cloth, threatened the livelihood of the traditional handworkers. They responded with some of the most severe attacks of the Luddite epoch. Although the machinery had been used for many years in many textile establishments, the severe economic conditions of the time brought matters to a head. More than the other instances of Luddite revolt, the attacks on cropping equipment were motivated by a deep fear of unemployment induced by technological change.[16]

Within a few years, the Luddite assaults came to an end due to the deployment of government troops; the execution, imprisonment, and exile to Australia of a number of the participants; and the general improvement in living conditions after the defeat of Napoleon. The succeeding decades of the nineteenth century also saw the replacement of the small manufacturing establishment by the large factory. Machine-smashing by riotous crowds was a likely form of labor protest when workers were scattered and lacking in permanent organizational linkages. In contrast, the factory served as a fertile ground for the development of labor unions and other organizational vehicles for pressing the interests of workers. Industrial sabotage did not come to an end, but it was generally superseded by unionization and more effective forms of worker protest.

Neo-Luddism

These early episodes of machine-smashing have led to the application of the "Luddite" label to anyone opposed to modern technology. But it is perhaps unfair to impute to the original Luddites a hostility to technology per se. As we have seen, most instances of Luddism were not motivated by a fear and hatred of new machinery; their grievances were those of people suffering from the low wages and unemployment caused by a generally depressed economy.

The machines were seen as convenient targets of their ire rather than the sources of it.

In recent times, technology itself has been attacked by individuals and groups who have been labeled as "neo-Luddites." In 1995, the *New York Times* and the *Washington Post* published a lengthy critique of modern society and the pivotal role of technology in creating and maintaining it. According to its author, a society based on modern technology brings some material comforts, but "all these technical advances taken together have created a world in which the average man's fate is no longer in his own hands or in the hands of his neighbors and friends, but in the hands of politicians, corporation executives and remote, anonymous technicians and bureaucrats whom he as an individual has no power to influence."[17] Regaining human freedom therefore required the total destruction of industrial society and the technologies that made it possible. This would not be a peaceful revolution, but one that required the destruction of factories, the burning of technical books, and the eradication of all of the components of an industrial civilization. This creed might have been dismissed as the agitated musings of a late twentieth-century Luddite, but its author was not just a misguided critic of the modern world. Shortly after the publication of the manifesto, it was discovered that its author was a brilliant mathematician named Theodore Kaczynski, dubbed by the media as "The Unabomber," an elusive figure who from 1978 to 1995 had been responsible for 16 bombings that killed three people and wounded 23 others.

Kaczynski was an extreme example; most expressions of neo-Luddism are aimed at excessive involvement in particular kinds of technology, especially those of the digital variety. A much milder criticism has come from a number of writers who have advocated for "digital detox." Few of them suggest jettisoning computers, tablets, and smartphones. Rather, they prescribe occasional respites from social media, Twitter feeds, email, and blogs. Occasional separation from digital devices, it is claimed, will lead to lower levels of stress, improved concentration, higher productivity, and more face-to-face human interaction.[18]

Whose Technology?

We have just seen how specific technologies have been used and resisted by particular groups in accordance with their own needs and concerns. These examples should help us to realize that technology does not proceed solely through its own momentum, as implied by technological determinism; its development is strongly influenced by existing social and political arrangements. Technological changes may take place because they advance the interests of a particular group. Conversely, some technologies may meet with stiff resistance because they threaten a group's interests. Technologies do not stand or fall solely on their intrinsic merits; the decision to develop and deploy a new technology is often shaped by the distribution of power in a society. Technological determinism "feels true" because most of us are "low inclusion" actors[19] who do not see the multiple trade-offs and negotiations, the priority-setting debates, and the social power deployed in the development of technology. We never see the technological failures, or the gap between the number of patents applied for, and the appearance, after many years, of new products and processes.[20] We end up with a "take-it-or-leave-it" experience of technology.

Social, economic, and political arrangements affect the course of technological change by influencing the kinds of investments that are made, the research projects that are funded, and the general priorities that are established. Large organizations, such as corporations and government agencies, often wield disproportionate influence over the process of technological change. As we will see in Part 7, the federal government is a major source of financial support for research and development, with the Department of Defense, the National Aeronautics and Space Administration (NASA), and the Department of Energy (primarily for nuclear research and development) accounting for a large share of these expenditures. Although we can only speculate about alternative outcomes, it seems likely that American technology would have diverged markedly from its historic path if financial resources had been distributed differently.

Perhaps with a different set of sponsors, technological development might have made greater contributions to the solution of a number of pressing social problems, such as environmental change, poverty, and crime. At the same time, however, it can be argued that certain kinds of problems are simply not amenable to technological solutions. Even with significant changes in the funding of research, technological solutions to many social problems will not be forthcoming. This is an important objection, and we will examine it in the next section.

What Technology Can Do—And What It Cannot Do

The growth of technology has brought dazzling changes to our lives. At the same time, we seem to be mired in problems for which there seem to be no solution. The continued existence of these problems is all the more frustrating when contrasted with the rapid progress of technology. For example, we can use all kinds of sophisticated medical equipment and techniques to preserve the lives of sickly infants who have been born many weeks premature, but we cannot seem to conquer the poverty that often results in sick infants. Why, it is often asked, is there such a gulf between technological progress and social progress? Why can technology not be applied as a solution for more, if not all, of our problems? If we can put a man on the moon, why can we not...?

The Technological Fix

These are troubling paradoxes, and in recent years we have searched for ways of finding technological solutions to a host of problems. The drug methadone has been widely used to eliminate addicts' cravings for heroin. As highway accidents continue to result in tens of thousands of deaths and hundreds of thousands of injuries each year, efforts have been mounted to develop and manufacture cars capable of protecting their occupants from the consequences of incompetent driving. Cities befouled by graffiti have turned to the use of new paints and cleaning solutions that resist the endeavours of spray-can artists. Overweight men and women spend billions of dollars annually on supplements, medications, and exercise apparatus in the hope of shedding excess pounds. College professors worried about plagiarized papers use computer programs to detect possible cheating.

The list of technologies that have been or could be applied to the alleviation of social problems is an extensive one, and examples could be endlessly supplied. What they have in common is that they are "technological fixes," for they seek to use the power of technology in order to solve problems that are nontechnical in nature. In this section, we will briefly examine a few of these technologies and consider the extent to which technology can alleviate these pressing problems.

Many technological fixes have been employed over time, although not always with the conscious understanding that technology was being used instead of some other method of achieving a desired end.

To take one example, at the beginning of the twentieth century the United States was undergoing severe growing pains; the urban population was expanding at a rapid rate, accompanied by congestion, pollution, and a host of other urban ills. In a nation steeped in the Jeffersonian belief that cities were inherently evil, and that the countryside was the best location for virtuous living, the conversion of the American populace into a race of unhealthy and disaffected city dwellers was viewed with alarm. A number of technologies did make urban life more tolerable, most notably those concerned with public health and sanitation, but these only served to ameliorate living conditions without addressing the real issue: the desire of many Americans to escape the city and return to a vaguely perceived rural idyll.

The pursuit of this goal gave a great impetus to the development of transportation technologies that would allow the solution of urban problems by reducing the need for cities, at least as places of residence. Instead of comprehensively addressing urban ills through planning and the development of social programs, Americans pinned their hopes on new transportation technologies. The first of these was the electric trolley. Through the construction of extensive networks of interurban electric lines, it was hoped, America's urban problems could be literally left behind as a new generation of workers could commute from their places of work to their rural or suburban homes.[21]

Figure 2.4 An electric streetcar in Cincinnati, Ohio, in the 1940s, looking west on 5th Street at Walnut Street. (Source: Metro Bus/CC BY 2.0)

In many American cities, the trolley was displaced by the automobile, as a great deal of automobile ownership was motivated by similar sentiments. Widespread automobile ownership promised an escape from the harsh realities of America's cities through individual commuting. As Henry Ford neatly summed things up, "We shall solve the city problem by leaving the city."[22] Ford's sentiments were taken to rhapsodical levels by one early twentieth-century journalist:

> Imagine a healthier race of workingmen, toiling in cheerful and sanitary factories, with mechanical skill and tradecraft developed to the highest, as the machinery grows more delicate and perfect, who, in late afternoon, glide away in their own comfortable vehicles to their little farms or houses in the country or by the sea twenty or thirty miles distant! They will be healthier, happier, more intelligent and self-respecting citizens because of the chance to live among the meadows and flowers of the country instead of in crowded city streets.[23]

It is hardly necessary to note that these hopes were not realized. The mushrooming growth of suburbs spawned by trolleys and automobiles did not create a harmonious social order based on rural values. All too often the legacy has been suburban sprawl, the deterioration of city centers, visual blight, air pollution, traffic fatalities, and many other ills. This is not to say that the automobile has been an unmixed curse; the benefits of personal mobility, privacy, and a sense of power have been too eagerly accepted to allow such a judgment. But the automobile, just like its predecessor the trolley, was hardly the technological panacea that was envisioned. The examples of the trolley and the automobile remind us that while some specific problems may be amenable to technological solutions, larger issues rarely admit of easy solutions through the application of technological fixes.

Why Technology Cannot Always Fix It

The main difficulty underlying the use of technology to solve social problems is that these problems are fundamentally different from technical problems. In the first place, social and technical problems differ in their specificity. If you intend to design an air conditioner, you at least know what your goal is: to keep a space cool. In many ways, this problem is similar to the far more grandiose objective of landing a man on the moon; although there may be daunting technical problems to overcome, at least the goal is clear and unambiguous. But what if your goal is to reduce crime? Crime, unlike air temperature, is a very diffuse concept, encompassing everything from forgery to murder. Even when a particular crime is singled out for treatment, its causes are likely to be manifold and not easily addressed by a single technology.

To make matters even more difficult, social problems are directly concerned with human motivations and behaviors. It is one thing to change the temperature of the air by inventing and installing an air conditioning system; it is quite another to attempt to change human behavior through the same kind of technological intervention. Human beings are wondrously intricate creatures whose actions are governed by extremely complex motivations. Trying to understand, let alone change, human actions is an exceedingly difficult task. And humans are likely to resist when attempts are made to change their behavior. Consequently, successful technological fixes tend to be those that get the job done without challenging the existing values and interests of the population. Hence, for example, electric cars, or seeding the ocean, to reverse climate change, are seen as desirable solutions because they do not require changes in our driving habits or pollution profiles.[24]

Or consider the 2014 Flint, Michigan, water crisis. The identification of lead and possibly bacteria in the water supply of Flint resulted from a change in the water supply due to budgetary constraints. It is estimated that between 6,000 and 12,000 children were exposed to high levels of lead in their drinking water, with possibly 100,000 residents exposed to high lead levels.[25] The technological fixes available: repairing the water supply and replacing utility and household lead pipes, were fairly obvious. But repairing Flint's water infrastructure required an economic and political commitment. We will review some of the specific political, ethical, and legal details of the case in a later chapter, but in the context of a discussion of a technological fix, the political will and organizational capacity to implement technological changes is part of making a technological fix work.

Figure 2.5 Lead pipes from Flint, Michigan.
(Source: Brand Diverse Solutions Steven Barber)

It is also apparent that technological solutions work best when they operate within closed systems—that is, when the issue to be addressed is sealed off from outside influences. Of course, no technology exists in isolation from the surrounding society. A transportation system based on private automobiles, for example, is the result of choices exercised within the economic and political realm, such as a government's decision to build a highway network. But within a given technology there are many specific matters that can be treated as purely technical problems. In these cases, it is possible to approach the problem directly and not worry about the influence of other factors. If your car fails to start one morning, you can be sure that the problem lies only with its components; you need not concern yourself with sunspot activity or a recent presidential election in Peru.

When a problem is not so easily isolated, a technological solution is much less likely. Today, nearly one out of six boys under the age of 10 in the United States is diagnosed with attention deficit hyperactive disorder (ADHD).[26] This behavioral problem undoubtedly has a neurological basis, at least for some children, and amphetamine-like stimulants such as Ritalin are routinely prescribed to alleviate the symptoms of ADHD. It is likely, however, that many children afflicted with the disorder have problems that go beyond the neurological. Dysfunctional relationships and actions within a family can create stresses that produce ADHD. Under these circumstances, the administration of a drug will be insufficient. As the ADHD website of the National Institute of Mental Health notes, "Sometimes, the whole family may need therapy."[27]

As a final point, it should be noted that no problem, technical or otherwise, is ever really "solved." Not only are most solutions incomplete, but they also generate new (and sometimes very different) problems. These unintended and often unforeseen problems may be even more difficult to solve than the original problem. This process has been dramatically illustrated by the rapid development of modern medical technologies, a topic that will be explored in greater depth in Chapter 7. Technical solutions such as the development of lifesaving drugs, organ transplants, and sophisticated diagnostic techniques have proliferated, but at the same time they have created a host of new dilemmas. Given the expense of many of these new technologies, it may be necessary either to spend more on medical care or to attempt to ration it. If these technologies are to be rationed, will this take place through the price mechanism, or will it be done according to some formalized procedure? In either case, serious ethical issues will have to be faced. Life-extending technologies have also raised vexing questions about the morality of prolonging a life under conditions that seem dismal indeed. Moreover, a longer individual life span leads to an aging population and the necessity for a wide range of adjustments to the society, the economy, and even the culture. Without belaboring the point, it should be apparent that no set of technologies will make our lives better without requiring the enactment of other changes, which may generate problems of their own.

The Appeal of Technocracy

These inherent limitations have not deterred a number of individuals and groups from trying to convert social problems into technical problems. There have been numerous flirtations with technocracy—the governance of society by engineers and other people with technical expertise, who attempt to develop policies based on technical and "scientific" principles. There is no denying that the technocratic vision is at first glance an appealing one. In a world too often governed by venal and incompetent politicians, there is something very attractive about a system of governance that supposedly bases itself on logic and the use of expertise. Moreover, where conventional political systems of all types seem endlessly involved with apportioning pieces of a small pie, adherents of some form of technocracy often promise a social and economic order that produces an ever-expanding pie through the application of the methods that have served technological development so well.

The promises and pitfalls of a technocratic approach to the solution of social problems are well illustrated by the theories of Scientific Management, as developed by Frederick W. Taylor (1856–1915) and his followers during the early decades of the twentieth century.[28] Scientific Management arose in an era marked by a profound paradox: industrial production was increasing at a rapid pace, but at the same time American society was racked by large-scale and potentially explosive conflicts between workers and management. Many cures for labor unrest had been proposed, but for Taylor, all of them missed the mark. Taylor had earned an international reputation as a metallurgical engineer, and his systematic studies on the cutting tools used for machining metal had resulted in major technological advances. If obdurate metals could be better controlled and shaped through the application of new technologies guided by scientific principles, why could the same thing not be done with workers?

To achieve this goal, Taylor and his colleagues developed a "scientific" regimen for studying work. It was premised on a naïve view of scientific objectivity, and of the "neutrality" of technologies. The main technique used for this task was the time-and-motion study through which workers were systematically observed and their work motions precisely timed. Through an analysis of these observations and measurements Taylor came up with a supposedly optimum set of motions for a given job, all of them subject to rigid time constraints. Equally important, the development and administration of these motions were the business of management exclusively, and any attempt by workers to go about their tasks independently would necessarily result in wasted motions and general inefficiency. A basic tenet of Scientific Management was that the planning and organization of work had to be separated from its actual execution. Only specially trained managers had the time and expertise necessary for the devising of optimal methods of production. The prime obligation of the workers was to do what they were told to do.

Although they had no power to plan and manage their own work, workers were supposed to benefit from the system. Because their work activities were now optimized, production would supposedly increase significantly. Workers would necessarily share in these higher returns, for Taylor also advocated that workers be paid according to piece rates rather than straight wages; the more they produced, the more they earned.

The technocratic spirit of Scientific Management is thus evident: the tasks and prerogatives of management rested not upon the exercise of raw power but on management's technical superiority in guiding the production process. At the same time, Scientific Management promised relief from continual squabbling over relative shares of the fruits of production; an optimal system of organization would result in more of everything for everybody. Taylor was not content with using Scientific Management as a solution for the problems of the workplace; its principles, he claimed, "can be applied with equal force to all social activities: to the management of our homes; the management of our farms; the management of the business of our tradesmen large and small; of our churches, our philanthropic organizations, our universities; and our governmental departments."[29]

The appeal of Scientific Management was not confined to the United States, or even to the capitalist world. Vladimir Lenin, the leader of the Bolshevik Revolution in Russia, expressed a deep admiration for American technology and American forms of industrial organization, and for Taylor's ideas in particular. Although he duly noted that Scientific Management embodied "the refined cruelty of bourgeois exploitation," Lenin made it clear that its basic principles and procedures could contribute to the realization of Soviet economic goals: "The possibility of building Socialism will be determined precisely by our success in combining Soviet government and the Soviet organization of administration with the modern achievements of capitalism. We must organize in Russia the study and teaching of the Taylor System and systematically try it out and adopt it to our ends."[30]

The Technocrat's Delusion

Although some of its elements, such as the use of time-and-motion studies, can still be found in contemporary managerial practices, Scientific Management in its pure form never took hold in the United States, the Soviet Union, or anywhere else. A number of technical problems impeded its use. Considerable skill was required for the administration of time-and-motion studies, and they were especially difficult to conduct in work settings not characterized by repetitious actions. But of equal or greater importance, both management and labour realized that the implementation of Taylor's system posed fundamental threats to their own interests. Most managers were highly reluctant to delegate their authority to the dictates of "scientific" procedures.[31] Workers, on the other hand, resented the loss of what little autonomy they had, and they widely believed—with considerable justification—that higher levels of productivity would result in the downward adjustment of piece rates, leaving them no better off than before the program had been enacted.

Scientific Management, like all technocratically inspired systems, ignored the distinction between technical and sociopolitical problems. As a technological fix, even if Scientific Management had generated the productive increases it promised—which is unlikely—it would still have been strongly resisted by those who had to submit to it. Scientific Management promised a conflict-free method of administration where no such thing was possible. Workers and managers had their separate interests, and each group was unwilling to entrust its fate to Taylor and his disciples.

The basic fallacy of Scientific Management, one shared by all other variants of technocracy, is that *administration* can replace *politics*. Administration is based on the application of rules that allow the realization of given ends. It is thus a manifestation of the rational spirit of applying the best means for the achievement of a particular

Machine shop ....
Order for ....Tires...........
Do work on Tire No ....
As follows and per blue print ....

	Templet	Size to be cut to	Depth of cut	Driving belt	Feed	Rate	Time this operation should take
Surface to be machined	.....	.....	.....	.....	.....	.....	....
Set tire on machine ready to turn....	.....	.....	.....	.....	.....	.....	....
Rough face front edge	.....	.....	.....	.....	.....	.....	....
Finish face front edge	.....	.....	.....	.....	.....	.....	....
Rough bore front...	.....	.....	.....	.....	.....	.....	....
Finish bore front...	.....	.....	.....	.....	.....	.....	....
Rough face front I. S.C.............	.....	.....	.....	.....	.....	.....	....
Cut out filled	.....	.....	.....	.....	.....	.....	....
Rough bore front I. S.F.............	.....	.....	.....	.....	.....	.....	....
Rough face back edge	.....	.....	.....	.....	.....	.....	....
Finish face back edge	.....	.....	.....	.....	.....	.....	....
Finish bore back ...	.....	.....	.....	.....	.....	.....	....
Rough bore back ...	.....	.....	.....	.....	.....	.....	....
Rough face back I. S.F.............	.....	.....	.....	.....	.....	.....	....
Cut out filled	.....	.....	.....	.....	.....	.....	....
Cut recess	.....	.....	.....	.....	.....	.....	....
Rough turn thread..	.....	.....	.....	.....	.....	.....	....
Finish turn thread..	.....	.....	.....	.....	.....	.....	....
Rough turn flange..	.....	.....	.....	.....	.....	.....	....
Finish turn edge....	.....	.....	.....	.....	.....	.....	....
Clean fillet of flange.	.....	.....	.....	.....	.....	.....	....
Remove tire from machine and clean face plate	.....	.....	.....	.....	.....	.....	....

Figure 2.6 A scientific management worksheet. Frederick Taylor believed that all kinds of work could be reduced to rationally derived actions, much as machining operations could be precisely timed through the use of this worksheet. (Source: *Shop Management* by Frederick Winslow Taylor, 1903)

goal. It does not, however, determine these ends. The relationship between means and ends will be one that we discuss many times throughout this book. Consider that the Internal Revenue Service (IRS) officials who administer the tax system are not the authors of the tax code. Around April 15, we may get angry about the perceived unfairness of the tax code, but it is pointless to blame the officials at the local IRS office, who are only executing a set of policies that they did not create.

Tax codes and other policies are formulated through choices made in the political arena. Neither technology nor administration can supply the values that form the basis of these choices. They cannot determine the ends, or the priorities, to which we apply our collective knowledge and power. They cannot tell us what we should do with our lives, nor can they help us to resolve the fundamental issue that all societies confront: how to fairly distribute life's necessities and luxuries. The resolution of these issues will always be marked by sizable differences of opinion and a good deal of conflict. The technocrat's hope that society can be run on the basis of engineering principles will always remain an illusion. For example, attempts at automating policing, through facial recognition and algorithms, are an example of a technocratic governance that reproduce preexisting assumptions and systems of bias, and are, of course, highly contested.[32]

To summarize, technological changes inevitably produce social changes, because technological changes themselves are the product of social forces working at different scales or social registers. Technological changes, in turn, do not affect everyone equally. Although many technologies produce widespread benefits, not everyone benefits to the same degree, and there are instances where particular individuals and groups lose out completely. A choice of technology is often a determination of who wins and who loses; it is therefore proper that affected parties have the opportunity to participate in the process. This issue will be taken up in greater depth in Part 7. At this point, we hope we have provided some sense of the limitations of technology without deflating the very real achievements. Technology and the procedures underlying its development have been immensely powerful in their own realm; outside this realm, however, they are less likely to be effective. Equally important, the methods that have been so successful in developing and applying new technologies cannot be transferred to the governance of society. Technological development may make some aspects of our lives better, but it can never substitute for a just and effective political and social system.

DISCUSSION QUESTIONS

1. Technological advance has often undermined established businesses. Most recently, the growth of internet-based e-commerce has posed a threat to conventional brick-and-mortar retail firms. Can you think of other business enterprises that the internet may damage or even destroy? Should anything be done to prevent this from happening?
2. Were the Luddites justified in mounting their attacks on machinery? How else might they have expressed their grievances? Would other kinds of actions have been more successful?
3. What examples of technological "fixes" can you think of? Have they been successful or not? What are your criteria for judging success and failure?
4. Political leaders at home and abroad are occasionally described as "technocrats." What are the implications of this description? Would you be more or less likely to vote for somebody who was described in this way?

Notes

1. Moses Abramowitz, "Resource and Output Trends in the United States Since 1870," *American Economic Review, Papers and Proceedings* 56 (May 1956): 5–23; John W. Kendrick, "Productivity Trends, Capital and Labor," *Review of Economics and Statistics* 37 (August 1956): 248–57; R. M. Solow, "Technical Change and the Aggregate Production Function," *Review of Economics and Statistics* 39 (August 1957): 312–20.
2. Jeffrey D. Sachs and John W. McArthur, "Technological Advancement and Long-Term Economic Growth in Asia," in Chong-En Bai and Chi-Wa Yuen (Eds.), *Technology and the New Economy* (Cambridge, MA: MIT Press, 2002): 157–85.
3. See Howard P. Segal, *Technological Utopianism in American Culture 1830–1940* (Chicago: University of Chicago Press, 1985).
4. Quoted in Herbert J. Muller, "Human Values and Modern Technology," in Edwin T. Layton, Jr. (Ed.), *Technology and Social Change in America* (New York: Harper & Row, 1973), p. 159.
5. MIT "Real-Time" Rideshare Research, "Rideshare History and Statistics" (January 24, 2015). Accessed from http://ridesharechoices.scripts.mit.edu/home/histstats/.
6. Business Insider, "Here's How Uber Got Its Start and Grew to Become the Most Valuable Startup in the World" (September 13, 2015). Accessed from

www.businessinsider.com/history-of-uber-and-its-rise-to-become-the-most-valuable-startup-in-the-world-2015-9.

7. See for example: BBC News, "France Cracks Down on Uber Service After Protests" (June 26, 2015). Accessed from at www.bbc.com/news/world-europe-33267581.
8. Sarah McBride and Dan Levine, "In California, Uber Driver Is Employee, Not Contractor: Agency," *Reuters* (June 18, 2015). Accessed from www.reuters.com/article/uber-california/in-california-uber-driver-is-employee-not-contractor-agency-idUSL1N0Z31BS20150617.
9. Davey Alba, "Florida Says Uber Driver Isn't an Employee After All," *Wired* (October 1, 2015). Accessed from www.wired.com/2015/10/florida-uber-decision-reversal/.
10. Helling, Brett, "Are Uber Drivers Independent Contractors or Employees?" Ridester (October 8, 2023). Accessed from www.ridester.com/uber-drivers-independent-contractors-employees.
11. W. F. Cottrell, "Death by Dieselization: A Case Study in the Reaction to Technological Change," *American Sociological Review* 16 (June 1951): 358–65.
12. Ibid., p. 362.
13. Langdon Winner, "Do Artifacts Have Politics?" *Daedalus* 109 (Winter 1980): 123–25.
14. Peter Blake, *Form Follows Fiasco: Why Modern Architecture Hasn't Worked* (Boston: Little, Brown, 1974), p. 59.
15. George Rude, *The Crowd in History: A Study of Popular Disturbances in France and England, 1730–1848* (New York: John Wiley & Sons, 1965), pp. 66–92; Malcolm I. Thomis, *The Luddites: Machine-Breaking in Regency England* (New York: Schocken Books, 1972).
16. See also: Sale, Kirkpatrick, *Rebels Against the Future: The Luddites and Their War on the Industrial Revolution: Lessons for the Computer Age* (NY: Basic Books, 1996).
17. Paragraph 128 of "The Unabomber Manifesto." This document is available on numerous websites.
18. Lori Osterberg, "7 Unexpected Benefits of a Digital Detox," *HuffPost* (October 22, 2015). Accessed from www.huffingtonpost.com/lori-osterberg/digital-detox_b_8361700.html.
19. Wiebe Bijker, "The Social Construction of Bakelite: Toward a Theory of Invention," in Wiebe E. Bijker, Thomas P. Hughes, and Trevor Pinch (Eds.), *The Social Construction of Technological Systems: New Directions in the Sociology and History of Technology*, Anniversary Edition (Cambridge, MA: MIT Press, 2012), pp. 155–182.
20. See the US Patent Office: www.uspto.gov/web/offices/ac/ido/oeip/taf/us_stat.htm for the number of patents applied for, and the number granted, over time. See also: www.inventiontherapy.com/how-many-patents-actually-make-money. It is estimated that about 10 percent of all patents actually earn enough money to warrant the cost of filing a patent. Just the "fact" of a new technology does not guarantee its up-take into society. We'll explore this in later chapters.

21. James C. Williams, "The Trolley: Technology and Values in Retrospect," *San Jose Studies* 3, (November 1977): 74–90.
22. James J. Flink, *The Car Culture* (Cambridge, MA: MIT Press, 1975), p. 39.
23. William F. Dix, "The Automobile as a Vacation Agent," *Independent* 56 (June 2, 1904): 1259–1260, quoted in ibid., pp. 39–40.
24. Daniel Sarewitz and Richard Nelson, "Three Rules for Technological Fixes," *Nature*, 456 (December 18, 2008). Accessed from www.nature.com/nature/journal/v456/n7224/full/456871a.html; and Philip Boyd and Chris Vivian, "Should We Fertilize Oceans or Seed Clouds? No One Knows," *Nature* (June 11, 2019). Accessed from www.nature.com/articles/d41586-019-01790-7.
25. Chris Clearfield and András Tilcsik, *Meltdown: Why Our Systems Fail and What We Can Do About It* (New York: Penguin Press, 2018), pp. 121–128; and "Flint Water Study Updates," *The Virginia Tech Research Team*," Flintwaterstudy.org.
26. "The Age of Unreason," *The Economist*, 416 (July 11, 2015), Special Report, p. 6.
27. National Institute of Mental Health, "Attention Deficit Hyperactivity Disorder (ADHD)." Accessed from chrome-extension://efaidnbmnnnibpcajpcglclefindmkaj/https://www.addrc.org/wp-content/uploads/2015/08/ADHD_Booklet_CL508_144426.pdf
28. For a biography of Taylor, see Robert Kanigel, *The One Best Way: Frederick Winslow Taylor and the Enigma of Efficiency* (New York: Viking, 1997).
29. Quoted in Samuel Florman, *The Existential Pleasures of Engineering* (New York: St. Martin's Press, 1976), p. 8.
30. V. I. Lenin, "The Immediate Tasks of the Soviet Government," *Izvestia* (April 28, 1918). Accessed from www.marxists.org/archive/lenin/works/1918/mar/x03.htm.
31. Daniel Nelson, *Managers and Workers: Origins of the New Factory System in the United States, 1880–1920* (Madison: University of Wisconsin Press, 1975): 75–76.
32. Ruha Benjamin, *Race After Technology: Abolitionist Tools for the New Jim Code.* (New York: Wiley/Polity Press, 2019).

PART 2

The Emergence and Development of Technologies

Much of the research, writing, and thinking about the relationship of technology and society centers on how particular technologies have affected some aspect of the latter. But how do new technologies come into being in the first place? Chapter 3 describes how social processes are at the core of technological innovation and examines two modes of technological change in that light—epochal transformations and less dramatic incremental ones. It describes how economics, politics, culture, and social arrangements have influenced the trajectory of technological change. A market-based economy is an important stimulus for technological innovation, but nonmarket forces—ranging from human curiosity to government institutions—are also stressed.

Scientific discoveries are often seen as the major source of technological advance, but Chapter 4 argues that this is at best a half-truth. Although science and technology have much in common, they do not interact in a straightforward manner. When they do interact, the commonly accepted relationship between scientific and technological advance may be inverted as technological advances propel advances in science.

Chapter 5 shifts the focus from the creation of technologies to the ways in which they spread. It pays particular attention to the pitfalls of transferring a technology from one social and cultural environment to another and from one business firm to another. Especially in the poor parts of the world, a technology may not be effective and may even be harmful. The chapter also takes a critical look at the patent system, which can be both a stimulus and an impediment to the diffusion of new technologies. Its influence on technological advance is examined in the final section.

Chapter 6 again takes up the idea of the "Social Construction of Technology" (SCOT) and situates it in relations to with special attention to the strengths and weaknesses of the classic diffusion model. The SCOT model has been central to critical thinking about new technology and with that model we will gain new tools for tracing the economic, cultural, and political factors which shape technologies. The chapter will use automobiles and competing alternatives such as bicycles, local public transport and national railway systems, to illustrate the complexity of technological development. We will also visit some long-lost technologies, such as the Apple Newton and other case studies, to look broadly at the processes pushing and pulling, and impeding, technological change.

Chapter 3

The Sources of Technological Change

What accounts for the emergence of particular technologies? Why do they appear when they do? What sort of forces generate them? How is the choice of technology exercised? To put it more concretely, why were digital computers developed only during the second half of the twentieth century even though their basic principles were understood more than a hundred years earlier? Why did photography undergo rapid development during the nineteenth century? What were the inventors of radio trying to accomplish, and how did their intentions differ from those of subsequent developers? These are some of the questions that this chapter will address as it considers some of the most basic issues in the study of technology.

Technological Change as a Social Process

As a starting point, it is important to keep in mind that technological change does not take place in a social vacuum. Technology is a human creation, and because humans are social creatures, technological change is necessarily a social process. In recent years, the study of technological change has been strongly influenced by a perspective known as "social constructivism," which we looked at in Chapter 1. According to this approach, technological change does not occur because new devices and processes demonstrate their clear-cut superiority over other ways of doing things. For social constructivists, the analysis must begin with the need to explain why certain technologies are assumed to work better than others.[1] As Wiebe E. Bijker has noted, social constructivism is predicated on a belief in "the malleability of technology, the possibility for choice, the basic insight that things could have been otherwise."[2]

To explain why things turned out the way they did, social constructivists describe how social structures and processes have affected choices of technologies. Since the presence of interest groups and unequal distributions of power are fundamental aspects of every society, social constructivists are particularly interested in delineating the main actors involved in the development and selection of particular technologies, and in noting how their actions reflect their positions in society. Accordingly, for scholarly practitioners of social constructivism, technological change is an inherently political process. New technologies do not succeed or fail solely on the basis of narrow technical merits. Rather, the achievement of technological "closure" (the point at which a particular technology is recognized as the accepted way of doing things, while others disappear or are marginalized) is closely tied to the presence of specific interest groups and their ability to affect the selection process.

Some form of social constructivism informs most contemporary studies of technological change, but considerable variation can be found in the relative emphasis put on social versus technical factors. Moreover, some students of technological change, most notably Thomas P. Hughes, have argued that the strength of the social constructivist approach may depend on the developmental stage of a particular technology. According to Hughes, social constructivism is most valid when a technology is at an early stage of development. Social, political, and economic forces are likely to exert the greatest influence when several alternative technologies emerge at about the same time.[3] Conversely, once a technology has become well established, it becomes difficult to deviate from the path that has been laid out by technical requirements.

To take one example, at the end of the nineteenth and beginning of the twentieth centuries, motorists could choose between cars with electric, steam, or internal combustion power plants. The triumph of the latter was not simply a matter of technical superiority, but was a reflection of the needs and expectations of the individuals who were the prime purchasers of automobiles.[4] Once automobiles powered by internal combustion engines became well entrenched, the adoption of another type of automobile engine became extraordinarily difficult because the basic elements of our personal transportation infrastructure were firmly in place, everything from the fuel used to the skills necessary for effective repair work. Under these circumstances, technical requirements will prevail.

The Great Breakthrough

One benefit of the social constructivist approach is that it challenges the belief that technological change largely stems from the insights and labors of a few supremely talented individuals. Popular histories of technology have often looked to individual genius as the chief source of technological advance; we have all heard or read stories of how the inspired labors of Thomas Edison or the Wright brothers produced epochal inventions that transformed the world. Histories written from this point of view are in essence biographies of great inventors whose brilliance is assumed to be the sole source of technological advance.

Other histories of technological advance have remained within this framework but have looked to a different kind of genius as the source of advance. The key players here are not the inventors of new technologies but rather the entrepreneurs who turn inventions into commercial successes by taking risks, moving into uncharted territory, and in general doing what hadn't been done before. There have been some individuals who have been both inventor and successful entrepreneur—for example, Edwin Land, the creator and moving force behind the Polaroid camera—but they have been rare. Entrepreneurs take other people's inventions and make them into commercial successes. From this perspective, the key figure in the development of the steel industry is not Henry Bessemer, the co-inventor of the iron-refining furnace that bears his name, but Andrew Carnegie, who laid the commercial and organizational foundations of the industry.

Arriving at a definitive determination of the relative importance of "great men and women" versus "social processes" in shaping the history of technology would require much more space than is available here. It can be noted, however, that an assessment of the relative contributions of the two must consider the fact that a great deal of technological change is the result of small, incremental changes. In contrast to the "heroic" approach to the history of technology, these involve the work of largely anonymous inventors, engineers, mechanics, and technicians.

Although their individual contributions may seem modest, in aggregate they have been an extremely important source of technological advance. These incremental changes often are the result of a learning process that occurs as a technology is used. Problems are identified and overcome, bugs are ironed out, and improvements are made. In many cases, the cumulative results of these efforts are technological advances at least as important as those that stem from fundamental breakthroughs. In industries as different as petroleum refining and building construction, the major source of productivity improvements has been a multitude of small technological improvements that have resulted in large cumulative gains.[5]

This process is nicely illustrated by Louis Hunter's narrative of how the impressive development of nineteenth-century steamboats was the result

> of plodding progress in which invention in the formal sense counted far less than a multitude of minor improvements, adjustments and adaptations. The story of the evolution of steamboat machinery in the end resolves itself in a large part into such seemingly small matters as, for instance, machining a shaft to hundredths instead of sixteenths of an inch, or devising a cylinder packing which would increase the pressure a few pounds, or altering the design of a boiler so that cleaning could be accomplished in three hours instead of six and would be necessary only every other instead of every trip. Matters such as these do not get into the historical record, yet they are the stuff of which mechanical progress is made.[6]

One can also witness the far-reaching consequences of numerous small improvements in the development of railroad technology. According to one calculation, if the traffic loads borne in 1910 had been carried by railroads employing the technologies of 1870, the additional costs would have amounted to $1.3 billion by the latter date. Fortunately, by 1910, American railroads had benefited from improvements in the size of cars and the power of locomotives, which in turn were the result of steady evolution.[7]

And so it goes today. Even though the railroad may be described as a "mature" industry, this pattern of incremental yet substantial technological development continues. American railroads have significantly lowered their costs through the implementation of a series of important but not epochal improvements to diesel-electric locomotives: turbochargers, computer-controlled fuel-injection systems, higher compression ratios, lower idling speeds, computerized controls to prevent wheelspin, and the substitution of alternators for direct-current generators. This cluster of advances culminated in much greater power, while also reducing energy costs.[8]

In a world where the ability to produce successful technological innovations is increasingly a requirement for a firm's success, if not its survival, each firm needs to tap every source of technological advance, no matter how modest it may seem. This is a point stressed by Anthony Athos and Richard Pascale in their book about American and Japanese management:

> Careful scrutiny reveals that despite the exalted status of "strategy" in the lexicon of American management, few great successes stem from one bold-stroke strategic thrust. More often, they result from one half-good idea that is improved upon incrementally. These improvements are invariably the result of a lot of "little people" paying attention to the product, the customer, and the marketplace.[9]

The "D" in R&D

The process of making a technology work is often summarized by the abbreviation R&D, which stands for research and development. "Research" calls to mind images of cutting-edge work in well-equipped laboratories, where great breakthroughs produce dramatically new technologies. Research can be an exciting, even glamorous activity, and we naturally look to it as the basis of technological progress. It is the source of fundamental change in technology, such as integrated circuits, cloning, and composite materials. Still, this sort of research rarely results in usable products. The realization of the potentialities created by research breakthroughs usually requires a lengthy process of development. Numerous problems must be resolved, and, equally important, the new material or device must be put into a form that allows it to be produced at a reasonable cost.

Here, again, we can see the importance of the slow, unspectacular improvements that turn a good idea into a working product or process. And here, too, we can often see a substantial outpouring of money. If basic research is an expensive process, development is often even more so. The development work that goes into preparing a new technology for actual production can entail massive expenditures for equipment, material, manpower, pilot plants, and the like.

A great deal of development work is oriented toward "scaling up"—that is, making the transition from a successful research result to large-scale production. It is one thing to invent a device or process that works under laboratory conditions, and quite another to make a product in large quantity or extensively employ a process in an industrial setting where commercial success is the goal. The development of penicillin provides an excellent illustration of the many facets of the scaling-up process.[10] Although the discovery of the bacteria-killing properties of penicillin initiated a major technological breakthrough, the development phase was certainly no less important.

Penicillin, a fermentation product of the mold *Penicillium chrysogenum*, was discovered in 1928 by Alexander Fleming, who observed that bacteria were destroyed in a culture that had been accidentally contaminated by the mold. Penicillium cultures grown in small quantities were the basis of laboratory and clinical research, but this process could not yield the large quantities of the drug needed for widespread therapeutic use. Large-scale production of penicillin was eventually done in huge fermentation vats, a process that required the solution of many technical problems. The key step was the development of a submerged fermentation process that allowed the mold to be grown directly in the nutrient medium. The success of this process in turn required several other improvements, such as new tank designs with special cooling systems and turbine mixers, which also had to be developed. The use of corn steep liquor (a by-product of corn starch production) as a culture medium increased yields tenfold, but it created a new problem. Penicillin requires air to grow, but severe foaming occurred when the culture was aerated. Antifoaming products therefore had to be developed to alleviate this problem. The extraction of penicillin from the moldy brew in the vats also created problems that were ultimately solved using freeze-drying, which was itself scaled up from a process first used to preserve blood plasma. As a result of this development work, production of penicillin had reached 650 billion units by the end of World War II, and the cost of a dose was 55 cents instead of the $20 it had been three years earlier.

Putting It All Together

The first chapter noted the importance of thinking of technologies as systems. This point is reinforced by a consideration of how particular technologies develop. Quite often, a technological leap forward takes place because of the availability of complementary technological developments that allow the resolution of fundamental problems. The history of the computer provides a good example of how complementary changes are essential for the translation of an idea into a workable technology. Beginning in the 1820s, Charles Babbage developed in England an "analytical engine" that contained a set of input devices, a processor, a control unit, a memory storage, and an output mechanism—the essential elements of today's computers. But Babbage's computer was operated by an exceedingly complex set of gears, cams, levers, and other mechanical linkages. Although he could draw on the talents of Ada Lovelace, who is often described as the world's first computer programmer, Babbage's ambitions were not fully realized for more than a century, when solid-state electronics, the cathode ray tube, and magnetic storage devices allowed the development of practical computers.[11]

Figure 3.1 Charles Babbage's computer.

Another illustration of the importance of complementary technological changes can be drawn from the history of one of the twentieth century's most important devices—the internal combustion engine. When these engines were first produced during the late nineteenth century, the spark that ignited the air–fuel mixture was timed to occur at the top of the compression stroke. This did not allow a sufficient time for the mixture to be completely ignited, and efficiency consequently suffered. When, after much experimentation, the spark was timed to occur before the piston reached the top of its stroke, the combustion process was improved, and much more power was consequently delivered.

After all this work, early engines were deficient in power because compression ratios were kept low to prevent the sudden detonation of the air–fuel mixture and resultant damage to the engine. Air-fuel mixtures were addressed in part by conducting careful inquiries into the combustion process and by reshaping the combustion chamber and piston crown. But this was not enough. To employ higher compression ratios successfully, it was necessary to modify not just the engine but also the fuel it used. In particular, the research of Thomas Midgley and Charles Kettering demonstrated that the addition of tetraethyl lead to gasoline allowed higher compression ratios and a subsequent improvement in both power and economy. It was thus through a combination of cumulative improvements in both the engine and the fuel it burned that the internal combustion engine reached an acceptably high level of efficiency.

This is not the end of the story. As we saw in the previous chapter, technological solutions can give rise to new problems. So it was with the case of tetraethyl lead. In addition to the immediate environmental damage it caused, leaded gasoline destroyed the active elements in a major emissions-control device, the catalytic converter. Beginning in the 1970s, lead had to be removed from gasoline and replaced by less destructive additives.

It is important to note that these changes in engine technology were made well after the internal combustion engine was first invented and was in widespread use. Lots of inventions require ongoing modifications over a lengthy period of time, and these in turn require ongoing sales revenues to pay for the needed improvements. Just as important, the experience of their customers may provide important guidance regarding where changes need to be made.

Under these circumstances, customers may be unwitting participants in the beta testing of new products. A tragic example of this occurred in 2016 when a Tesla Model S crashed into a truck, killing the Tesla's driver. The car was in Autopilot mode, which automatically controls steering and braking. Unfortunately, the sensors that controlled these functions failed to detect a truck that had made a left turn in front of the Tesla because the white truck did not stand out against the bright sky. Although Tesla had warned drivers to keep alert with their hands on the steering wheel, the temptation to yield control to a technology still being developed was too great.

Clearing Bottlenecks—The Reverse Salient in Technology Transfer

The commercial viability of a new technology may thus stimulate the development of complementary technologies. A bottleneck or reverse salient, as illustrated in Chapter 1, restricts the continued development of a particular technology and creates strong economic incentives to find new technologies that clear the obstruction. Many obstacles are surmounted using technologies that have been developed for use by a different industry or in a different sector. This can be seen today in the automobile industry, where emissions requirements have necessitated the use of microprocessors and computers for the regulation of spark advance and air–fuel ratios. Thus, a nineteenth-century technology, the four-stroke internal combustion engine, owes its survival to its marriage to a late twentieth-century technology.

The successful use of technologies from different realms can also be seen in the adoption of new ways of making things. This process is exemplified in the development of nineteenth-century metalworking industries. Although their products differed, the processes employed by these industries were basically the same: turning, boring, drilling, milling, and planing. At the same time, they all confronted similar technical problems of transmitting power, reducing friction, and controlling the rate of feed. This meant that a technological solution arrived at by one industry was often directly applicable to the problems of another industry.[12]

This process of technology transfer can also be seen in the early automobile industry when it made abundant use of the products and manufacturing techniques that had been developed by the bicycle industry during the 1880s and 1890s. Ball bearings, spoke wheels, drive chains, and the use of electrical resistance welding had been extensively employed for the manufacture of bicycles in the decades immediately preceding large-scale automobile production. One of the most novel and significant technologies entailed the use of stamped components to take the place of forgings. Their use eliminated a great deal of machining, with a consequent lowering of production costs. The cheap, mass-produced automobile thus owed much to technologies initially developed to make a product that it subsequently surpassed.[13]

This is hardly the only example of technologies developed by an established industry paving the way for a radically new one. Although the turbojet engine was a novel method of propulsion when it first appeared in the 1940s, it drew heavily on designs, components, and processes that had already been developed for steam turbines. In Edward Constant's summary, "All the work done on blade design, gas flow, shaft and bearing loads, temperature distribution, lubrication systems, governors, blade-cutting machines, test procedures and instruments, and countless other facets of design and production could be applied to gas turbine development."[14]

Interindustry transfers of technology do not happen automatically. The effective transfer of hardware, information, or simply (but importantly) the belief that a problem is solvable requires individuals and organizations that are capable of functioning in both worlds and have the incentives to do so. It also requires that these individuals and organizations are acceptable to the other individuals and organizations with which they interact, and that they are capable of speaking the same technical language. Technological innovation is, in Christopher Freeman's phrase, a "coupling process" that occurs at the interfaces between science, technology, and the market. This does not take place solely through intuitive flashes: "It is a continuous creative dialogue over a long period of research, experimental design, and development."[15]

Supply and Demand: The "Push" and the "Pull"

The first chapter outlined the deficiencies of technological determinism. As we have seen, social constructivists stress the importance of social, economic, and political forces in shaping technological change. And as was noted a few pages ago, contemporary scholarship has moved away from the "great person" approach to the history of technology. To be sure, many technologies owe their existence to the inspiration and hard work of individuals, including the unsung heroes who were responsible for the many improvements necessary for the realization of a new technology. But surely more is involved than their efforts. After all, human ability is presumably spread evenly throughout cultures and historical epochs, yet significant technological changes are not equally distributed over time and place. Thomas Edison produced 1,093 patents, and some of his inventions transformed the world. Had he been born in ancient Rome or dynastic China, he might have helped to design aqueducts or sections of the Great Wall, but it is unlikely that his talents would have changed the course of Roman or Chinese history. Geniuses require appropriate social settings for the realization of their talents.

What kind of social system is required if inventive ability is to flourish? Why does technological innovation occur in some places and times and not in others? In beginning to answer these questions, it is helpful to apply to technological change the concepts that have been so useful to economists: supply and demand. Everything from fundamental scientific breakthroughs to minor refinements push new technologies into the world. Still, simply having an available supply of new devices and techniques does not guarantee that they will be used. Many examples of technologies languished because they were "ahead of their time." The pneumatic tire was patented in 1845 and then forgotten until it was reinvented by John Dunlop in 1888. DDT was first synthesized in 1874, but it was not used as an insecticide until 1941. Several decades elapsed before the laser passed from being a laboratory curiosity to a practical device used for everything from supermarket scanners to instruments for microsurgery.

For a technology to make the transition from the potential to the actual requires not just that it exist; there must also be a desire for it, coupled with the ability to pay for it. Economists call this "effective demand." Seen in this light, technology is like any other good or service; it will not be produced unless some individual, group, or organization wants it and is willing to buy it. Technology is "pushed" by a variety of forces, but it also has to be "pulled" by effective demand. To understand why certain technologies have flourished while others have languished it is necessary to consider the configuration of a society and the way in which it determines the effective demand for particular technologies.

The most influential research on the importance of effective demand inducing technological change was done by Jacob Schmookler.[16] Schmookler examined a long series of patents in various industries and found that their emergence was closely related to the level of demand for the products of these industries. To take one straightforward example, inventors' interest in improving the horseshoe was strong when the horse was a primary means of transportation, but interest evaporated when the steam engine and the internal combustion engine began to displace it.[17] Although Schmookler's study was conducted more than fifty years ago, several studies have supported his thesis; commercial enterprises, anticipated profits and the perceived size of a market are major drivers of technological innovation.[18]

The history of photography in the nineteenth century is another example of the importance of demand inducing technological development.[19] For centuries, painters and scientists had made use of the *camera obscura*, a darkened room into which light was admitted through a pinhole, resulting in the projection of an inverted image of an outdoor scene on the opposite wall. Later developments substituted optical lenses for the pinhole, which made for a sharper image. Photographs were first produced during the 1820s and 1830s when pioneers such as Niepce, Daguerre, and Fox Talbot devised chemical emulsions that preserved the image on paper or a metal plate. But the rapid growth of photography cannot be attributed simply to the supply of these inventions. At this time, social changes were sweeping across Europe, resulting in the ascendance of a new social elite, the property-owning commercial and industrial bourgeoisie. The members of this group had a keen desire to flaunt their affluence by taking on characteristics of the old aristocracy. For the latter, a key artifact of their status was the painted portrait; we have all seen renditions of persons such as the Eighth Duke of Puddleswallop hanging in our local museums. But many of the rising bourgeoisie lacked the money or the time for such symbols of their importance, and, in any event, there were not enough skilled portrait painters to serve the needs of this growing group of people. Their aspirations were therefore met by the photographic portrait studio, where the subject posed with the trappings of upper-class status, such as rich draperies and elegant furniture.

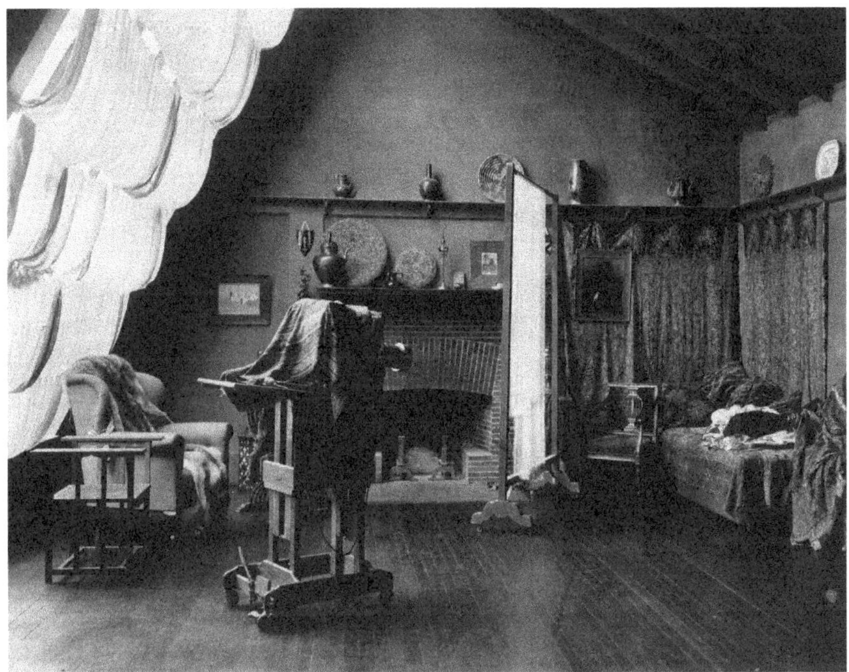

Figure 3.2 The interior of a nineteenth-century photographer's portrait studio.

In the later decades of the century, new and cheaper methods of photography emerged to meet the demands of poorer yet increasingly affluent people, such as immigrants to America who wanted portraits that could be sent back home. Today, the effective demand produced by a great mass of consumers has stimulated the development of a huge variety of photographic apparatus, ranging from ubiquitous mobile phone cameras to sophisticated digital cameras.

Good business practice is often implicitly based on the realization that successful technological development requires the presence of effective demand. One English study found that the key determinant of a firm's innovative success was an understanding of customer requirements. This meant that from its very inception, a new product or process had to be developed with an eye toward meeting the needs of actual or potential customers. This apparently is an infrequent occurrence; one study found that many inventions were commercial flops due to an inability to evaluate the true extent of demand. All too often entrepreneurs were so infatuated with an invention that they failed to realistically assess the nature of the market.[20] This kind of tunnel vision is a major reason that only about 3 percent of patent holders achieve any degree of commercial success.[21]

Belated Demand: The Question of Market Timing

At the same time, however, gauging the potential demand for a new product can be a tricky task. Many of today's "essential" technologies were not at first recognized as such. When in the late 1930s Chester Carlson attempted to interest established business machine manufacturers in his photocopying device—the first Xerox machine—they were of the uniform opinion that there was no point in employing complicated apparatus and chemicals simply to replace carbon paper. And even inventors can badly misjudge the ultimate consequences of their own creations. Alexander Graham Bell initially thought that the telephone he invented would primarily be used to transmit operas and other musical performances from the concert hall to the home.[22] Edison at first believed that one of the main applications of his phonograph would be to record the last words of dying men. We can also take note of the judgment of Howard Aiken, the director of the team that built one of the world's first computers, who in the early 1950s prophesied that in the foreseeable future, the total need for computers in the United States could be met by no more than a half-dozen machines. In similar fashion, before 1950, Thomas J. Watson, the president of IBM, believed there would be no commercial market for computers.[23]

The history of radio also shows how a technology may be put to uses not envisaged by its inventors. When equipment for sending and receiving radio waves was first developed during the late nineteenth century, no one imagined that it would be used for commercial broadcasts. The first transmitters and receivers were devised for purely intellectual purposes—to test the validity of James Clerk Maxwell's theories about the nature of electromagnetic waves. Only after the passage of more than a decade did some visionaries perceive a commercial use for radio apparatus, and then their imaginations were limited to the use of the radio for ship-to-shore communications. Broadcast was a problem to be reconciled in relation to the objective of point-to-point communications, such as telegraphy.

Decades passed before the idea of broadcasting to a mass audience emerged—before it became a solution, not a problem.[24]

What are we to make of these examples? They seem to refute the theory that technologies are primarily the result of "demand-pull," for the technological developments just described emerged in the absence of apparent demand for them. Perhaps we can salvage the demand-pull theory by distinguishing two different kinds of technological advances. The first kind consists of refinements and improvements to an existing way of doing things, while the second (and far less frequent) is the truly revolutionary breakthrough—the internet, digital computer, radio, telephones, and the like. In cases such as these, the very novelty of a revolutionary breakthrough makes it difficult to determine what its ultimate uses will be and who, if anyone, might be a future customer. By contrast, advances of the first kind occur within a known context; the basic technology is already in use, and there are likely to be people and business firms that want, and are willing to pay for, the new wrinkles that promise to improve an existing technology. Manufacturers of smartphones, for example, introduce new models on a regular basis even though the

Figure 3.3 ENIAC, a first-generation electronic digital computer, was programmed by plugging and unplugging cables.

enhancements may be trivial. Improvements of this sort therefore have a predictable market. Conversely, radically new technologies confront a great deal of uncertainty. They may satisfy a latent need, or they may create a new one. They may also sink without leaving a trace. They are flights into the unknown, and it is hazardous to guess what sort of a reception they will meet. If nothing else, they confirm the old Chinese saying that it is dangerous to make predictions—especially about the future.

How Market Economies Drive Technological Innovation

The forces that "push" and "pull" technological advance do not exist everywhere or in equal measure. In many places and at many times, the distribution of wealth and power retarded these forces, resulting in a slow pace of technological advance. Moreover, the particular technological advances that do occur usually reflect a society's general configuration of wealth and power.

In the European Middle Ages, the landowning aristocracy and Church officials controlled most of the wealth and wielded great power. The monastic orders often played an important role in land clearing, farming, and the construction of mechanical devices, but for the most part, the religious and secular establishment showed little interest in such matters. While the era gave rise to significant technological advances in water power, mechanical clocks, and weaponry, its most evident technological triumph was the great symbol of the traditional order: the Gothic cathedral.

As European history unfolded, the interests and demands of a growing merchant class led to the development of technologies that eventually surpassed even the soaring cathedrals. The great technological innovations that began in the mid-fifteenth century with improvements in shipbuilding and ocean navigation were closely associated with the rise of capitalism and the emergence of a market system. A market system organized around the principle of private property was of crucial importance for the stimulation and guidance of inventive and innovative abilities, as well as their application to production.[25]

One of the strongest accolades to the technological dynamism of capitalist society can be found, of all places, in *The Communist*

Manifesto. With unfeigned admiration, Karl Marx and Friedrich Engels note that

> The bourgeoisie, during its rule of scarce one hundred years, has created more massive and colossal productive forces than have all preceding generations together. Subjection of Nature's forces to man, machinery, application of chemistry to industry and agriculture, steam-navigation, railways, electric telegraphs, clearing of whole continents for cultivation, canalisation of rivers, whole populations conjured out of the ground—what earlier century had even a presentiment that such productive forces slumbered in the lap of social labour?[26]

In the time of Marx and Engels, and in our own era, a market economy driven by the activities of self-interested businesspersons has produced a highly receptive environment for technological innovation. There are several reasons for this. First, a market economy will stimulate inventive efforts, for it promises financial rewards to those able to meet the needs of consumers. For example, somebody invents a better mousetrap in the hope of selling it in the market. If the demand is there, eager customers will buy it. Everybody is better off: consumers have a better mousetrap, while the inventor gets rich and retires to Palm Springs, and nobody worries about how technological advance has lowered the quality of life for mice. Second, a market economy is characterized by the presence of numerous competitors. Under these circumstances, a producer is strongly motivated to develop and apply new technologies to make better products and to reduce production costs. Failure to do so may result in the eventual collapse of the enterprise, as the history of many once-successful firms demonstrates. Finally, a market system is particularly effective in eliciting the production of the auxiliary items necessary for technological innovation. A new technology will require special materials, components, and services. Because of its responsiveness to new sources of demand, a market economy is well suited to meet these requirements.

Obstacles to Innovation in Centrally Planned Economies

The advantages of a market economy in stimulating technological advance are further demonstrated by an examination of centrally planned economies. For decades, the economies of the Soviet Union and the People's Republic of China were organized through the mechanisms of central planning, but during the 1980s, it became painfully evident that these mechanisms were fatally flawed. One of the chief manifestations of that failure has been hindered technology. It cannot be denied that the Soviet Union produced some impressive technological achievements, most notably in its space and military programs, while China made significant progress in industrializing an impoverished country—no small feat. But taken as a whole their level of technological development did not significantly advance, while the United States, Western Europe, and Japan moved rapidly forward.

The flaws of centrally planned economies produced significant political repercussions. In China, dissatisfaction with the pace of technological and economic advance led to a retreat from centralized planning and a much greater scope for market-based economic relations. In the Soviet Union, the fear of falling even further behind the West motivated the Gorbachev regime to introduce some elements of a market economy and to attenuate the role of central planning. But these efforts were not enough, and within a few years the Soviet Union ceased to exist. In the countries of the former Soviet Union and in China, it has become evident that market-based reforms have made major contributions to economic and technological development, albeit at the cost of greater economic and social inequality, as the recent histories of the two nations have shown.

Many of the difficulties experienced by centrally planned economies in achieving technological advance have been the result of a basic tension between their system of economic management and the requirements of technological innovation. Centrally planned economies rest on the assumption that economic activities can be reduced to predictable routines. But the course of technological innovation is notoriously difficult to predict. The bureaucratic procedures that work tolerably well for the administration of routine productive tasks usually fail when they are applied to technological innovation. A planning agency can set goals and quotas for producing of established goods, and various ministries can oversee the actual operation of individual enterprises through routine bureaucratic administration. But these procedures work less well when innovation is the goal. Because innovation is an activity full of risk and unpredictability, and it cannot easily be accommodated to pre-programmed structures and activities.

To make matters worse, centrally planned economies attempt to motivate workers and managers through the allocation of rewards that create disincentives for technological innovation. A factory manager typically receives bonuses for fulfilling or exceeding quotas for established products, as given by the central plan. Creating an innovative product is not rewarded, for it has not been stipulated by the plan. The uncertainties that surround technological innovations create risks for those who seek to develop new products and new ways of doing things, but these risks are not matched by commensurate rewards for those who take them. Nonetheless, the features of the market itself can be a barrier to innovation, as short time frames (quarterly earnings expectations) and risk aversion on the part of financiers and shareholders can be barriers to innovation.

Noneconomic Sources of Technological Advance

It is not the intention here to convey the impression that only market forces can produce technological innovation; as we shall see, government institutions have become increasingly important sources of technological advance. And no inference should be made that the historical superiority of a market economy in promoting technological advance makes it a superior system in general. There is more to life than technological advance, and, as we have already seen and shall see some more, both capitalism and the market have produced technologies that have been detrimental to people and the environment.

Furthermore, technological innovation cannot always be traced to economic motives or even to the desire to address practical problems. To be sure, we tend to think of technology as the result of efforts to solve problems of this sort; after all, technology has already been defined as the product of knowledge that is used to get something done. The very word "technology" conjures up images of useful devices, and technology's practitioners—engineers, managers, and skilled workers—are often viewed as a serious bunch, sitting rigidly in front of computer terminals, making precise measurements, and, above all, applying their talents to the solution of practical problems that are usually tied to economic concerns.

The Spirit of Play and Curiosity

In fact, even the most practical of inventions may owe their origins to a spirit that seems more closely connected to play than to "productive" work. When Willis Whitney served as the first director of the research laboratory of the General Electric Company, he often asked his scientists and technicians there if they were "having fun." For Whitney, "fun" was working on problems that had stumped everyone. Pursuing these problems was nothing less than the most exciting thing that a person could do.[27]

Consider, too, one of America's most famous inventors, Benjamin Franklin. With typical American pragmatism he wrote, "Utility is, in my opinion, the test of value in matters of invention, and that a discovery which can be applied to no use, or is not good for something, is good for nothing."[28] Franklin's inquiries into the nature of electricity did result in one useful device: the lightning rod, which saved many a building from destruction. But his other inquiries had a less immediate payoff. Although Franklin devised several devices that helped him to learn more about the nature of electricity, decades passed before electricity had any practical value. Indeed, he was "chagrined a little that we have been hitherto able to produce nothing in this way of use to mankind."[29]

Later events proved him wrong, although he never shared in that knowledge. Still, the pursuit of useful innovations could not have been the prime motive for Franklin's inquiries. Franklin was an amateur in the literal sense of the word: a person who pursues an activity for the sheer love of it. For many years, the leisure-time pursuits of amateur scientists such as Franklin sustained research into the nature of electricity despite the absence of direct applications, yet these "idle" intellectual efforts were essential to the subsequent invention of a great variety of useful devices.

A century after Franklin died, a young man of 17 climbed a cherry tree and turned his imagination to possibilities that only a few had dreamed of. The year was 1899, and the young man was Robert Goddard, who was to be the inventor of the liquid-fuelled rocket. As he recalled in later years:

> It was one of the quiet, colourful afternoons of sheer beauty which we have in October in New England, and as I looked toward the fields at the east, I imagined how wonderful it would

> be to make some device which had even the possibility of ascending to Mars, and how it would look on a small scale, if sent up from the meadow at my feet.... I was a different boy when I descended the tree from when I ascended, for existence at last seemed very purposive.[30]

At that time, and for many years to come, Goddard could scarcely have imagined the products of that October vision: orbiting satellites for global communication links, weather prediction, global positioning systems, and the development of terrifying new weapons. Throughout his life Goddard continued to be energized by the dream of space travel for its own sake; practical consequences were at best a secondary concern.

Figure 3.4 The realization of a vision: Robert Goddard and his first liquid-fueled rocket, the progenitor of today's SpaceX rocket.

Cross-fertilization of the arts and technology also produce innovations, as artists apply new technologies to new forms of expression and make demands on existing technologies to achieve new effects. For example: 3D printing, laser cutting, and digital imaging all produce new artworks, new in their materials, ideas, and in interactivity, and artists make demands of the technologies to get them to do what they need.[31] The world of "DIY"(Do It Yourself) problem solving and innovation still brings new ideas to market with fresh approaches to accessible technologies, from life hacks to workarounds to new devices for solving local problems.[32] DIY approaches and the integration of arts and technology remind us that curiosity, pleasure, and the self-satisfaction of making something new, beautiful, and maybe useful, can still be sources of innovation even while much technological innovation is captured by big-scale and highly complex organizations and systems.

Government and the Promotion of Technological Advance

It also should be noted that even in a predominantly market-oriented, capitalist society such as the United States, not all technologies have been generated and shaped by market forces. In Part 7 we will take up this issue by examining the role of the government in promoting technological change. At this point, it will only be noted that technologies developed outside the constraints of the market system are less likely to be shaped by concerns about costs. Firms operating in a market environment know that minimizing costs is essential to success. Technologies that hold down production costs are quite appealing, as are technologies that expand sales by lowering the cost of the product itself. The Ford Motor Company during the glory years of the Model T is a striking example of a firm that owed its success to the use of new technologies to lower costs. This process has also been dramatically demonstrated in recent years as technological advances in the design and production of integrated circuits have led to sharp declines in the price of personal computers and a concomitant expansion of this market. At the same time, however, there are large sectors of the economy where prices and the costs of production arc secondary concerns. As a result, the pattern of technological development can be quite different. This has been particularly evident in the defense industry, where the presumed dictates of national security have spurred the deployment of staggeringly expensive military technologies. To take a particularly striking example, a single United States Air Force F-22 Raptor aircraft costs U.S. taxpayers an estimated $377 million (when R&D and testing costs are taken into account).[33] In a climate where only the presumed best will do, there are strong tendencies to "gold plate" weapons systems and to have little concern for cost constraints.

This tendency is not confined to the military sector. In recent years, there has been a mounting concern about the financial costs of modern medical technologies. Some medical technologies, such as antibiotics, have undoubtedly lowered the costs of medical care, but many others have had the opposite result. When faced with a choice between controlling medical expenses or saving lives and alleviating pain using sophisticated technologies, it is difficult for any individual or society concerned with the well-being of its members to put the former ahead of the latter. It is thus no surprise that about half the growth of health care spending that occurred during the second half of the twentieth century can be attributed to the use of new technologies.[34]

The case of military and medical technologies brings us back to the necessity to consider the political and social context of technological choice. Economic considerations, important as they are, are not the sole basis of decisions regarding the development, selection, and use of particular technologies. Nor does technology develop according to its own internal dynamics. Technologies are social creations, and any successful attempt at understanding why particular technologies are created, chosen, and used must consider their social context. Having made this basic point, we will pursue the matter a bit further by looking at the processes through which technologies spread and take hold. But before we do so, we need to consider one more source of technological change: scientific advance. This will be the topic of the next chapter.

QUESTIONS FOR DISCUSSION

1. To what extent were your prior ideas about new technology shaped by ideas about technological determinism? How does a new idea like social construction challenge your prior knowledge?
2. In times past, inventors like Edison, Morse, Tesla, and the Wright brothers were treated as celebrities. Can you name any contemporary inventors? Why do individual inventors appear to be less prominent today?
3. A person with a heart problem needs a pacemaker but is unable to pay for it. What, then, is the source of effective demand for this technology? Who ultimately pays for it, and why?
4. As a would-be inventor or entrepreneur, how would you go about ascertaining whether or not a new technology is likely to find a receptive market?
5. If you were the leader of a developing country, what sort of changes would you try to institute to accelerate the pace of technological innovation?

Notes

1. Wiebe E. Bijker, Trevor Pinch, and Thomas P. Hughes (Eds.), *The Social Construction of Technological Systems* (Cambridge, MA, and London: MIT Press, 2012).
2. Wiebe E. Bijker, "Understanding Technological Culture through a Constructivist View of Science, Technology, and Society," in Stephen H. Cutcliffe and Carl Mitcham (Eds.), *Visions of STS: Counterpoints in Science, Technology, and Society Studies* (Albany: State University of New York Press, 2001), p. 27.
3. Thomas P. Hughes, "Technological Momentum," in Merritt Roe Smith and Leo Marx (Eds.), *Does Technology Drive History? The Dilemma of Technological Determinism* (Cambridge, MA, and London: MIT Press, 1994), pp. 99–113.
4. Rudi Volti, *Cars and Culture: The Life Story of a Technology* (Baltimore: Johns Hopkins University Press, 2006), pp. 7–9; Gijs Mom, *Atlantic Automobilism: Emergence and Persistence of the Car 1895–1940* (New York and Oxford: Berghan Books, 2015), pp. 62–63.
5. Nathan Rosenberg, *Inside the Black Box: Technology and Economics* (Cambridge, England: Cambridge University Press, 1982), pp. 62–70.
6. Louis Hunter, *Steamboats on the Western Rivers* (Cambridge, MA: Harvard University Press, 1949), pp. 121–122. Quoted in Rosenberg, *Black Box*, p. 64.
7. Albert Fishlow, "Productivity and Technological Change in the Railroad Sector, 1840–1910," in *Studies in Income and Wealth No. 30: Output, Employment,*

and Productivity in the United States after 1800 (New York: National Bureau of Economic Research, 1966), pp. 635, 641.

8. Association of American Railroads, "Freight Rail Facts & Figures," (October 19, 2023). Accessed from www.aar.org/facts-figures.
9. Richard Tanner Pascale and Anthony G. Athos, *The Art of Japanese Management: Applications for American Executives* (New York: Warner Books, 1981), p. 306.
10. The following is based on John A. Heitman and David J. Rhees, *Scaling Up: Science, Engineering, and the American Chemical Industry* (Philadelphia: Center for the History of Chemistry, 1984), pp. 17–21.
11. Paul E. Ceruzzi, *Computing: A Concise History* (Cambridge, MA: MIT Press, 2012), pp. 5–6.
12. Peter George, *The Emergence of Industrial America: Strategic Factors in American Economic Growth since 1870* (Albany: State University of New York Press, 1982), p. 51.
13. David A. Hounshell, *From the American System to Mass Production, 1800–1932: The Development of Manufacturing Technology in the United States* (Baltimore: Johns Hopkins University Press, 1984), pp. 189–215.
14. Edward Constant, *The Origins of the Turbojet Revolution* (Baltimore: Johns Hopkins University Press, 1980), p. 82.
15. Christopher Freeman, "The Determinants of Innovation: Market Demand, Technology, and the Response to Social Problems," *Futures* 11 (June 1979): 211.
16. Jacob Schmookler, *Inventions and Economic Growth* (Cambridge, MA: Harvard University Press, 1966).
17. Ibid., p. 93.
18. Daron Acemoglu, *Introduction to Modern Economic Growth* (Princeton University Press, 2009), pp. 415, 430.
19. W. G. L. De Haas, "Technology as a Subject of Comparative Studies: The Case of Photography," *Comparative Studies in Society and History* 21 (July 1979): 367–70.
20. Christopher Freeman, *The Economics of Industrial Innovation*, 2nd ed. (New York: Cambridge University Press, 1982), p. 124.
21. Stephen Key, "In Today's Market, Do Patents Even Matter?" *Forbes* (November 13, 2017). Accessed from www.forbes.com/sites/stephenkey/2017/11/13/in-todays-market-do-patents-even-matter/?sh=1406480856f3
22. Freeman, *Economics*, p. 127.
23. Wilson Dizard, *The Coming Information Age: An Overview of Technology, Economics, and Politics* (New York: Longman, 1982), p. 33. See also Paul Ceruzzi, "An Unforeseen Revolution: Computers and Expectations, 1935–1985," in Joseph J. Corn (Ed.), *Imagining Tomorrow: History, Technology of the American Future* (Cambridge, MA: MIT Press, 1986).
24. Hugh G. J. Aitken, *Syntony and Spark: The Origins of Radio* (Princeton, NJ: Princeton University Press, 1985).
25. Robert Heilbroner, *Between Capitalism and Socialism: Essays in Political Economics* (New York: Random House, 1970), p. 162.

26. Karl Marx and Frederick Engels, *Manifesto of the Communist Party*, (1848). Accessed from www.marxists.org/archive/marx/works/1848/communist-manifesto /ch01.htm.
27. Elting E. Morison, *From Know-How to Nowhere: The Development of American Technology* (New York: New American Library, 1977), p. 126.
28. Quoted in Roger Burlingame, *March of the Iron Men: A Social History of Union through Invention* (New York: Grosset & Dunlap, 1938), p. 77.
29. Ibid., p. 77.
30. Barton C. Hacker, "Robert H. Goddard and the Origins of Space Flight," in Carroll W. Pursell, Jr. (Ed.), *Technology in America: A History of Individuals and Ideas* (Cambridge, MA: MIT Press, 1981), p. 233.
31. Randy Rieland, "7 Ways Technology is Changing How Art is Made," *Smithsonian Magazine* (August 27, 2014). Accessed form www.smithsonianmag.com/arts-culture/7-ways-technology-is-changing-how-art-is-made-180952472/; and "The Future of Art," Accessed from www.wired.com/brandlab/2018/07/the-future-of-art/
32. The Borgen Project, "DIY Innovations in the Developing World." Accessed from https://borgenproject.org/diy-innovations-in-the-developing-world/
33. David Axe, "Buyer's Remorse: How Much Has the F-22 Really Cost?" *Wired* (December 14, 2011). Accessed from www.wired.com/2011/12/f-22-real-cost/.
34. Daniel Callahan, "Health Care Costs and Medical Technology," *The Hastings Center* (2015). Accessed from www.thehastingscenter.org/Publications/BriefingBook/Detail.aspx?id=2178.

CHAPTER 4

Scientific Knowledge and Technological Advance

One of the most common beliefs about technology is that it is simply "applied science." There are certainly many examples that can be cited in support of this view. Modern medical practices have been strongly influenced by fundamental discoveries in biology. The development of the transistor depended on a thorough understanding of quantum mechanics. Synthetic materials have been made possible by research into polymer chemistry. But one should not be content to rest with these examples. When the full spectrum of technological advance is considered, it becomes evident that science does not always play the decisive role in the development of technology. Indeed, many are the times when technological advances have taken place without the benefit of scientific knowledge. Conversely, on some occasions scientific advance has depended on prior technological achievements. In this chapter we will look at the complex and shifting relationships between science and technology in order to come to a better understanding of how they differ, as well as the ways in which they have influenced each other.

The Historical Separation of Science and Technology

The definition of technology that was offered in the first chapter stressed that technology is based above all on the application of knowledge. But not all knowledge need be derived from scientific research. It is certainly true that today much of the knowledge required for technological advance is derived from scientific inquiries. Still, when the full history of technology is surveyed, it is apparent that most technologies have been developed and applied with little scientific input. For example, the ancient Greeks made important contributions to many sciences—most notably astronomy, optics, and acoustics—as well as producing major advances in mathematics. Greek technology also progressed through innovations in agriculture, building construction, mining, the refining of metals, and military equipment. Yet none of these innovations drew to any significant degree on Greek science. Moreover, the Greeks' technological achievements were far less impressive than their scientific achievements, again indicating the lack of connection between the two. This lopsided pattern of development continued with the Romans, although in reverse. Roman contributions to science were minor, while Roman engineering, manifest in in the construction of great aqueducts and roads that are still in use today, reached a high level of development through the use of existing principles.

The European Middle Ages were a time of slow but significant technological advance. Improved agricultural practices were introduced, and the power of wind and falling water was used for everything from grinding grain to polishing metal. An effective horse collar allowed the literal harnessing of another important source of power. Soaring cathedrals were built in many parts of Europe, where they continue to be a source of awe and inspiration. Again, these achievements owed nothing to the scientific inquiries of the time. In fact, the designers and builders of the cathedrals apparently did not even have knowledge of multiplication tables. Then, too, there was little that technology could have drawn on, for medieval science exhibited little of the dynamism of medieval technology.

At about the same time, blacksmiths in parts of the Middle East were using steel superior to anything made in Europe. The swords and other edge weapons that they made from the steel first produced in Damascus (the capital of present-day Syria) combined a hard cutting edge with the flexibility necessary for an effective weapon. Yet it was only late in the twentieth century that the metallurgical principles underlying Damascus steel were discovered. Although it was unknown to the swordsmiths of the time, minute quantities of impurities, vanadium especially, made an essential contribution to the unseen processes that gave the steel its desired qualities. Consequently, when the composition of imported iron ore changed, the steel made from it lacked the desired characteristics. Unable to draw on modern metallurgical knowledge, traditional swordsmiths could not make the necessary adjustments, and the "secret" of Damascus steel was lost for centuries.[1]

The disconnect between scientific and technological development continued through the succeeding centuries, but there were some points of tangency. During the fifteenth and sixteenth centuries the working methods of some "natural philosophers" (as scientists were so labelled in those days) benefited from an exposure to the practices of skilled artisans. In particular, the latter's commitment to careful observation, accurate measurement, hands-on practice, and an empirical approach to seeking the truth were eventually assimilated into the process of scientific inquiry.[2]

From the seventeenth century onward, scientific knowledge expanded at an unprecedented rate, but technological practices remained stagnant, largely unaided by these advances. This separation between scientific and technological progress is evident when France and England, two of the wealthiest and most powerful countries in Europe, are compared. By the early decades of the nineteenth century, France led the world in the generation of scientific advances, yet it lagged behind England in the practical use of steam power and the application of mechanical principles.[3] All in all, this historical record led one historian of science, Thomas Kuhn, to speculate that for the bulk of human history, technology has flourished in societies where science has remained undeveloped, and vice versa. It is possible that our era is unique in its apparent ability to simultaneously support scientific and technological advance, all the while forging strong connections between them.[4]

Studies of Contemporary Science–Technology Relationships

Even today, when the connection between science and technology is much stronger than it was in the past, a great deal of technological change takes place without substantial inputs from science. The relative unimportance of science for many technological developments was highlighted by a study that was conducted by the U.S. Department of Defense in the mid-1960s. Dubbed "Project Hindsight," this study assessed the extent to which pure scientific research had been essential to the development of 20 major weapons systems.

In conducting their study, the Hindsight researchers began with a weapon system and traced its history backward in order to determine the key "events" that produced the knowledge that had been essential to its creation and development. The results of the study gave little credence to the commonly accepted view that scientific knowledge is the primary basis of technological development. Of the 710 events surveyed, only two were the result of basic scientific research, a minuscule 0.3 percent of the total.[5] Scientific research that was specifically directed toward a particular military project was of greater importance, accounting for 6.7 percent of events, while 2 percent were the result of scientific research directed toward commercial or nondefense needs. The greatest portion of events, the remaining 92 percent, owed little to concurrent scientific research, and relied almost entirely on established scientific concepts and principles.

Nearly 40 years later, a group of researchers conducted a similar study dubbed "Project Hindsight Revisited," in which they surveyed the processes involved in the design of the Apache helicopter, the Abrams battle tank, the Stinger antiaircraft missile, and the Javelin antitank missile. Although the researchers did not attempt to determine the role of basic scientific research in the design of these weapons, their report noted that most of the relevant research had been done well before these projects were initiated, and that very little basic research had been done in order to address specific design issues.[6]

Although Hindsight showed that technology's connection to science had not been as straightforward as is often assumed, one should not draw sweeping generalizations from this study. The authors of the original Hindsight project, as well as a number of its critics, were quick to note that the long-term influences of scientific research were not captured by the study's methodology. Project Hindsight considered only the effects of scientific research conducted for the most part after 1945, thereby removing from consideration the immense body of scientific knowledge that had accumulated before that time. The study's researchers found that a median delay of nine years separated the completion of a scientific research project from its technological application, even when research efforts targeted at specific missions were included. It was therefore not surprising that basic scientific research had few technological consequences during the 20-year span covered by the study.[7] By taking a longer chronological view, another study, entitled "Technology in Retrospect and Critical Events in

Science (TRACES)," contradicted Project Hindsight by determining that a number of innovations, ranging from oral contraceptives to videocassette recorders, had been directly tied to prior scientific research.[8] But even here, the researchers were obliged to point out that the sequence from scientific discovery to technological innovation is not linear, and that "a better understanding needs to be achieved concerning the two-way influence of science and technology."[9]

How Technology Differs from Science

If nothing else, these studies show that the connection between science and technology is not adequately captured by the common belief that technology is simply applied science, and that scientific discoveries quickly and easily give rise to technological applications. Some technologies draw directly on scientific research, while others make little direct use of it. This is rather obvious. Of greater significance is the fact that science and technology are quite different in their basic natures. This makes the translation of scientific knowledge into technological application a difficult and complex process.

Whereas science is directed at the discovery of knowledge for its own sake, technology develops and employs knowledge in order to get something done. The content of the knowledge may be rather similar, but different motivations underlie its pursuit and application. Here, of course, we are on slippery ground; it is often extremely difficult to discern the motivations underlying a person's activities, and it may well be the case that a particular engineer may be driven by the same desire to understand something for its own sake that animates the work of a pure scientist. Motives are often mixed and complex.[10]

Much of the prestige accorded to science is the result of its supposed purity; science is thought to be an intellectual venture, free from political, organizational, and economic constraints. The perceived insulation of scientists from the demands of their patrons confers a sense of higher ethical standards; scientists are beholden to nothing but the internal demands of science. A great deal of recent scholarship has sharply questioned this assumption. As has been the case with some current studies of the history of technology, science has been described and analyzed as a social construction. From this standpoint, scientific inquiry is not a disinterested, fact-driven search for truth, but a human creation that has been shaped by cultural patterns, economic and political interests, and gender-based ways of seeing the world.[11] For uncompromising social constructivists, successful scientific outcomes may have more to do with negotiation, the support of designated authorities, and resonance with prevailing attitudes than theoretical elegance or experimental evidence. Under these circumstances, the idea that science is a "pure" intellectual endeavor cannot be supported.

The social construction of science approach remains controversial, and, in any event, no social constructivists believe that scientific facts and theories are purely social creations that have nothing to do with underlying realities. Moreover, social constructivism is a largely academic enterprise, and most laypeople still believe in the objectivity of science and the purity of scientific motives. These qualities give individual scientists a claim to autonomy not enjoyed by other employees. Scientists are thus in a particularly favorable situation. The assumption that scientific progress leads to material progress confers an aura of practicality on their work, while at the same time they are in a good position to resist the overt control of their work by their sponsors. This framework is central to the original charter of the National Science Foundation in the U.S., by Vannevar Bush in 1945.[12]

In contrast, most engineers work under tighter constraints. Their employers expect results that have immediate applications and fall within a narrowly defined range of possibilities. A scientist may abandon a theory or an experiment in order to pursue a line of inquiry that unexpectedly arises during the course of his or her research. An engineer, however, rarely has this opportunity; there may be some room for serendipity, but the bridge has to be built within a given time frame and under definite budget constraints. For this reason, what separates scientific and technological inquiries may not be the motivations of individual practitioners but the motivations of their employers and patrons.[13]

Scientific Knowledge and Technological Knowledge

Technology also differs from science in the type and depth of knowledge that is required. The ultimate question asked of scientific knowledge is "Is it true?" For technological knowledge, the key issue is "Will it work?" Technological problems can often be solved with no understanding of what is going on. As we have seen, throughout history many technologies were effectively applied even though the basic principles underlying their operation were poorly understood, if they were understood at all. A similar situation can be found today; superconductors—i.e., materials that conduct electricity with no resistance when temperatures are below –200°C (–320°F)—are beginning to be used in motors and other devices, even though the physics of the process remains something of a mystery. It is also instructive to consider the story of the great scientist Johannes Kepler (1571–1630), who developed and employed the calculus of variation in order to derive optimum dimensions of wine barrels—only to discover that these dimensions were already being employed by the coopers who actually built the barrels![14]

Many other technological innovations seem to fall into this pattern. Although scientifically derived principles may emerge after the fact, many technologies have been guided almost exclusively by trial and error, with the successful ones informed by an intuitive sense of the right solution, and not by scientific truths.[15] As Eugene Ferguson has observed, at the end of the nineteenth century there were no scientific principles that could be invoked during the design of the first motorcycles; the placement of the engine, fuel tank, and other major components could be determined only through the actual construction and operation of motorcycles, without the benefit of scientific principles or other forms of existing knowledge. Ferguson therefore makes the point that "there is often no *a priori* reason to do one thing rather than another, particularly if neither had been done before. No bell rings when the optimum design comes to mind."[16]

How Technology Facilitates Scientific Discovery

Although we tend to think of science as the leading factor in technological advance, the reverse often occurs: scientific knowledge and discovery may be a by-product of technological achievements. The most obvious examples of technological developments playing a vital role in fostering scientific advance can be found in the provision of devices and instruments. From early telescopes and galvanometers to today's electron microscopes and gene sequencers, the products of technology have steadily increased our ability to observe and analyze the phenomena that science presents as objects of inquiry. To take only a few recent examples, the scanning tunneling microscope has allowed a much better imaging of plant and animal cells; the Webb space telescope gives us new insights into the age and size of the universe and how it was created; particle accelerators have enabled physicists to obtain a better understanding of the basic constituents of matter; and magnetic resonance imaging, complemented by powerful computers, has played a key role in the rapid development of neuroscience, the study of the brain and how it functions. All in all, it is thus no exaggeration to claim that scientific "instruments shape research, determine what discoveries are made, and perhaps even select the types of individuals likely to succeed as scientists."[17]

Less obvious are instances where scientific advance was stimulated by a technology already in operation that defied accepted scientific explanations and theories. This process is exemplified by the story of how the steam injector contributed to the abandonment of a popular scientific theory regarding the nature of heat.[18] In the mid-nineteenth century, many scientists believed that heat was the result of the presence of a substance known as "caloric." According to this theory, when caloric combined with other materials those materials became hot. Also, caloric particles were supposedly self-repellent; thus, when sufficient quantities of these particles came into contact with water, their repulsive quality resulted in water turning into steam.

While this theory had its uses, it could not explain the operation of the steam injector that was patented by Henri Giffard in 1858. The injector used steam from the boiler to lift water into it, an operation that seemed to mimic perpetual motion for those who subscribed to the caloric theory. In fact, Giffard, who was well-trained in academic science, based his injector on the Bernoulli principle, which postulated that the pressure of a fluid (in this case steam) drops as its velocity increases. The operation of the injector was therefore the result of expanding steam producing a partial vacuum that sucked water into the boiler.

Giffard's injector was no perpetual motion machine; it used a quantity of heat that was equal to the quantity of work expended in raising water into the boiler, plus the losses due to radiation and contact with surrounding surfaces. Its operation therefore made sense only when the interconvertability of heat and work was understood. This idea rested on the kinetic theory of heat, and it followed the first law of thermodynamics (which stipulates that, quantitatively, energy cannot be created or destroyed). The kinetic theory of heat was formulated several years before Giffard's invention but had been slow in winning acceptance. The rival caloric theory had many adherents in the scientific community, and it took the apparent anomaly of the injector to convert many of them to the now universally accepted kinetic theory of heat.

The steam injector illustrates the often subtle interactions between science and technology. The operation of the injector provided a strong stimulus for the acceptance of one scientific theory and the rejection of another. At the same time, another scientific theory had been essential to the invention of the injector. But scientific theories by themselves were not enough; the design and effective use of the injector still depended on the experiments and modifications performed by practicing engineers, for no set of theories was powerful enough to guide its design. Again, we have an example of a technology that worked even though existing scientific principles did not completely explain its operation.

This example and many others that could be cited indicate that science and engineering are still separate enterprises, although there are certainly linkages between them. Scientific knowledge can result in technological advances, while at the same time new and extant technologies can create opportunities and motivations for new scientific inquiries. Many technological developments reach a plateau due to a lack of scientific knowledge, thereby generating a clearly perceived need for fundamental scientific research. The knowledge obtained through technological practices and applications is thus the raw material of many scientists, whose work centers on explaining technological practices at a deeper level.[19]

One example of this process is the invention of the laser. During World War II the United States and other countries were engaged in a major effort to develop radar as a means of detecting enemy ships and aircraft. While participating in the development of radar technology, scientists used the knowledge that they had gained to make significant advances in microwave spectroscopy, which allowed a more accurate determination of molecular structures. One of the main developers of microwave spectroscopy, Charles Townes, although nominally a physicist, continued to work on technologies for the generation of microwaves. In 1954, he and his co-workers created a device they called the "maser" (for "microwave amplification by stimulated emission of radiation"). In 1958, he and a former student published a paper that outlined how the principle of the maser could be extended into the region of infrared, visible, and ultraviolet light. These ideas were the foundation for the laser (the acronym for "light amplification by stimulated emission of radiation"). At first the laser was the classic example of an invention looking for an application. But in succeeding years, the laser became the basis for a host of technologies ranging from scanners used at checkout counters to devices used for the surgical rejoining of detached retinas. In short, the development of one technology (radar) gave rise to scientific advance (the determination of molecular structures through microwave spectroscopy) and at the same time provided a scientific foundation for an entirely new technology (the laser).[20]

As this example indicates, the relationship between science and technology, far from being linear, may be one characterized by considerable back-and-forth movement. This feedback between science and technology may be a fundamental source of their dynamism.[21] This complex reciprocal relationship can be seen in the highest

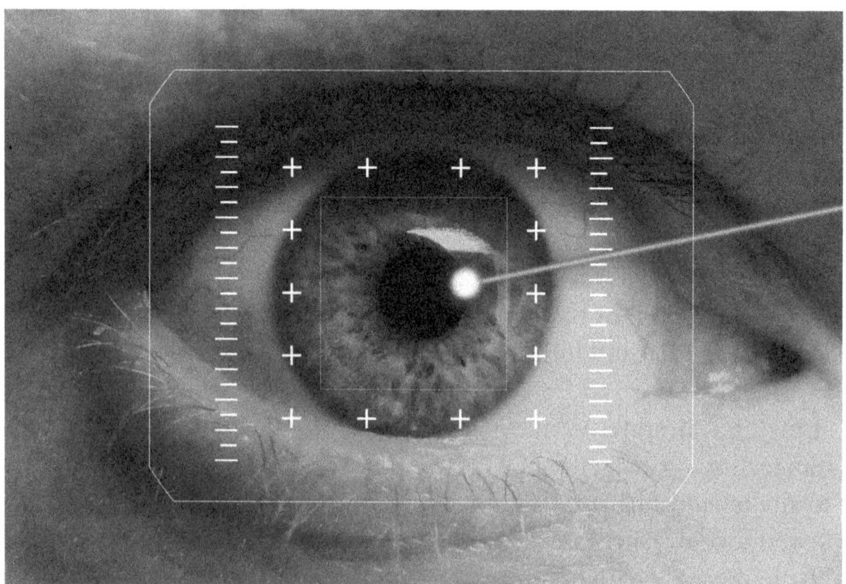

Figure 4.1 Laser eye surgery. One of the unanticipated uses of the laser is its use as a surgical instrument for the correction of faulty vision.

accolade for scientific achievement, the Nobel Prize. Although there is no prize for technology per se, a large portion of the prizes for chemistry, physics, and medicine have in fact been awarded for the invention of new devices and techniques. Some of them eventually resulted in commercially viable products, while others were used for further scientific inquiries.[22] In sum, when science and technology have gone their separate ways, as has been the case for most of human history, they develop more slowly than when they interact with each other, as they have done to an increasing degree during the present era.

Legitimating Science through Technology

There is a final and less immediately evident contribution that technology has made to scientific progress. Although an effort has been made here to demonstrate that science has not always been decisive in the development of new technologies, the opposite is widely believed. To a significant degree, this faith in the practical consequences of scientific research has given science the immense prestige and legitimacy that it enjoys today. Many areas of scientific inquiry have become increasingly expensive propositions. To take two admittedly extreme examples, the James Webb Space Telescope (the 2021 successor to the Hubble Space Telescope) cost nearly $10 billion, while Europe's Large Hadron Collider came with a price tag of $4.75 billion.[23] Expensive programs need sponsors with deep pockets. Although neither project is expected to generate technological spinoffs, many other high-priced research programs would die for lack of funding without the promise of some practical paybacks. Over the years, quite a lot of sophisticated and expensive biological research that has been justified on the grounds of its potential contribution to curing cancer—a hope that has yet to be realized. Biological research is hardly unique in this aspect, as scientists have become quite proficient in writing grant applications that stress the potential useful outcomes of their abstract inquiries.

Financial support, however important it is to the maintenance of scientific inquiry, is only part of the picture. The willingness of government agencies to grant money for scientific research and of citizens to have their taxes used in this manner is indicative of a widespread belief in the legitimacy of scientific research. This legitimacy is in large measure the product of the presumed ability of science to ultimately produce practical results. These ascribed powers of science have been analyzed by Langdon Winner:

> [The ultimate success of science] must be accounted to its fulfillment of Baconian ambitions—the delivery of power. Other modes of knowing have been able to give an intelligible, systematic, aesthetically pleasing picture of reality. If science had only been able to accomplish this and nothing more, it is likely that it would have been supplanted by yet another philosophy of inquiry. But in the West at least, the test is not so much *what do you know?* or *how elegant is your interpretation of worldly phenomena?* but rather, *what can you actually do?* This is the conclusive factor, the reason that, for instance, social science has never fully established its credentials in the halls of science.
>
> Science succeeds over rival ways of knowing—poetry, religion, art, philosophy, and the occult—not by its ability to illuminate, not even by its ability to organize knowledge, but by its ability to produce solid results.... In the last analysis, the popular proof of science is technology.[24]

This expected ability of science to "deliver the goods" is somewhat paradoxical, for science as a system unto itself responds rather poorly to economic needs. This has even been made into a virtue by many scientists who pride themselves on their insulation from the crass demands of the marketplace. As we have seen, the autonomy of scientists has been legitimized by the conception of science as a detached exercise in free inquiry. But it is also the case that the unpredictable nature of scientific discovery often precludes the possibility of useful discoveries being produced to order.

Scientific research, especially when directed at the discovery of basic principles, is an uncertain endeavor that cannot be guided by schedules and routinized procedures. This is illustrated by the response of one researcher who was offered more research funds by an officer of his company in the hope that the conclusion of a particular project could thereby be hastened. The researcher replied that it would be just as logical to expect that eggs could be made to hatch in half the normal time if twice as many hens were recruited to sit on them. Not only are the paths of scientific discovery full of twists and turns, but many of them terminate in dead ends. Of course, technology may also be incapable of solving the problems presented to it. If it were otherwise, we would have a cure for cancer by now. But most technology is directed toward the solution of specific problems, which narrows its scope and makes it a more predictable enterprise than science.

Scientific knowledge often ends up being extremely useful to technology, but in most instances that is not why it was produced. Scientists typically create knowledge for other scientists. Their efforts are focused on the testing of theories and the solution of problems that have been generated by previous scientific inquiries. If scientific knowledge is used for technological applications, it is because engineers and other technologists have appropriated it for their own use. In most places where science and technology meet, engineers and technicians "pull" knowledge out of science. Only in rare instances is knowledge directly relevant to technological application "pushed" by science itself.[25]

When knowledge is "pushed" from science into technology, it often happens indirectly. The transfer of knowledge from science to technology can be a subtle process, with scientific research motivating technological change by pointing out unseen problems and at the same time suggesting new opportunities. This happened in the 1930s when the evolving science of aerodynamics showed how the behavior of aircraft changed dramatically at high speeds. This research clearly indicated that conventional propeller-driven airplanes would encounter an insurmountable velocity barrier as they approached the speed of sound. At the same time, aerodynamic research indicated that proper streamlining could greatly increase the speed at which airplanes could fly, provided they had a different method of propulsion. In making these discoveries, aerodynamic researchers generated a powerful impetus for the development of jet engines that produced more power and did not have the inherent limitations of existing power systems.[26]

The Translation of Science into Technology

Today, many technologies make heavy use of the products of scientific inquiry. Much of this use, however, is indirect. A great deal of scientific information finds its way into technological practice through the education of engineers.[27] The findings of basic scientific research eventually appear in handbooks, university courses, and textbooks. Much of the scientific knowledge presented in these ways is eventually drawn on during the course of technological development.[28] Even here there can be problems. It has been argued that a significant amount of engineering education has been distorted by overreliance on science-based instruction. This has led to a devaluation of nonverbal thought, an excessive stress on mathematics, and an unwillingness to tackle problems that do not have a single unique solution.[29] Scientific thinking converges toward a single (if temporary) set of theories, while the history of technology is replete with examples of the old saying that there's more than one way to skin a cat. An excessive focus on the principles and methods of science may therefore restrict creativity and lead to an overly rigid approach to the solving of technological problems.

There is no getting around the fact that despite all that they have in common, science and technology usually operate in different worlds. If the two are to share in a productive symbiosis, they must be sustained by continual efforts to span the differences that separate them. In many cases, technological development has been stimulated by the presence of individuals and organizations that simultaneously participate in scientific and technological communities. Their primary role is to serve as translators, "decoding information generated in one system and transforming it into information usable in another."[30]

This process can be seen in the events that culminated in the invention of the vacuum tube, which in the pre-transistor era was an essential part of radio and television technology. The story began with Edison's invention of the light bulb.[31] While trying to determine why dark deposits were forming on the interior walls of the bulbs, Edison found that the needle of a galvanometer deflected when a wire probe was placed in a circuit between the galvanometer and the bulb's glowing filament. Edison did not understand what was producing the flow of electrical current through thin air, although he patented the modified bulb for use as a voltage indicator. (Many years later, the realization came that the current was produced by the migration of electrons from the negatively charged filament to the positively charged probe.)

Nothing practical came of Edison's discovery until John Ambrose Fleming renewed his acquaintance with the modified bulbs. During

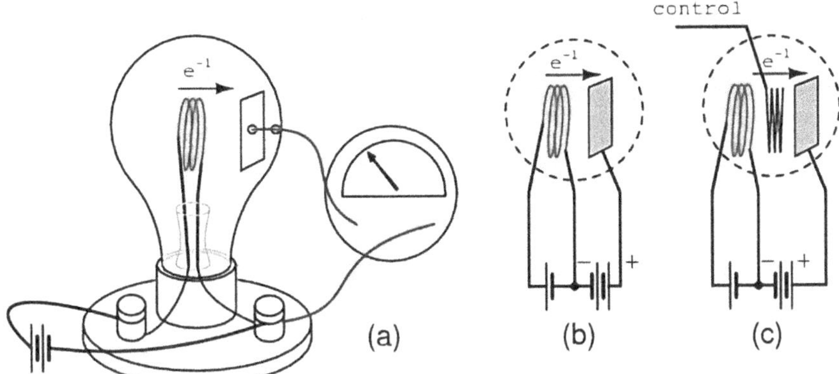

Figure 4.2 The Edison effect: Thomas Edison inserted a metal plate into a bulb and noted that when the power was turned on, a meter attached to the plate indicated that current was flowing through the air between the glowing filament and the plate. (Tony R. Kuphaldt/CC BY 4.0)

the 1880s and 1890s, Fleming had conducted a number of experiments using them; his sporadic efforts produced useful scientific knowledge, but no technological applications. Things began to change in 1899 when he became technical advisor to Guglielmo Marconi's Wireless Telegraph Company. At that time, the chief need of the infant radio industry was for a detector that could efficiently convert the weak oscillatory current of radio waves into direct current. In 1904, after a few years' work with other devices, Fleming came to the sudden realization that the specially equipped light bulbs with which he had previously worked might be used for this purpose. His hunch proved to be correct, and the "oscillation valve," as he named the device, began to be commercially used for the detection of radio signals a short time later.

Fleming had not been the only one to experiment with modified light bulbs, but he had been uniquely situated to act as a "translator" between science and technology. He was not an inventor like Edison or a full-fledged scientist like other experimenters. Rather, he was a scientifically trained engineer and teacher who was closely associated with the electrical industry and with engineering-training institutions. These separate but interrelated roles gave him the knowledge and the motivation to convert a scientific curiosity into a practical technology.

Figure 4.3 John Ambrose Fleming (left) with a vacuum tube diode (right).

The Commonalities of Science and Technology

Up to now, this chapter has stressed the differences between science and technology. At this point some mention should be made of the characteristics that they have in common. Both science and technology are based on the gathering of knowledge, and they both advance through the cumulative development of that knowledge. Isaac Newton is reputed to have said that he could see farther because he stood on the shoulders of giants. That is, his scientific discoveries were based on knowledge produced by earlier scientists. The same holds true for modern technology. Just as the scientific achievements of an individual chemist owe a great deal to the past research efforts of other chemists, the work of an aerospace engineer draws upon the accomplishments of other aerospace engineers.

More generally, science and technology have been nourished by a supportive culture at least since the days of the early Industrial Revolution. Although science provided few direct inputs into early industrialization, the values and attitudes of engineers and mechanics had much in common with those of scientists. As Peter Mathias has described this era:

> Together, both science and technology give evidence of a society increasingly curious, questioning, on the move, on the make, having a go; increasingly seeking to experiment, wanting to improve. So, much of the significance [of the cultural climate] impinges at a more diffused level, affecting motivations, values, general assumptions, the mode of approach to problem-solving, and he intellectual milieu rather than a direct transfer of knowledge.[32]

A key component of the shared culture of modern science and modern technology is their reliance on the rational thought processes described in Chapter 1. Although the development of both science and technology requires intuitive and other nonrational modes of thought, rationality is essential to the general methodology of science and technology. In general, a rational approach includes a propensity to challenge traditional intellectual authorities; a willingness to settle questions through observation, testing, and experimentation; and a desire to develop exact methods of measurement.[33]

Some of the basic elements of this mode of inquiry are described by Robert Pirsig in *Zen and the Art of Motorcycle Maintenance*, where he explains how even a clearly technological task like determining why a motorcycle won't start is addressed through the use of procedures that have much in common with scientific inquiry.[34] As a first step, a mechanic might formulate the hypothesis that the battery is dead; he or she will then try to honk the horn to see if the battery is working. If the horn honks, the mechanic concludes that the problem doesn't lie with the battery and proceeds to other parts of the electrical system. Should tests performed on these components show them to be in good shape, the mechanic may hypothesize that the problem lies with the fuel system and conduct tests (experiments) to check them out. And so it goes, with the formulation of a series of hypotheses and the conducting of experiments to test them. In the end the problem is isolated and perhaps fixed; if nothing else, you know what is wrong as you push your motorcycle along the side of the road.

Of course, one shouldn't take this analysis too far. Although both science and technology make heavy use of rational modes of thought, neither can be properly characterized as the embodiment of rationality. Scientific theories must be logically consistent and rationally articulated, but their ultimate source is human creativity and imagination—qualities often at a considerable distance from rational thought processes. At the other end of the scientific enterprise, the testing of these theories, there are no perfectly rational means of determining the criteria through which theories can be validated or disproved. Even empirically derived "facts" can be subject to interpretation, and general worldviews can strongly affect what is acceptable as "proof."[35] In similar fashion, a great deal of technological advance is also the product of nonrational thought. And, as was noted earlier, the benefit or harm of a particular technology cannot always be adjudged according to criteria based on rationally determined principles; a great deal hinges on values and ethical standards that have been derived through other means.

Other commonalities between science and technology can be noted. Mathematics is important to both as a kind of language and as an analytical tool. The practice of both science and technology very often requires university-based training that can stretch out for many years. Also, engineers and other technological practitioners employ organized knowledge that is presented and diffused through journals, books, blogs, and professional meetings that have many similarities to those found in the realm of science. And although engineers usually work for firms that try to retain exclusive use of innovative products and processes that were developed in-house, there can be a surprising willingness on the part of engineers to share their knowledge with engineers employed elsewhere, just as occurs between scientists.[36]

Although the sharing of information has long been a characteristic of science, in recent years an increasing number of scientific discoveries have come to be treated as proprietary information. This tendency has been particularly evident in biotechnology, where basic research is often essential for the rapid development of biological and medical technologies. Under these circumstances, the usual distinction between basic science and technological application no longer has much meaning.[37] In this field, and in a growing number of others, the distinction between science and technology has become so blurred that both can be subsumed under a single rubric, "technoscience." Since innovative, biologically based technologies can generate very large profits for the firms that develop them, these firms are likely to be reluctant to share their discoveries with the scientific community as a whole. It is not just for-profit private firms that have a reason for keeping scientific knowledge under wraps; universities are major players in industries based on cutting-edge technologies. Consequently, the research conducted in their laboratories may eventually generate substantial revenues. For universities and private firms alike, the lucrative coupling of basic research with technological application may seriously inhibit the sharing of new information, substances, and devices. These restrictions violate a basic canon of scientific culture—the free distribution of ideas and research findings—and, in the long run, they may result in a slower rate of progress for both science and technology.

Finally, at the core of the common culture of science and technology is a sense of optimism and progress within their own realms. Science and technology are dynamic enterprises that build on past successes, but they also make profitable use of their failures. An inadequate scientific theory may lead to the formulation of a better one, and a collapsed bridge is likely to provide valuable lessons that help to prevent future failures.[38] Above all, science is predicated on the belief that the world is knowable, while technology is animated by a conviction that it will always be possible to do something better. Both of these beliefs contribute to the dynamic, essentially optimistic spirits of science and technology.[39]

Although there are broad similarities between science and technology today, they are also different activities with different objectives. As noted above, the common-sense model of science and technology puts scientific research as a first step, followed by development, and eventually technologies for use. Stokes developed a different model, which was first of all far less simplistic than the Bush model, and also probably less self-serving for basic scientists.[40] There are two dimensions: do researchers explicitly search for use considerations, or try to get at fundamental principles? These are not mutually exclusive.

In the "Pure Basic" quadrant of Figure 4.4, particle physicists such as the late Niels Bohr are concerned with ascertaining the fundamental properties of matter, although a happy side effect of that has been technological innovations in support of their experimental works. In the "Use-Inspired Basic" quadrant, scientists are trying to

Considerations of Use: → Considerations of Fundamental Principles: ↓	No	Yes
Yes	Pure Basic (Bohr)	Use-Inspired Basic (Pasteur)
No	Descriptive (Peterson's Guides)	Pure Applied (Edison)

Figure 4.4 This chart is a representation of the model of different kinds of knowledge-seeking activities, as articulated by Stokes (1997). Is the primary goal theory construction and derivation of fundamental principles? Use? Each objective may or may not be prioritized, producing different kinds of research.

both discover basic principles and consider use. Most contemporary biotechnology is in the this quadrant, where scientists are trying to discover both the basic properties of living things (proteins, genomics, and so on) and prevent or cure disease. Advancing theory is part of the program for the two upper boxes.

For the "Descriptive" and "Pure Applied" quadrants, getting at generalizable truths or fundamental principles may not be an issue. In the "Descriptive" quadrant, the arts and some areas of the humanities might be classified there as a set of fields that make and interpret meaning and enrich human life. Generalizable theory is not really the objective; nor are developing reproduceable industrial products. A good example is birding guides (such as Peterson's[41]), which are not governed by an evolutionary theory of birds or anything generalizable. They are compilations of specific instances of birds, their behaviors, and their locations. It is not "useful" in an instrumental sense, such as producing new manufactured goods, but it is highly valued to the people who buy (and sell) the books and love to go birding. The knowledge-based activity has intrinsic value and meaning for users. Finally, the "Pure Applied" quadrant is the work of inventors, engineers, and technicists like Edison. Edison used the science of his time extensively, but wanted things to work and he himself didn't contribute to scientific literature or advance fields of knowledge, per se.[42]

One way to think about this framework is to consider the idea that all knowledge is *applied*. Is the knowledge applied to internal disciplinary questions, to generating generalizable knowledge, or is it employed to outward-looking applications, problem-solving, and new technologies? What we are trying to point out is that there are multiple relationships between scientific knowledge and technologies. Both are forward-looking fields of human action, and as we will discuss in Part 7, there are important policy questions to ask about the balance of research across the dimensions of theoretical and applied work, and between research outcomes and products which are instrumentally useful or with intrinsic or personal meaning and value.

This chapter began with the assertion that technology is not simply applied science. It is undeniable that technology today makes extensive use of scientific knowledge. But as we have seen, scientific knowledge often makes its way into technological practice in a very roundabout way. At the same time, a considerable amount of scientific advance stems from prior achievements in technology. Science and technology have evolved along separate paths that often intersect. At these points of intersection each has often contributed to the other's development. Both science and technology seem to do best when they remain in close contact, but this should not obscure the fact that in most instances they remain very different enterprises.

QUESTIONS FOR DISCUSSION

1. Summarize the relationships between science and technology outlined in this chapter. How do these relationships compare and contrast with your prior ideas about science and technology.
2. Monetary considerations aside, which would you find more personally satisfying: making a scientific discovery or inventing a useful technology? Why?
3. Quite a few research projects in chemistry, physics, and biology receive grants for millions of dollars, whereas most researchers in the arts, social sciences, and humanities consider themselves lucky to receive a few thousand dollars in grant aid. Why is this so? Does it represent a proper distribution of research funds?
4. Students in engineering programs typically take a substantial number of science and mathematics courses. Should some of these courses be eliminated and replaced with different kinds of courses? If so, which courses should be taken instead?

Notes

1. John D. Verhoeven, "The Mystery of Damascus Blades," *Scientific American* 284, 1 (January 2001), pp. 74–79.
2. Pamela O. Long, *Artisans/Practitioners and the Rise of the New Sciences, 1400–1600* (Corvallis: Oregon State University Press, 2011).
3. Thomas S. Kuhn, *The Rate and Direction of Inventive Activity* (Princeton, NJ: Princeton University Press, 1963), pp. 450ff.
4. Thomas S. Kuhn, "Comment on the Principle of Acceleration," *Comparative Studies in History and Society* 11, 4 (1969): 428. See also Thomas S. Kuhn, *The Essential Tension: Selected Studies of Scientific Tradition and Change* (Chicago: University of Chicago Press, 1977).
5. Chalmers W. Sherwin and Raymond S. Isenson, "Project Hindsight," *Science* 156, 3782 (June 23, 1967): 1571–1577.
6. Richard Chait, John Lyons, Duncan Long, and Albert Sciarretta, *Enhancing Army S&T: Lessons from Project Hindsight Revisited* (Washington, DC: National Defense University, Center for Technology and National Security Policy, 2007), pp. 23–24.
7. Sherwin and Isenson, "Project Hindsight," p. 1575. See also Wendy Faulkner, "Conceptualizing Knowledge Used in Innovation: A Second Look at the Science—Technology Distinction and Industrial Innovation," *Science, Technology, and Human Values* 19, 4 (Autumn 1994): 435, 454.
8. J. E. S. Parker, *The Economics of Innovation: The National and Multinational Enterprise in Technological Change*, 2nd ed. (London: Longman, 1978): 30.
9. Edwin Layton, "Mirror-Image Twins: The Communities of Science and Technology in 19th Century America," *Technology and Culture* 12, 4 (October 1971): 564–565.
10. Otto Mayr, "The Science–Technology Relationship as a Historiographic Problem," *Technology and Culture* 17, 4 (October 1976): 668.
11. See, for example, Bruno Latour, *Science in Action: How to Follow Scientists and Engineers through Society* (Cambridge, MA: Harvard University Press, 1987).
12. Vannvar Bush, *Science: The Endless Frontier* (US GPO, Washington, DC, 1945).
13. F. R. Jevons, "The Interaction of Science and Technology Today, or, Is Science the Mother of Invention?" *Technology and Culture* 17, 4 (October 1976): 731.
14. Roberto Cardil (MatematicasVisuales), "Kepler: The Volume of a Wine Barrel," *Convergence* (January 2012), DOI:10.4169/loci003499.
15. Ashish Arora and Alfonso Gambardella, "The Changing Technology of Technological Change: General and Abstract Knowledge and the Division of Innovative Labour," *Research Policy*, 23, 5 (September 1994).
16. Eugene S. Ferguson, "The Mind's Eye: Nonverbal Thought in Technology," *Science* 197, 4306 (August 26, 1977): 827.
17. Philip H. Abelson, "Instrumentation and Computers," *American Scientist*, 1 (1986). Quoted in Paul DeHart Hurd, "Technology and the Advancement of Knowledge in Science," *Bulletin of Science, Technology, and Society*, 14, 3 (1994): 126

18. Eda Fowlks Kranakis, "The French Connection: Giffard's Injector and the Nature of Heat," *Technology and Culture* 23, 1 (January 1982): 3–38.
19. Nathan Rosenberg, *Inside the Black Box: Technology and Economics* (Cambridge: Cambridge University Press, 1982), pp. 146–147.
20. Frederik Nebeker, "Charles Townes, the Maser, and the Relationship Between Engineering and Science," *Engineering Science and Education Journal* 4, 6 (December 1995): S41–S46.
21. Edwin T. Layton, Jr., "American Ideologies of Science and Engineering," *Technology and Culture* 17, 4 (October 1976): 688.
22. A list of all Nobel Prizes is available from www.nobelprize.org/prizes/lists/all-nobel-prizes/.
23. Casey Dreier, "How Much Does the James Webb Space Telescope Cost?" *The Planetary Society* (October 2021), accessed from www.planetary.org/articles/cost-of-the-jwst; Calum Roche, "How Much Money Did CERN's Large Hadron Collider Cost to Build and Who Paid for It? *AS* (July 2022), accessed from https://en.as.com/latest_news/how-much-money-did-cerns-large-hadron-collider-cost-to-build-and-who-paid-for-it-n/
24. Langdon Winner, *Autonomous Technology: Technics-Out-of-Control as a Theme in Political Thought* (Cambridge, MA: The MIT Press, 1977), pp. 24–25.
25. Hugh G. J. Aitken, *Syntony and Spark: The Origins of Radio* (Princeton, NJ: Princeton University Press, 1985), p. 316.
26. Edward Constant, *The Origins of the Turbojet Revolution* (Baltimore: Johns Hopkins University Press, 1980), pp. 15–16.
27. James M. Utterback, "Innovation in Industry and the Diffusion of Technology," *Science* 183, 4125 (February 15, 1974): 622.
28. Faulkner, "Conceptualizing Knowledge," p. 427.
29. John Rae and Rudi Volti, *The Engineer in History*, 2nd ed. (New York: Peter Lang, 2001), pp. 190–192.
30. Aitken, *Syntony, and Spark*, pp. 335–336.
31. George Shiers, "The First Electron Tube," *Scientific American* 220, 3 (March 1969): 104–112.
32. Peter Mathias, "Resources and Technology," in Peter Mathias and John A. Davis (Eds.), *Innovation and Technology in Europe: From the Eighteenth Century to the Present Day* (Oxford, England, and Cambridge, MA: Basil Blackwell, 1991), p. 37.
33. Peter Mathias, "Who Unbound Prometheus? Science and Technical Change, 1600–1800," in Peter Mathias (Ed.), *Science and Society 1600–1900* (Cambridge, UK: Cambridge University Press, 1972), p. 79.
34. Robert M. Pirsig, *Zen and the Art of Motorcycle Maintenance* (New York: Bantam Books, 1974), pp. 99–103.
35. John Ziman, *An Introduction to Science Studies: The Philosophical and Social Aspects of Science and Technology* (Cambridge, UK: Cambridge University Press, 1984), p. 104.
36. Ann Johnson, *Hitting the Brakes. Engineering Design and the Production of Knowledge* (Durham, London: Duke University Press, 2009).

37. Rebecca S. Eisenberg and Richard R. Nelson, "Public vs. Proprietary Science: A Fruitful Tension?" *Daedalus* 131, 2 (Spring 2002), p. 91.
38. See Henry Petroski, *To Engineer Is Human: The Role of Failure in Successful Design* (New York: St. Martin's Press, 1985).
39. Melvin Kranzberg, "Let's Not Get Wrought Up About It," *Technology and Culture* 25, 4 (October 1984): 742.
40. Donald E Stokes, *Pasteur's Quadrant: Basic Science and Technological Innovation*. (Washington, DC: Brookings Institute Press, 1997).
41. See: *Peterson Field Guides*, https://peterson-field-guides.harpercollins.com/home.
42. While he patented extensively, he himself wrote mostly product announcements and editorials describing new technologies, with only a handful of scholarly articles: https://edison.rutgers.edu/index.htm

CHAPTER 5

The Cultural Diffusion and Non-Diffusion of Technology

A technology is of no use unless it is put to use. This is a crushingly obvious statement, but it does serve to remind us that a technology that never makes it out of a corporate research and development (R&D) laboratory or an inventor's basement might just as well not exist. The past two chapters have examined some of the ways in which technologies have come into being; the task of this chapter is to describe how they are adopted in relation to cultural and organizational contexts. Our concern here is the diffusion of technologies—that is, the processes through which they spread from their initial sources into homes, factories, offices, and so forth. To begin, we will consider the diffusion of technologies from one country to another.

Technology on the Move

There is an understandable human tendency to believe that progress, technological or otherwise, is largely the result of efforts of one's own people. We exalt native scientists, entrepreneurs, and inventors, and see in their efforts the major source of our technological advance. The Soviet Union in the Stalin era carried this to ridiculous lengths when its historians claimed that their compatriots had invented everything from the steam engine to baseball, but many other countries have exhibited strains of technological nativism. The truth of the matter is quite different. Although indigenous skills and achievements are an indispensable part of technological advance, few technologically dynamic countries have been isolated from the influences of other lands. Continued technological advance requires an infusion of ideas, tools, and materials from other places, coupled with an ability to make good use of them.

Consider the world in the year 1500. There can be little doubt about what country had produced the most technological innovations up to that time. It was not England, nor was it France, Italy, or Germany. The country with the greatest heritage of inventive achievements was China. By 1500, Chinese technologists had produced not only the compass, paper, and gunpowder—the familiar textbook examples of Chinese inventive ability—but also had been responsible for such inventions as the segmental arch bridge, the chain drive transmission, the spinning wheel, watertight bulkheads for ships, printing, fishing reels, paddle wheel boats, the differential gear, earthquake detection devices, and of course the fine porcelain that takes its name from its country of origin.[1] Europe, too, could claim some significant technological advances by the beginning of the sixteenth century, most notably in mining and cathedral building, but on the whole, there could be little question where the greater degree of inventiveness had been residing.

Yet within less than two centuries, China's lead had dissipated, and the European continent exhibited the greater degree of economic and technological vitality. Much of this dynamism could be attributed to the successful adoption of Chinese inventions by European countries. Paper was an essential complement to printing; together they made possible the rapid circulation of new ideas and stimulated the growth of literacy, which in turn directly contributed to major historical changes such as the Protestant Reformation and the spread of capitalism. The magnetic compass aided transoceanic navigation and allowed the spread of European economic power throughout the world. And gunpowder helped to conquer and secure that world for the benefit of Europe.

The successful use of foreign technologies in early modern Europe was very much in accordance with long-standing historical patterns. Many of the most significant inventions used by the Western world up to that time did not originate there. The stirrup, which initiated major changes in medieval warfare and in so doing reshaped the social system, was not a Western invention, but diffused from Persia sometime after the eighth century, although it too probably originated in China.[2] Another vital innovation involving equine power was the horse collar; horses could not be used effectively as draft animals until it diffused from Asia to Europe at about the same time.[3]

Much of the economic progress of the Western world therefore can be traced to the willingness of its people to accept and make good use of technologies that originated elsewhere. By contrast, China was far less open to foreign inventions. Secure in the belief that they inhabited the "Middle Kingdom" (the literal translation of Zhong Guo, the name by which the Chinese called their country), the Chinese looked at the outside world with more than a trace of condescension. This attitude was particularly evident when China came into contact with the expansionary nations of Europe. During the late eighteenth century, English merchants and officials scoured the world in search of customers for the products of English factories, but they found no takers in China. The emperor made it very clear to one envoy of King George III that "there is nothing we lack.... We have never set much store on strange or ingenious objects, nor do we need any more of your country's manufactures."[4]

China's ruling elite persisted in this attitude. Toward the end of the nineteenth century the scholar Kang Youwei advocated the combination of Western techniques with traditional Chinese culture. Although he won the support of the emperor, the reform movement initiated in 1898 was aborted in a few months when the emperor was deposed and replaced by the reactionary dowager empress, Cixi (T'zu-hsi in the older Wade-Giles system of Romanization). The early years of the Chinese Republic (1912–1949) saw a considerable infusion of Western ideas and technologies, but these were eclipsed by Chiang K'ai-shek's New Life Movement, which stressed a return to Confucian virtues as the cure for China's weakness.

Even the emergence of a Communist state in 1949 did not result in a wholehearted acceptance of foreign ideas and ways of doing things. Although a considerable amount of technology transfer from the Soviet Union took place during the early 1950s, the Chinese were offended by Russian arrogance. The political rift between the two countries closed off China's main source of advanced technology, while at the same time, Maoist ideology contributed to an increasing aloofness to the outside world and stressed self-reliance in technological matters. This attitude has changed dramatically in the years following Mao Zedong's death in 1976. In the words of one Chinese journalist, "No nation has a monopoly on culture. Cultural exchanges between nations are inevitable and irresistible.... Whatever will facilitate China's material modernization and its cultural advancement is welcomed."[5]

In contrast to China's historical ambivalence and even hostility over the use of foreign technologies, Japan after the Meiji Restoration in 1868 pursued a policy of aggressive industrialization based on imported technologies. During its early stages, the Japanese industrial drive was propelled by the importation of locomotives, textile equipment, and other industrial goods. Foreign technical advice was also avidly sought; during the early years of the new emperor's reign, the salaries of foreign technicians absorbed 6 percent of the central government's budget, and 40 to 50 percent of the budget of the Ministry of Industry for the duration of its existence.[6]

As Japan's industrial enterprises began to emerge, their products were often outright copies of foreign products. At times, Japanese attempts to copy foreign artifacts produced comic results, as exemplified by the factory with a sawtooth roof that faced the wrong way, so no light shone through the windows, or the tailor who stitched together a Western-style suit complete with a patch on the pants. But with mounting experience, Japanese workers and managers were able to effect substantial improvements until some of their products were the equal of Western manufactured goods. An early example of this was the Toyoda automatic loom; it marked such an advance over existing looms that England's Platt Brothers & Co., a major manufacturer of looms, paid the 1929 equivalent of 1 million yen to obtain the patent rights.[7] Certainly no one laughs at Japanese technology today, although the accusation is occasionally still made that the country excels in the refinement of existing technologies but produces few completely new ones.

The importation of foreign technologies also plays a large role in the story of early American industrialization. One could practically date the start of America's industrial revolution to the year 1790, when Samuel Slater, who had recently emigrated from England, used his accumulated know-how to construct the first successful mechanized spinning factory in the United States. In the years immediately following, the United States was so deficient in indigenous technological capability that many implements and techniques had to be imported. American civil engineers found it necessary to secure from England drawings and examples of devices as simple as a wheelbarrow in order to successfully construct early canals.[8]

These were not simply instances of a young and economically immature nation relying on imported technologies due to temporary incapability. Modern, quintessentially "American" innovations such as automobiles, radios, jet airplanes, compact discs, and the World Wide Web all owe their existence to pioneering efforts that were undertaken in other lands. And along with imported implements, materials, and processes, American technological development has been spurred on by an even more valuable import: people. Although a large reservoir of "Yankee ingenuity" cannot be denied, many of America's greatest technological achievements were produced by immigrants. It would take many pages simply to list some of the most prominent of these, but a small sample can illustrate the debt that American technology owes to immigrant scientists, inventors, and engineers. One of the first plastics, Bakelite, was the work of a man born and educated in Belgium, Leo Baekeland. The mathematical and engineering work of Charles Steinmetz, a transplanted German, was essential to the development of the American electrical industry. Nicola Tesla (Serbian-born) made epic inventions in alternating current and long-distance transmission of electricity. The science of aerodynamics owes much of its early development to an immigrant from Hungary (via Germany), Theodore von Karman. The television camera was invented by a Russian exile, Vladimir Zworykin. This pattern continues today. As of 2018, immigrants and their children founded or helped to found 60 percent of the most valuable technology firms in the United States. These firms, which include Alphabet, Amazon, Apple, and Facebook, collectively are valued at $3.8 trillion.[9]

Equally important, inventors and entrepreneurs—both immigrant and homegrown—have benefited from the services provided by legions of unsung men and women who brought their skills from other lands and in so doing helped to push American technology to new heights.

Today, a considerable amount of technology transfer occurs when firms in the economically advanced nations of the world establish factories in developing countries, some of them engaged in the manufacture of highly sophisticated electronic equipment such as computers and smartphones. For critics, the appeal of these offshore

Figure 5.1 Two key figures in the history of American technology, one native born, the other an immigrant: Thomas Edison and Charles Steinmetz.

enterprises centers on the availability of cheap labor; their output consists of assembled products that require few skills on the part of the assemblers. As a result, these enterprises are little more than enclaves of modernity with few if any connections to the host country's economy and society. In contrast, a more optimistic assessment of these enterprises stresses their positive role in stimulating the development of workers' skills and technological advance. Governments that are committed to economic and technological advance know that labor gets more expensive as a country develops and workers demand their share of a rising national income. Under these circumstances, low-cost labor ceases to be a nation's main advantage. Consequently, countries like Taiwan, Korea, and China (as well as Japan in an earlier era) have made concerted efforts to upgrade the skills of their labor forces, so the quality of their employees makes up for the higher wages paid to these workers. No less important, these countries—China especially—have used multinational firms as sources of up-to-date technologies by requiring the transfer of these technologies as preconditions for allowing these firms to set up shop on their soil.[10] And although it is hardly a commendable business practice, many firms, especially Chinese ones, have gained a fair amount of technological know-how through outright piracy.

Clever Copyists

In calling attention to piracy and to less flagrant forms of technological borrowing, there is no intention to attribute the achievements of the United States, Japan, and today's China to technological plagiarism. It is not an easy matter to copy a technology that was produced somewhere else. Having an example of a machine or a material in front of you helps, but it hardly guarantees success. Life would be simpler and easier if technological advance were simply a matter of importing a prototype and then engaging in "reverse engineering." It does work occasionally, but all too often the process ends in frustration, as when the Chinese government built two jet airliners that may not have been direct copies of the Boeing 707 but had many similarities with it. The planes were overweight and very sensitive to changes in their center of gravity when loaded. The Y-10, as it was named, never went into series production nor was it flown in scheduled service.[11]

The very nature of technology makes copying a difficult enterprise. As was stressed in earlier chapters, technologies have to be considered as systems, and if one element is missing or deficient, even seemingly simple tasks such as copying a particular item may be impossible. A successful technology transfer requires numerous complementary inputs, many of which have to be developed and produced by the recipient.[12]

There are occasional possibilities for "leapfrogging" over whole stages of development through the importation of foreign technologies. For example, in many parts of the world, cell phones have given millions of people access to telephone service by eliminating the need for an expensive wired infrastructure. But this is an exception; more often than not, the successful use of advanced imported technologies depends on the preparation that comes through involvement with well-established technologies.

This brings us to a related point. The effective use of foreign technologies requires a labor force with a broad range of skills. It is essential that native managers, engineers, technicians, and ordinary workers are capable of making the best of opportunities that imported technologies present. To do so often entails significant modifications to an imported technology. Under these circumstances, the application of imported technologies is combined with indigenous innovative efforts to the point where the two are in fact part of a single process.[12]

A successful user of imported technologies must therefore have a reasonably high level of indigenous capability. As noted by Vernon Ruttan, "The ability to screen, borrow and adapt scientific knowledge and technology requires essentially the same capacity as is required to invent new technology."[13] This is demonstrated by one study that found that the countries most successful in using foreign technologies tended to be the ones with high levels of indigenous research and development.[14] It is also significant that countries with the highest innovation rate make the quickest use of technologies developed elsewhere.[15] A country (or an individual firm) that draws on technologies developed elsewhere is spared the expense of "reinventing the wheel," but making effective use of imported technologies—even if only direct copying is required—often requires a stock of indigenous skills that cannot always be easily found.

Adaptation and Adoption

In some cases, the diffusion process may result in a technology emerging as something fundamentally different. This is exemplified by the aforementioned transfer of gunpowder from China to Europe. The Chinese at first regarded gunpowder as a medicinal substance, and only after hundreds of years of experimentation did they begin to employ it for fireworks and on occasion for military rockets, bombs, and mines. By the end of the thirteenth century, the Chinese were using gunpowder to fire projectiles from vase-shaped guns, but development seems to have stopped at this point. A few decades later, Europeans began using gunpowder for weapons of steadily increasing destructive power, thereby transforming the nature of warfare and ultimately society as a whole.[16]

The changes made in the course of the diffusion of gunpowder from China to Europe were monumental, for the nature of the technology changed along with the purposes to which it was put. In most cases, however, the modifications are far less dramatic. A process of "fine tuning" takes place as the technology is transferred from one place to another. As noted above, many successfully diffused technologies succeed only because of alterations and improvements made during the process of diffusion. Some of these changes may be done in order to adapt an imported technology to local conditions. In the late nineteenth century, Japanese entrepreneurs set up textile factories that used ring-spinning machines that had been invented in Europe. Unlike in Europe, however, these machines were supplied with a combination of high-quality long-staple cotton and cheaper short-staple cotton. The thread that was produced was lower in quality, but it was less expensive, an important consideration in a nation that was still poorer than most European countries.[17] Another adaptive change occurred when the railroad, an English invention, crossed the Atlantic. Railroad systems were built in the United States with great haste because there were huge expanses of frontier to be opened up as quickly as possible. One consequence of these conditions was that curves were much sharper than in England. This in turn necessitated changes in locomotive design as American engines were equipped with pilot wheels placed ahead of the driving wheels to guide them through the curves.

Figure 5.2 The John Bull, from England, had to be fitted with leading wheels to guide it through the sharp curves of an American railroad.

Learning to Make Steel in Old Japan

The early development of the Japanese steel industry provides another illustration of a successfully diffused technology that required many adaptations wrought by native workers and technicians.[18] Although the Japanese had been making small batches of high-quality steel for centuries, their indigenous industry was incapable of making iron and steel in large quantities. Japanese steelworkers therefore looked to Europe for better ways of producing steel. The first Japanese attempts to draw on foreign techniques occurred in the 1850s when they attempted to produce cannons by using methods described in a Dutch book on steel production. Success was slow in coming. Numerous trials resulted in the production of cannons that burst the first time they were fired. Yet underlying these failures was a highly useful learning process. At first, the native ironworkers were incapable of melting even half of the required pig iron, but by the fifth attempt a complete cannon was cast, albeit not with complete success. Still, even this limited achievement required strenuous efforts to learn how to treat the raw material and to develop devices and techniques to bore the barrel of the cannon. Within about 20 years of the original attempt, the foundry had successfully cast approximately 200 cannons. In accomplishing this, the cannon founders had been able to draw on indigenous capabilities in such things as the production of firebricks for use in the reverberatory furnace in which pig iron was melted, the construction of water powered boring machines, and the employment of high-temperature furnaces for melting the iron.

Although their efforts represented an attempt at imitating foreign products and techniques, more was involved than mere copying. What began as an effort to imitate a foreign technology ended successfully only when indigenous capabilities were applied to the solution of production and design problems. The process continued in 1880 when foreign engineers were recruited to oversee an ambitious effort to increase steel production. Initially this too was a failure, largely because the foreign experts did not take into account the special characteristics of Japanese coal and iron ore. But even this failed attempt to produce steel through the importation of foreign materials and experts ultimately met with success 12 years later when a Japanese engineer and one of his students reopened the dormant blast furnace after solving the problems created by the use of native coke and iron ore.

The point of this brief narrative is simply that Japan's successful nineteenth-century attempt to create a modern iron and steel industry was the result of importing foreign equipment, techniques, and expertise while at the same time developing and making full use of the native capabilities that were required for their successful employment. This example shows that technological diffusion is not a one-way process; without active participation by the recipients of the foreign technology, an imported technology may end up an expensive failure. In turn, effective participation often is the product of previous efforts at indigenous technological transformation. The issue is not one of "imitation" versus "self-reliance," but of how to achieve an effective blending of the two.

Appropriate Technology

Up to this point, there has been an unspoken assumption that the use of technologies developed in advanced nations is essential to the technological and economic modernization of developing countries. Modifications may have to be made to these technologies, but, on the whole, technologies of foreign origin are a sound basis for indigenous development. This assumption can be challenged by noting that the countries receiving foreign technologies today are not like nineteenth-century Japan and the United States. Most are far poorer, not just in terms of wealth, but also in regard to human skills; this creates numerous problems in effectively using the sophisticated technologies that have been created for the developed nations of the world. Moreover, poor countries have large and growing populations, making the creation of jobs one of their prime needs. Although in recent years the rate of population growth has slowed in many poor countries, there remains a huge cohort of new entrants to the labor force. As a result, youth unemployment rates are still stubbornly high; according to a recent estimate by the International Labor Organization, the global youth (15 to 24 years old) unemployment rate is 13.6 percent.[20]

Unfortunately, many of the technologies that have been developed in the economically advanced nations have been designed to save labor, not to maximize its use. Labor is often saved by using technologies that allow the substitution of capital for labor, yet one of the distinguishing features of poor nations is that they are critically short of capital. And finally, is it reasonable to expect that the world as a whole can sustain the global diffusion of the technologies that have been developed by the rich nations of the world? It is hard to imagine China or India consuming resources at the rate the United States does. To take one notable example, in 2019 India had a ratio of about 20 cars for every 1,000 persons.[21] If India, a nation of 1.41 billion people, were to achieve a car-to-population ratio equal to the 439 cars per 1,000 persons found in the United States, the prospects for resource depletion and environmental pollution would be grim indeed.

Due to the drastically different circumstances of today's non-industrialized countries, technologies that have worked well in the rich nations where they originated may fail when they are taken to a different setting. Even worse, they may seriously distort the course of a nation's development and leave it poorer than when it started out by concentrating human and financial resources in a few privileged segments of the economy and society, leaving much of the population, especially those in rural areas, as poor as ever. To take one particularly notorious example, after Western experts introduced large tractors into Pakistan, farm owners replaced farm laborers with the new machines. In one region, 40 percent of these workers lost their jobs and thus were forced to migrate to the cities and live in dire poverty. Per-acre crop yields hardly increased at all.[22]

Figure 5.3 Moving water is a significant challenge in non-industrialized regions. (CC0 1.0)

According to critics of conventional technology transfers, what poor countries need is a set of technologies appropriate to the conditions prevailing in these countries.[23] Sometimes they are called "alternative technologies," sometimes they are called "intermediate technologies" (because they stand in an intermediary position between modern and traditional ways of doing things), and sometimes they are simply called appropriate technologies. These technologies, it is claimed, do more than boost production; they increase employment, help to redress the maldistribution of wealth and income, empower people, contribute to economic independence, and help to preserve the environment.

Examples of appropriate technologies in developing countries include hand-powered nut shellers, bicycle-powered pumps for irrigation and water distribution, and rolling drums for transporting water. Appropriate technologies less directly connected with production include shelters made out of concrete and canvas, solar-powered lights, and inexpensive laptop computers for schoolchildren.[24] The utility of these technologies seems evident, and few could take issue with the general goals of appropriate technology. The difficulty lies in their actual development and implementation.

Consider the first goal: increasing employment. As noted earlier, in poor countries there is good reason to select technologies that use relatively more labor than capital. But adopting a technology simply because it requires the employment of lots of workers is likely to be a poor strategy. If it makes sense to hire a few dozen workers with shovels rather than employing one operator of an earthmover, why not hire a thousand workers with teaspoons? This *reductio ad absurdum* reminds us that creating jobs is a laudable goal, but it is also necessary to take output and costs into consideration. In regard to the latter, large numbers of workers, even if they are poorly paid, may be more expensive than a small number of workers using modern technology. The resulting higher cost of production will cause higher prices for the product, which in turn will diminish sales and employment. Alternatively, purchases may not decline, but there will be less money spent on other goods and services, and fewer jobs in the firms that supply them. In sum, the use of highly labor-intensive production methods in one firm or industry may actually diminish employment in the economy as a whole.

Many economists would argue that a free market system, with its realistic reckoning of capital and labor costs, will naturally result in the use of productive technologies that optimize both labor and output. But the choice of technologies is not always governed by economic calculations. Technologies may be selected because they meet the requirements of powerful segments of the society, as when ultramodern hospital technologies are installed to serve the urban upper class of a country that lacks basic medical services for most of its people. At the same time, even practical, profit-oriented businesspeople in underdeveloped countries can be charmed by the technological sophistication of foreign technologies that do not make good economic sense. For example, according to one study of business firms in Indonesia, there were available intermediate technologies that lowered costs and maintained product quality, but they were often rejected in favour of advanced technologies of foreign origin. This was especially likely to happen in firms that had oligopolistic control over their market. Under these circumstances, the aesthetic appeal of sophisticated technologies and the prospect of managing machines instead of people overrode the economic advantages of more appropriate technologies.[25]

Finally, a consideration of alternative technologies that is confined to economic matters is incomplete, for the appropriate technology movement has been motivated by a deep concern with how technologies are connected to larger issues. In the final analysis, a technology can be adjudged appropriate or inappropriate only by reference to particular values, and these cannot be defined exclusively in economic or technological terms. The appropriate technology movement also has directed attention to the role played by powerful groups and institutions in the choice of technologies. Choices are likely to be seriously constrained when donors of foreign aid, international lending agencies, and multinational corporations occupy positions of dominance. It is also likely that the emergence and use of appropriate technologies is seriously hindered by the international distribution of R&D expenditures. In 2011, North America, Europe, and East Asia accounted for 90 percent of research and development expenditures; the remaining 10 percent was undertaken in the countries of Central and South America, Central Asia, the Middle East, and Africa.[26]

Appropriate technologies, however defined, are not a straightforward approach for reducing poverty and lack of access to technology. The transfer of sophisticated technologies from developed countries may cause more harm than good. Choices must be made, and in poor countries as well as in rich ones, the selection of technologies reflects many larger issues. One of the most important of these is *sustainability*. Today's technologies have enriched our lives in many ways, but can we continue to degrade the natural environment while also expecting the perpetual availability of energy supplies and other resources? In this context, technologies are appropriate when they save energy, require minimal resources, and relieve stresses on the environment. Examples of sustainable technologies include products that lend themselves to recycling, building designs that require little energy for heating and cooling, and road systems that are safe and convenient for cyclists and pedestrians.

Technology and Culture: Power, Culture, And Choice

Chapter 1 introduced the concept of technological determinism, the idea that technology acts as an independent force in human affairs. As we have seen, this perspective has a number of problems, one of them being that it simply takes technological change as a given and fails to consider how new technologies emerge and diffuse. In the first part of this chapter, we will revisit the concept of technological determinism by noting some of its corollaries and considering their strong and weak points. We will then assess the influence of cultural processes on technological choice. To repeat an earlier assertion, technology is a human creation, but at the same time it should be evident that individuals do not participate equally in initiating and directing technological change.

Technological Advance and Cultural Lag

For all of its shortcomings, technological determinism as an idea still exerts a lot of influence. One of the places where it shows up is in the theory of "cultural lag." Formulated by William F. Ogburn, whom we briefly met in Chapter 1, the theory of cultural lag is predicated on the belief that habits, thoughts, values, and social arrangements often fail to change at the same speed as technological innovation. Technology moves ahead, but many other things lag behind. On the face of it, this seems a reasonable concept, and we should have no difficulty in coming up with a number of examples. To take one, the expansion of public health services and modern medical technologies have lowered infant mortality, but people in some parts of the world continue to have large families, due in part to traditional expectations that many children will not survive infancy. Only after several generations, when expectations and practices have "caught up" with the level of technology, will birthrates begin to fall. To take another example—one that has been the source of considerable conflict, litigation, and governmental legislation—the ease of copying and transferring digitized books and music has been undermining traditional notions of intellectual property for years. Only now is the legal system beginning to catch up.

Appealing as it is on the surface, the theory of cultural lag has its shortcomings. In the first place, technological changes are simply taken as givens, and no attempt is made to delineate how they came to be in the first place. Although traditional histories of technology have conveyed the impression that inventions are carried forward largely by their own internal logic and are little influenced by external factors, this perspective leaves a great deal out of the story. As earlier chapters of this book have indicated, many things that are not technical in nature strongly influence the course of technological change. At the risk of repetition, suffice it to say that technology should not be viewed as an independent source of social change.

Second, many attempts to demonstrate the connection between technological and social changes present us with the problem of measuring and trying to connect different kinds of phenomena. Technological changes can often be more easily measured than sociocultural changes, making it hard to determine the precise connection between the two. We can easily tabulate the increase in the number of automobiles between 1900 and 1925, but to what extent did widespread car ownership contribute to changes in sexual activity during this period? In similar fashion, we can tabulate the number of television shows with violent programming, but determining the influence of these programs on individual behavior is fraught with difficulties.

Third, at what point can it be said that a society "catches up" with a given technological innovation? New productive technologies have allowed major alterations in the traditional divisions between "men's work" and "women's work," but customs and social arrangements have not completely taken account of this fact. To be sure, there have been significant changes, but at what point can it be confidently said that a complete accommodation has occurred? Then, too, some technological innovations have been followed by rapid adjustments, while in other cases these adjustments occur slowly, if at all. In regard to the latter, the killing power of modern weaponry far exceeds the spears and stones of our Neolithic ancestors, yet we still engage in warfare as a means of resolving our disputes. Obviously, something other than technology itself is affecting the extent of and speed at which particular cultural lags are eradicated.

Fourth, the theory of cultural lag assumes a unified society and culture, and ignores the economic, political, and cultural divisions to be found in all complex societies. Describing a social phenomenon as an example of cultural lag may therefore be a reflection of the values and preferences of particular groups or individuals.[27] The unwillingness of Amish farmers to use electricity could be labeled an instance of cultural lag, but this assumes that saving labor is a benefit of unquestionable value. Those convinced of the sanctity of physical work and the value of community would not agree.

Finally, there is a hidden value judgment contained in the concept of cultural lag. Implicit in it is the notion that technology is a progressive, dynamic element in human history, and that social and cultural arrangements are intransigent sources of obstruction. But is it not possible that a cultural lag may demonstrate the danger or inappropriateness of some technological innovations? Once again, it might be noted that just because something is technically possible, it does not necessarily follow that it should be done, or that people should submit to it and make the necessary alterations to their lives.

Technology, Globalization, and Cultural Convergence

The interplay of technological and cultural changes is especially relevant in the globalized era in which we live. As was noted in earlier, advances in transportation and communications have made individuals and the communities in which they live less insular by facilitating the movement of people, goods, cultural elements, and ideas all over the world. Globalization has brought yoga to Iowa, Delta blues to Japan, and pad thai to Ireland. Cultural change as a result of globalization is not a new phenomenon; the music we listen to, the clothes that we wear, the language that we use, and the devices we employ have been strongly influenced by importations from other lands. What has made things different is the speed at which these transfusions occur. As will be noted in the chapters reviewing issues in the military and war, there was a time when major battles were fought by combatants who were unaware that an armistice had been declared weeks earlier. Today, even trivial news items are broadcast to the world at the speed of light. It is obvious that technological advances in transportation and communications have helped to create a more interconnected and fast-moving world, but is a truly globalized civilization emerging? Or to put it slightly differently, are all of the world's societies and cultures moving together in the direction of a common model?

The idea that the use of similar technologies results in cultural and social sameness has been around for a long while in the guise of an approach known as convergence theory. Closely associated with a belief in technological determinism, convergence theory argues that the nations of the world are becoming more similar to one another—that is, they are converging—as they make use of the same technologies.

Technological advance promotes societal convergence in two ways. First, advances in transportation and communication bring the world together and expose its inhabitants to new ideas, products, and cultural elements. As these are adopted, the nations of the world and their populations begin to resemble one another to a greater degree. Second, and less obviously, some technologies seem to require similar social and cultural arrangements. For example, running a steel mill or operating an airline do not allow substantial culturally based variations in the procedures employed. For airlines, global airline safety is propagated by an insistence on English in air traffic control interactions, although cultural patterns such as deference to interpersonal hierarchy can impact cockpit communications and airline safety.[28] Organizational modes, attitudes, and behaviors all seem to converge on a common model. Langdon Winner asks the question: *Do Artifacts Have Politics?*[29] For example, solar power seems to suggest decentralization and the distribution of literal electrical power, as well as metaphorically, social power. Conversely, nuclear power seems to necessitate centralized control and a rather elaborated bureaucratic state to maintain safety and control of complex systems and dangerous materials. Do technologies *require* particular forms of society, culture, and governance? Or are there efficiencies that accompany technologies that match them to organizational and governmental forms? And similarly, if we are prioritizing particular political values, such as democracy, should we stay away from certain kinds of technologies which seem to be closely coupled to anti-democratic systems of governance? In philosophical terms, is a technology a necessary condition for a form of governance? Or a sufficient one? A sufficient condition is *causal*, while a necessary condition requires additional factors for an event, in this case a socio-political formation, to occur. Do nuclear power plants *require* strong governmental systems, or are they prioritized *by* strong governments to solve the problem of powering society? Are nuclear power plants uniquely compatible with and efficiently run by centralized administrative states, although other kinds of social organization would actually work to produce reliable power?

Many other examples of technologies producing convergence in social structure and modes of organization can be cited, but the effects of modern technologies are not limited to particular industries; according to convergence theory, the process is demonstrated by commonalities in educational systems, family interactions, organizational structures, and even preferences in food and clothing as they all move toward a universal pattern.[30]

Opposing Convergence

In geometry, convergence means that several lines are all moving toward a single point. Convergence theory, however, usually assumes that countries with relatively low technological levels will move toward a point already occupied by technologically advanced countries, most of which have been in the orbit of European culture (Japan being the notable exception). Since many advanced technologies have originated in Europe and North America, other countries will have to adopt Western institutions and culture if they are to make effective use of modern technologies. In other words, convergence really means Westernization. This equation of modernity with Western culture is embodied in the remark of a U.S. senator who, early in the twentieth century, expressed the hope that "we will lift Shanghai up and up, ever up, until it is just like Kansas City."[31]

This was a prospect that many Chinese did not view with pleasure. Throughout most of their history, the Chinese have been wary of foreign importations, and not even the adoption of Communist ideology altered this fundamental attitude. Especially when under the influence of Mao Zedong's ideology, many Chinese had misgivings about the adoption of foreign technologies. The underlying hostility toward the use of foreign technologies strongly asserted itself during the second half of the 1960s, when China was racked by the internal struggles that went under the name of the Great Proletarian Cultural Revolution. Among the victims of the often-bloody conflicts that occurred during this period were technical experts who were accused of being captivated by foreign technologies and opposed to the efforts of China's workers and peasants to develop indigenous technologies. Opponents of the use of foreign technologies also feared that these technologies would be accompanied by unacceptable foreign cultural patterns. Imported technologies, it was feared, would inevitably be accompanied by "bourgeois" habits and "decadent" activities.[32] Elements of this mentality remain today, as authorities have created "the Great Firewall of China" in the hope of blocking outside influences carried over the internet.

In developing countries and the global south can be found the apprehension that technologies developed in modern countries will result in the disruption of traditional social and cultural patterns. In some countries, these apprehensions have motivated a return to traditional cultural patterns. This has been most notable in Iran, where rapid economic modernization, led by Shah Reza Pahlavi's authoritarian government, produced cultural and social disruption that eventually resulted in the ouster of the Shah and a resurgence of Islamic fundamentalism.[33] Even in places where the reaction has not been as strong, many have mourned the passing of traditional patterns of life and thought, and their replacement by a social order that rejects established ways in favor of the latest novelties, and where a cold rationality intrudes into all human relationships.

This has been the fear. But is it really true that technological modernization has to be obtained as a complete package, and that it destroys all existing cultural patterns that are not favorable to it? Although it has been widely accepted, much of the evidence for convergence theory is impressionistic and anecdotal. Rigorous empirical tests are not abundant, and they do not always come to the same conclusion. In support of convergence theory, one study of Korean automobile workers found that exposure to industrial technology produced patterns of behavior and belief that were similar to the patterns exhibited by Italian and Argentinean automobile workers when their countries were at similar levels of industrialization.[34] On the other hand, another study found that the values of individual workers were much more affected by their nationality than the kind of work they did; Brazilian oil refinery workers held values and beliefs that more closely resembled those held by farmers in their country than those held by their occupational counterparts in India or Ghana.[35]

Robert Marsh took up the issue by hypothesizing that economically developed societies would exhibit less variation in the key characteristics of their social structures (form of governance, composition of the labor force, educational institutions, communications media, and so on) than less developed countries.[36] Economically developed countries have more specialization than less developed ones, making social interactions both essential and frequent. These interactions need to be mutually comprehensible and predictable, which necessitates more uniformity. To take a concrete example, a buyer and a seller of some commodity or service need to assume that their transaction will not be affected by differences in their gender, ethnicity, tribal membership, or other "particularistic" identities. This essential uniformity is made possible by similar institutional characteristics across developed countries, such as centralized governments, high levels of school enrollment for girls and women, and widespread consumption of newspapers, television, and other communications media. Marsh's research found that a few less-developed countries resembled more advanced countries in some of their characteristics, such as widespread educational opportunities for girls and women. But his essential finding was that, on the whole, these countries exhibited much more variation than was found in developed countries. In other words, many of the structural elements of a developed country undergo a process of convergence with other developed countries, resulting in a greater degree of uniformity in many key social, economic, and political categories than in less-developed countries. Equally important, the research demonstrated that the extent of uniformity in developed countries increased over time.

Although the case for convergence theory is plausible on both theoretical and empirical grounds, the role of technological change in promoting the convergence of societies is less certain. It would certainly be a stretch to think of technological change as a universal solvent that dissolves all prior cultural and social patterns so that they conform to the presumed dictates of modern technology. The effects of specific technologies or of technological change in the aggregate are complex in the extreme, and it is simplistic to expect that all of the features of a society will be summarily transformed in accordance with technology's requirements. In this matter, it is useful to consider the words of David Landes:

> Clearly there are no rigid compulsory relationships between a modern industrial economy and the entirety of its complex, multifaceted environment. Rather there is a wide range of links, direct and indirect, tight and loose, exclusive and partial, and each industrializing society develops its own combination of elements to fit its traditions, possibilities, and circumstances.[37]

Every society has its distinctive characteristics, and a particular technology will not necessarily produce the same effects for all that employ it. The actual consequences of a particular technology depend on why people have developed or adopted it in the first place.[38] Having invented gunpowder, the Chinese used it for rockets and primitive cannon, but fireworks became the main application. In contrast, in late medieval Europe, gunpowder was eagerly employed as the basis of increasingly destructive weaponry. The uses to which a technology is put depends on history, existing social arrangements, and the distribution of power.

Similarly, television in the United States as well as much of the world is primarily geared toward entertainment and the desire of advertisers to persuade the viewing public to buy their products. In contrast, authoritarian states have used television to trumpet the achievements of their regime, and a great deal of programming consists of stories that glorify the political leadership. Even when a diffused technology is used for the same goals as in its country of origin, the manner in which the technology is used may still reflect cultural differences. Consider baseball in Japan. The game uses the same equipment, bases are 90 feet apart, pitchers have the same repertoire of pitches, and an inning ends when both teams have made three outs. But Japanese baseball reflects some important cultural elements of that country. Players do not argue with umpires, base runners do not aggressively attempt to break up double plays, fans usually cheer primarily when cheerleaders (who are middle-aged men) tell them to do so, it is considered impolite for a pitcher to disagree with the catcher's pitch selection, and games can end in ties after twelve innings.

To take another example of only partial convergence, cars throughout the world use the same set of technologies and are similar in the layout of their controls. But as anyone who has driven outside his or her native land will attest, driving styles can vary greatly from place to place. In other cases, imported technologies may be accompanied by parallel behavioral and cultural changes, but these will be confined to a few enclaves that exert little influence over their surrounding society. Medical procedures at elite hospitals in poor countries may be similar to those in the developed world, but these are not available to the bulk of the populace. In sum, the introduction of new technologies may play a significant role in transforming a society, but only in conjunction with organizational, cultural, and other changes, and these usually unfold over a much longer period of time, if at all.

Fragmentation and Convergence: Local and Global Factors

While technological optimists thought the internet would bring us together, a very specific model of convergence, the facts of tribalization and fragmentation in the era of social media prove otherwise. The hope was that technological connectivity would allow people to interact with more geographically, and thus politically and culturally diverse, others, reducing bias and parochialism. It is pretty clear that most social media use reinforces distinctions and in-group identifications, a kind of virtual "localization" and production of electronic enclaves, sometimes referred to as "new tribalism," processes enhanced by algorithms and not just user actions.[39] But this concern over technology and social fragmentation is not new. There were concerns with both telephones and automobiles producing isolated Americans and insular social networks.[40]

As a heuristic, it's pretty clear that information and communications technology (ICTs) can maintain or strengthen already existing ties—such as face-timing a friend or parent after moving across country, but are far less likely to encourage new ties or strengthen weak ties.[41] Trust is most readily extended through interpersonal connections, not abstract relationships such as to experts or institutions, so, as noted in prior discussions of trust and dis/information, people tend to value the opinions of people they are already connected to, and information shared by trusted others.[42] These social-psychological tendencies are amplified by technology, resulting in the neo-tribalism lamented in modern societies. In its pure form, however, even fragmentation theory, as applied to ICTs, suffers from the same technological determinism as convergence theory and globalization theories. Local conditions and decision processes shape the acceptance and utilization of new technologies, and ICTs could have perhaps been designed to break down barriers, and may yet in the future. That is, the use of algorithms to reflect preferences back onto users could be disrupted.[43] In the meantime, enlightened users can make the effort to seek information and perspectives outside of algorithmic expectations. People can combat virtual tribalism and the metaphoric production of electronic enclaves and echo chambers.

The interplay of local and global forces emerges in the theory of the "McDonaldization" of society.[44] This is a more cultural and commercial, rather than technological, theory of convergence, where large global corporations have the power to erase cultural differences to produce highly fungible (meaning interchangeable) commodities. There are technological dimensions, however, as manufacturing processes support uniformity and efficiency in global commodity chains. In this model, the figure of *kokopelli* in Hopi and other Southwestern cultures in the US is stripped of his raciest representations (with a phallus) as a fertility symbol and sometimes dangerous trickster, and becomes simply a cute flute-playing southwestern motif, sold on t-shirts, trivets, and clever wall art. In Ritzer's model, culturally specific meanings are stripped away to reveal a neutralized "universal" cultural meaning for a commodity—it is reduced to "nothing"—or at least nothing controversial.[45]

Even McDonalds, however, is not as universal as might be imagined. There are quite a few regional variations in menu. For example, in India, the company does not sell beef or pork products on their menu, providing non-meat alternatives.[46] In Maine, during the summer, consumers can get the "McLobster Roll," which is lobster meat and a special sauce on a roll. Consumers can get the "McCrab" in coastal Delaware, Maryland, and Virginia, or poutine (cheese curds and brown gravy over French fries) across Canada. A meal is presented with a formal bow in a restaurant in Tokyo, with menu variations such as teriyaki burgers and edamame and corn as a side dish.[47] So as a counter to strict globalization or localization, the term "glocalization" reflects the influence of the local—local cultures and local supply chains—reasserting itself over commodities thought to be global.[48]

Let us consider two more examples of the ways that local cultures push back on convergence theory's technological determinism. It is easy to superficially dismiss these cases at first as merely "resistance" or irrational opposition to progress. But when looking at deeply held cultural values and practices that are central to social coherence and stability, it is clear that convergence theory has only partial applicability.

In Bali, in the 1960s, the "Green Revolution" was meant to transform agriculture through the application of Western science-based technologies. "Modern" farming technologies, pesticides, genetically improved monocultural crops, herbicides, and planting schedules were meant to improve productivity, primarily thought of as yield per acre, to improve food availability. Bali, as a center of rice production in Indonesia, was targeted by scientists, international development agencies, and the political leadership trying to modernize the image of Bali and develop rice as an export crop. However, it became rapidly clear that imposing Western farm organization and practice was decreasing local food security and disrupting environmental quality. Rice yields fell, pollution increased, and farmers lost livelihoods as they became indebted to purchase Western agricultural products.[49]

Water management and planting had been organized in *subaks*—groups of farmers in a watershed which, through a religious commitment to mutual aid, agree to self-regulate planting and irrigation schedules for rice paddies. Through these centuries-old practices, farmers have come to irrigation schedules which actually correspond to scientifically optimal water flows for planting and harvesting.[50] Using traditional planting schedules and rice varieties also protected the crops from the circulation of pests and plant diseases. Adding to this optimization, traditional rice-paddy production produces secondary food sources, such as ducks, frogs, and fish, that are part of the regional diet. In just a generation it became clear that convergence on a Western model of agriculture was harmful to the farmers of Bali, and many of the "green revolution" technologies and policies were abandoned.[51]

If you have ever chosen against a new technology (e.g., stepped away from social media, rejected a new gadget), you might have been accused by your friends, "What are you, Amish?" Like the Luddites of the nineteenth century, the Amish, or Old Order Amish and related Mennonite communities, are generally misunderstood and often misrepresented in the media. Their rejection of modern technology is straightforward, but with some unanticipated consequences. As an Anabaptist religious denomination, family, community, self-sufficiency, and humility are key values. They are effective farmers.[52] They seek to maintain independence from the complications of contemporary life. Most adults can do most tasks, although specialization and high skill, such as in woodworking or husbandry, is highly valued. Technologies are selected when they aid community self-sufficiency, so things that cannot be made or repaired in the local community are avoided. If driving through rural eastern Pennsylvania, you can identify an Amish farm by the lack of motorized farm equipment and the absence of power lines. Technologies are also avoided when they produce distinctions: displays of wealth and status are considered rude, or even sinful. And finally, the Amish highly value face-to-face communication, and regularly meet to discuss community issues, including adopting new technologies, through direct participatory democracy.

While an individual may not own a car or a cell phone, communities may. For example, a community may decide to have a cell phone, as community use is allowable, to reach restaurants and urban markets for fresh organic produce. Or a car may be owned by a group of community leaders, to provide transport to urban areas for health care, or so a group of Amish workers can get to a field or a barn raising site.[53] A modern plow rig may be re-engineered to be drawn by horses. But no question, the Amish life is a hard life, too, with rigid traditional gender roles, hard physical labor, and requiring strict religious discipline. Few people choose to join Amish and related communities, but their commitment to community and self-sufficiency, and to careful deliberation about the challenges they face, can serve as reminders that technological change is not a foregone conclusion.

Figure 5.4 An Amish farmer with a horse-drawn plow.

Efforts to Restrict the Diffusion of Technology

The exclusive possession of a particular technology can confer great advantages on those who have it. Individual firms or whole countries may therefore go to great lengths to block the diffusion of these technologies. During the sixteenth century, the glassmakers of Venice sought to prevent others from learning the secrets that went into the production of Venetian glass, and they even sent assassins to poison Venetian expatriates who had set up glassworks abroad.[54] England enacted a law in 1719 that forbade the emigration of skilled artisans, such as the aforementioned Samuel Slater, whose knowledge of British equipment and practices was a crucial element in the early development of the American textile industry.

Today, technologically advanced nations attempt to prevent the export of advanced technologies through export licensing requirements. Licenses may be denied if national security could be compromised by the transfer of a particular technology. The transmission of knowledge alone can also be deemed a threat. In 2012, the potential danger of scientific information falling into the wrong hands led to requests by the U.S. federal government to request two groups of researchers to temporarily withhold the publication of some research results involving the genetic manipulation of the virus responsible for bird flu. Although the research was conducted in order to aid in the prevention of future epidemics, it was feared that the data it generated could become the basis of a future bioterrorist attack.[55]

Past efforts to suppress the outflow of technology and the knowledge upon which it rests have rarely been successful, and there is little likelihood that contemporary efforts will fare any better. The transfer of plans, equipment, data, and products are extremely difficult to stop, even with rigorous attempts at policing. Also, as exemplified by the case of Samuel Slater, the movement of people has historically been the main vehicle for the transfer of technologies. This sort of movement is very hard to stop, especially in a free society.

Attempts to stem the outward flow of technology may even be counterproductive. A country that attempts to bar the export of technology may end up damaging itself more than anyone else. Technological development often requires the interchange of information from numerous sources. The attempt to seal off a technology will restrict the flow of information from both domestic and foreign sources, and thereby result in a slower pace of technological advance.

Finally, it is often the case that the most important "secret" is that a technology exists. During the late 1940s and early 1950s, there was a great deal of anguish in the United States after the Soviet Union exploded its first nuclear bomb. The general belief was that the Soviets had obtained our "atomic secrets," and two Americans were executed for their role in passing on these "secrets." But the Soviets had long known the most important thing: that a massive explosion could be produced through the fissioning of atomic nuclei. Once this fact was known, it was only a matter of time before the Soviets learned how to do it themselves, with or without the clandestine transfer of American technology. Knowing that a problem can be solved sometimes is the most important step in its solution.

Patents and the Diffusion of Technology

Individuals and managers in business firms naturally want technological diffusion to take place on their own terms. If it has developed a novel technology, it wants the benefits of that technology to accrue only to themselves, and not to others who might be able to copy an invention that likely required a great deal of effort and expense.

This is a legitimate desire, and it has been recognized as such by the establishment of the patent system. The possession of a patent confers exclusive use of an invention; it is a legal monopoly. Without the prospect of such a monopoly, it is believed, there would be a diminished motivation to invent, and in the long run society as a whole would suffer from a slower rate of technological advance.

Although it is aimed at restricting the sale or use of a particular technology to a single individual or firm, there are some ways in which the patent system may stimulate technological diffusion. The filing of a successful patent application makes the invention public, for the basic design and specifications are open to inspection by anyone willing to pay a small fee. In fact, the word "patent" is derived from the Latin verb *patere*, which means "to open." One of the chief justifications for awarding a patent is that the inventor has not attempted to keep his or her invention secret, but rather has revealed its workings to the public. In summarizing a number of previous decisions, the U.S. Supreme Court noted in 1933 that an inventor "may keep his invention secret and reap its fruits indefinitely. In consideration of its disclosure and the consequent benefit to the community, the patent is granted."[56]

Although direct copying is of course forbidden, access to the public record may give a clearer sense of how a technological problem may be addressed and may stimulate alternative approaches to its solution.[57] Then, too, patent holders often pass their inventions to other firms in return for a licensing fee. This arrangement may allow for the more rapid diffusion of a technology because the licensing agreement often facilitates the transfer of supplementary material necessary for the best use of the device or process.[58] In some cases, a licensing arrangement may result in a better product, as happened when DuPont developed waterproof cellophane after being licensed to produce the original version.[59]

At the same time, however, patent protection may inhibit technological advance. A well-entrenched firm may hold a patent in order to suppress an invention that could seriously shake up existing routines or even threaten the firm's existence. It seems as though everybody has a brother-in-law who knows a man who invented a way to make his car run on water straight from the garden hose, but "of course the big oil companies bought up all the patents." One shouldn't put much credence in these stories, but it is true that dominant companies have at times used their control of patents to restrict innovation, as happened with the design of light bulbs when General Electric held the key patents.[60]

The use of patents to directly suppress certain technologies poses less of a threat to innovation than the widespread use of patenting as a strategic weapon against potential rivals. Patents are supposed to motivate individuals and business firms to create innovative products and processes, but according to critics, their primary purpose is to allow patent holders to harass competitors, collect undeserved royalties, and bolster the value of financial portfolios.[61] Critics of the present patent system have also called attention to the monopoly power that has been conferred by a recent upsurge in the patenting of products and processes that in the past were not entitled to patent protection, most notably computer software, genes, and business methods.[62] Finally, and most fundamentally, the significance of patents in stimulating invention is by no means certain. According to critics, the correlation between a strong patent system and the rate of invention is weak at best.[63]

Although it may confer considerable power in the marketplace, the holding of a patent does not necessarily confer an overwhelming competitive advantage, since there may be considerable scope for "inventing around" the patent. Moreover, there are no industries where a key technology is covered by a single "master patent" that can be used to repulse potential competitors. There have been attempts to dominate an emerging industry through the wielding of a patent, most notably when, in 1895, George B. Selden was granted a patent for the automobile. What Selden held was a classic "submarine patent," a vague claim held in reserve for a later time when others had made the invention a practical reality. Although he had not built an actual automobile at the time when the patent was granted, through it, Selden and the subsequent holders of the patent were able to extract licensing fees from most automobile manufacturers. One manufacturer who refused to go along was Henry Ford, who successfully challenged the validity of the patent in court and won a great deal of public acclaim as a result.[64]

It also has to be recognized that a patent never confers ironclad protection for inventors; it may be little more than a "license to sue." In recent years, most patent holders have successfully waged court battles against individuals and firms who have been accused of infringing upon their patents, but both sides may incur sizable costs in

Figure 5.5 The patent model for the automobile that George Selden claimed to have invented. The patent office stopped requiring models like these in 1880. (National Museum of American History)

terms of time, talent, and money. As a result, many individuals and firms have been content to exploit their unpatented invention until the copiers are able to seize a large share of the market. In other cases, patented products and processes are quickly outmoded by the advance of technology, and a temporary monopoly position is soon lost. On the other side of the coin, parties who have been charged with patent infringement may feel that they are in the right, but they will nonetheless elect to pay royalties to the patentee in order to avoid an expensive and drawn-out courtroom battle.

Whether the patent system serves as an overall stimulus or hindrance to technological advance is a matter of considerable debate, and one not likely ever to be settled. Like many social institutions, as much as anything it owes its existence to inertia and the efforts of parties with a direct interest in maintaining it. Under these circumstances, the status quo usually prevails.

QUESTIONS FOR DISCUSSION

1. What sort of skills are required for the successful copying of a technology developed elsewhere? Why do you think that the Japanese have been so successful at making effective use of technologies that first appeared elsewhere? How might other countries duplicate their success?
2. Most economically advanced countries have technical assistance programs that are used to upgrade the technological levels of poorer countries. Under what circumstances might these programs be harmful? How might these programs be shaped to better meet the needs of poor countries?
3. If you were the leader of a poor nation, what general policies would you formulate in regard to the importation of technology? Where might you expect to find opposition to these policies?
4. Suppose that you are the CEO of a firm that has a choice between two new technologies: one that promises a modest profit with very little risk, and another that may yield a very high profit but at considerable risk. What would your choice be? Who in your company might support the first technology, and who might support the second?
5. Does the patent system encourage or discourage technological innovation? What would happen if it were abolished? Can you think of an alternative that would better serve the needs of both inventors and the general public?
6. Which twentieth-century technology has produced more changes in our lives, television or the automobile? What have their social consequences been? Are these consequences solely due to these technologies, or have these technologies interacted with other sources of change?
7. Globalization has enriched our cultural lives and has brought a number of economic benefits. At the same time, it has been the source of a fair number of disruptions, some of them, like lost jobs, quite serious. Should anything be done to restrain globalization? Is it even possible?

Notes

1. See Robert Temple, *The Genius of China: 3,000 Years of Science, Discovery, and Invention* (New York: Simon and Schuster, 1989).
2. Steven Till, "The Diffusion of the Stirrup into Medieval Western Europe." Accessed from http://steventill.com/2008/06/18/the-diffusion-of-the-stirrup-into-medieval-western-europe/.
3. Frances and Joseph Gies, *Cathedral, Forge, and Waterwheel: Technology and Invention in the Middle Ages* (New York: HarperCollins, 1994), pp. 45–47.
4. Gilbert Rozman (Ed.), *The Modernization of China* (New York: The Free Press, 1981), pp. 22–23.
5. Xin Xiangrong, "Open Policy, Import of Culture," *Beijing Review* 29, 46 (November 17, 1986): 4.
6. Angus Maddison, *Economic Growth in Japan and the USSR* (New York: W.W. Norton, 1969), pp. 16–17.
7. Japan Patent Office, "Sakichi Toyoda Wooden Hand Loom, Automatic Loom,"(October 7,2002), www.jpo.go.jp/e/introduction/rekishi/10hatsumeika/sakichi_toyoda.html
8. Elting E. Morison, *From Know-How to Nowhere: The Development of American Technology* (New York: New American Library, 1977), pp. 21–36.
9. Rani Molla, "The Top U.S. Companies Founded by Immigrants Are Now Worth Nearly $4 Billion," *Vox* (January 12, 2018). Accessed from www.vox.com/2018/1/12/16883260/trump-immigration-us-america-tech-companies-immigrants.
10. Henry S. Rowen, Marguerite Gong Hancock, and William F. Miller (Eds.), *Making IT: The Rise of Asia in High Tech* (Stanford: Stanford University Press, 2007), p. 10.
11. Global Security.org, "Y-10." Accessed from www.globalsecurity.org/military/world/china/y-10.
12. Paul Stoneman, *The Economics of Technological Diffusion* (Oxford and Malden, MA: Blackwell, 2002), p 75.
13. Richard G. Lipsey, Kenneth I. Carlaw, and Clifford T. Bakar, *Economic Transformations: General Purpose Technologies and Economic Growth* (Oxford: Oxford University Press, 2005), p. 87.
14. Vernon W. Ruttan, "Towards a Global Agricultural Research System: A Personal View," *Research Policy* 15, quoted in Edward J. Malecki, *Technology and Economic Development: The Dynamics of Local, Regional, and National Change* (New York: John Wiley & Sons, 1991), p. 142.
15. Tuvia Blumenthal, "A Note on the Relationship between Domestic Research and Development and Imports of Technology," *Economic Development and Cultural Change* 27, 2 (January 1979): 303–06.
16. Christopher Freeman, *The Economics of Industrial Innovation*, 2nd ed. (Cambridge, MA: MIT Press, 1982), p. 61.
17. Zhou Jiahua, "Gunpowder and Firearms," in *Ancient China's Technology and Science* (Beijing: Foreign Languages Press, 1983), pp. 184–191; Arnold Pacey,

Technology in World Civilization: A Thousand-Year History (Cambridge, MA: MIT Press, 1991), pp. 44–50.

18. Keijiro Otsuka, Gustav Ranis, and Gary Saxenhouse, *Comparative Technology Choice in Development: The Indian and Japanese Cotton Textile Industries* (London: Macmillan, 1988), pp. 23–24, in Morris-Suzuki, op. cit., p. 87.
19. The narrative is based on Nakaoka Tetsuo, "Imitation or Self-Reliance: A Lesson from the Early History of Modern Japanese Iron Manufacturing," in *Japanese Foundation Newsletter* 8, 4 (October–November 1979): 1–6.
20. International Labour Organization, "Global Employment Trends for Youth 2020." Accessed from www.ilo.org/global/publications/books/WCMS_737648/lang--en/index.htm.
21. Helgi Library, "Passenger Cars Per 1,000 People rose 2.93% to 19.7 vehicles in India in 2019." Accessed from www.helgilibrary.com/charts/passenger-cars-per-1000-people-rose-293-to-197-vehicles-in-india-in-2019.
22. Robert C. Toth, "Fitting Technology to Need Held Critical in Third World," *Los Angeles Times* (June 18, 1978), part 1: 1, 32.
23. The most influential book on this subject is E. F. Schumacher, *Small Is Beautiful: Economics as If People Mattered* (New York: Harper & Row, 1973). For some examples of current appropriate technologies, see "10 Cases of Appropriate Technology," at http://listverse.com/2010/06/12/10-cases-of-appropriate-technology/ (accessed on February 7, 2012).
24. Ibid.
25. Louis T. Wells, Jr., "Economic Man and Engineering Man: Choice and Technology in a Low-Wage Country," *Public Policy* 21, 3 (Summer 1973): 319–42.
26. National Science Foundation, "International Comparisons of R&D Performance." Accessed from www.nsf.gov/statistics/seind14/index.cfm/chapter-4#s2.
27. Ron Westrum, *Technologies and Society: The Shaping of People and Things* (Belmont, CA: Wadsworth, 1991), p. 57.
28. Carl Enomoto and Karl R. Geisler (2017). "Culture and Plane Crashes: A Cross-Country Test of The Gladwell Hypothesis." *Economics & Sociology*. 10(3). DOI: 10.14254/2071- 789X.2017/10-3/20
29. Langdon Winner, "Do Artifacts Have Politics," *Daedelaus* 109 (1980): 121–136.
30. Alex Inkeles, *One World Emerging? Convergence and Divergence in Industrial Societies* (Boulder, CO: Westview Press, 1998).
31. David Halberstam, *The Coldest Winter: America and the Korean War* (New York: Hyperion, 2007), p. 223.
32. Darryl E. Brock and Chunjuan Nancy Wei (Eds.), *Mr. Science and Chairman Mao's Cultural Revolution Science and Technology in Modern China* (Lanham, MD: Lexington Books, 2013).
33. Michael J. Mazaar, *Unmodern Men in the Modern World: Radical Islam, Terrorism, and the War on Modernity* (New York: Cambridge University Press, 2007), p. 132.

34. William Form and Kyu Han Bae, "Convergence Theory and the Korean Connection," *Social Forces* 66, 3 (March 1988).
35. Frederick Fliegel et al., "Technology and Cultural Convergence: A Limited Empirical Test," *Journal of Cross-Cultural Psychology* 10, 1 (March 1979): 3–21.
36. Robert M. Marsh, "A New Test of Convergence Theory," *Comparative Sociology* 6 (2007), pp. 251–94.
37. David S. Landes, *The Unbound Prometheus: Technological Change and Industrial Development in Western Europe from 1750 to the Present* (Cambridge: Cambridge University Press, 1972), p. 545.
38. Robert S. Merrill, "The Role of Technology in Cultural Evolution," *Social Biology* 19, 3 (Spring 1972): 246.
39. Brady Robards (2018). "Belonging and Neo-Tribalism on Social Media Site Reddit." In: Anne Hardy, Andy Bennett, Brady Robards (Eds.), *Neo-Tribes*. Cham, Switzerland: Palgrave Macmillan); Julie Frechette, "Public Opinion Formation in the Digital Age: How Tribalism and Political Fury are Manufactured through Social Media Microtargeting." (2019). *International Critical Media Literacy Conference*, 41; and Fukuyama, F. (2018). Against identity politics: the new tribalism and the crisis of democracy. *Foreign Affairs*, 97(5), 90–115.
40. Claude S Fischer, *America Calling: A Social History of the Telephone to 1940*. (Berkeley, CA: University of California Press, 1992).
41. Pew Research Center, "The Strength of Internet Ties," www.pewresearch.org/internet/2006/01/25/the-strength-of-internet-ties
42. Bhavtosh Rath, Wei Gao, and Jaideep Srivastava, "Evaluating Vulnerability to Fake News in Social Networks: A Community Health Assessment Model," *Proceedings of the 2019 IEEE/ACM International Conference on Advances in Social Networks Analysis and Mining (ASONAM)* (2019): 432–435.
43. Ruha Benjamin (Ed.), *Captivating Technology: Race, Carceral Technoscience, and Liberatory Imagination in Everyday Life*. (Durham, NC: Duke University Press, 2019).
44. George Ritzer, *The McDonaldization of Society: Into the Digital Age*, 10th ed. (Thousand Oaks, CA: Sage, 2018).
45. George Ritzer, *The Globalization of Nothing 2*. (Thousand Oaks, CA: SAGE Publications, 2007).
46. Joseph Bein-Kahn, "McDonald's menu items from abroad," *Stacker* (July 19, 2022), accessed May 8, 2023, at https://stacker.com/food-drink/mcdonalds-menu-items-abroad.
47. McDonalds turned 50 in Japan in 2021. www.mcd-holdings.co.jp/en/history; www.mcdonalds.co.jp/en/menu/side
48. Ritzer G. "Rethinking Globalization: Glocalization/Grobalization and Something/Nothing," *Sociological Theory*. 2003, 21:193–209.
49. "Indonesia: The Resistance of the Sacred in Bali to the 'green revolution' and the tourism industry," *WRM Bulletin* 237. Accessed from www.wrm.org.uy/bulletin-articles/indonesia-the-resistance-of-the-sacred-in-bali-to-the-green-revolution-and-the-tourism-industry.

50. Saran Shanmugam, "How the Bali Hindu agricultural practice proved superior to Green Revolution," *MyIndMakers*. Accessed from www.myind.net/Home/viewArticle/how-the-bali-hindu-agricultural-practice-proved-superior-to-green-revolution; and J. Stephen Lansing et al. (2017) "Adaptive Self-Organization of Bali's Ancient Rice Terraces," *PNAS Environmental Sciences*. 114: 6504–6509. www.pnas.org/doi/10.1073/pnas.1605369114
51. Bardini, Thierry. "A Translation Analysis of the Green Revolution in Bali." *Science, Technology, & Human Values*, 19:1994, pp. 152–68.
52. One of the authors was driving through rural Pennsylvania after a rain, and a farmer with a mechanized tractor and a farmer with horse drawn plow were working on opposite sides of the road. The horses were having a better day of it.
53. See: Ems, Lindsay (2002). *Virtually Amish: Preserving Community at the Internet's Margins*. Cambridge, MA: MIT Press; Kevin Granville and Ashley Gilbertson, "In Amish Country, the Future Is Calling," *The New York Times* (Sept. 15, 2017). Accessed from www.nytimes.com/2017/09/15/business/amish-technology.html; All Things Considered, "Amish Community Not Anti-Technology, Just More Thoughtful," *NPR* (September 2, 2013). Accessed from www.npr.org/transcripts/217287028.
54. Miroslav N. Jovanovic, *The Economics of European Integration*, 2nd ed. (Cheltenham, England: Edward Elgar, 2013), p. 211.
55. Dan Vergano, "Scientists Engineer Bioterror Fears," *USA Today* (January 26, 2012), pp. 1A–2A.
56. FindLaw, "United States vs. Dubilier Condenser Corp." Accessed from http://caselaw.findlaw.com/us-supreme-court/289/178.html.
57. J. E. S. Parker, *The Economics of Innovation: The National and Multinational Enterprise in Technological Change*, 2nd ed. (London: Longman, 1978), p. 310.
58. Ibid., p. 327.
59. Freeman, *Economics of Industrial Innovation*, p. 61.
60. Leonard S. Reich, "Lighting the Path to Profit: GE's Control of the Electric Lamp Industry, 1892–1941," *Business History Review* 66 (Summer 1992): 310, 312.
61. Stuart Macdonald, "When Means Become Ends: Considering the Impact of Patent Strategy on Innovation," *Information Economics and Policy* 16 (2004): 135–58.
62. Adam B. Jaffe and Josh Lerner, *Innovation and Its Discontents: How Our Broken Patent System Is Endangering Innovation and Progress, and What to Do about It* (Princeton: Princeton University Press, 2004).
63. "Intellectual Property: A Question of Utility," *The Economist*, 416, 8950 (August 8, 2015): 50–52.
64. James Flink, *America Adopts the Automobile, 1895–1910* (Cambridge, MA: MIT Press, 1970), pp. 318–28.

Chapter 6

Social Construction, Diffusion, and Technological Change

In Chapter 1 we discussed the idea of the social construction of technology, that technologies emerge in complex social processes, which are situated in networks of social power in specific contexts. This was presented in opposition to the idea of technological determinism, the broad idea that technological change is inevitable, often with the connotation that it is uncontrollable. Technological determinism comes in utopian and dystopian flavors. The first variety equates technological change with progress, that any and all technological change is an improvement over past versions of a technology, and that those who might be harmed by technological change are merely inconvenienced or experience the unhappy byproducts or side effects of inevitable progress. Conversely, there is the dystopian thread of technological determinism. These ideas are found in the jeremiad—the bitter list of woes produced by new technologies, such as found in the screeds of the "Unabomber" Ted Kaczynski and related neo-luddite lamentations about technological change. And much dystopian science fiction in print or film takes on a determinist and pessimist tone. Our objective is to provide a more analytic approach to studying the processes and outcomes of technological change, one less charged with assumptions about inevitability, whether positive or negative, but nonetheless attuned to issues of social power and inequality.

There are economic, cultural, and political factors that shape technological possibilities. The history of automobiles is our case study to look at competing alternatives, such as between propulsion types, or questions about individual automobiles in relation to local public transport and national railway systems. We will also visit cases such as the Apple Newton to look at the processes pushing and pulling, and impeding, technological change. But we will start with the bicycle.

What Does It Mean for a Technology to "Work"?

There are two primary premises that shape the theoretical model of the social construction of technological systems (SCOT).[1] The first is that *things could be otherwise*. The idea is that with careful historical study we can see many places in the development of new technologies where there might have been other choices made, with other technologies or variations in technologies that might have emerged. The second premise is a bit more esoteric: *deciding what constitutes "working" is a social process*. The very first bicycles, a kind of seated walking/riding machine that was often called a velocipede, were manufactured in Germany and France in the early nineteenth century. Note that these first "running machines" (*Laufmaschine* in German) were seen as less desirable than the other three- and four-wheeled human-powered vehicles invented in the 1820s through 1850s, because two wheels required human balance and skill to operate. Important to note is that there were numerous competing alternatives, variations, and varieties of bicycle-like inventions from the 1830s through the 1860s, with the French claiming invention of a velocipede in 1860, and the English the "boneshaker" (due to its rigid frame and rough ride) shortly thereafter.[2] The first organized intercity bicycle road race was in 1869; it was won by James Moore, who rode an average of 8 mph over the 10-hour race (including walking up steep hills), on a route between Paris and Rouen, France. The roads, and the bicycles, left a lot to be desired by modern standards, although this was apparently the first appearance of ball bearings in bicycle design.[3]

Figure 6.1 shows a line-up of "ordinary" bicycles from a race organized by the League of American Wheelmen in 1890. These were simply called "bicycles" because this was their "normal configuration" until what we now understand as a typical bicycle emerged in the later 1890s. At that time, the "high wheeler" began to be called the "ordinary" bicycle to distinguish it from bicycles like the one shown in Figure 6.2, which were called "safety" bicycles. The large front wheel of the high wheeler or "penny-farthing" bike allowed

Figure 6.1 "Ordinary" bicycles at the start of the L.A.W. Bicycle Race. The League of American Wheelmen mobilized to get cities and states to begin paving streets in the U.S. to facilitate their activities. (Source: Geo. Barker, 1890)

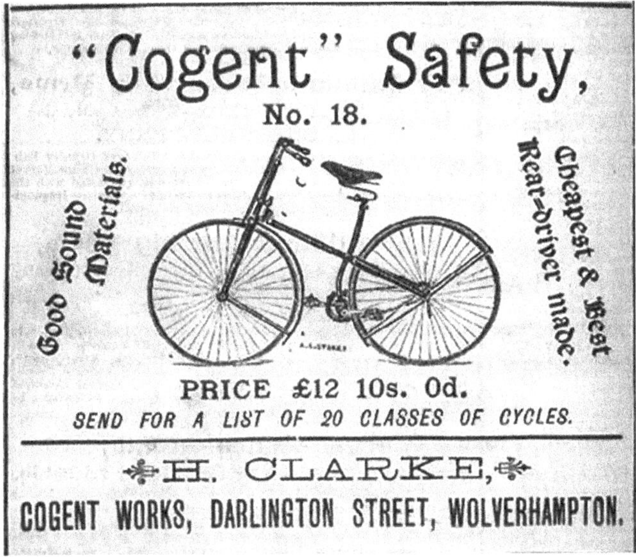

Figure 6.2 Advertisement for a Cogent Safety Bicycle, Barkers Wolverhampton Trade Directory, 1887. Note that this is identified specifically as a safety bicycle, as opposed to the ordinary high-wheel bicycles of the time.

for a greater leverage in the direct-drive from the pedals and gave a smoother ride on what would have been primitive urban and suburban roads. Pneumatic tires were (re-)invented in 1888 for bicycles.[4] Roads were largely unpaved, until the League of American Wheelmen helped push for road construction.[5] Gears and chains were not regularly incorporated into bicycle designs until the 1870s.[6] The high-wheeler bikes required a high degree of athletic ability, the province of "young men of means and nerve."[7]

What we now think of as a normal bicycle was initially designed for the less athletic, the elderly, and for women's exploding interest in bicycling. For women, bicycles were both a physical and symbolic revolution, requiring radical (for the time) changes in dress. Bicycles were symbolically and materially connected to suffrage in the United States and Europe, and growing general ideas about women's physical, political, and economic empowerment.[8]

The SCOT model asks us to consider the different relevant social groups, also including in this case "anti-cyclists," who interacted with bicycles. Each group had different priorities and experienced

Figure 6.3 A 1890s tintype of a woman with a safety bicycle. (Source: National Museum of American History)

different problems with the ordinary bicycle, including the group of primary users, young men. Archaeologist Michael Schiffer referred to these different priorities and problem definitions as performance characteristics.[9] Each group had different priorities—safety, speed, comfort, cost. Each group, whether women, older users, or young men, had different economic and social power with which to try to shift the technology to represent their preferences. Over time, closure shifted from the "ordinary" to the "safety" bicycle, reducing the production of the high wheelers over time.

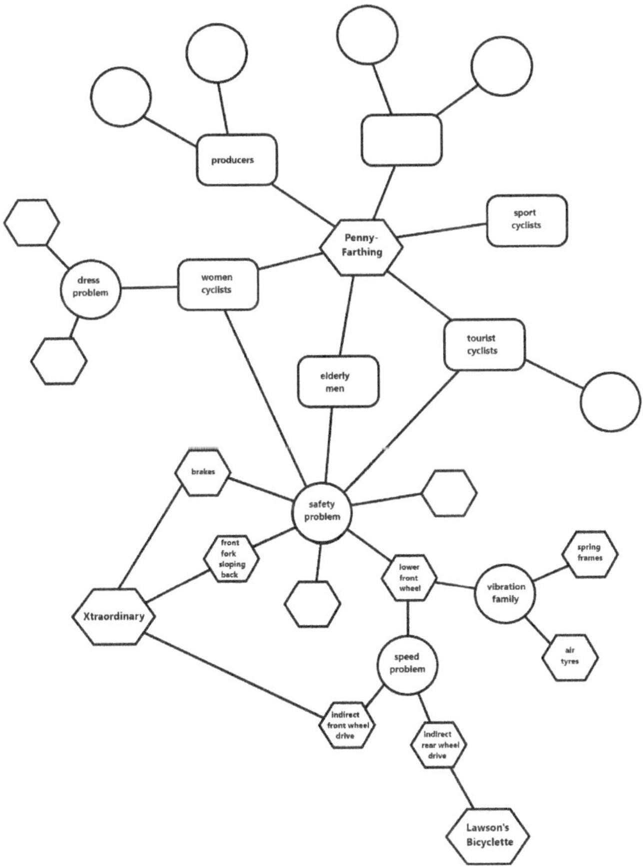

Figure 6.4 A sketch of relevant social groups, problem definitions, and competing alternatives in bicycle history. (Source: Pinch and Bijker, "The Social Construction of Facts and Artefacts: Or How the Sociology of Science and the Sociology of Technology might Benefit Each Other" *Social Studies of Science* 14, p. 418. Used with Permission.)

As can be seen in Figure 6.4, there is not some linear sequence of development for the bicycle. There were competing alternatives (not even all of them shown; more on the missing recumbent cycles below) and a pattern of user groups who experienced different problems with the "normal" configuration of the ordinary bicycle.

Closure is an important concept, because while it can change, producers and consumers need some relative stability in identifying what the "thing" is that they might adopt. Prior to closure, a technology—in this case, bicycles—has a great deal of *interpretive flexibility*. This is the way that different groups see a technology, in both its material functioning and symbolic meaning. Bicycling is to some a wasteful and distracting activity, and to others, a source of health and freedom. Especially when women began riding, the physical activity, the new clothing options such as bloomers, and the feelings of freedom produced quite the moral panic. There were concerns that bicycles would suppress women's fertility, lead to mania, and even be sexually enticing.[10]

The moral panic about bicycles represents the semiotic, or symbolic, ideas about a technology as "working," somewhat distinct from the material facts of working or non-working. The semiotic definition of a new technology may conflict with the rather straightforward material or technical workings of a technology. Interpretive flexibility narrows but never completely becomes fixed. Wheels might turn, bells ring, whistles whistle, but, as we will see again and again, if a new technology offends a cultural sensibility or proves morally suspect, it may fail and be perceived as non-working despite its mechanical adequacy.

While the historical examples surrounding the bicycle may seem quaint, in fact bicycling continues to be a bit open to interpretive flexibility as to what constitutes a "working" bicycle. This can range from novel frame designs for conventional bicycles, to tricycle and other alternatives, including unicycles and antique ordinary bicycles among their fans. One overlooked design option is the recumbent, which was available, at least as evidenced by a patent, as early as 1902.

Today there is a small and loyal market for recumbent bikes. What is often not appreciated is their biomechanical superiority to conventional bicycles: they can go further and faster than a conventional bicycle with less energy expended.[11] They are recommended for pregnant people, for people with back problems, and for older riders,[12] although they can be challenging to balance when in bicycle

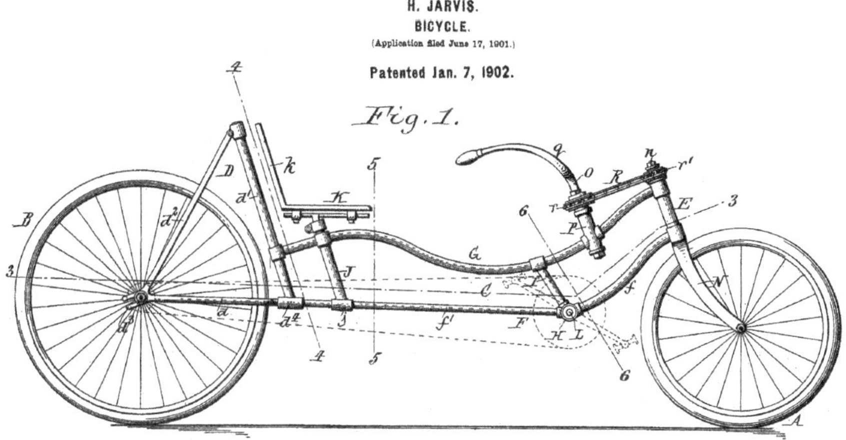

Figure 6.5 The Jarvis recumbent bicycle, patent filing 1902. Recumbent bicycles seem to emerge in the post-highwheeler phase of bicycle history.

rather than tricycle form. They are very efficient and won numerous bicycle races and trials prior to their being banned in international competition by UCI (Union Cycliste International) in 1934.[13] In the present, recumbents are largely interpreted as being "nerdy" and "not real" (meaning "non-working" in a semiotic sense) bicycles to competitive cyclists and most of the general public, despite their biomechanical superiority. They remain banned from international open competitions. Recumbent bicycles are a more recent case where the semiotic (non)working and material working are in conflict, the "suppression" of what is a logically superior technology achieved by the regulatory power of the *super actors* of the international cycling federations and a general cultural disparagement of the technology.[14]

Consumers, Organizations, and Technological Diffusion

Many of the same processes that characterize the diffusion of technology from one country to another also manifest themselves when new technologies are acquired by organizations within a country. Be they business firms, government bodies, charities, or schools, the same factors affect the speed with which new technologies diffuse and the extent to which they take hold. The *technological diffusion curve* is an important model for understanding the emergence, spread, and decline, of technologies.

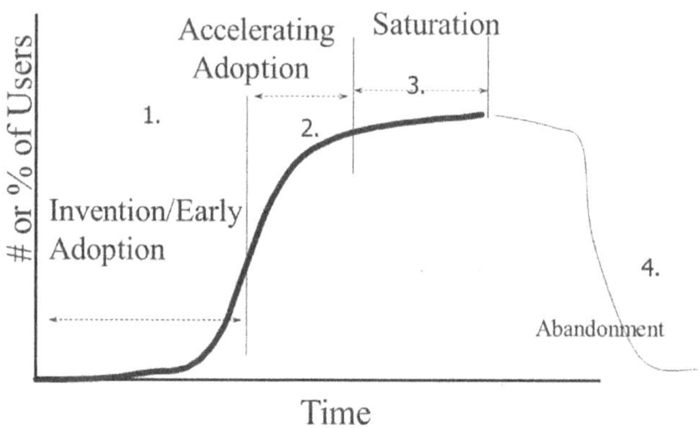

Figure 6.6 The Technological Diffusion Curve. (After C. Everett Rogers, The Diffusion of Innovations)

The technological diffusion curve was articulated by communications scholar Everett M. Rogers in 1962.[15] The curve, as identified by Rogers, has three primary phases: innovation and early adoption, accelerating adoption, and saturation. Also worth considering is an additional phase: abandonment. In each phase, different social groups, kinds of adopters, and communication patterns are in effect to either help or hinder the adoption of a new idea or technology.

In the first phase, inventors are developing new technologies, venture capitalists are screening potential investments, and the early adopters are often technological enthusiasts interested in the next "new new" thing. It must be emphasized that few technologies clear this phase. Much design and redesign are underway in the first phase. Here we can posit a heuristic: that it takes about 20 years to get a typical new technology from idea to innovation, to hit the first inflection point on the curve and have the potential for rapid adoption. The technology must move from prototype to a device that can be manufactured cost-effectively, which requires changes to both the design, and perhaps development of new manufacturing infrastructures.

According to Everett M. Rogers, the likelihood of a new technology's success rests on the following attributes:

- **Relative advantage:** The degree to which the innovation is perceived as being better than the existing way of doing things.

- **Compatibility**: The extent to which the innovation is perceived as being consistent with existing values and needs.
- **Complexity**: The ease or difficulty of understanding and applying the innovation.
- **Trialability**: The extent to which an innovation can be applied on a limited, experimental basis.
- **Observability**: The degree to which the results of an innovation are apparent.

When considering these attributes, it should be apparent that *perceptions* play an important role in the adoption or rejection of an innovation. The go/no-go decision is made by people, and these decisions are likely to be based on a combination of objective and non-objective factors. Performance characteristics and semiotic attributions shape technological choice. Moreover, individual decisions are usually forged in a social setting; once again we see that the technologies being adopted and used reflect social institutions and processes.

Closure, which we discussed in the context of social construction, needs to be established before the technology can "take off." The meanings of the technology and perceptions about its usefulness need to be rather firm in the buying public's mind, just like its actual physical configuration. It also has to be made convergent with potential infrastructures where it might be used. For example, in 1946, early microwave ovens were large commercial products that required a rewiring of a kitchen—an expensive and disruptive process. By 1967 they were portable devices, although still very expensive compared to today's low-cost countertop appliances.[16]

In addition to fulfilling at least some of the requirements just listed, effective technology transfer often requires a learning process as a new product or process moves from one environment to another, such as from a laboratory to a factory. Technologies are based on knowledge, but knowledge that is readily comprehensible in one setting may be a great mystery in another. As with the conversion of scientific knowledge to technological practice noted in the previous chapter, successful diffusion sometimes requires the services of a kind of "translator"—a person capable of functioning in both settings so that information can be transferred from one to the other.

As Hugh Aitken's study of the early days of radio indicates, the efforts of "translators" were crucial to the development of that technology and its adoption by commercial enterprises. The history of radio begins with the work of Heinrich Hertz, who in the 1880s "translated" James Clerk Maxwell's theory of electromagnetic radiation into a laboratory technology that allowed an empirical examination of the theory. Following Hertz, Oliver Lodge moved beyond the laboratory experiment through his development of an imperfect but workable method of using radio waves for actual communication. The first stage in the development of radio was essentially completed by Guglielmo Marconi, who took the embryonic technology and converted it into a practical (and highly lucrative) commercial business.[17] Throughout the course of radio's history, information was put into a new form, pushing the technology to the next stage of development. In Aitken's summary, "At each stage in the process of translation, information generated in one system was converted into a form that "made sense" in terms of another; and at each stage new information was blended with what was already known to create something essentially new."[18] This is a process we can see today when the decades-old technology of the web is used to stream music, movies, and other forms of artistic expression that have been around for much longer.

Accelerating adoption is the phase in which the technology has achieved relative closure and rapidly diffuses across new user groups. The rapid deployment of the cell phone is an example of this, but nearly all *commercial* technologies exhibit this. At saturation, substitutions, repairs and replacements, upgrades, accessories, and supporting devices are part of the economic activity. Consider again the microwave oven. They are exceedingly popular, but the market is not growing, per se. Most households interested in rapid cooking have microwaves. Nevertheless, ovens need to be replaced, and there are a plethora of specialized cookware and accessories, such as microwave popcorn makers, splash covers, egg poachers and so on, that are available as complements to the original microwave purchase.

Rogers' framework largely neglects the issue of abandonment. This is not the discard of a specific product by a user, due to breakage, disuse, or obsolescence. This is when a technology no longer is made or sold as a broad category. Studying abandoned technologies is as challenging as studying failed technologies: they are invisible and often difficult to track down. Segways were officially retired in 2020, having never achieved their commercial potential but being popular in closed environments, such as malls and airports, for security transport.[19] Segway's "failure" in the consumer market was somewhat offset, like Betamax, by a professional, niche, market. Similarly, pagers remain a commercial product: used in secure and specialized environments, such as hospitals or defence contractors, although replaced in the consumer market by the multifunctionality of cell phones. Abandoned technologies can also be repurposed—such as old Mac computers being turned into "Macquariums" (an aquarium made using an Apple Macintosh computer)[20]—and almost anything can become a collectible, including pagers, antique hand farm-implements, fire trucks, Apple Newtons, or Segways.

Figure 6.7 A 1923 Autocar Fire Engine, restored by Mr. William Croissant, now in the collection of Mr. R. James, affiliated with the Boalsburg Fire Company, one of its original service posts. (Photo by Mr. R. James, 2022)

Limits and Exceptions to the Diffusion Model

As hinted above, the technological diffusion curve works well for general consumer products, but it has limitations. Military products, and medical products, have formal restrictions on sales in specialized markets. Consider the Zamboni (the specialized tractor to groom ice rinks). For 80 years the Zamboni family has manufactured these machines,[21] but we do not expect this to be a general consumer good with one in every driveway. As an industrial good, there is a limited market. By managing innovation and new product development, just-in-time manufacture, and careful inventory management, the Zamboni technology lives in a state of "saturation," where their product is a generic trademark, the proprietary name for a general class of goods (like Kleenex or Xerox).

Similarly, luxury and status goods do not quite obey the technological diffusion curve. The exclusivity of fashion items, automobiles, and other goods means that companies deliberately "undershoot" saturation to maintain the extra value of products that are hard to come by. Even middle-tier products like specialized automobiles limit their total sales. For example, BMW produces Mini automobiles but does not expand production of the brand to meet demand. Because consumers do not want to find a substitute for owning a Mini, they will wait and pay a premium price for the vehicles.

Finally, the technological diffusion curve does have embedded within it some pro-innovation bias: people late to adopt a new technology are referred to as "laggards." It requires continual reminders that *most* technologies do not survive the first phase of the curve, to transform from inventions to innovations. As we see below, no new technology is a "sure thing."

A Risky Business

Some technologies, especially those involving military matters, are sponsored by governments, but much technological innovation is adopted by private businesses pursuing their economic goals. Firms may adopt new technologies in order to cut their costs, improve their products, bolster their profits, penetrate new markets, or achieve some combination of any or all of these. The speed and extent of diffusion are therefore strongly influenced by expectations of costs and benefits.

Expectations, however, are not always met. Along with the opportunities presented by a new technology come uncertainties that cannot be assessed through conventional economic analysis. No new technology is a "sure thing." Some end up embarrassing failures: ask Microsoft about its Zune media player, or Google about Google Glass. Even with a technology that is ultimately successful, its immediate advantages may be slight, and its benefits may be a long time in coming. On a more positive note, some new technologies may exceed expectations when they are complemented by other technologies. Under these circumstances, new products and processes are more than the sum of their parts, as when a combination of the laser and the computer gave birth to supermarket scanners, printers, and robotic surgery.[22]

It is often the case that the potential of an emerging technology is what matters, and this may be the hardest thing of all to assess. Many years may elapse before a technology passes from laboratory feasibility to commercial success, that is, from invention to innovation. Although some inventions made the transition rapidly—only one year in the case of Freon refrigerants—others have taken much more time. In the case of the mechanical cotton picker, it was 53 years, and for the zipper, 27 years.[23] Sometimes an invention appears on the scene, is quickly forgotten, and then re-emerges decades later, as happened with the pneumatic tire.[24] Even technologies we might characterize as "failures" may represent technologies ahead of their time or useful only in specific contexts. For example, consider Apple's Newton personal digital assistant or the Sony Betamax video recording system. The Newton had many of the features of a contemporary smartphone, but without materials advances, interface design improvements, miniaturization, batteries, and the availability of ubiquitous connectivity, its usefulness and adoption were limited.[25] The platform was developed in 1987 and the first products distributed in 1993. When connected to its successor, the iPhone, the Newton demonstrates the "20-year rule" mentioned above. Use infrastructure, technical problems, and manufacturing infrastructure take time to get worked out. Students who say the cell phone was invented with the first iPhone in 2007 do not know of this pre-history of the Newton.[26]

Context matters too, before judging a technology a failure. Betamax was preferred for professional video production, although a failure in the commercial marketplace.[27] As in examining the relationship between science and technology, ascertaining the success or failure of a technology depends in part on the time frame which is chosen.

The decision to adopt a technology represents an organization's assessment of the likely benefits accruing from the use of the new technology, weighed against the uncertainties that attend its use and the speed at which these uncertainties can be dissipated.[28] Given the nature of technological innovation, uncertainties will rarely vanish completely, but as information is gathered and experience gained, *uncertainties* can be converted into *risks*.[29] More than terminology is involved here; according to a distinction made by economists and game theorists, "uncertainty" indicates an inability to predict an outcome, whereas "risk" implies at least a rough notion of the probability of success and failure.[30] The ability to make calculations of this sort does not ensure accurate predictions of success or failure, but at least it enables decision makers to make informed judgments of their likelihood.

Not all risks are technological in nature. Although any new way of doing things can be presumed to have its share of bugs and other unanticipated problems, these may be minor when compared to the commercial uncertainties of a new venture. Uncertainties of this sort can be particularly problematic, for decision makers usually find it easier to assess the probability of technical success than the probability of marketing success.[31] "Will it fly?" is easier to determine than "Will anybody pay money to fly in it?" This was the question for the Anglo–French supersonic airliner, the Concorde. A stunning technological achievement that required the solution of countless complex problems, the Concorde was a commercial failure. An extraordinarily expensive aircraft to fly and maintain, it never generated revenues commensurate with the costs of its development and operation. No private business could have survived this sort of failure. Only the financial support of the governments of France and Great Britain, motivated by considerations of national pride as much as anything else, kept the Concorde aloft for over 50,000 flights, until it was retired in 2004.[32]

The NIH Syndrome

The inherently uncertain or at best risky nature of many technological innovations may explain the presence of one persistent block to the diffusion of new technologies—the Not Invented Here syndrome, or NIH for short. People and organizations exhibiting this syndrome are reluctant to make use of technologies that were invented elsewhere. The corrosive effects of the NIH mentality were bemoaned by Henry Ford II: "There's too much NIH—not invented here [in the automobile industry] . . . . Lots of times a guy brings something in, and unless the improvement is rather dramatic, nothing happens. The status quo is a hell of a lot easier than making changes."[33]

Internal resistance to technological innovation, while frustrating to some, is certainly understandable. As was noted, innovation is at best a risky process that can generate a great deal of discomfort, and this discomfort will be especially intense when the innovation has not been the work of those who are affected by it. It is also the case that internally generated technologies are likely to be more compatible with established organizational structures and activities. Moreover, an internally generated technology will probably have a "champion" within the organization who seeks its utilization, and management may be more inclined to pay attention to the ideas of their organization's own personnel.[34]

These are explanations for the presence of the NIH syndrome, but not excuses. A dynamic firm can ill-afford to ignore useful new technologies just because they were developed elsewhere. We have already seen that a receptivity to "foreign" technologies has been crucial to the development of Europe, the United States, and Japan, and that a resistance to them was a major reason for the economic retardation of imperial China. An individual firm that chooses to resist or ignore new technologies because of their source may ultimately find that security and pride have been purchased at a very great price. Its comfortable routines and organizational structures are retained, right up to the day that it goes out of business.

Connecting Social Construction and Technological Diffusion

The story of the bicycle we have presented is concerned with the competing alternatives *within* a technological framework. Unicycles, tri- and quad-cycles, uprights, safety, and recumbent bicycles were all competing to become the normal configuration of basic human-powered travel. If we expand the frame somewhat, we can also see that human-powered transport was part of a larger ecosystem of technological options for transportation: steam-powered trains, horse-drawn carriages, trolleys and trams, busses, and the emerging variants of automobiles.

Where to start? Like so many technologies, because multiple people recognize a need or an opportunity, there are often priority disputes about who was first. For example, Alexander Graham Bell's role as the inventor of the telephone is pretty common knowledge. But he was in close competition with Elisha Gray. There was literally a matter of a few hours difference in their patent filings: Gray's patent was at the patent office first, but Bell's was the first to be processed. It took a decade of bitter dispute in the courts and media of the time, including accusations of patent fraud, until Bell was enthroned as the inventor of the telephone, a victory in part enhanced by Bell's focus on transmitting speech. Both were deeply familiar with telegraph technologies which provided the backbone of telephony and shaped its emergence.[35] The point is that it is rare that only one person sees an opportunity for invention.

This was true, as well, for the automobile. And like the telephone is tangled up with telegraphy, the pre-history of the automobile is entangled with the history of bicycles, because many components, such as ball-bearings, axles, and pneumatic tires, were first patented for bicycles. And the necessary skills in part manufacturing and assembly emerged from bicycle makers as well as the rail industry, wagon-making, and industrial machine-making. Cars were manufactured one at a time, in numerous small workshops. There have been, in the United States, around 3,000 different car manufacturers. At the turn to the twentieth century, there were about 2,800. By 1930, there were fewer than 50.[36] Note here, too, that dedicated carriage manufacturers declined at about the same rate as car manufacturers emerged.[37] In Mannheim, Germany, Karl Benz patented a vehicle in January 1886, and in March 1886 Gottlieb Daimler and Wilhelm Maybach created their first car. By 1889, Daimler produced what we would recognize as a paradigmatic internal combustion engine.[38]

But electric cars were also in development. In 1899, driver Camille Jenatzy, in an electric car of Belgian design called *La Jamais Contente*, was the first to go 100 kilometers per hour (his official time was 105 kph, or 65 miles per hour). Electric "carriages" were part of the landscape of the global experimentation in the automobile industry.[39] In 1900, about 28 percent of the 4,912 cars produced in the US were electric.[40] At the turn of the twentieth century it would not have been clear whether to invest in gasoline or electric automobiles, or to invest in their competitors in public transport, such as trolleys and busses. And steam automobiles were not yet completely out of the picture.

So let's connect the social groups and their performance characteristics to the technological diffusion curve for contemporary automobiles. In the very earliest phase of the technological diffusion curve, 1880–1900, the social groups interested in the new automobiles were the emerging manufacturers, entrepreneurs, and of course potential consumers. Often forgotten as a relevant social group are the owners and operators of horse-drawn vehicles—whether individually owned carriages, trolleys and taxis, or cargo movers—which in New York City produced about 2.5 million pounds of manure, and 15,000 dead horses, annually.[41] Similarly, the emerging oil and gas and electrical industries were not, in the first phase of the diffusion process, power players in the automobile industry; But as the gasoline-powered automobiles came to dominate, they certainly gained an interest in developing the infrastructure for supplying energy as automobiles entered an accelerating adoption phase.

Figure 6.8 Approximate number of vehicles produced by U.S. automotive industry, 1886–1962. Time axis not to scale. (Data from Shaw (1987), p. 4; www.macsmotorcitygarage.com/the-year-in-cars-1942/ ; and https://en.wikipedia.org/wiki/U.S._Automobile_Production_Figures)

As we noted in Chapter 1, electric automobiles were a viable part of the emerging automotive landscape, but because they became associated with stereotypical Victorian femininity, they suited neither the male breadwinners who had purchasing power in upper- and middle-class homes, nor the many women who aspired to the ideals of independence symbolized by gasoline-powered vehicles. Electric cars were generally more expensive than their gasoline-powered competitors.[42] As Virginia Scharf shows, and Michael Brian Schiffer also demonstrates, women were the intended users of the electric automobile because electrics were clean, quiet, and didn't require a crank to start. Mrs. Henry Ford drove an electric car, as did Mrs. Thomas Edison. Mr. Henry Ford drove a Rolls Royce, and Thomas Edison a gasoline-powered car. But middle-class families could not afford two cars, and since men were both the ideological and economic leaders of households at the time, they chose cars which reflected their identities and interests, even if the vast majority of men never went on cross-country races. Women, similarly, wanted to participate in the idea of freedom and adventure that automobiles represented. Gasoline cars could carry their own fuel, were built more sturdily, and could travel further in rallies and races.[43] In terms of performance characteristics, electric cars were technically superior for short trips, and were cleaner and quieter, although not as effective as public transport for errands in urban areas in terms of cost and maintenance. Gasoline automobiles were more self-contained and perceived as more effective than electric vehicles, although also deficient in urban areas in terms of cost, maintenance, storage, and usefulness for short errands. There are no "pure" technical criteria which will tell you which alternative might win the race to become an accepted innovation.

Other things also had to come into place for the car to start to turn up the technological diffusion curve. After Henry Ford developed the assembly line for manufacturing the Model T in 1913, this led to a rapid expansion of reasonably priced vehicles. The gap between Daimler's 1886 internal combustion engine and Ford's start to mass production exemplifies the aforementioned "20-year rule" for technological diffusion. Only then could the automobile "take off" and move into the accelerating adoption phase.

Automobiles were relatively "big ticket" items, so purchases required the emergence and acceptability of consumer credit. It was rather taboo for people to have debts for anything beyond their mortgages, until the 1920s.[44] As a matter of cultural convergence, in the United States especially, automobiles came to signify freedom from landlords and freedom from both dense urban and isolated farm life. This association with freedom, encouraged by the press and emerging advertising industry, helped establish the public's desire for cars.

Note that the technological diffusion curve for automobiles is not smooth. The Great Depression (1929–1939) killed off the electric car manufacturers of that first generation—and many, many of the smaller gasoline-powered vehicle manufacturers as well. Technological "lock in"[45] and path dependency produced a final configuration for automobiles which defined the gasoline-powered vehicle as the "normal configuration" to the extent that, until Tesla popularized new electric vehicles in 2008, the pre-history of electric cars was nearly forgotten.[46] And it was never about the batteries, or not ever *just* about the batteries. "Range anxiety" plagued gasoline-powered cars users, too, who either had to carry fuel or ship it ahead in road races and long excursions. The disruptions of World War II also put a significant "blip" into what would have been the accelerating adoption of the automobile, as production was redirected toward wartime needs, and consumer discretionary spending was reduced and inhibited demand.

Behind this curve are also two changes in infrastructure. The first, noted above, is the development of mass production. While interchangeable parts had been used by Eli Whitney in 1798 for making muskets at his armory in Connecticut,[47] the actual assembly line was implemented by Henry Ford in 1913.[48] It facilitated producing a Model T, invented in 1908, in about 90 minutes rather than 12 hours. There were 78,440 Ford cars produced in 1912; 168,200 in 1913; and 306,162 in 1914. Other manufacturers only had incremental expansion of production in that time period until they, too, adopted mass production.

The second infrastructural change were the two waves of road-building. The first was facilitated as noted above, by the League of American Wheelmen, the bicyclists' organization. The Federal Highway Act of 1921 leveled the playing field for both electric and gasoline-powered vehicles. Until this time, road construction was locally controlled and frequently imagined as a way to get to trains as the primary mode of long-distance transport. Then the Federal-Aid

Highway Act of 1956 was signed by President Dwight David Eisenhower. He recognized the need for a transcontinental transportation system as a matter of national defense after his experiences in Europe in the Second World War. But again, note that there were some social groups opposed to the Federal Highway system.[49] Those displaced by freeways cutting through urban areas were particularly motivated. Disadvantaged groups such as ethnic minorities or those in older neighborhoods conducted "freeway revolts" in the late 1960s and early 1970s. With protests and legal action, they challenged the displacement of already marginalized groups by new highway construction, sometimes to great success.[50]

To be clear, the social groups most engaged in the development of a technology change as the technology moves across different phases of the diffusion curve. From inventors, entrepreneurs, and early adopters, to other groups perhaps impacted—positively or negatively—by the technology as it becomes more widespread, multiple social groups come into contact with technologies and can advance or impede development. Each of the different groups has varying priorities and identifies different problems and opportunities with the technology. They each decide whether or not the technology is "working" for them, in both semiotic and material terms. Every technology has trade-offs and prioritizing those trade-offs is a matter of choosing amongst values and preferences, which is not necessarily a technical process. Sometimes there are quantitative measures (e.g., cost-benefit analysis) that can help, but it still is up to the users, and competition among user groups, to identify which is the most pressing problem to solve and which technology to accept or decline.

One thing absent from the above representation of the automobile's technological curve is the possibility of abandonment—of the large-scale decision that a technology is *not working* anymore. While perhaps difficult to imagine, it is not impossible that we collectively will decide that gasoline-powered vehicles, and individual ownership of cars, don't work anymore. Looking at vehicle use across cultures gives us some indication of that possibility. For example, Japan, as well as China and other Asian nations, have prioritized trains for inter- and intracity transport. Japan has almost 3,000 kilometers (about 1,795 miles) of high-speed rail between major cities, with extensive trains and subway systems in urban areas. Besides prioritizing trains, there are also other initiatives to decrease automobility. France has decided to ban fossil fuels, both diesel and gasoline, in automobiles by 2040, with California setting a goal for 2035.[51] In the Netherlands, a car that produces typical emissions might face a 25% tax on the valuation of the car, while low- and no-emissions vehicles face a 15% and 4% tax rate, respectively. Similarly, Denmark has a value-added tax with additional taxes for luxury vehicles and inefficient vehicles that can sum up to 150% of the value of a vehicle. Of course, these high taxes discourage individual automobile ownership, but the primary purpose is to manage the externalities (see Chapter 8)—such as smog, traffic, road maintenance expenses, and parking and storage—that are not in the direct price of the vehicle and are borne by people beyond the individual owner.[52]

There are infrastructural changes required to support non-automotive transportation solutions. Besides investing in rail and other public transport, cities themselves may need to be re-designed. A rapid rail system between, say, Phoenix and Tucson, Arizona, (approximately 90 miles or about 145 km) doesn't make much sense when one gets off the train in either city to an extremely rudimentary public transport system for intracity transport. Nonetheless, there have been active freeway removal projects, including Madrid, Portland, and Seoul. Sometimes the highways are removed, sometimes they are covered with pedestrian greenspaces.[53] In all cases of highway removals and other projects that limit access by automobiles, pedestrian use is increased; noise, temperature, and smog is decreased; and local quality of life is much improved.

For many in the United States, because cars symbolize personal freedom, there is often a misidentification of actual travel patterns and needs. It is difficult to imagine a car-free future, particularly in rural areas and in the Western U.S., where city design is automobile-centric. The idea of being off-road sells SUVs, of which 90 percent never leave the pavement. Similarly, 90 percent of all driving is at less than 45 mph and for less than 30 minutes, perfectly in range for an electric car. But people often feel, and cars are sold on the basis of the idea, that we take lots of long trips on remote winding roads with the wind in our hair and thus "need" a longer-range vehicle than we actually do. We will explore this consideration, that the ideas about a technology, both real and imagined, can be a major determinant for the success or failure of an innovation, in the next section.

Culture, Acceptance, and Rejection

It is more than economics or mechanical functioning that shape the determination of a technology as "working" for a social group. Cultural and political factors, and as we will see here and in later chapters, organizational factors, also shape the detailed development of technologies and shape their reception in new contexts.

Consider the values that corn has in Mexico, both as central to a cuisine and to national identity. There are about 60 regional varieties of corn, which have been cultivated in Mexico for about 10,000 years. Displacement of these traditional varieties by GMO corn cultivation created protest, such as a hunger strike in 2013, because GMO production potentially erased traditional varieties, increased dependency on multinational corporations, and produced price increases and decreased quality of tortillas, a daily staple for Mexican families of all walks of life. A ban on GMO corn cultivation reflects a technological choice that prioritizes local economic self-sufficiency, food traditions, and the value of independence that outstrips whatever the economic benefits from adopting that technology might have for large-scale producers.[54]

Similarly, responses to reproductive technologies are highly influenced by culture, especially religious framings of new technologies. Consider three drugs involved in managing pregnancies. Levonorgestrel is known as "Plan B" emergency contraception; but it is also used to treat side effects of menopause and prevent osteoporosis and can be used to treat endometriosis. Like mifepristone, which is also used for abortions, it has numerous other uses, including having a preventive effect for some forms of breast cancer. And methotrexate is particularly useful for managing ectopic pregnancies but is extremely important in treating cancers like non-Hodgkin's lymphoma, or diseases such as lupus or arthritis. But the association with abortion has made access to these three drugs difficult for their non-reproductive uses.[55] These drugs have more than one use, but a very vocal minority has established closure around mainstream meanings which potentially limits access for other medical necessities.

Many other kinds of power are also part of the social construction process. Consider the case studies where labor politics are part of technological decision-making. As noted earlier (Chapter 2), changes to the production of the McCormick reaper were designed to disrupt skilled craft workers and unionization efforts, and after those changes had their effects, they were abandoned. But the conflict at the McCormick factory was central to a larger conversation in American public life, including the Haymarket Riots, on the control over labor.[56] In 2022, the retreat from work-from-home, which remains as technically feasible as it was during the height of the pandemic, reflects this continuing concern about controlling the productivity of workers.[57] Performance characteristics often go far beyond the merely technical and often engage issues of social power and control.

Finally, let's consider the example where the symbolic functioning of a technology completely overrides its mechanical adequacy. In late 1980s, before the rapid emergence of hand-held cellular telephones, phones in cars were a very high-prestige item. They were so prestigious that one of the authors (Croissant) knew someone who bought a kit that made it *look like* they had a cell phone in their car. With a fake antenna and assembly in the center console of the car, the *appearance* of being important enough and wealthy enough to have a cell phone was the basis for someone's designing and manufacturing, and buying, these kits.[58] While we can dismiss this as a trivial example, let it remind us about how important culture and beliefs are to the acceptance or non-acceptance of a technology, sometimes even overriding what actual functionality and benefits that may be obvious to designers but not to users.

Let's return briefly to the example of the Apple Newton. When first shipped in August of 1993, the designers fully defined it as a "working technology," and fully expected it to be successful. Something like 50,000 units were sold (at a price of $900, or $1600 in today's value) in the first four months. It lasted as a buyable product until 1998. By 1995, on both physical and semiotic dimensions, the Newton became framed as not-yet-ready for primetime. But fixing technical problems, like poor handwriting recognition and a poorly backlit screen, didn't allow it to overcome an emerging bad reputation as "non-working."[59] It is now interpreted either as a failure, or as a pre-emptive version of technologies that define success: iPads and iPhones. Social construction principles—such as context, competing alternatives, and infrastructure—are hashed out by designers and users and other relevant social groups as they decide what "works" for a technology.

Inclusion

One of the distinctions made in the SCOT model is between actors with a high and low inclusion, a term to describe the influence different groups have in early phases of technological change. As we noted in the early chapters, a great deal of technological change occurs out of the public view, and that is where actors with a high inclusion are generally found, because they are intimately involved in making key material technological choices. Henry Baekeland was highly included in the communities of electrochemistry and photographic chemistry but had little experience with synthetic resins. That low inclusion allowed him to see and experiment with phenol and formaldehyde in ways that traditional dye chemists were not in a position to see. This position allowed Baekeland to invent Bakelite—a moldable yet hard plastic resin.[60]

End users are generally low-inclusion actors, who might have a lot to say about the *meaning* of a technology as it comes into their world, but not so much about the design of the technology. Technological determinism *feels true* because most of us are low-inclusion actors in most technological design processes. We receive new technologies as products or systems which we can either take or leave, without understanding the complexity of the behind-the-scenes processes that brought the technology into consideration. "Leaving" a technology, choosing not to purchase or participate, can make one feel left behind. Not having a cell phone or choosing not to own an automobile in many places in the United States can leave someone both feeling left out, and literally left out of employment opportunities.

In the design and development phases, high-inclusion actors find a lot of technological choices. To paraphrase Wiebe Bijker, there's a lot of life inside a technology for high-inclusion actors. But even after design and development has stabilized and a technology has achieved relative closure due to the efforts of designers, engineers, and marketers, consumers and users make meaning, decide if a product "works" in their semiotic and material realms, provide feedback, don't follow the directions, use the technology for their own purposes, and sometimes organize and mobilize against new technologies. In establishing closure around an innovation, which is then identified as the "paradigmatic" frame, model, or normal configuration, various things can occur. As noted above in the case of recumbent bicycles, "super users" such as the International Cycling Association, can block technological innovation. Political and economic battles over standardization, such as between Edison and Tesla for direct current versus alternating current, can help shape technological lock-in.[61] Monopoly, oligopoly, or cartel conditions, such as the oil industry, can shape the cost and cultural desirability of potential technologies.

Social construction is a continuous process with moments of closure that can be temporary, and the inventors and translators, financiers and marketers, and of course end-users frequently underestimate the complexity of technological choice. We can connect inequalities in the design process with inequalities in outcomes—inclusion matters in understanding potential problems that users might have with a new technology. For example, both automatic handwashing systems and pulse oximetry machines are not good at engaging people with darker skin and thus can reproduce health inequalities.[62] Would the inclusion of more diverse designers and participants in the research and design process improve outcomes? Technological determinism does not engage the fact that technologies are designed, and designed by people situated in social worlds, and that technologies are adopted or not by people also situated in social worlds, perhaps very different from those of the designers.

Path Dependency and Sunk Costs: Prior Choices Shape Future Options

We need to look at one more factor that contributes to our ideas about technological determinism and reductions in technological choice. Path dependency also contributes to the lingering notion of technological determinism. Our material world is rather durable. It takes a great deal of time, money, labor, and often political will to make changes to infrastructures (e.g., to re-prioritize pedestrians over automobiles). We learned earlier of the controversy over the selection of alternating over direct current as the primary mode of electrical production.[63] These so called "current wars" laid the pathways for technological commitments both large and small. Multiple generations of devices, appliances, and systems were designed around the selection of alternating current, and landscapes transformed with transmission systems and power lines. But of course, direct current items (e.g., battery-powered devices) have always existed. The emergence of solar power and new local generating options, such as wind, and the development of stable and higher capacity batteries means that the balance of power, literally and figuratively, between alternating and direct current may yet again shift.

To turn to automobiles, the size of vehicular lanes and the amount of land used for transportation are important decisions in city planning. The size of vehicular lanes of course emerged from the pathways, streets and roads emerging when horse-drawn vehicles dominated. Every city planner now has to consider that while narrower streets are safer, they are not preferred by drivers who want the speed afforded by wider lanes. A decision to prioritize speed over safety (or vice versa) will shape a city nearly indefinitely as infrastructure becomes solidified around street formations.[64] These choices represent a path dependency, which like other features of our technological life, both constrains and enables future developments. A new vehicle would be expected to fit on existing roads, a new electronic device to plug into existing power grids. Those are examples where new technologies are constrained by past choices. But these choices also enable new technologies. With a stable infrastructure, the number of design choices is somewhat narrowed, and the predictability of infrastructure allows variation and elaboration of innovations.

Sunk costs are a similar phenomenon, but not identical to path dependency. Sunk costs reflect an economic commitment to a particular investment, financial or technological. That economic commitment can also come with psychological commitments that make it harder to walk away from an investment than might be rationally expected. Both individuals and organizations can become mired in sunk cost problems because it is very difficult to walk away from expenditures which are not going to pan out. Humorist Scott Adams gives us the following tongue-in-cheek example from a fictional list of sayings made by irrational people: "We've spent millions developing a water-powered pogo stick. We can't stop investing now or it will all be wasted."[65] This fictional person is not clear on the concept of a sunk cost. And for the record, powered pogo sticks are actually banned.[66]

One more serious challenge remains when looking to path dependency as proof of technological determinism. The possibility that path dependency forecloses growth or change even as it stabilizes technological systems gives us cases of leapfrogging, where technological change "skips over" commonly understood development patterns. A notable example is mobile phone use in Africa. In Tanzania, for example, 9 in 10 adults have a mobile phone.[67] Mobile phone use transformed banking, communications, and to a lesser extent, social network activities. This is despite the fact that landline infrastructure remains nearly completely absent.[68] The adoption of cell phones skipped over landline-based home phone usage, which was the typical pattern in the West. As with our model of social construction, there may be multiple technological options competing to solve a problem: it is not clear which option will "win" and become the dominant technology, and alternatives in the competition may come forward later or come to inhabit specific niches which in the future could become dominant. It is possible that something from "left field" can emerge, shift infrastructures, and truly disrupt or leap past existing technological systems as people find the new technology highly advantageous and worth disrupting old ways of doing things.

When we introduced the term "technocracy" back in Part 1, we argued that technologies are means to an end but do not, in themselves, tell us what ends we should be prioritizing. Social construction theory tells us that there are multiple potential user groups with competing performance characteristics. Different levels of power and the kinds of decision processes, such as Amish direct democracy in deciding about technology, could be available for technological decision making. To some extent, priority setting for new technology is generally left to the market and, as we will discuss in a later chapter, not always integrated as a process in political systems.

QUESTIONS FOR DISCUSSION

1. Can you think of other examples where there is a mismatch between the semiotic "working" of a technology and its material functioning? Did that mismatch help or hinder acceptance of a new technology?

2. Describe how closure works, and how it can be temporary. What does the concept of closure suggest for people wishing to accelerate adoption of a new technology?
3. Abandonment and resistance to technologies are not quite the same thing. At what phases of the technological diffusion curve do these orientations to technology occur? What do these processes tell us about the social identification of a technology as working or non-working?
4. On one hand, technologies are superseded as a matter of course with technological change (e.g., cell phones surpassed pagers). Yet in the United States, the automobile reigns supreme after a century since its invention, and despite an increasingly diverse set of technological alternatives. What material, economic, political, and cultural factors make some technologies more likely to be adopted for long-term acceptance while others never take hold or are passed over rather quickly?
5. What are low-inclusion and high-inclusion actors in a social construction process? How do their experiences with technological development different? Why might it matter?
6. How would you explain social construction and technological choice to someone else? Is there a technology or system that you are deeply familiar with, such as a hobby, which illustrates some of the same features as our other cases?

Notes

1. Wiebe J. Bijker, Thomas Parke Hughes, and Trevor Pinch (Eds.), *The Social Construction of Technological Systems: New Directions in the Sociology and History of Technology* (Cambridge, MA: MIT Press, 1987).
2. "History of the Bicycle: A timeline." Accessed from www.brown.edu/Departments/Joukowsky_Institute/courses/13things/7083.html; and Hans-Erhard Lessing and Tony Hadland, *Bicycle Design: An Illustrated History*. (Cambridge, MA: The MIT-Press, 2014).
3. Hugh Dauncey, *French Cycling: A Social and Cultural History* (Liverpool: Liverpool University Press, 2012); and "Evolution of the Bicycle (1817-1899)." Accessed from https://performancebikeblog.wordpress.com/2016/11/17/evolution-of-the-bicycle-1817-1899/.
4. There is a pre-history of pneumatic tires, attributed to R.W. Thompson in 1845, but forgotten because rubber was too expensive. Dunlop patented

a pneumatic tire in 1888. See: www.thehistorypress.co.uk/articles/from-discomfort-to-joy-dunlop-s-pneumatic-bicycle-tyre/; https://lemelson.mit.edu/resources/john-dunlop; and P. W. Litchfield, "Pneumatic Tires," *Transactions (Society of Automobile Engineers)* 8 (II) (1913): 208–223.

5. C. Reid, "Good Roads for America." in *Roads Were Not Built for Cars* (Washington, DC: Island Press, 2015), pp 143–158; see also: "A Smoother Ride," National Museum of American History, Behring Center; and P. P. Mason, *The League of American Wheelmen and The Good-roads Movement, 1880–1905*. Dissertation, 1957, University of Michigan.
6. See: Peter Suciu, "When Were Bicycle Gears Invented?" *Bike Radar* (December 14, 2012). Accessed from www.bikeradar.com/features/when-were-bicycle-gears-invented/#:~:text="In%201868%20and%201869%20there,built%20a%20three-speed%20system; and D. Karwatka, "Paul de Vivie and the Bicycle Derailleur." *Tech Directions*, 73 (2013): 10–11.
7. Trevor Pinch and Wiebe E. Bijker, "The Social Construction of Facts and Artefacts: Or How the Sociology of Science and the Sociology of Technology might Benefit Each Other" *Social Studies of Science* 14 (1984):199–441, p. 415, citing John Woodford, *The Story of the Bicycle* (London: Routledge and Kegan Paul, 1970), p. 47.
8. Hanna S. Ostroff, "How the 19th-Century Bicycle Craze Empowered Women and Changed Fashion," *Smithsonian Sparks* (May 17, 2018). Accessed from www.si.edu/stories/19th-century-bicycle-craze; see also https://dangerouswomenproject.org/2016/10/07/dangerous-women-bicycles/
9. M. B. Schiffer, "Studying Technological Change: A Behavioral Perspective." *World Archaeology*, 36 (2004): 579–585.
10. Jonathan Jarry, "The Moral and Medical Panic Over Bicycles," (October 2, 2020). Accessed from www.mcgill.ca/oss/article/history-did-you-know/moral-and-medical-panic-over-bicycles; and Joseph Stromberg, "Bicycle Face: A 19th-Century Health Problem Made up to Scare Women Away from Biking," *Vox* (March 24, 2015). Accessed from www.vox.com/2014/7/8/5880931/the-19th-century-health-scare-that-told-women-to-worry-about-bicycle
11. Carlo Capelli, Luca Paolo Ardigò, Federico Schena, and Paola Zamparo, "Energy Cost and Mechanical Efficiency of Riding a Human-Powered Recumbent Bicycle, *Ergonomics*, 51, 10 (2008): 1565–1575.
12. Michael Watson, "Benefits of Recumbent Bikes for Seniors." Accessed from www.caringseniorservice.com/blog/recumbent-bikes.
13. Jan Trenson, "Things They Are A-Changing (and It's Not the UCI)," *Recumbent News* (April 17, 2020). Accessed from www.recumbent.news/2020/04/17/things-they-are-a-changing-and-its-not-the-uci/; and also "The History of the Recumbent Bicycle," Accessed from https://bentupcyclingjournal.blogspot.com/p/history-of-recumbent-bicycle.html.
14. Hassaan Ahmed, Omer Masood Qureshi, and Abid Ali Khan, "Reviving a Ghost in the History of Technology: The Social Construction of the Recumbent Bicycle." *Social Studies of Science*, 45, 1 (2015): 130–136.

15. Everett M. Rogers, *Diffusion of Innovations*, 5th ed. (New York: The Free Press, 2003), pp. 15–16.
16. Aggieland Appliance Repair, "Microwave Ovens—An Accidental Invention!" Accessed from www.aggielandappliancerepair.com/microwave-repair-college-station-texas/microwave-ovens-an-accidental-invention/; and Amanda Davis, "A History of the Microwave Oven," *IEEE Spectrum* (May 2, 2016). Accessed from https://spectrum.ieee.org/a-history-of-the-microwave-oven.
17. Hugh G. J. Aitken, *Syntony and Spark: The Origins of Radio* (Princeton, NJ: Princeton University Press, 1985), pp. 329–35.
18. Ibid., p. 335
19. Rachel Treisman, "After Nearly Two Bumpy Decades, The Original Segway Will Be Retire," *NPR* (June 23, 2020). Accessed from www.npr.org/2020/06/23/882536320/after-nearly-two-bumpy-decades-the-original-segway-will-be-retired-in-july; Mark Wilson, "Exclusive: Segway, the Most Hyped Invention since the Macintosh, Ends Production," *Fast Company* (June 23, 2020). Accessed from www.fastcompany.com/90517971/exclusive-segway-the-most-hyped-invention-since-the-macintosh-to-end-production.
20. See https://en.wikipedia.org/wiki/Macquarium for more info on Macquariums.
21. For more info visit https://zamboni.com.
22. Richard G. Lipsey, Kenneth I. Carlaw, and Clifford T . Bakar, *Economic Transformation: General Purpose Technologies and Economic Growth* (Oxford: Oxford University Press, 2005).
23. See John Enos, "Invention and Innovation in the Petroleum Refining Industry," in Richard Nelson (Ed.), *The Rate and Direction of Inventive Activity* (Princeton, NJ: Princeton University Press, 1962).
24. John Thomson, "What We Should Know about Tires: A Historical Background." Archived at https://web.archive.org/web/20150306211250/http://www.jags.org/TechInfo/2001/05May01/tires/historyoftires.htm
25. Mat Honan, "Remembering the Apple Newton's Prophetic Failure and Lasting Impact," *Wired* (August 5, 2013). Accessed from www.wired.com/2013/08/remembering-the-apple-newtons-prophetic-failure-and-lasting-ideals.
26. Ibid; and Jeremy Reimer, "Remembering Apple's Newton, 30 years On," *Ars Technica* (June 1, 2022). Accessed from https://arstechnica.com/gadgets/2022/06/remembering-apples-newton-30-years-on.
27. Laurence Murphy, "How Betamax Bit the Dust—And Other Tales of Forgotten Tech," *The Conversation* (January 4, 2016). Accessed from https://theconversation.com/how-betamax-bit-the-dust-and-other-tales-of-forgotten-tech-50619; and Joshua M. Greenberg, *From Betamax to Blockbuster: Video Stores and the Invention of Movies on Video* (Cambridge, MA: MIT Press).
28. Donald A. Schon, *Technology and Change: The New Heraclitus* (New York: Dell, 1967), pp. 24–25.
29. Richard R. Wilk, *Economies and Cultures: Foundations of Economic Anthropology* (Boulder, CO: Westview, 1996), p. 63
30. Pablo A. Guerron-Quintana, "Risk and Uncertainty," *Federal Reserve Bank of Philadelphia* (2012). Accessed from www.phil.frb.org/research.

31. Freeman, *Economics of Industrial Innovation*, p. 154.
32. See: British Airways, "Celebrating Concorde." Accessed from www.britishairways.com/en-us/information/about-ba/history-and-heritage/celebrating-concorde.
33. Henry Ford II, quoted in James C. Jones, "Dr. Schweitzer's Dilemma: Not Invented Here," *Ward's Auto World* (November 1974): 41.
34. Rogers, *Diffusion of Innovations*, pp. 414–17
35. David A. Hounshell, "Elisha Gray and the Telephone: On the Disadvantages of Being an Expert," *Technology and Culture*, 16, 2 (1975): pp. 133–61; and Everson, A. Edward, *The Telephone Patent Conspiracy of 1876: The Elisha Gray-Alexander Bell Controversy and Its Many Players* (Jefferson, NC: McFarland & Company Publishers, 2000).
36. See Beverly Rae Kimes, *Pioneers, Engineers, and Scoundrels: The Dawn of the Automobile in America* (Warrendale, PA: Society of Automotive Engineers, 2005).
37. Glenn R. Carroll, Lyda S. Bigelow, Marc-David Seidel, and Lucia B. Tsai, "The Fates of De Novo and De Alio Producers in the American Automobile Industry, 1885–1981," *Strategic Management Journal* 17 (1996): 117–137; and Joyce Shaw Peterson, *American Automobile Workers, 1900–1933* (Albany, NY: State University of New York Press, 1987).
38. Mercedes-Benz Group, "The First Automobile: 1885–1886." Accessed from https://group.mercedes-benz.com/company/tradition/company-history/1885-1886.html; and SuperCars.net, "1886 Daimler Motorized Carriage." Accessed from https://www.supercars.net/blog/1886-daimler-motorized-carriage.
39. Michael T. Hannan, Glenn R. Carroll, Elizabeth A. Dundon, and John Charles Torres, "Organizational Evolution in a Multinational Context: Entries of Automobile Manufacturers in Belgium, Britain, France, Germany, and Italy." *American Sociological Review* 60, 4 (1995): 509–28.
40. For the census data: www2.census.gov/prod2/decennial/documents/05457254v10ch2.pdf; and Cadie Thompson, "How the Electric Car Became the Future of Transportation," *Business Insider* (July 2, 2017). Accessed from www.businessinsider.com/electric-car-history-2017-5; and Department of Energy, "Timeline: History of the Electric Car." Accessed from www.energy.gov/timeline/timeline-history-electric-car.
41. Electric Vehicle Association, "Electric Vehicle History." Accessed from https://eaa-1967.clubexpress.com/content.aspx?page_id=22&club_id=222684&module_id=241560.
42. Megan Barber, "Before Tesla: Why Everyone Wanted an Electric Car in 1905," *Curbed* (September 22, 2017). Accessed from https://archive.curbed.com/2017/9/22/16346892/electric-car-history-fritchle
43. Virginia Sharff, *Taking the Wheel: Women and the Coming of the Motor Age* (Albuquerque, NM: University of New Mexico Press, 1992); Michael Brian Schiffer, *Taking Charge: The Electric Automobile in America* (Collingdale, PA: Diane Publishing, 1994).

44. John P. Watkins. "Corporate Power and the Evolution of Consumer Credit," *Journal of Economic Issues*, 34, 4 (2000): 909–932; and Charles E. Persons, "Credit Expansion, 1920 to 1929, and Its Lessons," *The Quarterly Journal of Economics*, 45, 1 (1930): 94–130.
45. James Foreman-Peck, "'Technological Lock-In' and the Power Source for the Motor Car." *Oxford Economic and Social History Working Papers*. St.Antony's College, Oxford.
46. M. Armand and J.-M. Tarascon, "Building Better Batteries," *Nature* 451 (2008): 652–657.
47. See www.eliwhitney.org; and Robert S. Woodbury, "The Legend of Eli Whitney and Interchangeable Parts," *Technology and Culture*, 1, 3 (1960): 235–53.
48. Kat Eschner, "In 1913, Henry Ford Introduced the Assembly Line: His Workers Hated It," *Smithsonian Magazine* (December 1, 2016). Accessed from www.smithsonianmag.com/smart-news/one-hundred-and-three-years-ago-today-henry-ford-introduced-assembly-line-his-workers-hated-it-180961267; and *History*, "Ford's Assembly Line Starts Rolling." Accessed from www.history.com/this-day-in-history/fords-assembly-line-starts-rolling.
49. Interstate Highway system: David Karas, "Highway to Inequity: The Disparate Impact of the Interstate Highway System on Poor and Minority Communities in American Cities," *New Visions for Public Affairs* 7(2015): 9–20; Wendell Cox and Jean Love, *The Best Investment a Nation Ever Made: A Tribute to the Dwight D. Eisenhower System of Interstate and Defense Highways*. (Collingdale, PA: Diane Publishing, 1996); and *U.S. Department of Transportation*, "History of the Interstate Highway System." Accessed from www.fhwa.dot.gov/interstate/history.cfm August 4, 2022.
50. Raymond A. Mohl, "The Interstates and the Cities: The U.S. Department of Transportation and the Freeway Revolt, 1966–1973," *Journal of Policy History*, 20, 2 (2008): 193–226; and Raymond A. Mohl, "Stop the Road: Freeway Revolts in American Cities," *Journal of Urban History*, 30, 5 (2004): 674–706.
51. Reuters Staff, "France to Uphold Ban on Sale of Fossil Fuel Cars by 2040." Accessed from www.reuters.com/article/us-france-autos/france-to-uphold-ban-on-sale-of-fossil-fuel-cars-by-2040-idUSKCN1TC1CU; and Alvin Powell, "California Dreaming? Nope," *The Harvard Gazette* (September 9, 2022). Accessed from https://news.harvard.edu/gazette/story/2022/09/what-to-expect-from-california-gas-powered-car-ban.
52. Institute for Local Self-Reliance, "Kilometer Tax—Netherlands." Accessed from https://ilsr.org/rule/road-pricing/2424-2.
53. Carey L. Biron, "Neighborhoods United: Highway Removal Gains Steam in U.S. Cities," *Reuters* (April 12, 2021). Accessed from www.reuters.com/article/us-usa-cities-transportation/neighborhoods-united-highway-removal-gains-steam-in-u-s-cities-idUSKBN2BZ0VF; and Alissa Walker, "Six Freeway Removals That Changed Their Cities Forever," *Gizmodo* (May 25, 2016). Accessed from https://gizmodo.com/6-freeway-removals-that-changed-their-cities-forever-1548314937.

54. Elizabeth Fitting, *The Struggle for Maize: Campesinos, Workers, and Transgenic Corn in the Mexican Countryside* (Durham, NC: Duke University Press, 2011); and Erick Castellanos and Sarah Bergstresser, "The Mexican and Transnational Lives of Corn: Technological, Political, Edible Object," in Ronda L. Brulotte, Michael A. Di Giovine (Eds.), *Edible Identities: Food as Cultural Heritage* (London: Routledge, 2014); and most recently: Ernesto Hernández-López, "Racializing Trade in Corn: México Fights Maíz Imports and GMOs," *Journal of International Economic Law*, 25, 2 (2022): 259–276.
55. Caroline Hopkins, "The 'Abortion Pill' May Treat Dozens of Diseases, but Roe Reversal Might Upend Research," *NBC News* (June 25, 2022). Accessed from www.nbcnews.com/health/health-news/abortion-pill-may-treat-dozens-diseases-roe-reversal-might-upend-resea-rcna34812.
56. History Bits, "Haymarket Riots (Chicago)." Accessed from www.historybits.com/haymarket-riots.htm; and ThoughtCo., "The Haymarket Riot," Accessed from www.thoughtco.com/1886-haymarket-square-riot-chicago-1773901; and Langdon Winner, "Do Artifacts Have Politics" *Daedelus* 1(1980): 123–125.
57. Hannah Hickok, "Why Remote Work Has Eroded Trust among Colleagues," *BBC* (March 18, 2021). Accessed from www.bbc.com/worklife/article/20210315-why-remote-work-has-eroded-trust-among-colleagues; and Ian Bogost, "Hybrid Work Is Doomed," *The Atlantic* (July 6, 2022). Accessed from www.theatlantic.com/technology/archive/2022/07/work-from-home-rto-flexibility/661495.
58. Murilee Martin, "In 1987, a Fake Cellphone Antenna for Your Car Made You Look like a High Roller," *Autoweek* (July 22, 2015). Accessed from www.autoweek.com/car-life/but-wait-theres-more/a1873646/1987-fake-cellphone-antenna-your-car-made-you-look-high-roller; and Alex Perrone, "The Car Phone, The Coolest Accessory for Your Car… in 1995," *Endurance* (January 11, 2022). Accessed from www.endurancewarranty.com/learning-center/rewind/throwback-thursday-car-phone-1995.
59. William Gallagher, "Newton's August 1993 Launch Set the Stage for What Would Become the iPad and iPhone," *AppleInsider* (August 3, 2023). Accessed from https://appleinsider.com/articles/18/08/02/newton-launched-august-2-1993-setting-the-stage-for-what-would-become-the-ipad-and-iphone.
60. Wiebe Bijker, *Of Bicyles, Bakelites, and Bulbs: Toward a Theory of Sociotechnical Change*, (Cambridge, MA: MIT Press, 1997).
61. Allison Lantero, "The War of the Currents: AC vs. DC Power," *U.S. Department of Energy* (November 18, 2014). Accessed from www.energy.gov/articles/war-currents-ac-vs-dc-power.
62. Taylor Synclair Goethe, "Bigotry Encoded: Racial Bias in Technology," *Reporter* (March 2, 2019). Accessed from https://reporter.rit.edu/tech/bigotry-encoded-racial-bias-technology; and Michael W. Sjoding et al., "Racial Bias in Pulse Oximetry Measurement," *New England Journal of Medicine*; 383 (2020):2477–2478.

63. Allison Lantero," The War of the Currents: AC vs. DC Power," *U.S. Department of Energy* (November 18, 2014). Accessed from www.energy.gov/articles/war-currents-ac-vs-dc-power.
64. See: National Association of City Transportation Officials, "Urban Street Design Guide." Accessed from https://nacto.org/publication/urban-street-design-guide/street-design-elements/lane-width; and Angie Schmitt, "Compelling Evidence That Wider Lanes Make City Streets More Dangerous," *StreetsBlogUSA* (May 27, 2015). Accessed from https://usa.streetsblog.org/2015/05/27/compelling-evidence-that-wider-lanes-make-city-streets-more-dangerous.
65. Scott Adams, The Joy of Work: Dilbert's Guide to Finding Happiness at the Expense of Your Co-Workers (New York: HarperCollins, 1999).
66. Speed Society, "Gas Powered Pogo Stick from 1960's, Dangerous Toys Banned!" Accessed from https://news.speedsociety.com/gas-powered-pogo-stick-from-1960s-dangerous-toys-banned.
67. Laura Silver and Courtney Johnson, "Majorities in Sub-Saharan Africa Own Mobile Phones, but Smartphone Adoption Is Modest," *Pew Research Center* (October 9, 2018). Accessed from www.pewresearch.org/global/2018/10/09/majorities-in-sub-saharan-africa-own-mobile-phones-but-smartphone-adoption-is-modest.
68. Pew Research Center "Cell Phones in Africa: Communication Lifeline," (April 15, 2015). Accessed from www.pewresearch.org/global/2015/04/15/cell-phones-in-africa-communication-lifeline.

PART 3

How Technology Affects the Health of Earth and Its Inhabitants

The following four chapters illustrate how technological change can create both problems and solutions. Chapter 7 describes how technological innovations have, at times, been the source of two general categories of environmental damage: pollution and depletion. The record isn't altogether depressing, however, as the chapter presents several examples of successfully addressing these problems through the development of new technologies. As is often the case, technological fixes are only part of the story, however. Any fix must be carried out amidst a set of social and economic forces, which, along with governmental policies, can help or hinder the solution of environmental problems.

Chapter 8 looks at tools for environmental assessment. Building on the social construction framework, there are tools for environmental, economic, and social assessments, known variously as value chain or life-cycle analysis. We'll consider the idea of *externalities,* the secondary or side-effects of an activity that may not be reflected in the costs. We will look at value chain analyses of cars and other technologies. The 'costs' of a car don't simply start or stop at the bumper. We'll also look at the economic idea of future value and how to evaluate the worth of things that might be priceless. What, for example, are the different ways to value a landscape or a species?

Chapter 9 turns from environmental health to human health. It describes the many ways in which medical technologies have cured our ills and extended our lives while at the same time creating new problems and challenges. The most serious of these is escalating health care costs. Many medical procedures can be described as "halfway technologies" that extend lifespans and alleviate pain but leave patients in a state of permanent—and expensive—dependence on medical care. On another front, sophisticated diagnostic technologies give physicians more data, but at the expense of marginalizing important diagnostic skills based on personal relationships with patients, and often are overused, leading to excess costs.

Chapter 10 looks into the development of medical technologies based on advances in genetics and assesses recent developments in the genetic engineering of food crops, once again noting that technological advances are often mixed blessings. This theme is continued with an exploration of recent developments in genetic technologies that are being applied to humans now and may be employed in the future. Perhaps no set of technologies has a greater potential for transforming our lives for good and for ill. The benefits and possible pitfalls of these technologies are discussed, along with an examination of the legal and ethical issues they raise.

CHAPTER 7

Technology, Energy, and the Environment

Although technological advance has been blamed for a variety of ills, its most obvious and long-lasting negative consequence has been the degradation and even the destruction of the natural environment. For most of human existence, people left the environment pretty much as they found it. But beginning with sedentary agriculture and accelerating with industrialization, the use of new technologies has at times left a ruined environment as its legacy. There is a strong possibility that environmental problems will worsen as more nations make greater use of industrial technologies. Today, the United States, with fewer than 5 percent of the world's population, consumes about 25 percent of the world's resources. If the poor countries of the world, which contain the majority of Earth's population, used resources at the current U.S. level, the consequences for the environment would be dire. This will not happen in the near future, but the general trend is inescapable. For example, according to the U.S. Energy Information Administration, global consumption of energy from 2010 to 2050 is projected to increase by 50 percent, with Asia expected to account for the largest portion of the increase.[1] The rich countries of the world should not begrudge the efforts of poor countries to pursue better lives for their people, but for all countries, rich and poor alike, the sustainability of Earth's environment must be an essential goal.

The deleterious consequences of technological advance can be grouped into two broad categories: pollution and depletion. In the former, the environment is damaged by the addition of harmful substances, resulting in polluted skies, acid rain, poisoned soil, contaminated water, and climate change. In the case of depletion, the unchecked application of technology leads to the permanent loss of resources, deforestation, and the extinction of plant and animal species. On many occasions, of course, the two go together; a mining operation could extract all of the available ore and leave a devastated environment in its wake. Either separately or together, pollution and depletion threaten the sustainability of the environment.

But there is some cause for optimism. Technology is not just a source of environmental problems; it can also be part of the solution. In this chapter, we will look at some of the environmental consequences of technological advance, review some technologies that can alleviate the degradation of the environment, and briefly consider how governmental policies affect the natural environment.

Figure 7.1 Beijing's Forbidden City can barely be discerned through the thick smog enveloping the area. (Source: Brian Jeffery Beggerly/CC BY 2.0)

Figure 7.2 The development of technological fixes to prevent or mitigate climate change is known as geoengineering. One example, seen here, is a proposed fleet of wind-powered vessels that would generate many small seawater droplets to increase the reflectivity of clouds by the Twomey effect, which in turn would block some of the sunlight that is the source of the greenhouse effect. This is known as marine cloud brightening. (Used with permission from Stephen Salter; illustration by John MacNeill)

Fossil Fuels, Air Pollution, and Climate Change

The blanket of smog choking many of the world's cities makes the consequences of the use of fossil fuel all too evident. Potentially more dangerous in the long run is the accumulation of the main product of carbon combustion: carbon dioxide (CO_2). The use of fossil fuels (petroleum, coal, and natural gas) has vastly extended the amount of available energy. At the same time, cars, power plants, factories, and other sources emit immense quantities of carbon dioxide from their use of fossil fuels, which hit a record high of 36.8 billion tons in 2022.[2]

CO_2 is produced when the carbon in gasoline, diesel fuel, coal, or natural gas unites with oxygen during the process of combustion. This creates quantities of CO_2 significantly greater than the original carbon; burning 1 pound of gasoline generates more than 3 pounds of CO_2.[3] To look at it in another way, a car traveling 100 miles produces about 100 pounds of CO_2.

CO_2 is a clear, odorless gas that by itself presents no threat to health. It is the normal product of any animal's respiration, and without it, beverages like beer and soda pop would be unpalatably flat. The danger presented by CO_2 emerges only when large quantities of it accumulate in the atmosphere along with two other products of an industrial society, methane (CH_4) and nitrous oxide (N_2O). As light rays from the sun pass through the atmosphere containing these gases, the radiant energy warms the surface of the earth, but most of the heat cannot pass back through the atmosphere. This is known as the "greenhouse effect." Just as a greenhouse is warmed by the light of the sun on a cold winter's day, Earth's temperature begins to rise as greenhouse gases accumulate in the atmosphere. Concentrations of these three greenhouse gases have increased to a marked degree as industrialization has advanced. In 1750, at the dawn of the Industrial Revolution, concentrations of CO_2, the most important greenhouse gas, stood at 280 parts per million (ppm); by 2022, that figure had climbed to 417 ppm.[4]

Continued warming due to increasing CO_2 emissions seems likely. As noted by the United Nations Intergovernmental Panel on Climate Change, average temperatures rose about 0.8°C (1.4°F) since 1880. Taking a shorter time span, the last decade (2010–2020) has been the warmest ten-year period since at least the 1880s.[5] Projections of future trends indicate substantially greater increases, but as a general principle, the reliability of any kind of prediction diminishes as it moves further into the future. Regarding Earth's climate, predicting the course of global climate change is a complex exercise because the many variables have complex interactions. Besides temperature change, increased energy in the atmospheric system may lead to increased storm severity, increased fire and flood occurrences, and even wider and more unpredictable winter temperature swings. Predictions are further complicated by the need to estimate increases in the production of greenhouse gases resulting from economic growth and consequent energy use in the developing countries in the decades to come.

For all the uncertainties surrounding global climate change, it is certain that significant changes to Earth's climate are now taking place. Future temperature increases will not be uniform, and some regions may even end up with colder climates due to changes in ocean and atmospheric currents. Rainfall patterns would shift to the advantage of some areas and to the detriment of others. In conjunction with increases in CO_2 and other greenhouse gases, global climate change could increase agricultural yields in certain regions but reduce them elsewhere. It also seems certain that climate changes will result in the loss of some plant and animal species, and the relocation of others. One of the consequences of the latter could be the spread of insect-borne diseases to previously unaffected parts of the world. Most troubling, significant warming is leading to partial melting of glaciers and the northern polar ice cap. It is also causing water to expand in volume as the oceans warm. Over the long term, warming oceans will result in a rise of sea levels, and potentially catastrophic floods in low-lying areas. At least five Pacific Islands no longer exist.[6] One particularly vulnerable place is the nation of Bangladesh; according to a United Nations report, a three-foot rise in sea level would inundate a quarter of its coastline, resulting in the displacement of 30 million people.[7]

Atmospheric scientists are learning more and more about the complex interactions that govern Earth's climate, but many unanswered questions remain. In the meantime, the build-up of greenhouse gases could reach a potentially catastrophic tipping point in a few decades. Reducing the emissions that contribute to climate change will require the expenditure of billions—if not trillions—of dollars over the decades to come, but the costs of doing nothing will likely be even greater. Effectively addressing the causes of climate change also requires taking some painful actions today to forestall problems that lie well in the future, a difficult course of action for business and governmental leaders who generally have short-term horizons. Moreover, as noted above, decisions (or nondecisions) must be made amid considerable uncertainty, although the most fundamental issue, the reality of human-induced climate change, can no longer be questioned.

A Planet under Stress

While the burning of carbon-based fuels is the largest cause of global climate change, deforestation also contributes. Particularly serious has been the loss of forested areas in Africa and Latin America, much of it occurring in tropical regions. According to the World Bank, since the beginning of the twentieth century about 3.9 million square miles (10 million square km) of the world's forests have been lost. Just in the past 25 years deforestation has destroyed 502,000 square miles (1.3 million square km) of woodland, accounting for an area bigger than South Africa.[8] For a 12-month period ending in July 2019, fires set deliberately for farming and cattle raising in Brazil's Amazon rainforest wiped out about 9,762 square kilometers (3,769 square miles) of rainforest. Deforestation affects climate change in two ways: the burning of forests and the rotting of fallen trees release CO_2, while the loss of trees removes an important source of CO_2 capture. Through these processes, deforestation contributes about 17 percent of human-caused greenhouse gases.[9]

In addition to affecting the skies above, human activities are generating massive amounts of refuse, a significant portion of which is hazardous. According to the U.S. Environmental Protection Agency, in 2017 Americans tossed out 267.8 million tons of municipal solid wastes, which comes to 4.51 pounds per person per day. To this can be added a much greater volume of industrial and other waste products.[10]

One component of waste that his grown exponentially in recent years is plastic. In 2017, discarded plastic products accounted for 35.4 million tons, or 13.2 percent of total refuse.[11] There are several varieties of plastic, and most can be recycled if kept separate. This is rarely done, and producers of plastic materials find it cheaper to use virgin materials rather than recycled ones. The substantial increase in plastic production has generated a great amount of waste; every nine minutes, discarded plastic weighing as much as a blue whale (about 300,000 pounds) ends up in the ocean, where it harms marine life through entanglement, ingestion, and exposure to toxic substances. Some of us might be ingesting as much as 5 grams worth of plastic every week, the equivalent of a credit card.[12] Plastic refuse also breaks up into microplastics (pieces less than 5 millimeters long) that absorb and emit chemicals, many of them toxic, for indefinite periods of time.[13]

Some of the materials discarded by households and industrial enterprises reside in inadequate disposal facilities that threaten adjacent communities. Already, some areas have been so badly contaminated that they had to be abandoned for a number of years while the mess was being cleaned up. The most toxic sites have qualified for special cleanup funding. There are 1,337 of these Superfund sites in the United States, some of which have had that designation for nearly 40 years. In 2013, the EPA found that roughly 17 percent of the U.S. population, including 18 percent of all children in the U.S. younger than five, lived within 3 miles (5 km) of a Superfund site. Despite an obvious need for special attention, federal funding for the Superfund program declined by nearly half between 1999 and 2013 (from about $2 billion to about $1.1 billion), leaving much of the cleanup costs to the states.[14]

Transitioning from Fossil Fuels

While pollution, global climate change, the massive generation of trash, and the other negative consequences of industrial production cause problems on one front, depletion of the resources that provide energy and raw materials poses another set of challenges. Fossil fuels have been essential to the process of industrialization since at least the middle of the nineteenth century, and petroleum-based fuels power the world's transportation systems. Supplies of fossil fuel are necessarily finite, although many uncertainties attend efforts to estimate their size, especially in the case of petroleum. Not too long ago, many experts believed that we had reached "peak oil" and inevitably faced a future of declining petroleum extraction.[15] In recent years, these concerns have been mitigated by the development of technologies that followed extraction of petroleum from shale deposits through hydraulic fracturing or "fracking." Whether this trend will continue is open to debate, due to the health and environmental consequences of the process.[16] Although fuel prices were low in the middle of the second decade of this century, the increasing costs of employing new technologies, environmental damage, and especially concerns about climate change may reinstate peak oil as the dominant way of looking at petroleum supplies.[17]

Neither the peak oil thesis nor its antithesis can be settled once and for all. Determining the amount of oil still remaining underground entails considerable guesswork, and there is not even universal agreement on how to define petroleum "reserves." The amount of oil that can be economically extracted from a known reservoir depends on the prevailing market price. The price of oil will increase as supplies dwindle, so the rate at which oil is extracted will depend on what users are willing to pay. At the same time, high oil prices will motivate the development of new extraction technologies, and a consequent increase in supplies and lower prices. The complex interaction of supply, demand, and technology make projections of future supplies of oil a hazardous enterprise. Oil shortages and price shocks will likely still occur, just as they did during the energy crises of 1973–1974 and 1979, but it is important to note that they were triggered by political actions such as by OPEC or the Russian invasion of Ukraine, and not by an abrupt decrease in the world's petroleum supplies.[18]

Feeding the World's Population

In the 1960s, some books, notably *The Population Bomb*, confidently predicted widespread famine in the immediate years to come.[19] Although there were localized incidents of famine due to wars and political upheavals, this apocalyptic vision never came to pass. But this does not mean that concerns about future food supplies can be ignored.

At the most basic level, agriculture requires sunlight, soil, and water. The first does not pose a problem, but soil resources are under threat in many parts of the world. Since 1945, 1.2 billion hectares of agricultural land have been degraded by moderate to extreme soil erosion, the equivalent of the total land area of China and India. This trend has not abated; somewhere between 24 and 75 billion tons of soil are being lost annually around the world.[20] In the past, the loss of topsoil could be offset by moving to virgin lands, but this option is no longer available in most parts of the world. Not only is new land unavailable, erosion and land degradation are removing about 1 percent of existing arable land annually. In many parts of the world, the loss of topsoil has increased the costs of farming, but in some places, the consequences have been far more serious. Severe loss of topsoil has converted some regions into deserts, bringing widespread famine in its wake. Elsewhere, topsoil loss coupled with climate changes induced by global climate change have dire consequences for agriculture, even as the continual growth of the world's population calls for the production of more food. Due to climate change, overpopulation, deforestation, excessive tapping of water tables, and poor farming and grazing practices, a large portion of the earth is vulnerable to desertification. It is estimated that one-third of the planet's land surface is already degraded, threatening the health and livelihoods of more than 1.5 billion people in 167 countries.[21] It is also important to note that 25 million of the world's refugees, 58 percent of the total number, are fleeing degraded lands.[22]

Prolonged droughts are one of the major causes of desertification. More generally, questions can be raised about the state of the world's water supplies. Once again, climate change is creating major problems; as soils become drier, they absorb more water when it rains, resulting in less inflow into lakes, rivers, and reservoirs. Water from these sources is essential for irrigated agriculture, and their diminishment can seriously affect agricultural production.[23]

Having inadequate water supplies for agriculture is only part of the problem; 2.1 billion people lack access to safely managed supplies of drinking water because 80 percent of wastewater flows back into the environment without being treated. Children are major victims of unsafe water supplies; 340,000 children under five die every year from diarrheal diseases.[24]

Is Technology the Problem or the Solution?

The previous pages have presented a depressing catalogue of environmental ills. Pollution, climate change, species extinction, and resource depletion pose multiple threats to our standard of living and, perhaps, even to our continued existence. The reliance on fossil fuels has left our atmosphere contaminated by various pollutants, and Earth is threatened by climate change. Irresponsible agricultural practices have poisoned groundwater with pesticides and other chemicals while depleting the topsoil. Industrial processes produce thousands of tons of toxic wastes, and nuclear power plants leave behind radioactive wastes that will pose potential health hazards for thousands of years. In the opinion of some critics, the technological advances of the past few centuries seem to have produced only temporary benefits that will ultimately be overwhelmed by the consequences of environmental stress. In the long run, it may be argued, technology generates more harm than good.

Is a modern economy, supported by advanced technologies, doomed to extinction, leaving us with no choice but to retreat to the simpler technologies of the past, perhaps even back to the era of hunting and gathering? Or can technology itself provide us with the solutions to technologically induced problems? In addressing these questions, it is useful to first gain some historical perspective. If nothing else, some acquaintance with the past should convince us that damage to the environment is not solely a phenomenon of modern times and modern technologies. Entire civilizations have collapsed due to excessive land clearance, overgrazing, withdrawal of groundwater, and eventual desertification. Epidemic diseases of catastrophic proportions have been spread by the careless disposal of household wastes. There are some scholars who believe that a major factor in the collapse of the Roman Empire was sterility and premature death brought on by the widespread use of lead in pipes and utensils. Air pollution existed long before the invention of fossil-fuel consuming engines; those suffering from twenty-first-century smog can find a seventeenth-century counterpart in John Evelyn, who decried the growing use of coal for energy and warmth:

> In London we see people walk and converse pursued and haunted by that infernal smoake. The inhabitants breathe nothing but an impure and thick mist, accompanied by a fuliginous and filthy vapour, which renders them obnoxious to a thousand inconveniences, corrupting the lungs and disordering the entire habit of their bodies, so that catarrs, phtisicks, coughs and consumption rage more in that one city than in the whole earth besides.[25]

Some Technological Fixes of the Past

These examples of past environmental ills provide little comfort, for they can be seen as a preview of what might be in store for us. But there are other examples of environmental threats that have been successfully countered. One example is the deforestation that had become a serious problem in England during the sixteenth century. By that time, vast quantities of wood had been consumed by the demands of an expanding population and the growth of shipbuilding, construction, and iron manufacturing (which required large quantities of charcoal). Within a century, the depletion of timber was perceived as a serious problem, as seen in the complaint of one contemporary writer that "at this present, through the great consuming of wood as aforesaid, and the neglect of planting of woods, there is so great a scarcity of wood through the whole kingdom."[26]

England's forests were never fully restored, but fuel shortages were alleviated by burning coal in the place of wood. Although there were misgivings about the noxious vapors given off by burning coal, it came to be widely used for domestic heating and as a source of heat for the production of beer, sugar, bricks, soap, glass, and iron. More than simply a substitute for wood, by the end of the nineteenth century coal had become the basis of industrial civilization, as the rich coal deposits of Britain significantly contributed to that country's unique position as "the Workshop of the World." Much of the industrial age was the era of coal, as coal-fired steam engines powered factories, hauled railroad trains, generated electricity, and propelled ships to distant destinations.

Yet just when coal had established its primacy as the most important energy source for industrial society, hard questions were being asked about the continued viability of coal-based technologies. By the end of the nineteenth century, it was becoming evident that stocks of coal, while still large, were being depleted at ever-increasing rates. The projection of established trends seemed to offer indisputable proof that the day of reckoning was not far off: Britain was running out of coal. In the words of the contemporary English economist, W. Stanley Jevons, "There is no reasonable prospect of any relief from a future want of the main agent of industry. We must lose that which constitutes our particular energy."[27]

Coal was king, and in Jevons's estimation, as well as those of other informed students of the British economy, there was no hope that anything could take its place. In Jevons's gloomy appraisal, "All things considered, it is not reasonable to suppose or expect that the power of coal will ever be superseded by anything better."[28] His pessimistic assessment of potential substitutes is exemplified by his quick dismissal of petroleum as a fuel: "Its natural supply is far more limited and uncertain than that of coal, and an artificial supply can only be made by the distillation of some kind of coal at considerable cost. To extend the use of petroleum, then, is only a new way of pushing the consumption of coal. It is more likely to be an aggravation than a remedy."[29] Natural gas, another possible substitute, was an equally forlorn hope. Jevons approvingly quoted the assessment of an American steel executive: "Of late years the supply of gas has been decreasing...and it would seem that before many years this fuel would cease to be a factor in the large operations of a steel works."[30]

One can smile at the remarkable wrongheadedness of these assessments, but it is easy to be wise after the fact. After all, the true extent of oil and gas reserves were only dimly perceived at the time Jevons was coming to his gloomy conclusions. The third edition of Jevons's book that contains the passages quoted above was published in 1905, four years after the Spindletop field demonstrated the vast oil reserves of East Texas. There, a single well produced twice as much oil as the entire state of Pennsylvania, until then the center of the American oil industry.[31] And it was not until three decades later that the immense oil deposits of the Middle East began to be explored.

The essential point here is that a problem such as resource depletion can often be solved by the use of substitutes, just as coal substituted for wood, and oil replaced a great amount of coal. This does not happen easily or automatically, of course; it requires the invention, development, and application of many new ways of doing things. The large-scale employment of petroleum fuels required a host of new technologies: seismic exploration devices, casings and bits for rotary drilling, new compounds for the cementation of bore holes, drilling mud, and so on. Equally important, the use of a new source of energy must be complemented by the emergence of new energy-using technologies. In the early 1900s, the coal-fired reciprocating steam engine was a proven technology, while the gasoline-fueled internal combustion engine was cranky and unreliable, and the diesel engine had scarcely emerged from the laboratory. The rapid strides made by these new engine types in the ensuing years was both a product of the availability of new fuels as well as a stimulus to their accelerated extraction.

Alternatives to Fossil Fuels

The example of the widespread substitution of petroleum-based energy for coal-based energy applies to many other diminishing resources. It demonstrates that at one level of analysis, the cure for depletion and environmental damage can be found in new technologies that successfully address the problems generated by old technologies. Still, this is not the end of the matter. Historical analogies have their place, but like all analogies, they must be treated with caution. The fact that technological solutions have been found in the past is no guarantee that they will be found in the future.

Nuclear Energy: From Panacea to Peril

The limitations of technological solutions to energy shortages are all too apparent when nuclear energy is examined. Beginning in the 1950s, nuclear energy was heralded as the next stage in the evolution of energy sources. In the words of a former head of the U.S. Atomic Energy Commission, nuclear power would produce electricity that was "too cheap to meter" (it has been argued that this phrase has been misleadingly taken out of context).[32] In any event, nuclear power has not been an energy panacea. In 2018, U.S. nuclear plants produced 807 billion kilowatts of electricity, 19.4 percent of total generation,[33] but its future is uncertain at best. In addition to costing twice as much per megawatt as a coal-fired plant and five times as much as one using natural gas,[34] the generation of nuclear power requires considerable expenditure of energy for uranium mining and transportation, equipment manufacture, plant construction, maintenance and administration, and waste disposal. When all of these energy inputs are taken into account, nuclear plants make a much smaller contribution to net energy supplies than their output indicates, and a number of years will go by before the cumulative contributions of a nuclear facility exceed the energy that went into its construction and operation.

Second, the product of nuclear plants is electricity, and electricity has its limitations. Electricity is not a source of primary energy; it is a means of transmitting energy, and as much as 10 percent of the energy used to generate electricity is lost in transmission. This means that, if possible, it is always better to produce energy close to where it will be used. Although rooftop solar panels are not the most efficient way of generating electricity, they may outperform a large power plant situated hundreds of miles away when transmission losses are considered. Also, electrical energy cannot be used for all purposes. Although fossil fuels are an important source of electrical power, the substitution of nuclear plants for conventional generating facilities would diminish but not come close to eliminating the need for energy derived from fossil fuels. Nuclear power is sometimes offered as an effective way of slowing global warming through the replacement of power plants fueled by coal or gas. Again, its contribution would be significant, but limited. Depending on the type of conventional power plant being replaced, tripling the present generating capacity supplied by nuclear power would reduce greenhouse gases by between 11 and 26 percent.[35]

Finally, accidents at nuclear plants are rare but potentially catastrophic.[36] In 2011, an earthquake and subsequent tsunami caused a partial core meltdown and hydrogen explosion at the nuclear facility near Fukushima, Japan. In addition to the massive economic losses incurred by the region and the nation as a whole, the disaster resulted in hundreds of immediate deaths and will lead to the eventual loss of thousands more due to greater-than-normal incidences of cancer. The potential catastrophic failure of the Zaporizhzhia nuclear power plant in Ukraine after attacks during the Russia-Ukraine War shows that the vulnerability of nuclear energy to intentional damage cannot be overlooked.[37]

Less dramatically, nuclear wastes also present serious long-term problems. The processing of ores into nuclear fuel has left a residue of hundreds of millions of tons of radioactive waste, while thousands

Figure 7.3 Although nuclear power plants do not produce greenhouse gases, they present other dangers, as was demonstrated when a tsunami led to a catastrophic explosion at the Fukushima Daiichi nuclear power plant in 2011. The severely damaged No. 3 reactor building, shown here, housed 460 tons of nuclear fuel in a storage pool in its upper floors. (ZUMA Press, Inc./Alamy Stock Photo)

of tons of radioactive materials remain as by-products of civilian and military nuclear programs. In addition, large quantities of low-level radioactive wastes are generated through medical and industrial applications of nuclear technologies. In the United States, these dangerously radioactive materials now reside in 131 "temporary" storage sites while efforts to open a permanent waste facility near Yucca Mountain, Nevada, have been stymied by technical problems and political opposition.[38]

Problems with the disposal of nuclear wastes are not confined to the United States, and they will only intensify with the expansion of nuclear power. If global production of nuclear energy went from its current annual output of 350,000 megawatts to 1 million megawatts, the resultant waste material would fill a disposal facility the size of Yucca Mountain in only three-and-a-half years.[39] Alternatively, nuclear plants could use a "closed cycle" process through which nuclear wastes are recycled into nuclear fuel. This method, however, entails additional costs. More ominously, it results in the production of large quantities of plutonium that could be the basis of a nuclear weapon if only a small amount fell into the hands of terrorists or hostile states.

Renewable Sources of Energy

Lest it be thought that nuclear energy is being singled out as an especially problematic source of energy, it should also be noted that other alternative energy sources have their own shortcomings. The energy of the sun offers a fantastically large source of potential energy; the solar energy that strikes Earth in 40 minutes equals the world's annual consumption of energy.[40] The trick, of course, lies in effectively using even a modest fraction of that energy.

On a positive note, the cost of solar energy has continued to fall rapidly. The International Renewable Energy Agency (IRENA), an association with more than 150 member countries, notes that since 2010 the cost of generating power from the sun has fallen by 73 percent. Consequently, the cost of photovoltaic electricity has dropped to $0.10 per kwh. In comparison, the cost of electricity generation based on fossil fuels typically falls in a range of $0.05 to $0.17 per kwh, in some circumstances making solar competitive with fossil-fuel power plants.

Wind power, a very old technology, has been increasing in importance in recent years. When sited in the proper location, the most efficient wind turbines can produce electricity for as little as 2.35 cents per kWh.[41] As with solar energy, this price reflects the operating cost of generating electricity; when construction costs are included the price is necessarily higher, but putting an exact value on these costs is complicated when interest rates and subsidies are figured in.

As with solar energy, the use of wind power will increase as more efficient turbines are developed and the costs of carbon-based sources rise, although complaints about noise and the degradation of landscape aesthetics will limit or prevent the construction of wind turbines in some areas. Wind power also shares with solar cells the problem of episodic operation. This means that some way to store energy is needed for times when the sun doesn't shine, and the wind doesn't blow. Wind and solar also suffer from a fundamental mismatch: areas of maximum sunlight or wind are usually far from the places that consume the most electricity. Redressing this situation will require substantial upgrading of electrical transmission lines, the design of which goes back a hundred years.[42] Batteries are the obvious choice for the storage of electrical energy. Lithium-ion batteries are now ubiquitous, powering everything from smartphones to cars, but they are not suitable for large-scale installations. For these, a recent development, the vanadium-flow battery, holds out a great deal of promise, but more development is needed.[43]

At present, the most important source of renewable energy is hydropower, which accounts for about 10 percent of electrical generation in the United States. In recent years, however, it has come under considerable criticism because dams alter the ecology of river systems, often to the detriment of the plants and animals that live in them. It is unlikely that massive new hydropower projects will appear in the years to come.

Other sources of renewable energy might take on more significance as reserves of fossil fuels are depleted and their environmental consequences become unacceptable. Among these are biomass (plants and other organic sources), geothermal, methanol (methyl alcohol) produced from coal and gas, ethanol (ethyl alcohol) produced from plants, and nuclear fusion (although not for many decades, if ever). In evaluating these alternatives, it is essential to keep things in perspective.

Figure 7.4 Wind turbines and nuclear power plants are alternatives to the use of fossil fuel, but each carries disadvantages as well as advantages. (Source: Jeanne Menjoulet/CC BY 2.0)

In 2022, renewable sources of energy accounted for only 21.5 percent of electricity generated in the United States. Wind turbines generated 10.2 percent while, for all its recent advances, solar contributed a mere 3.4 percent. For the immediate future, none of them can rival petroleum and natural gas as relatively cheap and convenient sources of energy. Fossil fuels will continue to be of central importance for many decades. The key issue will be using them efficiently.

Doing More with Less

The generation and use of massive quantities of energy is the cornerstone of industrial society. Much of the expansion of production (and of course consumption) that has taken place since the Industrial Revolution has come through the development of technologies dependent on external sources of energy. Countering the depletion of fossil fuels as well as their adverse consequences will require the use of new sources of energy. Still, it must be remembered that new sources will generate new problems, as will the employment of effective pollution-control strategies; in both cases, there are few, if any, cost-free technological fixes. In the early twentieth century, the replacement of the horse by the automobile was widely applauded because it promised a far cleaner urban environment—in those days horses in New York City deposited 2.5 million tons of manure annually.[44] One might speculate on which source of pollution is preferable, the emissions of cars or of horses.

Instead of substituting one source of pollution for another, a better course of action is to try to reduce pollution by cutting back on energy use. Here, it is useful to distinguish *conservation* from *curtailment*. The former implies doing the same sorts of things with fewer inputs of materials and energy, while the latter implies an actual loss of output and consumption. Fortunately, most of the advances in recent decades have in fact come from conservation. In the United States, energy intensity (the ratio of energy used to productive output) peaked during the second decade of the twentieth century and has been declining ever since.[45] The total amount of energy used has increased, but the emergence of more fuel-efficient sources of energy (primarily the shift from wood and coal to oil and natural gas), changes in the structure of the economy, and the development of more energy-efficient technologies allowed the rate of economic growth to outstrip by a comfortable margin the rate of energy use.

A portion of the improved output-to-energy ratio can be attributed to the growth of services (which use relatively little energy) and the relative decline of industrial production (which uses a lot of energy). It should be pointed out, however, that the growth of the service sector has been accompanied by increasing energy use in developing countries, most notably China, that have become major producers of manufactured items bought by economically advanced nations.[46] The rest of the improvement in energy intensity has been the result of increased efficiency. New technologies along with better management of heating, lighting, and ventilation systems have reduced energy costs by billions of dollars.[47]

Relatively modest changes in everyday items such as refrigerators and freezers have produced large benefits. Refrigerators sold today use only one-quarter the energy of those made in 1974. Were it not for this substantial improvement, operation of the 150 million refrigerators and freezers in the United States would require the generation of an additional 40,000 megawatts of electrical power.[48] Similar savings have been gained by replacing incandescent bulbs with light emitting diodes (LEDs), which use 75 percent less energy while lasting 25 times as long as incandescent bulbs. According to the U.S. Department of Energy, by the year 2027, widespread use of LEDs will save the equivalent annual electrical output of 44 electric power plants, each generating 1000 megawatts.[49] Devices such as smart thermostats similarly can have impactful reductions in consumer cost, by hundreds of dollars per year, and about 10% less usage for most users.[50]

A multitude of other energy conservation measures can also be applied to commercial and residential buildings. Many of them require nothing in the way of advanced technology. Intelligent site selection and proper orientation of buildings and the windows in them can result in structures staying warmer in the winter and cooler in the summer while making use of natural light. Adequate sealing and insulation prevent the loss of hot air in the winter and cool air in the summer, while the provision of thermal storage in buildings allows the use of electricity for air-conditioning to be shifted to off-peak hours, thereby obviating the need to build new generating facilities.[51] At the same time, there are many electronic technologies that can produce significant energy savings. Microprocessor-based integrated controls for heating, ventilation, and cooling can keep temperatures within narrow limits that are optimal for different rooms in a building, so during the summer a computer room can be kept cooler than a storage room.[52]

Additional gains could come through more intensive recycling efforts. One noteworthy example comes from a prodigious user of energy, the aluminum industry. Aluminum smelters consume 3 percent of the world's electrical supply, and the electricity used to produce one ton of aluminum would run a typical U.S. household for nearly 18 months. Much less energy is required to convert aluminum beverage cans back into raw aluminum, making these cans a valuable part of the waste stream, as is demonstrated by the fact that industrial sources recycled 63.6 percent of aluminum cans while consumers recycled 49.8 percent in 2018.[53]

More Miles to the Gallon (or kilowatt)

Even greater savings follow when the fuel consumption of private vehicles is reduced. Gasoline, which is primarily used by cars and light trucks, accounts for about 45 percent of the petroleum refined in the United States.[54] American automobiles are also responsible for 20 percent of CO_2 production in the United States, an amount greater than the total emissions of CO_2 from all but three of the other nations of the world.[55] Improving automotive technology to produce greater efficiency would thus be a major step in arresting both resource depletion and global climate change.

There are essentially three ways of enhancing the fuel economy of cars and trucks: reducing weight, improving aerodynamics, and making engines and accessories perform more efficiently. All of these have contributed to the substantial gains in fuel economy scored by American automobiles after the first energy crisis in 1973. For the 1975 model year, American cars averaged about 13.2 miles per gallon (mpg); by 2013, fuel mileage had nearly doubled to an average of 25 mpg.[56] An even more impressive advance—if it can be achieved—will come with the 2025 model year, when each automobile firm's fleet will be required to average 54.5 mpg.

The largest share of this improvement was affected by reducing weight. Cars became smaller and lighter, but the shift to front-wheel-drive designs for many models maintained interior space. Better aerodynamics also helped, while under the hood, fuel injection, turbochargers, computerized engine controls, variable valve timing and lift, and detail improvements allowed engines to squeeze more mpg while actually increasing performance.[57] Although the internal combustion engine is not likely to be replaced anytime soon, other parts of the automobile's powertrain could undergo significant changes, such as the widespread use of continuously variable transmissions that allow engines to run at their most efficient rotational speeds. Even more radical concepts, such as cars powered by fuel cells, have already been sold to customers in small numbers. It is possible that in a decade or two these vehicles will be on the road in substantial numbers. Several automobile manufacturers now produce battery-powered electric vehicles and more will be offered in the years to come. Even more successful at present have been hybrid vehicles that use both an internal combustion engine and an electric motor to optimize fuel consumption. Hybrids are now a common sight on the world's highways; by the middle of 2018, Toyota had sold nearly 2.3 million Prius hybrids in the United States since their launch in late 2000.[58]

Figure 7.5 The Volkswagen XL1 experimental car was designed to travel up to 240 mpg on one liter of diesel. Some of the technology it embodies will eventually be used for cars sold to the public. (Source: RudolfSimon/CC BY-SA 3.0)

Fuel-efficient cars will become more attractive as rising corporate average fuel economy (CAFE) standards required by the federal government will mandate more miles per gallon of gasoline. It is far from certain that the days of the conventional automobile are numbered, and it is likely that new automotive technologies will be able to at least partially compensate for the environmental problems caused by rising car populations worldwide.

Economic Systems, Government Policies, and the Environment

Just as technological advances affect energy supplies and energy-conservation measures, so too are there a substantial number of innovative technologies that can help to control toxic wastes, pollutants, and other undesirable by-products of modern technology.

Many innovations can be used to substitute for hazardous materials and processes, make more efficient use of potentially toxic materials, and counteract existing hazards. New technologies attempt to undo or recuperate damaged earth systems. The "Great Pacific Garbage Patch," also known as the Pacific trash vortex, is actually two patches of thousands of acres of plastic, estimated at a trillion pieces. These plastic materials start as debris and garbage from lost shipping containers, fishing nets, and circulating waste and end up, as noted earlier, as toxic microplastics. Technologies for cleanup have been proposed, such as scoop and recycle systems, although prevention will be the better option.[59] Or consider drone technology to reseed and replant acres of deforested lands. While of course reforestation is an important project, it discounts the value of old-growth and irreplaceable trees such as long-growing sequoias. Tree growth takes time, and firing seeds or seedlings into the landscape does not immediately fill in the lost ecosystem, i.e., the complexity of the destroyed forests.[60]

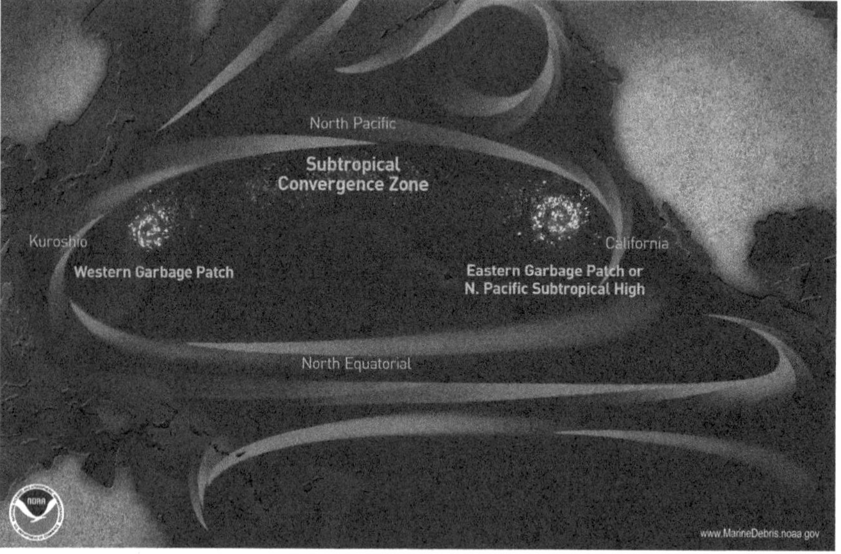

Figure 7.6 The Great Pacific Garbage Patch is a large collection of trash in the North Pacific, between Hawaii and California, formed by ocean currents. It consists of both larger pieces of trash, such as fishing nets, and tiny microplastics, making most of the Patch invisible when sailing through it. While difficult to determine its size due to there being no exact borders, some estimates put it at around 620,000 square miles.

So, as important as these technological advances are, one would be overly optimistic to think that new and improved technologies are the solution to all of our environmental ills. As Chapter 1 argued, technological fixes have their place, but when major problems are involved they must be complemented by social, political, and cultural changes. Some of these changes may occur spontaneously, but many of them must be encouraged by laws, regulations, and other government interventions. Even when a technological fix can be developed, it is often necessary to stimulate its use through the formation and implementation of effective policies.

During the energy crisis brought on by the OPEC oil embargo in 1973, some scholarly observers came close to panic. As they saw things, the massive reduction in oil supplies was the latest symptom of a massive crisis. Combined with population growth and environmental destruction, the contraction of energy supplies was forcing major changes in industrial societies. Democratic procedures would have to be replaced by powerful, even authoritarian, government mandates. One influential economist looked toward a dangerous future and reluctantly concluded that "the passage through the gauntlet ahead may be possible only under governments capable of rallying obedience far more effectively than would be possible in a democratic setting."[61] His assumption, one shared by many others, was that in the face of a massive crisis, the only salvation lay in a government's ability to force people to sacrifice their individual interests in favor of the collective good. A highly centralized government was inescapable, for only such an authority could bring sufficient expertise to bear on the shaping of the necessary policies and at the same time stifle individual actions contrary to these policies.

There may be a surface plausibility in these ideas, but they are easily refuted by an examination of the performance of centrally planned economies like the Soviet Union and the People's Republic of China during the years of Maoist ascendency.[62] Inefficient and wasteful production along with massive environmental destruction were painfully evident in these and other centrally directed economies.

As will be noted in greater detail in Part 7, the imperfections of a market system can be alleviated by intelligent regulations and tax policies, government-sponsored research programs, and educational efforts. The consequences of governmental policies have been particularly evident in the development and application of energy technologies, although the results have not always been benign. Critics of the nuclear industry argue that atomic power would never have gained a foothold without loan guarantees, limits on liability in the event of accidents, and financial support for the disposal of wastes.[63] Renewable energy has also depended to a significant degree on government subsidies. From 1992 to 2014, wind power installations received a tax credit of 2.2 cents per kWh of electricity generated for their first 10 years of operation, while until recently, the solar industry was supported by cash grants that amounted to 30 percent of the construction costs of new projects.[64] The oil and gas industry does not receive direct subsidies, but according to critics, tax breaks of a sort that only an accountant or lawyer could understand and love have been worth $70 billion over a 10-year period.[65]

Less controversial is governmental support of scientific and technological research relating to the environment. Providing economical and environmentally friendly sources of energy will require an increased pace of research and development (R&D) efforts on a variety of fronts. Although private industry has been an important source of energy-related R&D funding, governments have also been essential, if erratic, providers. Federal government-sponsored R&D expenditures on non-fossil sources of energy rose sharply in the wake of the energy crises in the 1970s, but only temporarily. R&D for nuclear energy continued at a high level in the 1980s, but R&D for other sources of energy underwent a sharp decline and then more or less leveled off from the mid-1980s onward.[66]

The federal budget allocation for renewable energy R&D during the fiscal years stretching from 2009–2018 came to $9.37 billion.[67] In comparison, during that period, $9.98 billion went for fossil fuel R&D and $13.71 billion was allocated for nuclear R&D.[68] There is no guarantee, of course, that higher levels of R&D support would have resulted in an array of technological breakthroughs on the renewable energy front or for the benefit of the environment as a whole. Still, it does seem likely that we would be in better shape today if R&D on conservation and renewable sources of energy had at least as generous as expenditures on fossil and nuclear energy. Moving forward, the 2022 Federal budget for the Department of Energy added over $12 billion dollars in research to achieve a goal of net-zero carbon emissions by 2050.[69]

At the same time, it should be recalled that the availability of a potentially superior technology does not ensure that it will be used. The development of technologies that could help us use energy more efficiently, reduce pollution, and make fewer demands on a fragile environment is only part of the solution. Individual choices, corporate and governmental policies, the general distribution of income and power, and the willingness to forsake short-term advantages for longer-term benefits are at least as important as any combination of technological fixes. There are 20 firms which generate one third of all carbon emissions. Willingness to regulate and intervene in the market will be essential to future sustainability.[70] These will be the ultimate determinants of our ability to maintain and even improve our material standard of living while at the same time preserving the environment that has sustained us through centuries of carelessness, neglect, and exploitation.

QUESTIONS FOR DISCUSSION

1. What do you consider to be the greatest environmental threat facing the world today? What sort of measures need to be taken to counter it? What will be the major sources of resistance to these measures?
2. The long-term environmental consequences of CO_2 emissions cannot be known for sure. Earth's atmosphere is an extremely complex system that is driven by a great multitude of variables. Predictions of climate trends are, and probably always will be, based on data that are not completely adequate. Are we justified in enacting laws to enforce significant reductions in the use of fossil fuels on the basis of imperfect scientific information? At what point can we decide that the assessment of risk is exact enough to warrant taking firm actions?
3. In the text, a distinction was drawn between conservation and curtailment. Are the energy-saving measures taken in the past 15 years primarily examples of conservation or curtailment? Will future energy-saving strategies be based on the former or the latter?
4. A considerable improvement in the fuel economy of automobiles has been the result of "downsizing." Yet all other things being equal, smaller cars are not as safe as larger ones in the event of an accident. Can a substantial savings in fuel justify the likelihood

of more traffic-related injuries and fatalities? At the same time, more fuel-efficient automobiles also produce fewer pollutants, leading to fewer pollution-induced deaths and illnesses. Is it possible to construct a balance sheet that takes into account all of these factors in order to determine if smaller cars improve or threaten our physical health?

5. A major issue in environmental analysis is sustainability. Is our present economy and society sustainable over the long run? In what ways do present technologies undermine sustainability? Are there any that promote it?

Notes

1. U.S. Energy Information Agency, "EIA Projects Nearly 50 Percent Increase in World Energy Usage by 2050, led by Growth in Asia," (September 24, 2019). Accessed from www.eia.gov/todayinenergy/detail.php?id=41433.
2. International Energy Agency, "Global CO2 emissions rose less than initially feared in 2022 as clean energy growth offset much of the impact of greater coal and oil use," (March 2, 2023). Accessed from www.iea.org/news/global-co2-emissions-rose-less-than-initially-feared-in-2022-as-clean-energy-growth-offset-much-of-the-impact-of-greater-coal-and-oil-use.
3. Newton H. Copp and Andrew W. Zanella, Discovery, *Innovation and Risk: Case Studies in Science and Technology* (Cambridge, MA, and London: MIT Press, 1993), p. 299.
4. Rebecca Lindsey, "Climate Change: Atmospheric Carbon Dioxide," *National Oceanic and Atmospheric Administration* (May 12, 2023). Accessed from www.climate.gov/news-features/understanding-climate/climate-change-atmospheric-carbon-dioxide.
5. NASA Earth Observatory, "Global Temperatures." Accessed from https://earthobservatory.nasa.gov/world-of-change/global-temperature.
6. GVI, "Climate Change and Rising Sea Levels: 5 Pacific Islands That No Longer Exist," (September 28, 2022). Accessed from www.gviusa.com/blog/disappearing-land-5-pacific-islands.
7. Anuj Chopra, "How Global Warming Threatens Millions in Bangladesh," *U.S. News and World Report* (March 26, 2009). Accessed from www.usnews.com/news /energy/articles/2009/03/26/how-global-warming-threatens-millions-in-bangladesh.
8. Sarah Derouin, "Deforestation: Facts, Causes & Effects," *Live Science* (November 6, 2019). Accessed from www.livescience.com/27692-deforestation .html; and Alaa Elassar, "Amazon Deforestation Rate Hits Highest Level in Over a Decade," *CNN* (November 19,2019). Accessed from www.cnn .com/2019/11/19/americas/brazil-deforestation-amazon-2019-trnd/index .html.

9. Ross W. Gorte and Pervaze A. Sheikh, "Deforestation and Climate Change," *Congressional Research Service* (March 24, 2010). Accessed from http://forestindustries.eu/sites/default/files/userfiles/1file/R41144.pdf.
10. Environmental Protection Agency, "Facts and Figures on Materials, Wastes, and Recycling." (December 3, 2022). Accessed from www.epa.gov/facts-and-figures-about-materials-waste-and-recycling/national-overview-facts-and-figures- materials.
11. Ibid.
12. Brooke Kato, "You're Eating a Credit Card's Worth of Plastic a Week—and It's Killing Your Gut," *New York Post* (March 30, 2022). Accessed from https://nypost.com/2022/03/30/youre-eating-a-credit-cards-worth-of-plastic-a-week-and-its-killing-your-gut; and Casey Decker, "Claims You Eat a Credit Card's Worth of Plastic per Week Need Context," Verify (April 22, 2022). Accessed from www.verifythis.com/article/news/verify/science-verify/claims-you-eat-a-credit-cards-worth-of-plastic-per-week-need-context/536-748d91cf-00f1-4de6-8e01-776694ce6fdf for clarification.
13. Monterey Bay Aquarium, "Plastic Pollution." Accessed from www.montereybayaquarium.org/act-for-the-ocean/plastic-pollution.
14. Jessica, Morrison, "Polluted Sites Linger under U.S. Cleanup Program," *Chemical and Engineering News* (April 3, 2017). Accessed from https://cen.acs.org/articles/95/i14/Polluted-sites-linger-under-US-clean-up-program.html
15. Kenneth S. Deffeyes, *Hubbert's Peak: The Impending World Oil Shortage* (Princeton, NJ: Princeton University Press, 2001).
16. Kristina Marusic, "After a Decade of Research, Here's What Scientists Know about the Health Impacts of Fracking," *Environmental Health News* (April 15, 2019). Accessed from www.ehn.org/health-impacts-of-fracking-2634432607.html; and Christina Nunez, "How Has Fracking Changed Our Future?" *National Geographic*. Accessed from www.nationalgeographic.com/environment/article/how-has-fracking-changed-our-future.
17. Michael T. Klare, "Peak Oil Is Dead . . . Long Live Peak Oil," *Huffington Post* (January 9, 2014). Accessed from www.huffingtonpost.com/michael-t-klare/peak-oil-is-dead_b_4567978.html.
18. See: www.federalreservehistory.org/essays/oil-shock-of-1978-79 on the Iranian Revolution (1978–1979); On the Russia-Ukraine War see: Pippa Stevens, "These Charts Show How Russia's Invasion of Ukraine Has Changed Global Oil Flows," *CNBC* (May 31 2022). Accessed from www.cnbc.com/2022/05/31/these-charts-show-russias-invasion-of-ukraine-has-changed-global-oil.html; and Bassam Fattouh, Andreas Economou, and Ahmed Mehdi, "Russia-Ukraine Crisis: Implications for Global Oil Markets," *The Oxford Institute for Energy Studies*. Accessed from www.oxfordenergy.org/publications/russia-ukraine-crisis-implications-for-global-oil-markets.
19. Paul Ehrlich, *The Population Bomb* (New York: Ballantine Books, 1968).
20. David R. Montgomery, *Dirt: The Erosion of Civilizations* (Berkeley, Los Angeles, and London: University of California Press, 2007), pp. 4, 171–74.

21. James F. Reynolds, "Desertification," *Reference Module in Earth Systems and Environmental Sciences* (2013). Accessed from www.sciencedirect.com/topics/social-sciences/desertification.
22. April Reese, "What's Behind Desertification?" *Population Reference Bureau.* Accessed from www.prb.org/whatsbehinddesertification.
23. Wilson Da Silva "The Long Dry: Global Water Supplies Are Shrinking," (December 13, 2018). Accessed from https://phys.org/news/2018-12-global.html.
24. United Nations, "Water." Accessed from www.un.org/en/sections/issues-depth/water.
25. Carlo M. Cipolla, *Before the Industrial Revolution: European Society and Economy, 1000–1700*, 2nd ed. (New York: W.W. Norton, 1980), p. 138.
26. Quoted in Ibid., p. 288.
27. W. Stanley Jevons, *The Coal Question: An Inquiry Concerning the Progress of the Nation, and the Probable Exhaustion of Our Coal-Mines*, 3rd ed. (London: Macmillan and Co., 1906), p. 9.
28. Ibid., p. 187.
29. Ibid., pp. 184–85.
30. Ibid., p. 185.
31. James R. Chiles, "Spindletop," *American Heritage of Invention and Technology* 3, 1 (Summer 1987): 34.
32. Thomas Wellock, "Too Cheap to Meter": A History of the Phrase," *U.S. Nuclear Regulatory Commission* (January 1, 2023). Accessed from www.nrc.gov/reading-rm/basic-ref/students/history-101/too-cheap-to-meter.html.
33. "What Is U.S. Electricity Generation by Energy Source?" U.S. Energy Information Administration. Accessed from www.eia.gov/tools/faqs/faq.php?id=427&t=3.
34. Calculated from figures in John M. Deutch and Ernest Muniz, "The Nuclear Option," *Scientific American* 295, 3 (September 2006): 78.
35. Mark Z. Jacobson, "The 7 Reasons Why Nuclear Energy Is Not the Answer to Solve Climate Change," *Heinrich Böll Foundation* (April 26, 2021). Accessed from https://eu.boell.org/en/2021/04/26/7-reasons-why-nuclear-energy-not-answer-solve-climate-change.
36. Adam Piore, "Planning for the Black Swan," *Scientific American* 304, 6 (June 2011): 50.
37. Anthony Burke, "Attacks on Ukrainian Nuclear-Power Plants Challenge Treaties," *Nature* (November 4, 2022). Accessed from www.nature.com/articles/d41586-022-03580-0.
38. Matthew L. Wald, "What Now for Nuclear Waste?" *Scientific American* 301, 2 (August 2009): 46–53.
39. Deutch and Muniz, "The Nuclear Option," pp. 82–83.
40. Dominic Dudley, "Renewable Energy Will Be Consistently Cheaper Than Fossil Fuels By 2020, Report Claims," *Forbes* (January 13, 2018). Accessed from www.forbes.com/sites/dominicdudley/2018/01/13/renewable-energy-cost-effective-fossil-fuels-2020/#5d10d77f4ff2.

41. U.S. Department of Energy, "2014 Wind Technologies Market Report Highlights" (August 2015). Accessed from www.energy.gov/sites/prod/files/2015 /08/f25/2014-Wind-Technologies-Market-Report-Highlights.pdf.
42. Natural Resources Defense Council, "Renewable Energy for America." Accessed from www.nrdc.org/energy/renewables/solar.asp.
43. James Conca, "Energy's Future—Battery and Storage Technologies," *Forbes* (August 26, 2019). Accessed from www.forbes.com/sites/jamesconca/2019/08/26/energys-future-battery-and-storage-technologies/#481d5ec544cf.
44. James J. Flink, *The Car Culture* (Cambridge, MA: MIT Press, 1975), p. 34.
45. The Economist "Power Slide," (January 11, 2009). Accessed from www.economist.com/blogs/dailychart/2011/01/energy_use.
46. Chris Nelder, "The Worst Way to Measure Energy Efficiency," *Slate* (March 13, 2013). Accessed from www.slate.com/articles/technology/the_efficient_planet /2013/03/energy_intensity_is_a_misleading_metric_in_the_fight_against_climate _change.html.
47. Arthur H. Rosenfeld and David Hafemeister, "Energy-Efficient Buildings," *Scientific American* 258, 4 (April 1988): 78.
48. Eberhard K. Jochem, "An Efficient Solution," *Scientific American* 295, 3 (September 2006): 66.
49. U.S. Department of Energy, "LED Lighting." Accessed from www.energy.gov/energysaver/save-electricity-and-fuel/lighting-choices-save-you-money/led-lighting
50. Nest, "Saving Energy Starts with Your Thermostat." Accessed from https://nest.com/thermostats/real-savings; and Dan Seifert, "Nest Thermostat Review: More Simple than Smart," *The Verge* (December 1, 2020). Accessed from www.theverge.com/21725036/google-nest-thermostat-2020-review.
51. Rosenfeld and Hafemeister, "Energy Efficient Buildings," p. 81.
52. Rick Bevington and Arthur H. Rosenfeld, "Energy for Buildings and Homes," *Scientific American* 263, 3 (September 1990): 78.
53. The Aluminum Association, "The Aluminum Can Advantage: Key Sustainability Performance Indicators," (September 2019).
54. U.S. Energy Information Administration, "Gasoline Explained" (October 29, 2015). Accessed from www.eia.gov/energyexplained/index.cfm?page=gasoline_home.
55. John DeCicco, Rod Griffin, and Steve Eertel "Putting the Brakes on U.S. Oil Demand" (Environmental Defense, 2003). Accessed from www.edf.org / documents/3115_OilDemand.pdf, p. 3.
56. U.S. Environmental Protection Agency, "Light-Duty Automotive Technology, Carbon Dioxide Emissions, and Fuel Economy Trends: 1975 through 2013" (2013). Accessed from www.fueleconomy.gov/feg/pdfs/420r13011_EPA_LD_FE_2013 _TRENDS.pdf.
57. Ben Knight, "Better Mileage Now," *Scientific American* 302, 2 (February 2010): 50–55.
58. Car Salesbase, "Toyota Prius." Accessed from http://carsalesbase.com/us-car-sales-data/toyota/toyota-prius/47.

59. See "The Ocean Cleanup," at https://theoceancleanup.com; and "Great Pacific Garbage Patchh" at ttps://www.nationalgeographic.org/encyclopedia/great-pacific-garbage-patch.
60. See Mast Reforestation at www.mastreforest.com; and James Conca, "Drones Can Reforest The Planet Faster Than Humans Can," *Forbes* (Sep 30, 2020). Accessed from www.forbes.com/sites/jamesconca/2020/09/30/drones-can-reforest-the-planet-faster-than-humans-can.
61. Robert L. Heilbroner, *An Inquiry into the Human Prospect* (New York: W.W. Norton, 1974), p. 110.
62. Judith Shapiro, *Mao's War against Nature: Politics and the Environment in Revolutionary China* (New York: Cambridge University Press, 2001).
63. Doug Koplow, "Nuclear Power: Still Not Viable without Subsidies," *Union of Concerned Scientists* (February 2011). Accessed from www.ucsusa.org/assets/documents/nuclear_power/nuclear_subsidies_report.pdf. See also Union of Concerned Scientists, "Production Tax Credit for Renewable Energy," Accessed from www.ucsusa.org/resources/production-tax-credit-renewable-energy.
64. Diane Cardwell, "Energy Tax Breaks Proposed Despite Waning Support for Subsidies," *The New York Times* (January 26, 2012). Accessed from www.nytimes.com/2012/01/27/business/energy-environment/clean-energy-projects-face-waning-subsidies.html?_r=2&pagewanted=1&ref=windpower.
65. Seth Hanlon, "Big Oil's Misbegotten Tax Gusher," *Center for American Progress* (May 5, 2011). Accessed from www.americanprogress.org/issues /2011/05/big_oil_tax_breaks.html. It should be pointed out that many of these quasi-subsidies are also collected by firms in other sectors of the economy.
66. Government Accountability Office, "Budget Trends and Challenges for DOE's Energy R&D Program," (March 05, 2008). Accessed from www.gao.gov/products/gao-08-556t.
67. Corrie Clark, "Renewable Energy R&D Funding History: A Comparison with Funding for Nuclear Energy, Fossil Energy, Energy Efficiency, and Electric Systems R&D," *Congressional Research Service* (June 18, 2018). Accessed from https://fas.org/sgp/crs/misc/RS22858.pdf.
68. Ibid.
69. International Energy Agency, "Department of Energy, Federal Fiscal Year 2022 Budget," (November 18, 2022). Accessed from https://www.iea.org/policies/13738-department-of-energy-federal-fiscal-year-2022-budget.
70. See Tess Riley, "Just 100 Companies Responsible for 71% Of Global Emissions, Study Says," *The Guardian* (July 10, 2017). Accessed from www.theguardian.com/sustainable-business/2017/jul/10/100-fossil-fuel-companies-investors-responsible-71-global-emissions-cdp-study-climate-change; and more recently: Matthew Taylor and Jonathan Watts, "Revealed: The 20 Firms behind a Third of All Carbon Emissions," *The Guardian* (October 9, 2019). Accessed from www.theguardian.com/environment/2019/oct/09/revealed-20-firms-third-carbon-emissions.

CHAPTER 8

Tools for (Environmental) Assessment

Quite a number of disciplines have devised tools for evaluating technologies and making comparisons among technological options. In this chapter we will look at several of these and connect these tools to some of our other resources and ideas. By the end of this chapter, we will have woven together the social construction of technology, tools such as value chains or life-cycle analyses, and the technological diffusion curve to give us a sense not only of how products come into being, but how those production processes have changed and might be transformed, over time. As the title suggests, environmental analyses are one of the most obvious kinds of assessments of the consequences of producing a technology, but not the only ones.

Case Study: Origins of a Mug

The ancient ceramic chocolate pot (Figure 8.1) is of Mayan origin, from somewhere around AD 600 to 900. It was made by hand in a local workshop along with other ceramic pieces, with locally sourced clay. It would have been fired in a small kiln or furnace after being manufactured by hand, with a small batch of related pieces. When broken or lost, like its lid, it may have continued to be a treasured object. If it had been completely broken, its remains would have been tossed into a midden (trash dump) or been ground up and reincorporated into raw clay as "grog" to stabilize a new batch of ceramics. We can imagine that it was a treasured gift, perhaps made or purchased as a special present for a wedding or special event. Archaeologists can tell a great deal about the source of the

Figure 8.1 This is a traditionally crafted Mayan "chocolate pot," from about AD 750. Glyphs describe the owner and place of origin, and that this is a vessel for keeping and preparing cacao. (Photo from the Minneapolis Institute of Art).

Figure 8.2 A commercially produced beverage mug.

materials through chemical analysis. Traces of manufacture, including even fingerprints, can be found in ancient clays. The steps to manufacture are inferred from the remains of workshops unearthed through archaeological exploration. We can't know for sure whether this chocolate mug remained within a few miles of its point of origin or travelled through the Mayan empire across the Yucatan Peninsula, or exactly what it experienced before it arrived as a treasured artifact in a private collection.

Yet whatever it is we don't know about this ancient chocolate pot, we intuitively know that its production story was probably very different from a typical mass-produced mug made today (Figure 8.2). Ceramic manufacture for these kinds of products is generally by a mold-pressing system, which makes identical samples in batches of several dozen at a time, with minimal human contact. The bulk clay can come from nearly anywhere in the world—wherever it is cost-effective to procure and transport. Industrial-sized kilns finish this and thousands of other products. This industrial, manufactured good is shipped anywhere in the world, in palleted boxes packed into the ubiquitous shipping container. It may at some point have travelled by plane, train, or truck. After it is damaged or separated from the set that it came with, perhaps it too, remains a useful object—holding paper clips or pens on a desk. But it is not likely to be ground into grog for remanufacture in new ceramics, and ultimately ends up in a landfill, perhaps after a cycle of sales at thrift stores.

In each case, different people in social groups are involved in the process. In a complex society, there might be more intermediate steps, such as shipping and transport, with more people involved. In each case, there may or may not be information flows. For example, if the mug is defective when unwrapped new, does the information make it back to the producer—as a product return, online review, or email? Or does the information simply stop circulating when the user tosses the defective product into the trash? The material flows between the two cases are of course different: one local, with some capacity for feedback, the other potentially global, with distinct gaps between manufacturing, use, and end-product waste management.

So here we are taking the idea of social construction to look at not only the invention of a technology and its relationship to competing alternatives, and the arrangements and power dynamics amongst various social groups, to the *actual* construction process for specific technologies. If nothing else can convince you that technological determinism is false, it should be the study of the specifics of producing technologies, and the decisions, systems, and consequences of bringing technologies into being.

Matters of Scale

The Mesoamerican chocolate mug[1] exemplifies what is probably on the smaller scale of pottery production. Archaeologists have studied these smaller scale operations across civilizations, and these kinds of studies are among the first to articulate the idea of the life history of an artifact. The *chaine operatoire* is a concept that integrates the knowledge of a group, the gestures and technical sequences, and the objects produced by a group. The term has been extensively applied to prehistoric groups and lithic (stone) tools. For example, Pierre Lemonnier, an archaeologist, is concerned with the most local scale—the chain of gestures, movements, materials choices, and the implied decisions and culture that go into physically producing a stone shard or arrow/spear point.[2] Archaeologist Michael B. Schiffer articulated a similar framework, although more focused on space and at a somewhat more extensive and generalizable scale. But he was concerned with roughly the same question—how does a material artifact get from raw materials to use.[3]

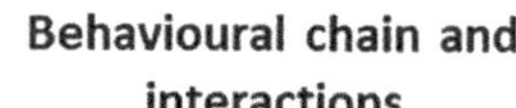

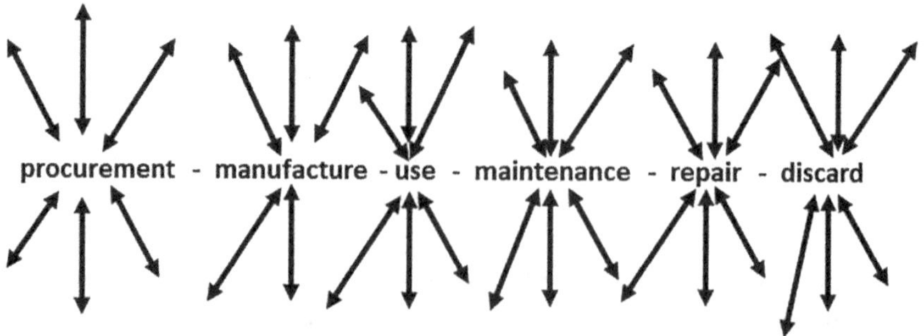

Figure 8.3 This is a representation of the information and material flows as an object comes into being and moves through its life history. (Source: From Ian Hodder, "Human-Thing Entanglement: Towards an Integrated Archaeological Perspective," *Journal of the Royal Anthropological Institute* 17 (2011): 154–177. Used with permission.)

For pre-historic societies, or in contemporary non-industrial societies and many kinds of craft work, an artifact may not pass through many different hands over the course of its life history. The maker-user could have collected the raw materials, fabricated, used, reused, and discarded the product all in one rather small area. It is a rather straightforward model, although, as we see, as social complexity increases, the scale and scope of any one of the stages in a product can dramatically increase and have broad impacts well beyond a single participant in any stage of the chain, or any single end-user.

More general approaches to the life history of a product emerged with the development of what is called "world systems theory" in American sociology.[4] Scholars Terence K. Hopkins and Immanuel Wallerstein define a commodity chain as "a network of labor and production processes whose end result is a finished commodity."[5] Gary Gereffi extended the commodity chain to a global orientation,[6] establishing a framework for evaluating multiple inputs and outputs for making and distributing products. We also see terms like cradle-to-grave accounting, generally focused on the value of an asset in business calculations, or in comparisons between products in environmental fields. Which is more environmentally impactful: bio-based or petrochemical-based plastic bottles? Life-cycle assessments are often focused on environmental and social impacts, and value-chain analysis generally focus on human dimensions, such as on where value is added to products as they move through various phases of manufacture. For simplicity, we will use commodity chain analysis as our term of choice.

All of the frameworks get at the same idea, although with different emphasis and scales. Even simple questions such as where raw materials come from, or where the product goes when broken or obsolete, can rapidly expand into complex networks of exchanges and impacts. Does engagement in a global value chain improve or worsen local poverty? Does it shift patterns of gender and poverty by providing access to new markets for women? In an area with a lot of hydroelectric power, are electric vehicles a better cost and environmental option than in areas where electricity is produced through coal or natural gas combustion?[7] We will investigate two case studies, smartphones and automobiles, to look at how there are social and environmental consequences entangled in each phase of a commodity chain, allow us to evaluate a technology's consequences beyond just the evaluation of its impacts on users.

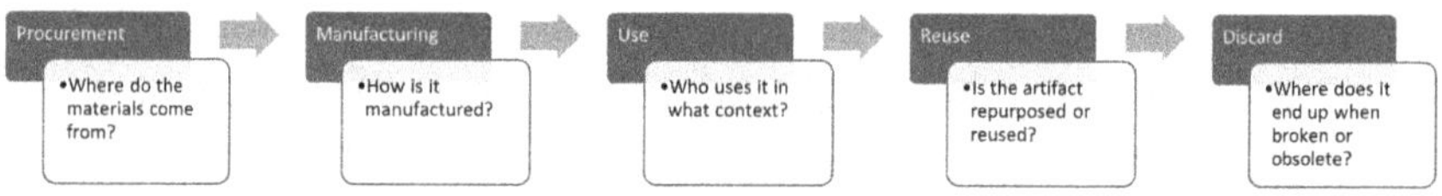

Figure 8.4 Some starting questions for a basic commodity chain analysis.

Case Study: The Phone in Your Pocket

Probably the most overused student essay topic in recent years, smartphones are a nearly ubiquitous feature of life in contemporary societies, and while physically small, the global reach of the commodity chains which bring these devices into being and into our pockets is exemplary as a case study.

As of 2023, the primary materials in a cell phone include gold for circuitry, copper, and materials such as gallium arsenide for the integrated circuits, and lithium for batteries, plus petrochemicals for the plastic components. These materials are mined or extracted. Gallium Arsenide (GaAs) and Gallium Nitride (GaN) are two very important compounds in microelectronics, light-emitting diodes (LEDs), and solar cells. At present, Gallium is mined largely in Russia, China, and Ukraine,[8] with increasing production in Canada, Japan, and Singapore. It is often found in conjunction with bauxite (aluminum ore) mining,[9] but not actively mined in the United States. All mining has severe consequences on the environment, and potential health impacts. For example, expansion of bauxite mining in Malaysia is largely unregulated and producing health effects due to toxic dusts, plus causing deforestation and pollution of water supplies, and creates the potential for flooding and erosion as landscapes are disrupted.[10]

Similarly, lithium and gold mining have had intensive environmental and social impacts, although both have changed over time. Lithium, used extensively for electric batteries, has been extracted from pegmatite ores through conventional mining activities.[11] This is how it's done in Australia and Canada, and it is proposed to be done for a controversial new mine in the United States.[12] However, technological innovation has shifted to extraction of lithium from brines, with a very different environmental profile, largely in Chile and Argentina.[13] The process changes the landscape and ecosystems, including decreasing flamingo populations and related species

Figure 8.5 A Gallium mine hillslope in Pulaski County, Arkansas, which was active from 1947 to 1983. Arkansas is one of two states where this metal has been produced in the U.S., although no mining is currently active. (Photo from the Arkansas Geological Survey)

biodiversity.[14] The production of one metric ton of lithium requires 400,000 liters of water, which stresses water supplies for humans and watersheds.[15] While lithium may be extracted globally, more than half is refined China.[16] Thus, while lithium is seen as essential to green technologies which rely on electricity, it is not impact-free, even though it does not add directly to greenhouse gas emissions.[17]

The rush for gold after its discovery in Sutter's Mill at the Nevada–California border in 1848 was undeniably transformative for the West in the United States, even though of relatively short duration (1848–1855). Much has already been written about how gold drew the attention of the nation westward, incentivizing the railways, intensifying the Indian Wars, and pushing the center of population further west into the continental United States.[18] Crude methods of blasting away at hillsides with streams of water led to rapid environmental degradation,[19] eventually harming the emerging farming industry of California. In the present, gold is mined in both massive-scale mining operations,[20] and smaller-scale, often "wildcat" or "artisanal" small-scale operations that disrupt ecosystems[21] and have been implicated in the displacement and murder of indigenous populations,[22] and with other negative outcomes for human and environmental health.[23]

All of these primary elements—gold, gallium, or lithium—plus a host of others, require extensive extraction processes, which use acids and other complex and toxic chemicals, and a great deal of energy. There are risks of leaks and chemical accidents in these extractive processes, as well as general workplace exposures.[24] From these raw materials, components are refined, transported, and manufactured. As of this writing, 90% of electronics components are manufactured in Shenzhen China and its nearby region.[25] The Chinese government and private investors have developed a comprehensive manufacturing infrastructure, which aggregates raw materials refinement, component manufacture, and finished product assembly, and an experienced workforce. This region, with a population of at least 13.44 million people, has an estimated per-capita GDP (gross domestic product) of 203,489 yuan (US $31,000). But there may actually be nearly 20 million people in the region, which would suppress the calculated wage and stress the infrastructure.[26]

This concentration of manufacturing capacity provides efficiency and economies of scale, but also becomes the seat of potential civil unrest. For example, in 2012, workers in the FoxConn factory in Shenzhen had a rooftop strike. The plant, which manufactured Apple electronics, experienced working conditions so bad that people were committing suicide.[27] One solution to this problem was to add nets around the buildings.[28] In 2022, this labor unrest re-emerged in the context of COVID-19 lockdowns, which again centered on Foxconn plants and the ways in which government lockdowns intensified already prison-like labor conditions.[29]

As might now be clear, global commodity chains are very sensitive to geopolitical issues, such as the autonomy of Taiwan, or China's own internal politics. These vulnerabilities have led to initiatives for building electronics components and batteries within the United States, although often by foreign manufacturers.[30] European nations have responded as well to the vulnerability of industry concentration and dependency.[31] However, no matter the location, chip manufacturing is energy and water intensive. For example, while the new fabrication plants in Arizona and elsewhere are heralded as a new opportunity for employment and technological development, these new plants come online in the midst of a drought emergency.[32] And despite water-recycling systems and attention to green manufacturing, chip manufacturing is still energy-intensive.[33]

After manufacturing, cell phones and related electronics are shipped globally, and the distribution process has its own complexity. In general, oil and petrochemicals appear in multiple places in contemporary commodity chains—in product manufacture, in packaging, and in transport. Fossil-fuel impacts on the environment were reviewed more extensively in Chapter 7. While global ocean shipping only produces about 3% of total greenhouse gas emissions in any given year,[34] shipping accidents produce spills of oil and waste water (bilge) which contribute to local waste problems. To focus in on the effects of shipping on wildlife, we should also consider acoustic (noise) pollution, which affects ocean wildlife such as whales. Whale strikes caused about 35.5 percent of the recorded North Atlantic right whale deaths between 1970 and 1999, accelerating the possibility of extinction for the highly endangered species, with perhaps fewer than 350 left in 2022.[35] Technologies, such as drone and sonar spotting, and sociotechnical solutions, such as speed controls, may have some benefit in solving this problem, but in the meantime this consequence remains salient. At each phase of the commodity chain, a layered network of impacts add complexity to analyses of technological systems.

In Part 5, especially Chapters 16 and 17, we will look more extensively on the effects that the *use* of various forms of ICTs (information and communication technologies) have on individual users and social institutions. Remember, these consequences are unevenly distributed; globally, the benefits of cell phone access do necessarily not line up with the people bearing the environmental costs of producing the cell phones. Countries such as Chile bear extensive environmental costs for lithium production, although they do benefit from high levels of smartphone access. As we discuss later, in the use phase of a commodity chain for electronics there are uneven positive and negative impacts on student learning, social connectedness, mental health, and changes in employment. And let's not forget the energy production for cloud systems and cell phone use. For the emerging cryptocurrency market, server energy use surpasses total usage of smaller countries such as Denmark or Chile.[36]

Typically, the next phase of a commodity chain might be refurbishment or reuse. The market for used mobile phones is significant—approximately $52.34 billion in 2021 and expected to nearly triple by 2030.[37] But even then, electronics have a finite lifespan, and when broken, or totally phased out and unsupported, not much can be done to repair significant damage or obsolescence. The toxic effects of electronics waste are very well known. The average U.S. user replaces a cell phone approximately every 18 months, with about 120 million phones discarded every year. Globally, it is estimated that about 5 billion phones became waste in 2022, or were at least decommissioned for later disposal.[38] Phones may spend some time stashed in a drawer, but they eventually join the vast volume of e-waste. This waste is not only toxic, with plastic and poisonous materials leaching out of degrading devices, but this waste is concentrated in places that will accept international e-waste and otherwise lack recycling capabilities, such as West Africa.[39]

Recycling is rapidly becoming integrated into the discard process for consumer electronics. Not just the recycling of devices for resale, but the development of mechanisms to break down e-waste into reusable component materials is also spreading widely. As technologies for managing e-waste have improved, in some cases it has become more environmentally and cost-effective to rely on recycling and what's been called "urban mining" to remove electronic materials from waste streams or recover them from landfills.[40] In the case of gold, waste-stream intercepts may be more cost-effective than primary mining.[41]

So, at each phase of production, transport, use, reuse, and discard, there are different social groups facing differential social, economic, and environmental consequences. Commodity chain analysis in its various forms is a useful framework for looking at the environmental, health, employment, and related social consequences of the multiple phases of the production of cell phones and many other electronic devices. Costs and benefits of technologies, and the potentially iniquitous distribution of effects are always more than just the consequences for end users. Care must be taken in drawing equivalences between technologies and making sure that we are comparing "apples to apples" rather than "apples to oranges."[42] Otherwise, bias and error can creep in, allowing for ineffective, and possibly distorting, evaluations of new technologies. For example, while lithium mining has significant local environmental impacts, it does not have the global climate change or geopolitical impact of the petrochemical industry and fossil fuel industry to which it compares.[43]

Case Study: Automobiles, Over Time

Our second case study is to illustrate the issue that commodity chains vary over time, with the maturation of the technology along the technological diffusion curve. We looked at the technological diffusion curve in Chapter 6, with some significant attention to the emergence of automobiles. Here, we will briefly connect the commodity chain to the technological diffusion curve. Each framework is a kind of life history of an artifact or product, and they are interrelated but not the same. The diffusion curve is about the car as a broad category, while the commodity chain focuses on a specific car, or cohort of cars, in the context of their production.

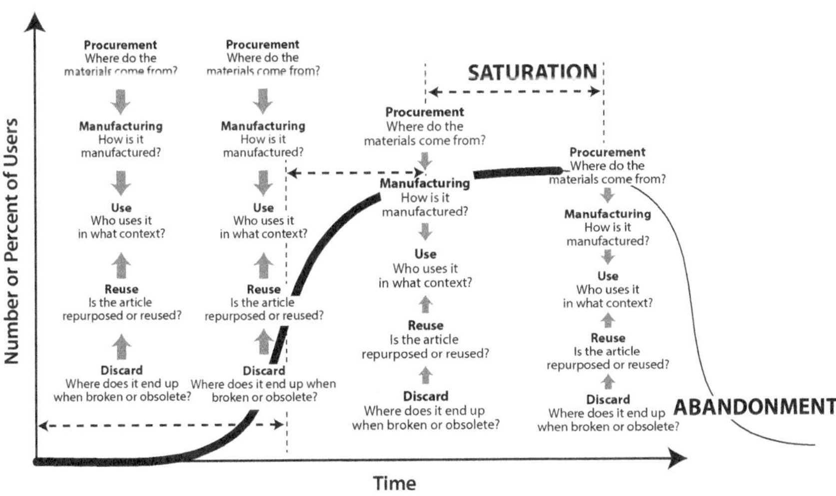

Figure 8.6 Combining the technological diffusion curve with the commodity chain gives us a sense of how production chains change over time as a technology becomes stabilized and enters the consumer marketplace. We can expect that as it approaches abandonment, the commodity chain might "revert" to more localized or bespoke production present at the early phases of the invention of the technology, before manufacturing infrastructures developed.

Reviewing the technological diffusion curve, at different periods in the invention of a technology there are different actors with different interests. In the invention and early adoption phase, inventors, developers, venture capitalists and financiers all have their eye on a new technology. Early users are significantly different from later adopters, with a great deal of enthusiasm for almost anything new. And people who may face negative impacts, real or imagined, often take notice of technologies in this first phase. Recall also that it's generally about 20 years to go from "bright idea" to the inflection point of the curve—closure in material and symbolic registers. For the automobile, it's just over 20 years from the 1886 Daimler prototype until the 1908 Model T. Accelerating adoption intensified in 1913 when Henry Ford figured out mass production. This inflection point is where "closure" occurs, where the meaning of the technology, and its physical form, is relatively stable so that the product can "move up" the technological diffusion curve and into mass, industrial production. In addition, the semiotic dimension, the cultural identification of the product as an object to be desired, has stabilized. People learned to *want* cars. The remaining three phases of the curve similarly have diverse social groups concerned about the manufacture, distribution, and use of the technology.

So, in 1886, the emerging automobile was designed and built, basically one at a time, in workshops such as wagon makers or machine makers. Industrial machine makers, and bicycle makers, were among the first firms that had the technical capacity and supply chains available to build early automobiles. For example, the Benz corporation had to invent the manufacturing machinery at the same time they were refining automobile designs. While there was a prototype of the assembly line, crews of workers stayed with one vehicle, rather than the vehicle moving to different worker stations. Total production in 1899 was 572 vehicles, with a workforce of 50.[44]

Henry Ford is credited with applying the idea of mass production with interchangeable parts to automobile construction in 1913. Other firms adopted the assembly line as finances allowed. But the simplicity and affordability of the Model T means that by 1927, 15 million had been produced.[45] The Model T was the emblematic, normal configuration of a car, establishing closure on the basic expectations of how a car should look and work. To produce over 100,000 cars in a year required a significant supply chain for parts, stable supply of manufacturing equipment, and a reliable and knowledgeable workforce.

While World War I and II disrupted demand, and deflected manufacturing resources and personnel to the war efforts, nonetheless by 1950 the stability of supply and demand for automobiles led to a kind of market saturation and a localization of the commodity chains that made automobiles. "Detroit" had arrived. Its origins as a central source for flour-milling in the U.S. Midwest in the nineteenth century provided some of the backbone for industrial machines which were necessary for car production. Parts, knowledgeable labor, and supply chains for machine tools were already available. Nearby communities, such as Gary, Indiana, were major suppliers of steel for parts, and raw materials were available through the strength of the U.S. steel industry,[46] centered around supply chains of iron ore around Lake Superior and coal from Pennsylvania and West Virginia.[47] Ford was a major supplier of steel for its own products.

While the relative decline of the U.S. steel and automobile industry are fascinating case studies in themselves, in short what the 1970s brought was rapid globalization and international competition for automobile markets and low-cost steel from international sources. By the 1990s, the commodity chains for automobile manufacture had globalized. Cars made by U.S. companies might be manufactured in Mexico and Canada, while international manufacturers have strongholds in Tennessee, South Carolina, Kentucky, Michigan, and several other states.[48] While distributed around the U.S. and globally, there are nonetheless regional clusters of automobile manufacturers that reflect the economic importance of accessible components, labor, manufacturing infrastructure, and supply chains for economic efficiency.

Change and Disruptions to Commodity Chains: Fluidity versus Rigidity

As the automobile case study illustrates, there are often multiple sources of pressure which transform commodity chains. The phase which the technology might inhabit—early development, accelerating adoption, saturation, or abandonment—places demands on a commodity chain. Economic competition from other firms or nations is of course an obvious source of change. Technological change can occur *within* the specific providers in the commodity chain itself, in search of efficiency or in response to regulation. Social movements and competition can shape the overall landscape of manufacturing. For example, regulations for fuel efficiency place demands on materials selection, such as for aluminum or plastic or low-weight alloys, which propagate through components suppliers and manufactures, and to raw materials providers. Social movements and related media and cultural pressure shape demand, say for electric cars or green or ethical manufacturing, thus shaping commodity chain development. These kinds of changes are generally more smoothly integrated into commodity chain development, unlike changes that are the result of shocks.

Like other forms of infrastructure, commodity chains are often most noticed when disrupted. They are relatively invisible when working well. The COVID-19 pandemic is an example which illustrates the disruptions and the sometimes nonlinear or disproportionate effects of commodity chain disruptions. For example, shipping backlogs affect not only consumers, but manufacturers of finished products, and manufacturers of components from raw materials. Ideas such as resiliency, re-localizing supply chains, and other business and policy methods are central to the recovery process.[49] Commodity chains can be very rigid, as stable times reinforce existing patterns and relationships, while in times of shock, fluidity might be a better option.

Industry concentration or resource concentration can contribute to the rigidity of a supply chain. In the case of extremely powerful firms and industry concentration, powerful firms can define and narrow the supply chain. For example, Walmart is a very powerful retailer. They are so large in the market that they generate near monopsony (a single buyer) conditions, placing high demands for very low costs from firms who make goods that they might sell. These demands for low costs from suppliers suppress competition and forces smaller suppliers out of the market.[50] While that might seem to be "just capitalism," it reduces the number and variety of suppliers and may reduce commodity chain sustainability. Walmart's low prices make it harder for small retailers to compete, because of Walmart's economies of scale. And as a very large employer, Walmart's national-level wage structure can depress earnings, employment, and local rents.[51] Walmart distorts the valuation of both goods and labor through its economic power and may increase commodity chain vulnerability as well as decrease small-town or small-supplier economic sustainability.

Other Impacts: Boundaries, Externalities and the Hard-to-Measure

The history of capitalist economic development has been marred by numerous examples of pollution, contamination, and heedless exploitation of natural resources. This is to be expected; a pure market system, engine of economic growth though it may be, is by its nature a threat to the environment. Markets generally do a good job of coordinating production and consumption, but they are not effective in dealing with consequences that lie outside the transactions of individual buyers and sellers. The problem lies in what economists identify

Figure 8.7 Meme of lane expansion. For its meme history, see: https://knowyourmeme.com/memes/one-more-lane-bro-one-more-lane-will-fix-it.

as "negative externalities." These occur when the cost of a commodity sold by one party and bought by another does not take into account the costs borne by other parties, such as the damages suffered by downstream residents when a paper mill dumps its wastes into a river.

In similar fashion, a transaction may not consider the loss of an irreplaceable resource to future generations. The price of a commodity may reflect some depletion costs, but time lags may prevent adequate corrective actions. In many cases, the price of a dwindling resource will increase at too slow a rate to signal its rapid depletion. The result is an overshoot-and-crash situation. This situation resembles a car driven by someone with slow reflexes; by the time the driver perceives a problem ahead and decides to apply the brakes, it is too late to avoid a collision.[52]

There are no measures or metrics for cultural change, and thus these transformations are hard to incorporate into commodity chain analysis. We briefly discussed the moral panic around women and bicycling. By both symbolizing and enhancing freedom, bicycles were seen as not only promoting liberty, but licentiousness. Other ideas change too. Consider the idea of who "owns" the streets, who morally, and as a matter of policy, is prioritized as a user. We take it for granted that street design is based on the needs of drivers. But how did this come to be? The crime of jaywalking was invented as a policy initiative of automakers to make the streets safer for cars.[53] Much has already been written about the United States as a "car culture."[54] It has shaped the landscape, house designs, city and street planning, and even ideas about time. It shapes our imagination about what we need for the future, whether as individuals with the automobile ownership being a marker for adulthood, or collectively in how future cities and homes might be designed. There are no hard and fast rules, however, as to whether culture will be an accelerating or decelerating force in changing commodity chains, or how commodity chain transformations will affect culture.

Valuations

When a commodity chain analysis is reduced to a single point in time and focus on user outcomes, we have the classic "cost-benefit analysis." The *Consumer Reports* (*CR*) model of product reviews and ratings encapsulates this approach. Collecting and analyzing end-user experiences to provide advice to consumers on which of a fixed set of options is the most cost-effective solution, in terms of desirable features and reliability, is useful. But *CR* hardly encapsulates the entire range of consequences of a technological choice, especially when taking a systems approach. People who don't *use* a technology still face consequences from its production. Ultimately, as useful as commodity chain analysis might be, it is, like all tools, subject to design choices and requires human judgement as to priorities and attention to context for implementation. How far do we extend the attention to labor outcomes, or to environmental consequences? How far into the future do we assess health consequences? These modelling decisions are simultaneously analytical and sociopolitical choices. Ultimately, assessing technology is engaging multiple systems of valuation. We'll start with economic valuation, but also end up considering the question of how to evaluate the worth of things that might be priceless. What, for example, are the different ways to value a landscape or a species? How about a human life? The continuation of a local language? The value of a safe walking environment?

One long-standing economic and policy question is the value of labor. Is labor to be valued according to its market price? If there are lots of interchangeable and low-skill workers, any given laborer faces pressure to accept the lowest wage. Or is the value of labor to be measured by the value added in the production process. If a worker spends an hour on building a chair and is paid $15 per hour, but the chair is sold for $250, who collects that profit? The entire system of capitalism relies on the difference between these two values, which is extracted by the owners of productive enterprises, the capitalists, in the form of profits. Very often, laborers argue for a larger share of that profit, and unions and collective bargaining are one mechanism for addressing that gap. But "quiet quitting" (doing only the bare minimum required at one's job) reflects a perceived devaluation of a worker's labor and is an individualistic attempt to recapture part of that extracted surplus.[55] Government policy, especially taxation of profits and income, but also business regulations about hours, minimum wages, and health and retirement benefits, are important mechanisms to address that difference between labor market and value-added pricing for labor. Of course, employee-owned business skip this problem entirely.[56] The relevance of questions about the value of labor will be more clear in later chapters when we talk about technological change and work.

One of the authors has a friend who inherited his grandfather's drill press. What is the value of that machine tool? It has a use-value, the value that it might have in making products for the market, in connection with a position in a commodity chain. The drill press, as an antique, probably has an exchange value, its value to collector. As a family heirloom, the drill press has intrinsic or cultural value that may not be amenable to setting a specific price: it is "priceless." The difference in profits between using the drill press to make money or keeping it on display as an object with intrinsic or cultural value—as a family heirloom—represents an opportunity cost. That is the potential cost of doing, or not doing, something in relation to another activity. For example, the time and money spent on higher education reflects an opportunity cost in relation to entering the labor market immediately following high school. It is thus a matter of the present value of money in relation to its potential future value, and the time frame selected. One might be able to make money immediately out of high school, but long-term earnings for an advanced degree generally outpace lifetime earnings for only a high school degree. That, however, does depend on field, location, and qualifications.[57] And yes, compared to high school diplomas, skilled trades are a significant benefit to lifetime earnings.[58] But what about the intrinsic value of education outside of the labor market? Satisfaction in facing intellectual challenges, enjoyment in learning new things, critical thinking, connections to friends, and gaining a sense of independence are among many of the non-monetary values of higher education. How do you know something, anything, is "worth it" as an undertaking?

Similarly, what is the value of, say, endangered North Atlantic right whales?[59] Since they have no commercial value as a species due to whaling bans, how do we find a way to include its worth in calculations of the consequences of regulatory policies? If fitting commercial shipping in the North Atlantic with sensors and requiring speed controls to save whales will cost some specific number of dollars in technology and productivity, how do we "set" the value of any given whale, or the viability of the species in such a way as to incentivize whale-saving activities? Assessing the intrinsic or cultural value of things, as opposed to calculated economic values, is very difficult to incorporate into commodity chain analysis, which despite its strengths as a tool, is only as good as the assumptions built into it.

Figure 8.8 How do we assess things that are priceless? (Source: CALVIN AND HOBBES © 1995 Watterson. Reprinted with permission of ANDREWS MCMEEL SYNDICATION. All rights reserved.)

QUESTIONS FOR DISCUSSION

1. The continual extraction and use of natural resources does not result in their being completely "used up," but eventually the costs of extracting a diminishing resource exceed the value of the resource; barring improvements in extraction technologies, the resource is as good as gone. This may not happen for a long time, perhaps not until we as individuals are long departed from this earth. Is there anything immoral about using large quantities of Earth's resources for our own benefit? Do we owe anything to future generations? If so, by how much should we restrict our use of resources? How should these restrictions be mandated?
2. What is your most prized possession? What is it *worth* to you if someone tried to buy it from you? Is it replaceable?
3. One of the decisions to be made in applying a value-chain analysis is to consider the boundaries of the system in time, and how far in space to trace the consequences of developing a technology. How do electric car batteries exemplify this choice? How does the effectiveness and desirability of electric vehicles change in comparison to public transportation or traditional internal combustion engines if you keep a narrow scope of study? What about if you look broadly at the environmental and social impacts of batteries in comparison to the environmental and social impacts of oil extraction and burning fossil fuels?
4. What are the environmental and social impacts of the entire life cycle of a cell phone from procurement and production, to use and discard? Does knowing that change how you intend to

interact with the life cycle? What could be done, technologically and otherwise, to make the standard life cycle more environmentally sustainable?

5. How can we account for the value that open space, stable wild species, and healthy ecosystems have now and in the future? What kinds of new technologies can be imagined engaging those values? What technologies might we have to abandon?

Notes

1. Chocolate is called *xocolatl* in Aztec, while the Mayan word for "cacao" is *KaKaWa* and the chocolate drink is called *chocolhaa* (bitter water). See www.iaf.gov/content/story/kakawa-chocolhaa-xocolatl-cacaoit-all-tastes-like-chocolate.
2. See Frédéric Sellet, "Chaine Operatoire; The Concept and its Applications," *Lithic Technology* 18, 1/2 (1993): 106–12; and Pierre Lemonnier, "Introduction," in Pierre Lemonnier (Ed.), *Technological Choices: Transformation in Material Cultures Since the Neolithic.* (London: Routledge, 1992), pp. 1–35.
3. Michael B. Schiffer, "Behavioral Chain Analysis: Activities, Organization, and the Use of Space," *Fieldiana. Anthropology* 65 (1975): 103–119.
4. I. Wallerstein, *The Modern World-System: Capitalist Agriculture and the Origins of the European World-Economy of the Sixteenth Century* (Cambridge: Cambridge University Press, 1974).
5. Hopkins and Wallerstein, 1986: 159; from T. K. Hopkins and I. Wallerstein, "Commodity Chains in the World-Economy Prior to 1800. *Review* 10, 1 (1986): 157–170. As quoted in Joonkoo Lee, "Global Commodity Chains and Global Value Chains," *Oxford Research Encyclopedia of International Studies.*
6. Gereffi and Korzeniewicz, (Eds.), *Commodity Chains and Global Capitalism* (Westport, CT: Greenwood/Praeger, 1994).
7. Examples of different approaches: Ashim Joshi, Raghav Sharma, and Bivek Baral, "Comparative Life Cycle Assessment of Conventional Combustion Engine Vehicle, Battery Electric Vehicle and Fuel Cell Electric Vehicle in Nepal," *Journal of Cleaner Production* 379 (2022): 134407; For value chains see: Simon Bolwig, Stefano Ponte, Andries Du Toit, Lone Riisgaard, and Niels Halberg, "Integrating Poverty and Environmental Concerns into Value-Chain Analysis: A Conceptual Framework," *Development Policy Review* 28 (2010): 173–194; For a cradle-to-grave assessment: Iris Vural Gursel, et al., "Comparative Cradle-To-Grave Life Cycle Assessment of Bio-Based and Petrochemical PET Bottles," *Science of The Total Environment*, 793 (2021): 148642.
8. Joaquim Sanz, Oriol Tomasa, Abigail Jimenez-Franco & Nor Sidki-Rius, "Gallium (Ga) [Z=31]. In *Elements and Mineral Resources* (Basel, Switzerland: Springer Nature Switzerland).

9. "Gallium Statistics and Information," by National Minerals Information Center, for access to commodity summaries. Accessible from www.usgs.gov/centers/national-minerals-information-center/gallium-statistics-and-information.
10. N. H. Abdullah et al., "Potential Health Impacts of Bauxite Mining in Kuantan," *Malaysian Journal of Medical Sciences* 23, 3 (2016): 1–8.
11. Sebastián Ibarra-Gutiérrez, Jocelyn Bouchard, Marcel Laflamme, and Konstantinos Fytas, "Project Economics of Lithium Mines in Quebec: A Critical Review," *The Extractive Industries and Society* 8 (2021): 100984. See also: www.nrcan.gc.ca/our-natural-resources/minerals-mining/minerals-metals-facts/lithium-facts/24009
12. Katelyn Weisbrod, "Snapshots, Hotshots and Moonshots: Images of Climate Change in 2022," *Inside Climate News* (December 31, 2022). Accessed from https://insideclimatenews.org/news/07112021/lithium-mining-thacker-pass-nevada-electric-vehicles-climate/ 12/29/2022.
13. Maeve Campbell, "In Pictures: South America's 'Lithium Fields' Reveal the Dark Side of Our Electric Future," *EuroNews.Green* (January 2, 2022). Accessed from www.euronews.com/green/2022/02/01/south-america-s-lithium-fields-reveal-the-dark-side-of-our-electric-future.
14. See: Jorge S. Gutierrez, et al., "Climate Change and Lithium Mining Influence Flamingo Abundance in the Lithium Triangle," *Proceedings of the Royal Society B* 289, 1970 (2022).
15. Victoria Flexer, Celso Fernando Baspineiro, and Claudia Inés Galli, "2018 Lithium Recovery from Brines: A Vital Raw Material for Green Energies with a Potential Environmental Impact in Its Mining and Processing. *Science of The Total Environment* 639 (2018): 1188–1204.
16. Alex Scott, "Challenging China's Dominance in the Lithium Market," *C&EN* (October 29, 2022). Accessed from https://cen.acs.org/energy/energy-storage-/Challenging-Chinas-dominance-lithium-market/100/i38#.
17. Carlito Baltazar Tabelin, et al., "Towards a Low-Carbon Society: A Review of Lithium Resource Availability, Challenges and Innovations in Mining, Extraction and Recycling, and Future Perspectives," *Minerals Engineering*, 163 (2020): 106743.
18. Patricia Nelson. Limerick, "The Gold Rush and the Shaping of the American West." *California History* 77, 1 (1998): 30–41. See also: /https://americanexperience.si.edu/wp-content/uploads/2015/02/The-Gold-Rush-and-Westward-Expansion.pdf.
19. R. F. Dasmann, "Environmental Changes Before and After the Gold Rush," Chapter 5, pp. 105–121 in J. Rawls, and R. Orsi (Eds.), (1999). *A Golden State: Mining and Economic Development in Gold Rush California* (Berkeley, CA: University of California Press). See also: https://education.nationalgeographic.org/resource/after-gold-rush 12/29/2022.
20. Kemel Toktomushev and Saipira Furstenberg, "Understanding Gold Mining and Social Conflicts in Kyrgyzstan," (June 30, 2021). University of Central

Asia, Institute of Public Policy and Administration (IPPA) Working Paper No. 63.

21. José Luis Rivera-Parra, Bernardo Beate, Ximena Diaz, and María Belén Ochoa, "Artisanal and Small Gold Mining and Petroleum Production as Potential Sources of Heavy Metal Contamination in Ecuador: A Call to Action," *International Journal of Environmental Research and Public Health* 18, 6 (2021): 2794.
22. Jeff Tollefson, "Illegal mining in the Amazon hits record high amid Indigenous protests." *Nature* 598 (2021): 15–16. See also: Scott Wallace, "An Illegal Gold Rush is Igniting Attacks on Indigenous People in the Amazon," *National Geographic*. (July 6, 2021). Accessed from www.nationalgeographic.com/history/article/war-on-indigenous-amazon-communities-in-brazil.
23. L .T. Ogundele, O. A. Oluwajana, and A.C. Ogunyele, et al., "Heavy Metals, Radionuclides Activity and Mineralogy of Soil Samples from an Artisanal Gold Mining Site in Ile-Ife, Nigeria: Implications on Human and Environmental Health," *Environmental Earth Sciences* 80, 202 (2021).
24. Philip Edelman, (1990). "Environmental and Workplace Contamination in the Semiconductor Industry: Implications for Future Health of the Workforce and Community," *Environmental Health Perspectives* 86(1990): 291–295. For an example of a company's MSDS (Material Safety Data Sheet) identifying the safety issues of GaAs see: www.sigmaaldrich.com/US/en/sds/ALDRICH/651486.
25. Briony Harris, "The Astonishing Rise of Shenzhen, China's Gadget Capital," *World Economic Forum* (November 9, 2017). Accessed from www.weforum.org/agenda/2017/11/inside-shenzhen-china-s-gadget-capital; and Will Yakowicz, "5 Things to Know About Doing Business in Shenzhen," *Inc.* (Februrary 24, 2015). Accessed from www.inc.com/will-yakowicz/shenzhen-city-of-electronics.html.
26. He Huifeng, "Shenzhen's Population Size, Key to Gauging Economic Development, Remains a Matter of Debate," *SCMP* (November 26, 2020). Accessed from www.scmp.com/economy/china-economy/article/3111332/shenzhens-population-size-key-gauging-economic-development.
27. Malcolm Moore, "'Mass Suicide' Protest at Apple Manufacturer Foxconn Factory," (January 11, 2012). Accessed from www.telegraph.co.uk/news/worldnews/asia/china/9006988/Mass-suicide-protest-at-Apple-manufacturer-Foxconn-factory.html.
28. Brian Merchant, "Life and Death in Apple's Forbidden City," *The Guardian* (June 18, 2017). Accessed from www.theguardian.com/technology/2017/jun/18/foxconn-life-death-forbidden-city-longhua-suicide-apple-iphone-brian-merchant-one-device-extract
29. Amy Chang Chien et al., "In a Challenge to Beijing, Unrest Over Covid Lockdowns Spreads," *The New York Times* (November 24, 2022). Accessed from www.nytimes.com/2022/11/24/world/asia/china-unrest-covid-lockdowns.html.
30. Don Clark and Kellen Browning, "In Phoenix, a Taiwanese Chip Giant Builds a Hedge Against China," *The New York Times* (December 6, 2022).

Accessed from www.nytimes.com/2022/12/06/technology/tsmc-chips-factory-phoenix.html; and see Adam Bruns, "Valley of the Sun," Site Selection Magazine (November 2022). Accessed from https://siteselection.com/issues/2022/nov/valley-of-the-sun.cfm.

31. Willy Shih, "Intel's $88 Billion European Expansion Is Part of a New Phase in the Globalization of the Semiconductor Industry," *Forbes* (March 15, 2022). Accessed from www.forbes.com/sites/willyshih/2022/03/15/intels-european-expansion-regionalization-of-the-semiconductor-industry/?sh=363b509c27c9.
32. Justine Calma, "Water Shortages Loom over Future Semiconductor Fabs in Arizona," *The Verge* (August 18, 2021). Accessed from www.theverge.com/22628925/water-semiconductor-shortage-arizona-drought.
33. Pádraig Belton, "The Computer Chip Industry Has a Dirty Climate Secret," *The Guardian* (September 18, 2021). Accessed from www.theguardian.com/environment/2021/sep/18/semiconductor-silicon-chips-carbon-footprint-climate.
34. Jackie Northam, "Shipping Industry Is Pressured to Cut Pollution Caused by Merchant Fleet," *NPR* (December 1, 2021). Accessed from www.npr.org/2021/12/01/1060382176/shipping-industry-is-pressured-to-cut-pollution-caused-by-merchant-fleet.
35. L. I. Ward-Geiger, G. K. Silber, R. D. Baumstark, and T. L. Pulfer, "Characterization of Ship Traffic in Right Whale Critical Habitat," *Coastal Management* 33, 3 (2005): 263–78; and S. Taylor and T. R. Walker "North Atlantic Right Whales in Danger," Science 358 (2017): 730–31.
36. Paul Kirvan, "How Much Energy Do Data Centers Consume?" *TechTarget* (April 25, 2022). Accessed from www.techtarget.com/searchdatacenter/tip/How-much-energy-do-data-centers-consume; and for crypto energy use: Jon Huang, Claire O'Neill, and Hiroko Tabuchi, "Bitcoin Uses More Electricity Than Many Countries. How Is That Possible? *The New York Times* (September 3, 2021). Accessed from www.nytimes.com/interactive/2021/09/03/climate/bitcoin-carbon-footprint-electricity.html.
37. Numbers from a report released from Custom Market Insights titled, "Global Refurbished and Used Mobile Phones Market Size/Share Worth 146.43 Billion by 2030 at an 11.45% CAGR: Custom Market Insights (Analysis, Outlook, Leaders, Report, Trends, Forecast, Segmentation, Growth)."
38. See: "Importance of Cell Phone Recycling," at https://getrepowered.org/importance-of-cell-phone-recycling; and Victoria Gill, "E-Waste: Five Billion Phones to Be Thrown Away in 2022," *BBC News* (October 14, 2022). Accessed from www.bbc.com/news/science-environment-63245150.
39. See: S. Needhidasan, M. Samuel, and R. Chidambaram, "Electronic Waste - An Emerging Threat to the Environment of Urban India. *Journal of Environmental Health Science and Engineering* 12, 1 (2014): 36; and see: E. C. Merem, et al., "Analyzing the Environmental Risks from Electronic Waste Dumping in the West African Region," *Journal of Health Science* 11, 1 (2021): 1–16.

40. R. P. de Oliveira, J. Benvenuti, and D.C.R. Espinosa, " A Review of the Current Progress in Recycling Technologies for Gallium and Rare Earth Elements from Light-Emitting Diodes," *Renewable and Sustainable Energy Reviews* 145 (2021): 111090.
41. Shahjadi Hisan Farjana and Wen Li, "Integrated LCA-MFA Framework for Gold Production from Primary and Secondary Sources," *Procedia CIRP*, 98 (2021): 511–516.
42. Malek Kamal Hussien Rabaia, et al., "Environmental Impacts of Solar Energy Systems: A Review," *Science of The Total Environment*, 754 (2021): 141989; and V. Raja Sekhar and P. Pradeep, "A Review Paper on Advancements in Solar PV Technology, Environmental Impact of PV Cell Manufacturing" *International Journal of Advanced Research in Science, Communication and Technology* 8, 1 (2021): 485–492.
43. See https://sciencing.com/factories-cause-air-pollution-5169.html for plastics as a petrochemical industry output.
44. "Corporate History," *Mercedes-Benz*. Accessed from www.mercedes-benz.com/en/innovation/milestones/corporate-history.
45. Mary Bellis, "History of the Automobile: The Assembly Line," *ThoughtCo* (September 24, 2018). Accessed from www.thoughtco.com/history-of-car-assembly-line-4072559.
46. Natasha Ishak, "Haunting Photos of Gary, Indiana—'The Most Miserable City in America,'" *All That's Interesting*. Accessed from https://allthatsinteresting.com/gary-indiana.
47. "A Brief History of the American Steel Industry," *National Material Company* (January 24th, 2018). Accessed from www.nationalmaterial.com/brief-history-american-steel-industry.
48. List of automobiles manufactured in the United States: https://en.wikipedia.org/wiki/List_of_automobiles_manufactured_in_the_United_States
49. Sanjoy Kumar Paul, Priyabrata Chowdhury, Md. Abdul Moktadir, Kwok Hung La, "Supply Chain Recovery Challenges in the Wake of COVID-19 Pandemic," *Journal of Business Research* 136 (2021): 316–329.
50. OECD, "Monopsony and Buyer Power." Accessed from www.oecd.org/daf/competition/44445750.pdf
51. Justin Wiltshire, "Walmart Is a Monopsonist That Depresses Earnings and Employment Beyond Its Own Walls, but U.S. Policymakers Can Do Something About It," *The Washington Center for Equitable Growth* (March 29, 2022). Accessed from https://equitablegrowth.org/walmart-is-a-monopsonist-that-depresses-earnings-and-employment-beyond-its-own-walls-but-u-s-policymakers-can-do-something-about-it/; and Mark Stater and Michael S. Visser, "Implications of Big Box Retail Location on Regional Profits, Consumer Utility, and Land Rents," *Review of Regional Studies* 38, 1 (2008): 8–28.
52. Kimon Valaskakis, et al., *The Conserver Society: A Workable Alternative for the Future* (New York: Harper & Row, 1979), p. 166.

53. Joseph Stromberg, "The Forgotten History of How Automakers Invented the Crime of 'Jaywalking,'" *Vox* (November 4, 2015). Accessed from www .vox.com/2015/1/15/7551873/jaywalking-history.
54. James Flink, *The Car Culture* (Cambridge, MA: MIT Press, 1976).
55. Kara Dennison, "Quiet Quitting Is the Natural Response to Toxic Workplaces," *Forbes* (September 28, 2022). Accessed from www.forbes.com/sites/karadennison/2022/09/28/quiet-quitting-is-the-natural-response-to-toxic-workplaces/?sh=5248ea4c7429.
56. Rachel Merfalen, "Employee Ownership: A Good-For-Business Tool That Creates Equity and Opportunity for Workers," *San Diego Workforce Partnership* (May 26, 2021). Accessed from https://workforce.org/news/employee-ownership-a-tool-for-san-diego-to-create-equitable-economic-opportunity.
57. Anthony P. Carnevale, Stephen J. Rose, and Ban Cheah, "The College Payoff: Education, Occupations, Lifetime Earnings," *The Georgetown University Center on Education and the Workforce*. Accessed from www2.ed.gov/policy/highered/reg/hearulemaking/2011/collegepayoff.pdf.
58. Jon Marcus, "Long Disparaged, Education for the Skilled Trades is Slowly Coming into Fashion," *The Hechinger Report* (December 31, 2021). Accessed from https://hechingerreport.org/long-disparaged-education-for-the-skilled-trades-is-slowly-coming-into-fashion.
59. Barbara Moran, "North Atlantic Right Whale Population Continues to Shrink, According to Latest Report," *WBUR* (October 24, 2022). Accessed from www.wbur.org/news/2022/10/24/north-atlantic-right-whale-population-decline.

CHAPTER 9

Medical Technologies

Some may doubt the overall benefits of technological advance, but almost everyone would agree that improvements in medical technologies have made lives better. Who would want to go back to the not-so-good old days, when vast numbers of children died in infancy, epidemic plagues wiped out millions, and pain and suffering were an inevitable part of everyday life? Not too long ago, medicine's ability to cope with sickness was severely limited at best. In some cases, medical intervention only made bad situations even worse. The knowledge base of medical technology was pathetically underdeveloped, rendering the many therapies ineffective or worse. As Lewis Thomas describes this unhappy situation:

> Bleeding, purging, cupping, the administration of infusions of every known plant, solutions of every known metal, every conceivable diet including total fasting, most of these based on the weirdest imaginings about the cause of disease, concocted out of nothing but thin air—this was the heritage of medicine up until a little over a century ago.[1]

More recently, technological advance supported by an expanding knowledge base has extended the lives of many, while eliminating or at least reducing a great deal of agony. How could anyone quibble with such positive results? And yet, it should be apparent by now that, in aggregate, technological advance is never an unmixed blessing. Problems are solved or at least alleviated, but at the cost of a new set of difficulties, many of them unforeseen. Medical technology is no exception. As we shall see, for all its unquestioned benefits, technological advance has generated some vexing problems. Understanding the nature of these problems is important in its own right, and at the same time it may help us come to a deeper understanding of how technological advance can take with one hand as it gives with another.

In assessing the significance of modern medicine, it is important to bear in mind that not all historical improvements in mortality and morbidity are the result of advancing medical technologies. During the past 200 years, dramatic advances have taken place in medical care: immunization for the prevention of infectious diseases, antiseptic surgery, anesthetics, antibiotics, and the transplantation of organs. Breakthroughs in safe vaccine production have likely saved 10 million lives since 1962, with approximately 1.5 million childhood deaths from polio avoided.[2] But most of the gains in life span and improvements in the physical quality of life have been due to better nutrition, sanitation, and personal hygiene, along with higher standards of living in general.[3]

Recent decades have seen the development and application of many new medical treatments, but it is not at all certain that they have always had their intended effect. In fact, there may even be a negative correlation between the use of some of these treatments and patient health. American medicine is considered especially aggressive. This is strongly suggested by a recent study of Medicare patients conducted by researchers at Dartmouth University's Medical School. The study found considerable regional variations in the treatment of patients with serious illnesses during the last two years of their lives, with aggressive medical treatments being much more common in some parts of the country than in others. Paradoxically, however, patients in these regions were less satisfied with their care, ended up no better in terms of day-to-day function, and died at a slightly higher rate than patients in regions with less aggressive care. There are a number of reasons for this anomaly, but chief among them is the higher concentration of medical specialists in regions that have the most aggressive medical practices. Medical care in regions with large proportions of specialists is characterized by an excessive use of tests and procedures, extended periods of hospitalization, and a fragmented approach to patient care that sharply contrasts with the holistic approach of primary-care physicians.[4]

None of this is to say that technological advances in medicine have been of no value. Many people are alive today because of kidney dialysis, computerized tomography (CT), and antibiotics.

Modern medical technologies have generated new possibilities and new hopes. Instead of accepting sickness and death as the inevitable product of fate or God's will, we increasingly expect that cures will be found, and that they will be made available when we need them. But along with rising expectations come rising costs. We do not have to suffer and die quietly (and inexpensively); new medical technologies allow life to be extended for long periods of time. As a result, medical advances and their associated costs have the potential to expand without limit, for as long as people are subject to sickness and death, there will be no lack of demand for new medical interventions. And unfortunately, the cost of medical care is sometimes weighed against other possible expenditures. You may decide to defer the purchase of a new car to make a down payment on a house. Deferral of a quadruple bypass operation or insulin rationing[5] are dismal choices when they are all that stands between you and life or death.

While the costs of medical care have been escalating, in the U.S., a mix of government and private insurance programs pay for the bulk of medical expenditures, a trend that has been reinforced by the implementation of the Affordable Care Act ("Obamacare"). In 1960, one-half of U.S. healthcare costs were directly borne by patients and their families. The years since then have seen a major expansion of private insurance plans, as well as the creation and massive expansion of government-supported Medicare and Medicaid programs. Private insurance plans require either employer or individual contributions in some combination: in 2020, the average cost was $1,152 for a family policy, per month.[6] In 2021, of the $4.3 trillion that went for personal health care expenditures, private insurance plans paid out $1.2 trillion (28 percent of the total). Other payers accounted for $596.6 billion (14 percent of expenditures). Government programs have assumed a significant part of financial burden; Medicare and Medicaid were responsible for $900.8 billion (21 percent of the total) and $734 billion (17 percent of the total), respectively. The out-of-pocket total was $433.2 billion billion (10 percent).[7]

In other economically advanced democracies, health care costs are much lower for far more effective health care systems, due to the prevalence of universal health care.[8] Where health care is considered a human right, rather than a privilege associated with employment or a matter of private expenditure, costs are less, outcomes are better, health disparities fewer, and access to primary care—the foundation of health access—are improved.[9] Removing the profit motive from medicine makes preventive care more fiscally attractive than late, expensive treatments and improves trust in medicine.

More medical treatments are available than was the case 50 years ago. Under these circumstances, there has been a strong tendency to provide increasing amounts of medical care and for a country's medical costs to increase with no apparent endpoint. As one analyst put it, "When benefits do not have to be weighted against costs—when the only criterion is that there be some benefit—the number of good things that can be done in medical care is, for all practical purposes, unlimited."[10] All of these "good things" have created a situation in which 40 to 50 percent of the increase in medical costs over recent decades can be attributed to new medical technologies and the increased use of preexisting treatments.[11] Although it is difficult to do so, especially when life and well-being are at stake, choices eventually have to be made. Increasingly sophisticated medical technologies offer many benefits, but they can also generate costs that threaten to spiral out of control. Items of high-technology medicine should not be applied indiscriminately. Like any other claim on financial resources, they need to be evaluated in terms of the benefits they provide and the costs they incur. A useful starting point for evaluating the appropriateness of particular medical technologies has been provided by Bryan Jennett, a British neurologist. According to his analysis, a medical technology can be deemed inappropriate for a number of different reasons:[12]

- **Unnecessary**: The patient has a condition too advanced to respond to treatment, as would be the case with attempting an extended treatment of metastatic cancer.
- **Unsafe**: Complications outweigh the probable benefit, as with some invasive investigations and dangerous therapies.
- **Unkind**: The quality of life after treatment is not good enough or its duration long enough to have justified the intervention.
- **Unwise**: It diverts resources from activities that would yield greater benefits to other patients.

These are useful analytical categories, but in actual practice it may be very difficult to determine which of these categories apply to particular cases. Doctors still need to have the ability to make accurate prognoses, and compassion always has to be combined with informed judgment.

New Medical Technologies: Choices and Trade-Offs

It is not always easy to scrutinize a particular medical technology in terms of costs, benefits, and general appropriateness—especially when it applies to us or someone close to us. Still, it is worth making the effort, if only to better comprehend the dilemmas often posed by the advance of medical technologies. The brief studies that follow should help to make the key issues more concrete.

The Case of Kidney Dialysis

Sophisticated technologies may be of limited importance to the population as a whole, but for many individuals they are literally the difference between life and death. The benefits of recent technological advances are nowhere more evident than they are to people suffering from kidney disease. In a healthy individual, the kidneys regulate the body's acid–base concentration, maintain proper water balance, and concentrate metabolic wastes that are subsequently excreted as urine. Complete or substantial kidney failure (known as "end-stage renal disease") usually results in incapacitation and premature death. Many are afflicted: Renal diseases are the United States' fourth-largest killer, right behind cardiovascular diseases, cancer, and pneumonia.

In the early 1940s, a Dutch physician used a bathtub and parts salvaged from a foundry to construct the first device to successfully take the place of the kidneys. The inventor, Dr. Willem Kolff, later immigrated to the United States, where his device inspired the development of more refined versions at a number of hospitals. At that time, cleansing the blood by means of an artificial kidney (a process known as dialysis) was confined to short periods of time. Use of dialysis as a substitute for functioning kidneys became possible in the early 1960s through equipment improvements and the invention of a connecting tube that obviated the need to tap into a new artery and vein every time that a patient was hooked up to the machine. This made long-term dialysis a practical proposition, but at the same time, it generated a host of nonmedical problems that have yet to be resolved.

Figure 9.1 An early Kolff-Brigham Artificial Kidney (est. 1946). Kidney dialysis has become a common procedure that has saved many lives while at the same time contributing to rising medical costs. (Source: Rob Koopman/CC BY-SA 2.0)

When dialysis first became an accepted medical practice, the number of patients who could potentially benefit from it far exceeded the number of available machines. It was therefore necessary to select some patients for dialysis and to reject others. To make these choices, the nation's pioneering dialysis institution, the Seattle Artificial Kidney Center, established an Admissions and Policy Committee to screen applications and determine who would get dialyzed and who would not. The committee was intended to reflect the community as a whole, being initially comprised of a lawyer, a minister, a housewife, a labor leader, a government official, a banker, and a surgeon, as well as two physician-advisers.[13] The members of the committee made their first selective cut by accepting patients only from Washington State, and by eliminating children along with adults over the age of 45. Having done this, the committee then applied a set of criteria that took into account the personal characteristics of prospective patients. According to one report, these included the "sex of patient, marital status, and number of dependents; income and net worth; emotional stability, with regard to patient's ability to accept the treatment; educational background; nature of occupation; past performance and future potential; and names of people who could serve as references."[14]

As might be expected, making life-or-death decisions on the basis of the presumed economic worth of the patient generated a fair amount of indignation in some quarters. According to two critics, a psychiatrist and a lawyer, the published accounts of the selection criteria "paint a disturbing picture of the bourgeoisie sparing the bourgeoisie, of the Seattle committee measuring persons in accordance with its own middle-class suburban value system: Scouts, Sunday school, Red Cross. This rules out creative nonconformists, who rub the bourgeoisie the wrong way but who historically have contributed so much to the making of America. The Pacific Northwest was no place for a Henry David Thoreau with bad kidneys.[15]

The problem of controlling access to a potentially lifesaving technology was mitigated as the number of machines increased and they were made easier to operate. It even became possible for a patient to routinely dialyze himself or herself at home. But formidable cost problems remained. In 1970, a twice-weekly dialysis cost $3,000 to $5,000 and required start-up expenditures of $9,000 to $13,000. For most patients, these expenses produced grave financial pressures, and for some, they were impossible to meet. The problem was solved in 1972 when the U.S. Congress authorized payment for dialysis treatments through the federal government's Medicare insurance program.

The process whereby this policy was enacted illustrates the capricious way in which technologies are sometimes supported. The government's assumption of payments for dialysis treatments was in large measure the result of intense lobbying, an effort that was made all the more effective by connecting an abstract ailment with actual people. In one instance, the vice president of the National Association of Patients on Hemodialysis and Transplantation not only testified but dialyzed himself in the presence of members of the House Ways and Means Committee.[16] The provision of Medicare payments for dialysis was only a small portion of a larger piece of legislation; only 30 minutes of debate on the floor of the Senate took place, culminating in a lopsided vote in support of the measure at a time when close to half of the senators were absent. The conference committee of the House and the Senate took only 10 minutes to discuss the measure, which was only slightly modified as a result.

Both houses subsequently accepted the conference committee's report, and the bill including the dialysis provision was signed into law by President Nixon two weeks later.[17]

Payment for dialysis through the Medicare program addressed a real need, but at considerable cost. Initial estimates of annual expenditures of $135 to $250 million were far too low. In 2014, annual hemodialysis costs in the United States came to $42 billion; $34 billion of this sum was absorbed through the Medicare budget. The rest was either covered by Medicaid, private insurance, or paid by the patient.[18] Of course, it is not some abstract entity that we call the federal government that pays for most of the costs of dialysis; these expenditures are ultimately borne by individual taxpayers, the vast majority of whom are not afflicted by end-stage renal disease. There is nothing inherently unreasonable or unfair about this. A catastrophic organ failure could happen to any of us, and we shouldn't begrudge helping people who have been less fortunate than ourselves. Still, questions remain about the equity of paying for dialysis and not for other serious ailments, especially since people suffering from end-stage renal disease comprise 1 percent of Medicare enrollees while absorbing 7 percent of the Medicare budget.[19]

Today, about one-third of the patients receiving dialysis are over the age of 65. As the population ages, it can be expected that the demand for dialysis will increase apace, putting further financial strains on the health care system. This may not occur in some countries where government insurance programs do not always cover dialysis for patients with other serious medical problems. In Great Britain, the government-run National Health Service always has operated under tight financial constraints. Consequently, certain treatments have been restricted, dialysis included. During the early years of dialysis treatment, patients over the age of 55 were rarely given the opportunity to receive dialysis because it was believed that their physical constitutions would not allow them to survive and flourish under a regimen of daily dialysis. One British physician candidly stated that people over the age of 55 were not suitable candidates, for they were all "a bit crumbly."[20] This may be true in a statistical sense, but in fact, there is greater variation in overall levels of health among the elderly than there is in any other age group. Using age as a means of limiting access to an expensive medical technology is administratively convenient, but it dodges the issue of making choices on the basis of more relevant criteria. In any event, in Great Britain, increasing numbers of people over the age of 65 are now receiving dialysis. As in the United States, dialysis takes a disproportionate share of the overall medical budget; the cost of providing dialysis for 0.05 percent of the population with end-stage renal disease absorbs about 1 to 2 percent of the National Health Service budget in the UK.[21]

Is the American policy of giving universal access to dialysis the correct one, or is the former British policy more defensible when financial resources are limited? There are no easy answers. At the same time, the problems of distributive justice posed by the use of artificial kidneys are only a preview of what will have to be faced as new and even more expensive medical technologies become available and medicine gains the ability to successfully address hitherto untreatable conditions.

Replacing Broken Hearts

The dilemmas engendered by the advance of medical technologies are further illustrated by the prospect of using new technologies to counteract the number one killer in the United States—heart disease. In recent decades, many victims of clogged arteries and weakened hearts have gained a new lease on life through coronary artery bypass grafts. This procedure is now routine in many hospitals, but once again, questions of cost and benefits have to be confronted.

Open-heart surgery became feasible in the early 1960s with the development of devices that could take over the heart's functions during surgery. By 1970, the efficacy of these procedures was attested to in several papers appearing in medical journals. Over the following years, bypass surgery became a common surgical intervention. In bypass surgery, a segment of a vein is removed from the leg or chest and spliced into one or more of the five coronary arteries that transport blood from the heart. In extreme cases, all five arteries may receive this treatment. This procedure was effective in reducing early mortality, but only for a limited amount of time and primarily for patients with generally poor health conditions. Negative findings, however, did not alter the incidence of bypass surgery; since 2017, more than 160,000 of these operations were performed annually at a total cost of more than $6.7 billion.[22]

For patients with less severe blockages, percutaneous transluminal coronary angioplasty, commonly known as "balloon angioplasty," has been widely employed. This procedure threads a tiny balloon into a coronary artery to break up a blood clot that is obstructing the flow of blood to the heart. This procedure has saved the lives of many men and women who would otherwise have succumbed to heart failure. The total cost of approximately one million annual angioplasties amounted to more than $10 billion in 2014.[23]

Although many lives have been saved through these procedures, questions have been raised about their possible overuse. One study of angioplasties performed on a sample of 828 Medicare patients found that only about one-third of the patients benefited from the procedure. It was deemed completely inappropriate for 14 percent, and it could not be determined whether or not the remaining patients were helped by the procedure.[24] These unimpressive results were mirrored in another study conducted in 2007, when 2,287 patients with stable coronary artery disease were divided into two groups. About half underwent balloon angioplasty, while the others received nonsurgical medical therapy that entailed the administration of statins (to reduce cholesterol), beta-blockers, diuretics (to lower blood pressure), and aspirin or other blood thinners (to prevent excessive clotting). The latter group had more chest pains (which could be controlled by medication), but there was hardly any difference between the two groups in the frequency of strokes and heart attacks.[25]

Although this study seriously called into question the massive employment of balloon angioplasty with its attendant financial costs and patient discomfort, it seems to have had limited influence on actual practices. The ratio of balloon angioplasty operations to nonsurgical interventions hardly changed after the publication of the study. In summarizing their results, the authors of the study noted that more attention needed to be paid to nonsurgical treatments: "These findings represent a significant opportunity for improvement and a limited effect of an expensive, highly publicized clinical trial on routine clinical practice."[26]

For many cardiac patients, even effective and appropriate angioplasties or bypass operations are of no use because the heart has been so weakened as to be barely functional and is in danger of failing completely. The transplantation of another person's heart offers one solution, but demand exceeds the supply. Back in the mid-1960s, the implantation of artificial hearts seemed to offer a way out of this impasse. At that time, a consulting group reporting to the federal government assumed that the major technical impediments were well on the way toward solution, and that before too long, many people would have a diseased heart replaced by a mechanical one. According to this optimistic scenario, recipients of these hearts would be able to return to the work force, adding $19 billion to the gross national product over a 10-year period. It was even asserted that the taxes paid by artificial heart recipients would more than offset the federal government's expenses in supporting the program.[27]

In the 1960s, these hopes seemed close to realization. Artificial hearts began to be used for brief periods of time while patients were awaiting a heart transplant, and efforts to develop a permanent artificial heart accelerated. An apparent breakthrough occurred in 1982 when the Jarvik-7 artificial heart was used for the first time. Unfortunately, the first recipient of the device, a Seattle dentist named Barney

Figure 9.2 The AbioCor Replacement Heart (in use 2001–2009). (Source: Album/Alamy Stock)

Clark, underwent great suffering as a result of the immaturity of this technology. Seizures, severe nosebleeds, pneumonia, kidney disease, gout, epididymitis, and an intestinal ulcer followed the implantation of the artificial heart.[28] Clark's postoperative existence for 112 days could hardly justify the expense and pain of the operation. Subsequent implants met with barely any greater success. One patient lived for 620 days, but during that span he suffered four strokes and a series of infections that severely eroded his physical and mental capacities.[29] Despite these setbacks and the bad publicity that attended the use of the Jarvik-7, about 90 more patients received the device in subsequent years. In 1990, the Food and Drug Administration (FDA) banned the Jarvik-7 as a permanent replacement for a human heart because most of the recipients lived for only a few months after its implantation. Artificial hearts are the epitome of high-tech, high-cost medical technology. They can save lives, if only temporarily, but they come with substantial price tags. Should a widespread program of heart replacement become feasible, the costs of implanting artificial hearts, along with subsequent therapy and follow-up care, would surely run into many billions of dollars. A small portion of these expenditures might be recovered when some recipients return to the labor force, but patients might not have this option. Whatever the exact figures might be, extensive use of artificial hearts would make a significant contribution to a national medical bill that already absorbs more than 17.9 percent of the national income of the United States.

Extensive use of these technologies would engender some very difficult choices. The primary beneficiaries would be those who could afford an artificial heart either through insurance coverage or personal resources. Alternatively, some sort of rationing system could be set up, as had occurred during the early years of dialysis. This is clearly a distasteful prospect for the great majority of citizens. If rationing were to be implemented despite public resistance, what sort of criteria might be used? Ominously, criteria could include those invoked by the Seattle Admissions and Policy Committee five decades ago, including the "worth" of a prospective recipient, that is, how much the person had contributed or was likely to contribute to society. Whatever the criteria invoked, the decision to use an expensive medical technology in a world of finite resources necessitates making decisions. In many cases this can be akin to "playing God," a role that few humans are eager to assume.

The selection or exclusion of recipients is not the only equity issue that the use of expensive medical technologies pushes to the forefront. In a world of necessarily limited resources, there is no escaping the fact that an expenditure in one area of health and medicine means the foreclosing of an expenditure in another. The tens of billions of dollars that would be required for the large-scale implantation of artificial hearts could be used to produce significantly higher levels of overall well-being through other means. To cite one example, at least a portion of the funds absorbed by an artificial heart program could instead have been used for antitobacco education campaigns. According to the Centers for Disease Control and Prevention, more than 480,000 deaths can be attributed to cigarette smoking and exposure to cigarette smoke. Smoking has been identified as a major contributor to heart disease; consequently, even a relatively small reduction in tobacco use would produce life-extending benefits considerably in excess of those offered by artificial hearts.[30] But as we have seen, technological fixes can be easier to implement than changes in individual behavior.

In general, the use of high-tech medicine is understandable; physicians are reluctant to take responsibility of deciding when to withhold treatment. Families often want to try any and all interventions to extend life. It would be foolish to deny the value of many advanced medical technologies. But whatever the outcome, there is no getting around the fact that their use contributes to the escalation of medical costs that have been painfully evident in recent decades. The contribution of modern medical technologies to ever-higher medical expenditures is particularly evident during the last year of life, when an average of about 25 percent of Medicare expenditures are incurred.[31] For some critics, the treatment of aged patients with advanced medical technologies is a misallocation of resources; as they see it, much of the money spent on expensive care would be better applied to the elimination of poverty among both the elderly and nonelderly, improving nutritional awareness, or promoting fitness programs.

Diagnostic Technologies

The escalation of medical costs is one of the most problematic consequences of the advance of medical technologies, but hardly the only one. To take another example, a variety of diagnostic technologies have substantially advanced physicians' ability to determine the source of medical problems. At the same time, however, they have altered the nature of medical practice, sometimes with unfortunate results.

For centuries, medical diagnosis had been based on the observation of external symptoms, patients' narratives of their illness, and the application of dubious theories that explained illness in terms of such things as imbalances in four bodily "humors." A major reorientation began in the 1820s when doctors began to use a simple tube to hear the heartbeat and other sounds within the patient's chest. Dubbed the "stethoscope," the new instrument allowed more accurate diagnosis by amplifying and focusing the sounds produced by the circulatory system. Within a few decades, other instruments for apprehending hitherto hidden parts of the body were invented, such as the ophthalmoscope for observing the interior of the eye and the laryngoscope for the throat. A series of sophisticated diagnostic technologies such as X-ray machines and electrocardiographs followed. By the beginning of the twentieth century, physicians were

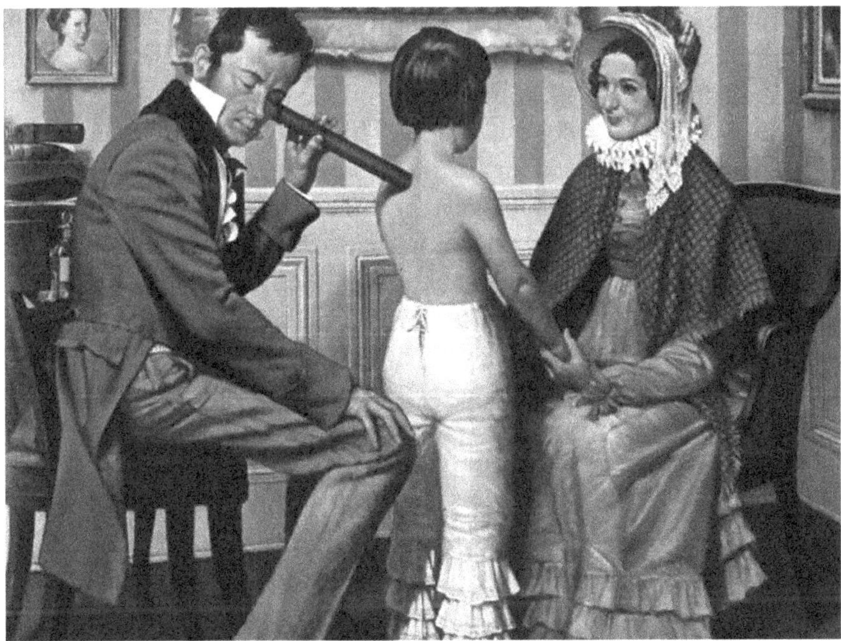

Figure 9.3 Rene Laennec holds an example of his invention, the stethoscope. Dr. Laennac was also a flautist. This device allowed him to maintain a "dignified" distance between himself and his patients.

using instruments for measuring everything from blood pressure to the electronic currents generated by the nervous system. The accuracy of diagnosis was further advanced by the development of laboratory tests for such disparate medical issues as high cholesterol levels and the presence of the organism that causes syphilis.

The use of diagnostic tests exemplifies the tendency of technologies to create their own needs. Many of the tests routinely performed today seem to be unnecessary; according to one study of a teaching hospital, 47 percent of tests performed could have been eliminated with no evident loss in the quality of patient care.[32] One such example is the fetal heart monitor. It has had few discernable benefits; their main consequence seems to have been an increase in the number of deliveries using a cesarean section.[33] To be fair, the increase in lab tests is not solely the result of the availability of these tests. Some of the increase is directly attributable to the practice of "defensive medicine." Doctors are justifiably nervous about the threat of malpractice suits, so the safest course seems to be the overuse of tests to avoid accusations of not having used every available diagnostic tool. To make better decisions about diagnostic technologies, the American Board of Internal Medicine Foundation has established a campaign to better inform physicians and patients about the advisability and inadvisability of a variety of diagnostic tests.[34]

The overuse of tests can produce more than financial pain. According to a past president of the Association of American Physicians, "As our interventions have become more searching, they have also become more costly and more hazardous. Thus, today it is not unusual to find a fragile elder who walked into the hospital, [and became] slightly confused, dehydrated, and somewhat the worse for wear on the third hospital day because his first 48 hours in the hospital were spent undergoing a staggering series of exhausting diagnostic studies in various laboratories or in the radiology suite."[35]

It can be reasonably argued that the physical and financial discomfort produced by diagnostic tests is outweighed by the greatly expanded ability to understand and treat illnesses. In recent years the use of artificial intelligence (AI) for medical diagnosis promises to revolutionize cancer detection. Through the use of a component of AI called Machine Learning, it is now possible to distinguish between healthy tissue, precancerous cells, and cancerous cells with high degrees of accuracy, in some cases exceeding the success rate of trained pathologists.[36]

The benefit of these as well as longstanding diagnostic technologies cannot be denied. But there is another, more subtle consequence of sophisticated diagnostic technologies—they have encouraged the treatment of specific disorders rather than human patients. In many cases, a physician, by examining an X-ray plate or a microscope slide, can produce an accurate diagnosis even when the patient is not physically present; indeed, the physician does not even have to be personally acquainted with the patient. Outsourcing, both local and international, is concerning for quality control and depersonalization questions, although it provides an opportunity for resource-limited areas, such as rural areas, to have access to timely results.[37]

In going about their diagnostic work, few physicians today trouble themselves with obtaining lengthy narratives from their patients regarding their patients' ailments. The physician's personal diagnostic skills are often slighted in favor of more "objective," scientifically based knowledge attained through the use of sophisticated tests and instruments. This is a process that began with the stethoscope; although a very simple device, it changed the nature of the interaction between doctors and their patients. As Stanley Joel Reiser explains, the use of the stethoscope "helped to create an objective physician, who could move away from involvement with the patient's experiences and sensations, to a more detached relation, less with the patient but more with the sounds from within the body."[38]

A similar story could be told of the sphygmomanometer, the ubiquitous instrument used for the measurement of blood pressure. After its introduction in the late nineteenth century, physiologists praised the instrument's capacity to produce objective, quantitative, and precise data. Many practicing physicians saw things differently, however, fearing that the sphygmomanometer would "intervene between patients and doctors, dehumanize the practice of medicine, and replace the delicate art of medicine with mere technical proficiency."[39]

The use of increasingly sophisticated diagnostic technologies illustrates the mixed blessings of technological advances in medicine. On the positive side, far more data can be gathered and analyzed, allowing physicians to identify illnesses that might otherwise have been misdiagnosed, or to catch a disease in its early stage when it can be more easily treated. Standard tests also help to promote a standardized medical practice, removing the biases and blind spots of individual practitioners. Consequently, medicine is more "scientific" and less tied to the variable skills of individual practitioners. At the same time, however, other kinds of medical skills are in danger of being lost. Not every aspect of a functioning organism can be reduced to "objective" data. Treating illnesses often requires more than the ability to process information of this sort and use it as the basis of treatment. Judgment, experience, and even "intuition" can be of vast importance for a successful medical practice. Ideally, the use of sophisticated diagnostic technologies frees doctors from tedious work, leaving them with more time for personalized interactions with patients and more accurate diagnoses. Unfortunately, this is not what always happens. A fixation on the use of modern diagnostic technologies can easily result in the denigration of diagnostic skills that historically rested upon more personal relationships with patients. Consequently, a focus on "objective" data such as blood pressure or hormone deficiency to the exclusion of all else can lead a physician to miss some crucial clues regarding the source of a patient's illness.

This has a significance beyond diagnosis. Sick people usually have something physically wrong with them, but organic disorders may be only part of the problem. Illnesses often have psychological components as well, and the successful treatment of these requires more than the application of medical techniques; communication, the ability to build a bond of trust with the patient, and some measure of empathy are also necessary. An overreliance on "objective" data can obscure and even drive off these qualities, to the detriment of medical care as a whole. As Reiser notes, a physician must resist the tendency to place an excessive reliance on diagnostic technologies and the information they provide: "Accuracy, efficiency, and security are purchased at a high price when that price is impersonal medical care and undermining the physician's belief in their own medical powers."[40]

Medical Technologies and Medical Ethics

While contemporary medical training has emphasized the objective, "scientific" dimension of medical care, it has also conferred the belief that the highest duty for members of the medical profession is to preserve life. On the face of it, it is hard to take issue with this principle. At the same time, however, advances in medical technology have rendered the creation and preservation of life increasingly problematic. At all stages of life, medical practitioners along with their patients and their patients' families may have to grapple with major ethical dilemmas that are closely related to the application of contemporary medical technologies. These and other dilemmas have given rise to the field of medical ethics, a discipline that has emerged in recent decades in an effort to systematically address the issues that have attended advances in medical technology.[41]

New Ways of Making and Sustaining Babies

Technological advances have created ethical issues that arise long before birth. One important threshold was crossed in 1978 with the birth, in England, of the first baby to have been conceived through in vitro fertilization (IVF). Since that time, hundreds of thousands of babies have been born worldwide through the use of this procedure, although the exact number is difficult to determine. IVF entails bringing sperm and egg together in a petri dish (the term *in vitro* literally means "in glass"). The sperm can be supplied by the husband or partner of the woman whose eggs are to be fertilized, or it can be provided by a donor, anonymous or otherwise. After the fertilized egg is cultured for three to four days and develops into an embryo with at least four cells, the embryo is transplanted into the uterus of the egg donor or a surrogate. Because each procedure entails the fertilization of more than one egg, IVF results in the production of many surplus embryos every year. These can be kept viable for long periods of time, but many are simply discarded. Alternatively, the embryos can be a source of stem cells, the basis for an emerging technology that will be discussed in the following chapter.

IVF is most commonly used to circumvent blockages of a woman's fallopian tubes or a man's insufficient sperm production. A different reproductive issue occurs when a normal pregnancy is impossible or undesired. Under these circumstances, a surrogate mother can be implanted with an embryo that has resulted from the fertilization of another woman's egg. Alternatively, a surrogate mother can be artificially inseminated with sperm that fertilizes her own egg. The latter process led to a court battle in the 1980s when a surrogate mother refused to give the baby she had carried through pregnancy to the couple who had contracted for her services. Although the case was eventually resolved largely in favor of the couple who had hired the surrogate, the present legal status of surrogate motherhood has been described as being "in a state of flux and confusion." Some states ban the practice entirely, while others enforce surrogacy contracts. Other states allow them under certain circumstances, while others have no applicable laws at all. In the last case, it is up to the courts to determine parenthood when these contracts are subjects of disputes between the various parties involved in the process.[42]

More generally, new reproductive technologies have changed accepted notions about a fundamental constituent of human society, the family. Whether through artificial insemination, in vitro fertilization, or surrogate motherhood, the ability to bypass the usual methods of procreating children has opened up the definition of a family. It is now possible for a child to have several parents: two fathers—the sperm donor and the man taking on the conventional social role of father; and four mothers—an egg donor, a nucleus donor, a surrogate to carry the embryo through pregnancy, and the person who ends up occupying the social role of mother.

While controversy still surrounds nontraditional reproductive methods, a different set of issues comes to the fore with the birth of babies who have been born prematurely. On the one hand, the ability to preserve the lives of prematurely born infants has been all to the good, as many "preemies" have gone on to lead healthy and productive lives as a result of these interventions. Unfortunately, some prematurely born infants are afflicted with major medical problems that result in lack of function, continual pain, and early death. As with the end-of-life issues about to be discussed, the value of preserving the lives of infants destined to a lifetime of pain and sickness can be debated at considerable length. And, as with the case of many modern medical technologies, it is at least open to question if the hundreds of thousands of dollars sometimes spent on preserving the life of one severely premature infant might better have been used for prenatal care, which continues to be inadequate for many expectant mothers and unevenly available.[43]

Some of the medical problems that potentially afflict all newborns, premature or otherwise, can be forestalled through the use of prenatal and postnatal testing. Many of these procedures do not entail complicated ethical or financial issues. For example, the genetically based illness phenylketonuria (PKU) causes seizures, intellectual disabilities, and hyperactivity by affecting the ability of the body to metabolize a particular enzyme. It can be successfully treated through dietary restrictions, but only when there is an awareness that the disorder is present. To this end, newborns are customarily screened for possible PKU through an easily administered blood test, and a potential individual and social problem is averted.

Other tests are more problematic and give rise to a number of ethical questions. The target of these tests is the genetic abnormalities associated with particular illnesses. From the late 1960s onward, amniocentesis (testing of the amniotic fluid in the uterus) has allowed the detection of genetically linked abnormalities such as Down syndrome and spina bifida. Unlike PKU, however, treatment of these conditions is difficult or nonexistent. Should a serious disability of this sort be discovered, prospective parents may want to decide if the fetus should be aborted or be allowed to survive, a difficult decision for many. Amniocentesis also makes it possible to determine the sex of a fetus, thereby bringing another ethical issue to the fore: Is the desire for a boy or girl a sufficient basis for aborting a fetus of the "wrong" sex?[44]

The decision to abort a medically abnormal fetus can be circumvented through the use of IVF in conjunction with a process known as preimplantation genetic diagnosis (PGD). This is done by analyzing the DNA of one cell that has been extracted from an embryo to determine if there are chromosomal abnormalities associated with diseases such as cystic fibrosis and Tay-Sachs. Only an embryo without this abnormality would then be transferred to the uterus. PGD obviates the need to abort an abnormal fetus, but the disposal of embryos is still unacceptable to those who believe that life begins at the moment of conception. Ethical issues of this sort will become more common in the near future as the ability to detect potential problems increases. Further complicating matters will be the emergence of prenatal tests that may not be definitive, providing only statistical probabilities of future disorders. What might expectant parents decide if a test indicates that there is a 50 percent chance that their child will be stricken with cancer before the age of 20? The rapidly developing science of genetics will surely raise many issues of this sort, some of which will be explored in the next chapter.

When Does Life End? When Should It?

While medical technologies have raised a host of ethical issues regarding how life is to be created, they also have been the source of new ethical dilemmas that emerge at the other end of the life span.

One centers on the very conception of when life has become unsustainable. In the not-too-distant past, the failure of one or more vital organs was tantamount to death. Today, however, organ failure and even multiple organ failures can be circumvented by devices such as feeding tubes, respirators, and dialysis machines which serve as partial organ replacements. Even a major and irreversible loss of brain function can no longer be taken to be an unequivocal indication of death.

The issue of determining when death occurs received a great deal of national attention from 1998 onward when family members took opposing sides over the removal of a feeding tube that was sustaining a 34-year-old woman named Terri Schiavo. Although she had been in a "persistent vegetative state" for more than eight years her parents strongly resisted her husband's efforts to terminate their daughter's life through the removal of the tube. What had been a family matter was then thrust into the political arena when it was taken over by politicians with opposing values about the prolongation of life through artificial means. In 2005, after many court hearings, legislation that was overturned by the Supreme Court, political posturing, and a great deal of acrimony, the tube was removed, and Terri Schiavo died at the age of 41.

A feeding tube is not a particularly sophisticated item of medical technology, but in some circumstances it helps to maintain organic functions for an indefinite period of time. Whether or not the maintenance of these bodily processes is the same thing as preserving a life lies at the center of the ethical debate over the use of modern medical technologies for this purpose. This issue is not easily resolved, and it will become more prominent in the years to come as emerging technologies make it possible to preserve organic functions for longer periods of time and under increasingly difficult circumstances. At the same time, end-of-life issues are becoming increasingly salient as the elderly comprise a large and growing portion of the total population of the world's industrially developed nations. As with costly medical technologies in general, measures to extend the lives of elderly patients are in competition with alternative uses of scarce medical resources. Economic calculations cannot substitute for ethically based judgments, but both will have to be taken into account as advanced medical technologies solve one set of problems while giving rise to others.

Halfway Technologies

The dilemmas presented here demonstrate the decidedly mixed blessings bestowed by medical technologies that are primarily palliative. A kidney dialysis machine does not cure the disease that led to kidney failure in the first place, just as a heart bypass operation comes into play only after the damage has been done. Many of the apparent breakthroughs of recent years are not as efficacious and cost-effective as the dramatic advances of the past. For example, relatively inexpensive vaccination campaigns completely eradicated smallpox as a human disease, significantly contributing to longevity and a better quality of life. In contrast, despite decades of intensive efforts, we have made only marginal progress in overcoming cancer, the second-largest cause of death in the industrial world. The limitations of curing, rather than preventing cancer in the first place, leaves unchallenged the toxic industries implicated in causing cancers.[45] Many of today's technological advances in medicine have extended life spans and eased pain but have done little to address the causes of disorders themselves.

For a patient suffering from organ failure, the availability of dialysis or a heart transplant can be the difference between life and death. But for the society as a whole, the escalating costs of medical care driven by these "halfway technologies" constitute a formidable problem. Like the technological fix, without contextualization and complexity halfway technologies are only stop-gaps rather than cures. No one is comfortable with putting a dollar value on a human life, but the years to come may see the emergence of more explicit efforts to balance the costs of new technologies with their benefit to society as a whole. Far from offering across-the-board advantages, the development of new medical technologies will force us to grapple with increasingly difficult choices concerning their application. Financial considerations aside, new and emerging medical technologies will also intensify existing ethical quandaries while creating some new ones as well. As we shall see in the next chapter, the rapid development of genetic technologies will allow the successful treatment of many illnesses and infirmities, and may allow the complete eradication of some others. But to repeat a now familiar theme, these advances will give rise to a new set of issues that will demand our attention.

QUESTIONS FOR DISCUSSION

1. As noted in this chapter, about one-quarter of Medicare and Medicaid expenditures are incurred during the last year of life. Given the increasing financial burdens of both programs, should anything be done about this situation? If so, what?
2. When the demand for a new medical technology exceeds the supply, what should be used to determine who gets it? A lottery? The ability to pay? The "merit" of the recipient? Might it be better to limit the development of new technologies in order to forestall the need to make these choices?
3. Attempts at developing devices to completely replace human hearts have not been completely successful, but many technologies require a long development period before they are successfully employed. Should R&D for artificial heart technologies continue? If so, how should it be financed? Are there some ongoing medical programs that are more deserving of financial support than the development of an artificial heart?
4. Who should make the decision to terminate life support when a person is in a permanent vegetative state, and has left no instructions concerning the indefinite prolongation of his or her life? Should government officials, elected or otherwise, have a role in making this decision?
5. Some medical procedures can be characterized as "halfway technologies," while others are at least close to being completely effective. What accounts for the difference? What "halfway technologies" of the past have become much more effective today? Why has this advance occurred?

Notes

1. Lewis Thomas, *The Medusa and the Snail: More Notes of a Biology Watcher* (New York: Bantam Books, 1980), p. 133.
2. For polio, see www.sciencedaily.com/releases/2017/03/170303163208.htm, and more specifically www.cdc.gov/globalhealth/newsroom/topics/polio/index.html
3. Thomas McKeown, *The Role of Medicine: Dream, Mirage, or Nemesis?* (Princeton, NJ: Princeton University Press, 1979).
4. John E. Wennberg, et al., "Tracking the Care of Patients with Severe Chronic Illness," *The Dartmouth Institute* (April 2008); and Consumer Reports, "Too Much Treatment?" Accessed from www.consumerreports.org/cro/2012/04/too-much-treatment/index.htm; and Aaron Carroll, "The High Costs of

Unnecessary Care," *JAMA* (September 27, 2017). Accessed from https://jamanetwork.com/channels/health-forum/fullarticle/2760210.

5. See: Samantha Willner, Robin Whittemore, and Danya Keenea, "Life or Death": Experiences of Insulin Insecurity among Adults with Type 1 Diabetes in the United States," SSM - Population Healthwww.ncbi.nlm.nih.gov/pmc/articles/PMC7352063/ and www.sc.edu/uofsc/posts/2021/11/11_conversation_insulin.php#.YoKd06jMLcs and https://fortune.com/longform/insulin-cost-diabetes-treatment/
6. Anna Porretta, "How Much Does Individual Health Insurance Cost?" *eHealth* (October 01, 2022). Accessed from www.ehealthinsurance.com/resources/individual-and-family/how-much-does-individual-health-insurance-cost.
7. Centers for Medicare & Medicaid Services, "NHE Fact Sheet 2021." Accessed from www.cms.gov/research-statistics-data-and-systems/statistics-trends-and-reports/nationalhealthexpenddata/nhe-fact-sheet.
8. Matthew McGough, et al., "How Does Health Spending in the u.s. Compare to Other Countries?" *PETERSON-KFF Health System Tracker* (February 9, 2023). Accessed from www.healthsystemtracker.org/chart-collection/health-spending-u-s-compare-countries-2; and Johns Hopkins Bloomberg School of Public Health, "U.S. Health Care Spending Highest Among Developed Countries," (January 07, 2019). Accessed from https://publichealth.jhu.edu/2019/us-health-care-spending-highest-among-developed-countries.
9. See: "Communities in Action: Pathways to Health Equity," from www.ncbi.nlm.nih.gov/books/NBK425844; and "Racial/Ethnic Disparities in Obstetrical Outcomes and Care: Prevalence and Determinants," from www.ncbi.nlm.nih.gov/pmc/articles/PMC2847630.
10. Louise B. Russell, *Technology in Hospitals: Medical Advances and Their Diffusion* (Washington, DC: The Brookings Institution, 1979), p. 3.
11. Daniel Callahan, "Health Care Costs and Medical Technology," *The Hastings Center* (December 14, 2015). Accessed from www.thehastingscenter.org/Publications/BriefingBook/Detail.aspx?id=2178.
12. Adapted from Bryan Jennett, *High-Technology Medicine: Burdens and Benefits* (Oxford: Oxford University Press, 1986), p. 174.
13. Renee C. Fox and Judith P. Swazey, *The Courage to Fail: A Social View of Organ Transplants and Dialysis* (Chicago: University of Chicago Press, 1978), p. 230.
14. Shana Alexander, "They Decide Who Lives, Who Dies," *Life* 53 (November 9, 1962), quoted in ibid., p. 231.
15. Fox and Swazey, *The Courage to Fail*, p. 233.
16. Ibid., p.
17. Ibid., p.349.
18. University of California, San Francisco, Schools of Pharmacy and Medicine, "The Kidney Project: Statistics" (2018). Accessed from https://pharm.ucsf.edu/kidney/need/statistics.
19. Ibid.
20. Henry J. Aaron and William B. Schwartz, *The Painful Prescription: Rationing Health Care* (Washington, DC: The Brookings Institution, 1984), p. 35.

21. Adnan Sherif and Keshwar Baboolal, "Update on Dialysis Economics in the UK," *Peritoneal Dialysis International* (March 31, 2011). Accessed from https://pubmed.ncbi.nlm.nih.gov/21364210.
22. Michigan Health Lab, "Heart Bypasses Cost Billions, but There's Opportunity for Improvement"(August 24, 2017). Accessed from https://labblog.uofmhealth.org/industry-dx/heart-bypasses-cost-billions-but-theres-opportunity-for-improvement.
23. Centers for Disease Control and Prevention, Health, United States, 2013, Table 116 "Cost of Hospital Discharges with Common Hospital Operating Room Procedures in Nonfederal Community Hospitals by Age and Selected Principal Procedure: United States, Selected Years 2000–2011." Accessed from www.cdc.gov/nchs /data/hus/hus13.pdf#116. See also: "A Study of Cost Variation for Percutaneous Coronary Interventions (Angioplasty) In the U.S.," *Blue Cross Blue Shield* (July 16, 2015). Accessed from www.bcbs.com/the-health-of-america/reports/study-of-costtion-percutaneous-coronary-interventions.
24. Shannon Brownlee, "The Perils of Prevention," *The New York Times Magazine* (March 16, 2003). Accessed from www.nytimes.com/2003/03/16/magazine/the-perils-of-prevention.html.
25. Thomas H. Maugh II, "Cardiologists Rush to Angioplasty despite Evidence for Value of Drugs," *Los Angeles Times* (May 11, 2011). Accessed from http://articles.latimes.com/2011/may/11/news/la-heb-angioplasty-medical-therapy-05112011.
26. Ibid.
27. Diana B. Dutton, *Worse than the Disease: Pitfalls of Medical Progress* (Cambridge: Cambridge University Press, 1988), p. 99.
28. Ibid., p. 118.
29. Gideon Gil, "The Artificial Heart Juggernaut," *Hastings Center Report* 19, 2 (March/April 1989): 24.
30. Centers for Disease Control and Prevention, "Smoking and Tobacco Use, "November 15, 2019). Accessed from https://www.cdc.gov/tobacco/data_statistics/fact_sheets/fast_facts/index.htm.
31. Henry J. Kaiser Family Foundation, "10 FAQs: Medicare's Role in End-of-Life Care," (September 26, 2016). Accessed from http://kff.org/medicare/fact-sheet/10-faqs-medicares-role-in-end-of-life-care.
32. U.S. Congress, Office of Technology Assessment, "The Changing Health Care System," in Schwartz, *Dominant Issues*, op cit., p. 286.
33. Alexander Friedman, "Why Do Doctors Cling to Continuous Fetal Heart Monitoring?" (2011). Accessed on February 23, 2012, at www.ncbi.nlm.nih.gov/pmc/articles /PMC1464043. On c-section rates see: www.cnn.com/2018/10/11/health/c-section-rates-study-parenting-without-borders-intl; and on maternal mortality, including racial disparities within the U.S., see: www.cdc.gov/nchs/data/hestat/maternal-mortality/2020/maternal-mortality-rates-2020.htm; and for comparisons see: www.commonwealthfund.org/publications/issue-briefs/2020/nov/maternal-mortality-maternity-care-us-compared-10-countries.

34. See: The America Board of Internal Medicine Foundation, "Choosing Wisely," from www.choosingwisely.org.
35. David E. Rogers, "On Technologic Restraint," *Archives of Internal Medicine* 135 (1975), p. 1395, cited in Stanley Joel Reiser, *Medicine and the Reign of Technology* (Cambridge: Cambridge University Press, 1978), p. 161.
36. Amber Dance, "AI Uses Images and Omics to Decode Cancer," *The Scientist* (May 1, 2019).
37. For radiology outsourcing, see: Marty Stempniak, "Experts Warn Hospitals about the Perils of Outsourcing Radiology Services," *Radiology Business* (August 23, 2021). Accessed from https://radiologybusiness.com/topics/care-delivery/healthcare-quality/hospitals-outsourcing-radiology-services and https://healthmanagement.org/c/it/issuearticle/outsourcing-and-radiology.
38. Reiser, *Medicine and the Reign of Technology*, p. 38.
39. Hughes Evans, "Losing Touch: The Controversy over the Introduction of Blood Pressure Instruments into Medicine," *Technology and Culture* 34, 4 (October 1993): 803.
40. Reiser, *Medicine and the Reign of Technology*, p. 231.
41. Richard Sherlock, "Medical Ethics," in Carl Mitcham (Ed.), *The Encyclopedia of Science, Technology, and Ethics* (Detroit: Thomson Gale, 2005), p. 1184.
42. Carla K. Spivack, "The Law of Surrogate Motherhood in the United States," *The American Journal of Comparative Law* (2010), p. 114.
43. Wanda Barfield, "Addressing Health Inequities Among Pregnant People," *CDC* (February 16, 2022). Accessed from www.cdc.gov/hearher/resources/news-media/addressing-health-inequities.html; and Samuel L Dickman, David U Himmelstein, Steffie Woolhandler, "Inequality and the Health-Care System in the USA," *The Lancet* 389, 10077 (2017): 1431–1441.
44. See the legal status of this issue in the US: https://scholarship.law.cornell.edu/facpub/1399; and the plateau and gradual decrease in sex-selection abortions in China and India: Guo, Zhen; Monica Das Gupta, and Shuzhuo Li, "'Missing Girls' in China and India: Trends and Policy Challenges," *Asian Population Studies* 12, 2 (2016):135–155.
45. Samantha King, *Pink Ribbons, Inc: Breast Cancer and the Politics of Philanthropy*. (Minneapolis: University of Minnesota Press, 2008); and Robert Proctor, *Cancer Wars: How Politics Shapes What We Know and Don't Know about Cancer*. (New York: Basic Books, 1996).

CHAPTER 10

Genetic Technologies

As was noted in the previous chapter, many of the dilemmas that surround the use of medical procedures such as dialysis and the implantation of artificial hearts are a consequence of their status as "halfway technologies." They offer the prospect of survival and some relief from a potentially fatal disease, but they do not cure it, and certainly don't prevent it. But we may be on the threshold of medical advances equal to vaccination or antiseptic surgery. In recent years, much has been learned about the structure and mechanisms of cellular reproduction, giving rise to the prospect of eventual cures for a host of ailments. Yet when we consider the possible impact of technological breakthroughs in this area, we are confronted with a familiar theme: every advance is both a solution and a problem.

The Genetic Fix

The fundamental scientific insight underlying many emerging medical technologies is that the growth and reproductive mechanisms of each cell are governed by the basic unit of heredity, the gene. Humans have about 20,000 genes arrayed along the 46 chromosomes found in the nuclei of every cell, other than red blood cells (which have no nuclei) and sperm and egg nuclei (which have 23 chromosomes). The human genome consists of long chains of deoxyribonucleic acid (DNA), which collectively contain more than 3 billion letters that comprise the human genetic code. These letters (also called "bases" or "basc pairs") are in effect instructions for making proteins, the essential building blocks of life. Some of the DNA sequences that make up a genetic code may be faulty and can give rise to a variety of disorders. Genetic screening or altering an organism's genetic code offer revolutionary new ways of diagnosing, curing, and even preventing the emergence of many disorders.

Gaining the ability to successfully treat genetically based disorders has been the work of many decades. People had long been aware that parental traits often reappeared in their offspring, but they could only speculate on the mechanism that made this happen. An important step forward came with experiments conducted during the latter half of the nineteenth century by Gregor Mendel. His systematic study of the propagation of peas revealed rules of inheritance through which traits were passed from generation to generation, although the biological processes that drove them were still unclear.

While Mendel was conducting his experiments, other scientists had been taking advantage of improvements in microscopy to obtain a better understanding of the constituents of plant and animal cells. First came the discovery of the cell nucleus, followed by the use of improved staining techniques to reveal the presence of chromosomes in cell nuclei. During the first decade of the twentieth century, Thomas Hunt Morgan found that these chromosomes were the site of the key agent of heredity, the gene. Morgan and his associates performed experiments with fruit flies that helped them to unravel the process of trait transmission, but little was known then about its molecular basis. The puzzle began to be solved in the 1940s when DNA (deoxyribonucleic acid) was identified as the key

"Brother Mendel! We grow tired of peas!"

Figure 10.1 A cartoon illustration of monks at a dinner table. The caption: "Brother Mendel! We grow tired of peas!" (Source: John Churchill Chase)

constituent of genes. This set the stage for one of the most important scientific discoveries of the twentieth century. In 1953, James Watson and Francis Crick[1] determined that the molecular structure of DNA consists of bases attached to pairs of sugar-phosphate backbones that form intertwined spirals, the now-famous double helix. This discovery laid the foundation for rapidly expanding insights into the functioning and reproduction of cells.

Another milestone was reached in the 1970s when researchers learned how to separate and isolate portions of DNA. This allowed the development of procedures for transferring genetic material from one organism to another. As has been the case with a great deal of contemporary scientific research, new technologies have created greatly expanded opportunities for advancing scientific knowledge. In the case of genetics, the development of polymerase chain reaction machines gave researchers the ability to obtain and analyze virtually unlimited amounts of DNA, while at the same time identifying mutations associated with particular diseases.

Continued advances in laboratory equipment and techniques made possible the discovery of the distinctive sequence of bases that characterized the DNA of a few individual genes. This gave rise to the possibility of determining the sequences for all of a particular organism's genes. By the mid-1980s, the 150,000 bases that made up the genome of a virus had been sequenced, but the possibility of doing the same thing for the more than 3 billion bases of the human genome seemed remote at best. But by the early 1990s, new technologies in the form of automated sequencers and mapping machines, along with faster and more powerful computers, were propelling advances in basic science. The eventual determination of the human genetic map had become a distinct possibility.

Discovering Genes and Patenting Them

Up to the early twenty-first century, the sequencing effort had been conducted by researchers working as part of the Human Genome Project, an international effort launched in 1990 and coordinated by the U.S. Department of Energy and the National Institutes of Health. By the end of the decade, a private company, Celera Genomics, also was engaged in decoding the human genetic code. In early 2001, both enterprises separately published "rough drafts" of the human genome that were in general agreement with each other. The complete sequence was published in 2003, exactly 50 years after the discovery of the molecular structure of DNA.[2]

Although Celera Genomics made a major contribution to human genetics, its involvement in genetic research has highlighted some of the issues raised by the coupling of scientific discoveries with commercial motivations. Even before the mapping of the human genome, the potential monetary value of genetically engineered organisms was underscored when in 1980, the U.S. Supreme Court in a 5-4 decision ruled that a human-made organism (in this case, a bacterium that could digest petroleum) was entitled to full patent protection.[3] The next stage appeared with the mapping of the human genome. Researchers were now able to home in on specific genes and determine their function. Armed with this information, biotech firms successfully filed patent applications that gave them commercial control over the application of that gene. For example, when a biotech firm held a patent based on their discovery of a gene implicated in certain kinds of cancer, they then had an exclusive right to market laboratory tests, diagnostic devices, and therapeutic products based on that gene.

The rationale for allowing genes to be patented was the same as for the patenting of mechanical devices or new chemical compounds, that the possibility of obtaining a patent would serve as a powerful stimulus for innovative research. At the same time, however, critics pointed out some serious problems inherent in the patenting of genes. In addition to giving rise to monopolies, critics noted that the patenting of genes would likely retard scientific advance by inhibiting the free flow of information. Patent documents could provide essential information, but to directly employ the gene usually required the payment of permission fees to the holder of the patent. Moreover, genetic research often required numerous steps and stages, each of which may have been controlled by a different patent, while fees were sometimes assessed for the use of the information covered by each patent. Even the process of simply finding out who owned a patent could entail substantial costs. Critics also pointed to the practice of "patent stacking," whereby a single gene sequence was covered by separate patents covering key elements of the gene, such as expressed sequence tags and single nucleotide polymorphisms, along with the gene itself.

The patenting of genes was the source of considerable revenue. Commercial entities such as pharmaceutical firms were profiting handsomely from gene patents, but so were public and not-for-profit institutions such as universities and research organizations. Allied against these for-profit and nonprofit organizations were public-interest organizations and many components of the medical establishment, including the American Medical Association, which has a long history of opposing the patenting of medical and surgical procedures.

After a number of lower court rulings, and with billions of dollars at stake, the U.S. Supreme Court took on the patenting of genes in 2013.[4] The case centered on Myriad Genetics, a Salt Lake City firm that held patents on two genes, BRCA1 and BRCA2, linked to ovarian and breast cancer. Their ownership of these patents allowed Myriad Genetics to claim royalties from any medical provider performing tests for the presence of these genes. In the case of *Association for Molecular Pathology, et al. v. Myriad Genetics, et al.*, the Court unanimously ruled that as "naturally occurring" entities, genes were "products of nature," and as such were not subject to being patented.[5] Patients should know that at this time, they do not have a legal claim to intellectual property (IP) derived from tissues removed from their bodies. As "discarded" materials, the IP derived from cell lines, such as John Moore's leukemia cells or Henrietta Lack's (HeLa) cervical cells, belong to the researchers which establish the cell lines as valuable research products.[6] We will explore some of the ethical and legal challenges again in Part 7.

Bioengineering on the Farm

A controversial application of genetics has been the development and cultivation of genetically modified (GM) crops, also known as genetically modified organisms (GMOs). Genetic modification is nothing new, of course; humans have been using selective breeding for millennia in order to improve naturally occurring plants and animals. But selective breeding can be a hit-and-miss affair that entails thousands of genes and may take many generations to achieve an intended result. Genetic engineering makes it possible to regularize and accelerate the process by transferring a single gene or set of genes to obtain a desired characteristic.

In some quarters at least, genetic modification of food crops has become especially attractive as a way of dealing with present and future food shortages. Although there is no imminent crisis in the global production of grain as a whole, some parts of the world have been, and will continue to be, grain-deficit regions.[7] In other parts of the world, increasing affluence is triggering a growing appetite for meat, which demands considerably more grain (and water) per capita than does a largely vegetarian diet. Finally, the effects of climate change can be debated, but they are not likely to be wholly benign as far as grain production is concerned.

Most genetic interventions do not increase crop yields by making individual plants more prolific. Rather, the costs of cultivation may be lowered through new means of controlling weeds, pests, and diseases. It is now a common practice to sow fields with crop varieties that have been genetically modified to have increased resistance to herbicides so the application of weed killers will not affect the food crop. Use of these crops may have the added benefit of allowing the application of safer herbicides that need to be used less frequently. It is also possible to eliminate or reduce the use of pesticides through the cultivation of corn and cotton varieties that have been genetically modified to produce a pesticide derived from *Bacillus thuringiensis*, a naturally occurring soil bacterium.[8]

Genetic modification can also be used to increase the nutritional value of foods. One notable example has been the development of strains of rice that prevent certain kinds of blindness by supplying increased quantities of vitamin A. Another example of a useful genetic modification is the creation of crop varieties capable of tolerating long periods of drought. Somewhere over the horizon is the production of "edible vaccines"—food crops that prevent certain diseases when consumed. Genetically engineered crops may also be used to promote "conservation tillage," a mode of energy-efficient cultivation that conserves the soil and the organisms in it by eliminating or sharply reducing the need for regular plowing.[9] On the far horizon are possible technologies that promote more efficient photosynthesis, drought resistance, nitrogen fixation, and other qualities that have the potential to significantly increase food production.[10]

GM crops now constitute a substantial portion of the global food economy; in 2018, 17 million farmers in 26 countries were cultivating GM food crops on 191.7 million hectares of farmland, a 113-fold increase over the first use of GM crops in 1996.[11] In some countries and for some crops, GM varieties dominate. In Brazil, the second largest user of GM crops and the world's largest producer of soybeans, 91 percent of that crop has been GM.[12] Argentina, India, and Canada are also major planters of GM crops, but the United States is by far the largest user of GM crops, with 169 million acres—about half of total cropland acreage—sown in 2013.[13] At this point, only four types of GM crops are cultivated in significant numbers—soybeans, canola, cotton, and corn—but they dominate the worldwide market. In 2018, GM crops accounted for 78 percent of the soybeans, 76 percent of the cotton, and 30 percent of the corn.[14] With figures like these, it is apparent that the consumption of food based on GM crops is hard to avoid. In the United States, at least 60 percent of processed foods found in American supermarkets contain one or more GM ingredients, usually corn, soy, or canola.[15]

Although GM crops are confined to only a few countries, their widespread use in the United States and a few other countries has generated substantial revenues. The 2023 genetically modified crop market will be about $22.29 billion, and growth is expected in the future. One major change in farming practices is that GM crop seeds are purchased every year, rather than the traditional practice of seed saving. This changes financial risks for farmers. In high income countries (such as the U.S. and Canada), less than 10% of the seed is saved year-to-year.[16] Most of these revenues have gone to the few firms that dominate a highly oligopolistic industry. In 2020 the top four companies controlled more than 50 percent of the global seed market. And not only seeds themselves, but the genetic traits specifically are subject to patent and intellectual property protections. Bayer controls 98% of the trait markers for herbicide-resistant soybeans, for example.[17]

Opposition to GM Crops

As previously noted, the United States accounts for more than half of the world's croplands sown with GM seeds. GM crops are far less common in many other places. Some European countries have

Figure 10.2 A field of GMO corn. (Source: Lindsay Eyink/ CC BY 2.0)

put a moratorium on their use, and opposition to such crops can be found throughout the world. At the most fundamental level, critics of GM crops share a concern that changing the genetic makeup of plants is a dangerous venture into uncharted territories. Alterations to an organism's genome, which in nature may take millions of years, can now take place in a matter of days. And, according to critics, the introduction of laboratory creations into the natural world may have dangerous unintended consequences. Opponents of GM crops point to the possible presence of new allergens or toxins that could afflict many consumers of these foods. Concerns have also been voiced about the danger of cross-pollination (sometimes referred to as "outcrossing") of GM plants with naturally occurring plants, which could result in the emergence of "super weeds" with a resistance to commonly used herbicides. There have also been cases in which genetic material from corn used for animal feed has shown up in corn intended for human consumption.[18]

Arguments against the use of GM crops can also be based on the economics of agricultural production. In the industrially developed world, the "farm problem," at least until recently, has stemmed from too much production, not too little, resulting in low food prices and depressed incomes for farmers. Increasing production through the use of GM crops exacerbates the problem and puts even more pressure on small farmers who are most vulnerable to falling prices. Overproduction is, of course, not a problem in most poor countries, where simply keeping up with population growth is a major challenge for agriculture. But even there, the use of GM crops raises some important issues. Seeds for GM crops are more expensive than naturally occurring varieties, as are fuel, fertilizer, and herbicide and pesticide costs. Extensive cultivation of GM crops would likely benefit wealthy, large-scale farmers and intensify rural economic inequalities. Less problematic is the alleged development and sale of GM seeds with "terminator technology" that prevents them from propagating new seeds. Consequently, each new crop has to be sown with seeds purchased from the firm that markets them. In fact, "terminator technology" is not currently employed by seed producers. Moreover, commercial crops are not usually propagated with seeds from the previous season's crop; new crops are generally sown with commercially obtained seeds, both GM and otherwise.[19]

A more serious concern, only marginally related to the use of GM crops increase, is the extension of monoculture—the

Figure 10.3 In 2013, protests against Monsanto, producer of GMOs and Roundup herbicide, occurred around the world. The above photo is from the Vancouver, BC, Canada action. (Photo by Rosalee Yagihara/CC BY 2.0)

cultivation of only a few varieties of plants over a wide area. This practice poses the danger of massive crop failures because it eliminates the natural firewalls that prevent the spread of a pest or disease. To be sure, massive crop failures are a rarity in modern farming, but the main thing that prevents their occurrence is the extensive use of insecticides and other pesticides, which create problems of their own.

Regarding the real and imagined dangers of GM crops, recent research has not demonstrated that GM crops pose clear-cut dangers to human health, but, as with all novel and potentially far-reaching technologies, a large dose of caution is certainly in order. As David Ehrenfeld's study of rBST (recombinant bovine somatotropin—sometimes known as rBGH—recombinant vine growth hormone) argues, there's not much evidence for any negative *direct* impacts to human health.[20] But rBST is bad for cows[21] and the environmental benefits of higher-yield cows (more product per cow means fewer cows) comes at the expense of bovine health and industry concentration and farm community decline, along with increased presence of antibiotics into the milk supply and environment. Because rBST stresses cow physiology to improve productivity, they get mastitis and illnesses and are prescribed antibiotics more extensively.[22] Those antibiotics are problematic for human health primarily through proliferation of antibiotic-resistant bacteria. We can understand the consequences to human health of rBST and similar products as distributed and tertiary effects that are hard to measure. Ehrenfeld argues that there is a spiral of consequences, from (in this case small) direct individual health effects upwards and outwards in scale and scope through community and ecosystem effects of intensified production enhanced by genetic technologies.

At the very least, continued research into possible negative consequences is essential. In the words of one plant pathologist, "We need to devote the research to risks now, rather than deal with repercussions later."[23] GM crops exemplify the limitations of technological fixes. At present and for the near future, existing agricultural technologies are sufficient for feeding the world's population. The tragic existence of millions of malnourished people is not the result of inadequate food production, but of warfare, unfair trade practices, misguided governmental policies, and the maldistribution of income and productive resources. When GM crops disrupt long-standing land- and water-management strategies and destabilize communities, they may even make hunger worse.[24] Genetically based agricultural technologies have the potential to improve food production and nutritional standards, but they shouldn't be viewed as a substitute for needed economic and political reforms.

Human Applications: Genetic Mapping and Screening

While recently developed technologies are used to alter the genetic makeup of plants and other organisms, some technologies are being used to get a better understanding of natural genetic endowments and their consequences. Most notable is the use of pre- and post-natal screening to discover potential abnormalities in infants. It is hard to take issue with technologies that prevent early deaths and crippling diseases, but as will be noted next, even these benefits can have a dark side. Also problematic is the genetic screening of adults that does not have a curative intent.

Perhaps the most accepted use of genetic mapping is in the realm of law enforcement. Identifying the perpetrators of crimes, always a major concern for police forces, has been significantly augmented through DNA testing. As has now become familiar to viewers of televised crime, both real and fictional, a DNA sample obtained from a suspect can be matched with DNA evidence taken from a crime scene to determine guilt or innocence. Sometimes it is not even essential to have a suspect's DNA sample; a close match with a relative's DNA may provide an important insight into the identity of an at-large murderer or rapist.

Tests of this sort require a prior record of an individual's genome or at least a portion of it. In recent years, many police forces have been accumulating DNA records for large numbers of people. In many jurisdictions, anyone arrested is required to submit a DNA sample, usually obtained through a simple mouth swab. Even though a suspect may be exonerated, his or her genome will remain on file, most likely in a centralized data bank maintained by the FBI. On the other side of the coin, some individuals convicted of various crimes have been exonerated when subsequent DNA tests indicated that they were in fact innocent; 367 convictions had been overturned from the first use of DNA evidence in 1988 to 2018 and 20 residents of death row were shown to be innocent.[25]

Whether or not the mandatory submission of a DNA sample can be successfully challenged by invoking the U.S. Constitution's prohibition of "unreasonable search and seizure" is still being debated. Meanwhile, the collection of DNA samples by firms in the private sector has become a source of concern to civil libertarians, although such collection is not yet widespread. As might be expected, physicians and other medical practitioners have taken the lead; having a patient's genetic information can be an invaluable contribution to diagnosis and treatment. The growing use of electronic medical records increases the danger that these records, many of which contain sensitive information, could be accessed by individuals with malicious intent.[26] More benign but still problematic is the legal right of employers and insurance companies to require individuals to authorize the release of their medical records, which may contain information about genetically based disorders. The Genetic Information Nondiscrimination Act, passed by the U.S. Congress in 2008, forbids insurers from denying health insurance or charging higher premiums because an individual's genetic profile indicates a higher risk of certain health problems. This stricture does not, however, apply to life or disability insurance.[27] The U.S. Equal Employment Opportunity Commission has ruled that employers may not use genetic information when making decisions about hiring, promotions, layoffs, and other aspects of employment. At the same time, however, an employer may make a conditional job offer to a prospective employee and then require him or her to authorize the release of his or her health records as a condition of employment. Clearly, the emergence of DNA testing has created several issues and problems in balancing improved medical care with the protection of individual privacy, a conundrum that will reappear when we consider in a later chapter how advances in electronics also pose potential threats to personal privacy.

Cloning, Present and Future

The patenting of genes, the growing of GM food crops, and the threats to privacy posed by digitized genetic records are prominent examples of the legal and ethical issues raised by recently developed biological technologies, but they are hardly the only ones. Another prominent source of controversy has been the practice of cloning organisms. A clone is an organism that has been derived from another organism through asexual reproduction.[28] Because all the clone's genetic material originated with the "parent" organism, it is a genetic duplicate. All navel oranges are clones,[29] and during the 1950s, biologists successfully cloned frogs, but for many years it was thought that the cloning of higher organisms was impossible, or at least unlikely.

Complacency over cloning was shattered in 1997 when a team of researchers at the Roslin Institute in Scotland announced that they had successfully cloned a female sheep during the previous year.[30] The genetic material of the sheep known as "Dolly" originated in the nucleus of a cell that had been part of the mammary tissue of an adult sheep that had been transplanted into the enucleated eggs of sheep of a different breed. Since these nuclei carried full complements of chromosomes (unlike sperm and eggs, which each contain only half the genetic material of ordinary cells), the eggs did not require fertilization. Instead, an electric shock was applied, stimulating some of the eggs to begin the process of cell division that results in an embryo. Thirteen out of 277 of these eggs were implanted into the uteruses of sheep who served as surrogate mothers. Of these, 12 miscarried, but Dolly was carried to term and came into the world on July 5, 1996.

The ultimate goal of the Roslin team was to genetically modify the cells of sheep udders so that the cells would produce commercially valuable drugs. The cells would then be cloned, allowing the reproduction of sheep that produced the drug along with their milk. It was anticipated that cloned sheep eventually could also be used for conducting research into genetically borne diseases. These, of course, are not the only possible applications of cloning. Sheep are not fundamentally different from other mammals, humans included, and there is no scientific reason that the procedures used to produce Dolly could not also be applied to the cloning of people.

The prospect of human cloning has raised a host of ethical questions. Would it be prudent to circumvent the normal process of reproduction, which ensures the healthy genetic diversity of a population? Will members of a segment of the population be especially inclined to clone themselves, and are they necessarily the kind of people who should be perpetuated? Will people with special abilities allow themselves to be cloned for a fee so that prospective parents will be able to have a child with inborn talents? Will people with damaged organs clone offspring so that they are guaranteed future organ replacements that pose no danger of rejection? Might it be possible to eventually clone individuals from the cells of dead people? Right now, this is a far-fetched prospect, but so was the cloning of mammals a few decades ago. Some might argue that such an outcome would be desirable, as it would preserve the genetic endowments of geniuses in the realms of science, music, the visual arts, and other exalted human endeavors. But at the same time, of course, it could also allow a ruthless dictator to attain a kind of immortality. Who would decide?

Concerns over human cloning have led several governments to ban any efforts in this direction, and the U.S. government has barred the use of federal funds for human cloning research. Even so, cloning is a technology that will be difficult or even impossible to contain. Clandestine human cloning could take place anywhere, and eventually it might even be encouraged by governments beguiled by the prospect of creating a cohort of "super citizens." In any event, neither tacit nor direct government support is essential for human cloning. There are many individuals wealthy enough to underwrite such a venture, and their ranks include persons whose outsized egos are likely to make their physical duplication an attractive proposition.

It is important to note, however, that governments or individuals will be disappointed if they expect that cloning will create identical copies of the original organism. Even "identical" twins, who are natural clones, are not really identical. There are many cases, for example, of one twin being left-handed while the other is right-handed. Animals cloned in laboratories often differ considerably in both appearance and behaviour from the animal from which they were cloned.[31] These differences stem from biological processes that are still not well understood, along with environmental influences that begin in the womb and continue to shape an organism throughout its life. Without getting into the age-old debate about the relative importance of "nature versus nurture," it simply can be stated that genetic endowments are not all-powerful, and that an organism's environment is a major influence on its appearance, behavior, and its genetics through *epigenetic* processes. An organism's environment, in addition to being a major influence on its appearance and behavior, may also alter its genetic makeup.[32] Putting to rest the ancient debate of the significance of "nature" vs. "nurture," the great lesson of epigenetics is that the two work in tandem to shape the appearance and behaviour of an organism. For example, cloned pets are often not physiologically or behaviorally identical to their original sources, often leading to disappointment.[33]

Stem Cells and Future Therapies

At this point, the cloning of humans is only a hypothetical issue.[34] Of much greater current salience is a promising but controversial medical technology, treating serious diseases using embryonic stem cells. These cells appear during the early phase of embryo formation and have the quality of "pluripotency," which means that they have the capability to become any of the 220 kinds of cells that constitute the human body, everything from bone cells to neurons. Stem cells in bone marrow are now being successfully used to treat leukemia, but the real promise of stem-cell therapies lies in the future. With further research it is hoped that stem cells can be used to regenerate organs that have been damaged by Parkinson's disease, Alzheimer's disease, and many other ailments.

Although promising, therapies using stem cells lie at least a decade in the future. The source of stem cells is still a source of controversy; current stem-cell research and application are based on stem cells obtained from embryos that have been created through in vitro fertilization (IVF) but have not been implanted into a uterus for further development. Alternatively, it may be possible to extract stem cells from embryos that received the genetic material of a patient to be treated with these cells. In either case, the use of embryos has galvanized opposition to stem-cell research. For individuals and groups who hold that life begins at conception and that an embryo is a human being, the destruction of an embryo through the extraction of stem cells is akin to murder.

Adhering to this point of view, President George W. Bush in 2001 prohibited the federal funding of stem-cell research unless it used existing stem-cell lines derived from embryos that were the product of IVF and where the sperm and egg donors had provided informed consent. Originally thought to constitute approximately 60 stem-cell lines, only about 15 of them proved suitable for research purposes.[35] Other restrictions on embryonic stem-cell research created difficulties for researchers. According to the government's rules, research using stem cells had to be kept entirely separate from other kinds of research receiving federal funding; even the shared use of a refrigerator was forbidden.

In 2009, these restrictions were lifted by the Obama administration, a move upheld in a number of court cases. Even so, nonembryonic stem-cell research subsequently received considerably more federal funding than research involving embryos.[36]

Embryonic stem-cell research continues to be controversial. Polls indicate that a majority of Americans are in favor of it, but support or opposition tends to break along political and cultural divisions; Republicans and frequent church attenders were more likely to oppose embryonic stem-cell research, while Democrats and less frequent churchgoers tended to support it.[37]

Although most supporters of this research do not believe that an embryo constitutes a human life, its use as a source of stem cells still triggers many issues and problems. Currently, the embryos used for stem-cell research are the surplus products of IVF that are destined to be discarded when they are no longer needed. However, the supply of these embryos will likely not be sufficient if stem-cell research results in therapies that have large numbers of potential beneficiaries. This would give rise to two alternative means of dealing with supply limitations, both of which may be distasteful to many. On the one hand, the products of IVF would continue to be the sole source of stem cells, but because the number of embryos is necessarily limited, their stem cells would go to the highest bidder, the winners of a kind of lottery, or would be allocated in a manner reminiscent of the early days of kidney dialysis. On the other hand, an emerging technology is being used to reprogram specialized adult cells so they will act as embryonic stem cells. Nonembryonic stem cells can be derived from many kinds of tissues, including adipose tissue and bone marrow. Known as "induced pluripotent stem cells," these cells are currently used for drug development, but as with stem cells in general, their application to specific disorders remains an ongoing effort.[38]

The Perils of Genetic Intervention

Stem-cell research, mammal cloning, GM foods, and the sequencing of the human genome have given rise to a host of ethical concerns that touch upon some of the most basic issues of human existence.

On a somewhat less abstract level, governments are faced with the task of creating appropriate public policies to encourage the development of beneficial technologies while preventing the emergence of harmful ones, or at least mitigating their worst consequences. In hammering out these policies, it will be important to guard against an excessive enthusiasm for seeking in the human genome the answer to all our medical, psychological, and social problems. Genes are central to the process of biomedicalization,[39] which focuses on the individual body at the expense of looking at the community systems in which the human body is embedded. For example, poverty reduction programs have a positive impact on health outcomes, and conversely, healthier populations can prevent ill health.[40]

At the level of the individual, there is no doubt that many diseases have a genetic component, but genetic defects are not always clear-cut; a gene that malfunctions under one set of environmental, nutritional, or other external conditions may be perfectly adequate in different circumstances.[41] It cannot even be said with complete confidence that the presence of a seemingly harmful gene is an unmixed hazard; genetic "imperfections" provide the raw material of genetic variability upon which natural selection operates.[42]

So, a search-and-destroy campaign waged against "defective" genes is likely to be a quixotic affair, for it may not always be possible to identify them. Each of us carries between 5 and 10 defective genes that have no consequence for us individually, but could at some future time become a problem for our progeny.[43] The emergent understandings of epigenetics are showing the ways in which DNA and other cellular nucleic acids (RNA, mRNA) can be modified through environmental exposures. Many illnesses are now understood as a consequence of epigenetic mutations, including cancers and neurological, reproductive, and autoimmune diseases, among others.[44]

Finally, and most importantly, an overemphasis on the role of genes in determining physical and behavioral qualities could lead to a revival of the now discredited eugenics movement. Initiated in the late nineteenth century by Charles Darwin's cousin, Francis Galton, the eugenics movement had many adherents, including leading scientists, politicians, and intellectuals.[45] Its basic postulate was that biological inheritance was the major source of human differentiation; if some people had superior strength, intelligence, and even morality, it was because their ancestors exhibited these qualities and passed them down to their descendants. Although its focus was purportedly on the qualities of individuals, eugenics easily lent itself to racist notions of human differentiation. According to most eugenicists, the white race occupied the pinnacle of human development, and all others were arrayed at lower places of the human hierarchy. The sexes, too, could be divided according to their general abilities, with women on average presumed to have lower intellectual capabilities than men, although they were deemed to be superior in the "lesser" qualities of empathy and nurturance.

Eugenics aimed at being far more than a scientifically based way to account for human variation; it was also a call to action. According to eugenicists, human progress depended upon the perpetuation of individuals with superior genetic endowments and the limitation or even elimination of those with inferior qualities. According to the program envisaged by "positive eugenics," individuals whose physical, mental, and moral qualities marked them as "superior" were encouraged to have children who would inherit and perpetuate these qualities. This, however, was a difficult task. Much easier was the promotion of "negative eugenics," which aimed at preventing the transmission of "undesirable" qualities from one generation to another. In its mildest form, negative eugenics motivated the promotion of birth control for low-income groups. It was also associated with more draconian measures, most notably the sterilization of "defectives." In the United States, more than 60,000 compulsory sterilizations of "defectives" were performed between 1907 and 1963.[46] The most horrific examples of negative eugenics took place in Nazi Germany, where 350,000 sterilizations were performed in the name of improving the German *Volk* (people) while millions of men, women, and children were killed because they were adjudged to be members of "inferior" races."[47]

These ghastly actions undermined much of the support for negative eugenics, but support for positive eugenics continued well into the 1960s. Although it was shorn of its overtly racist underpinnings, eugenics was still predicated on a concern that humanity was threatened by the deterioration of its biological foundation. Genetic endowments were taken to be the major source of human variation, and as one eminent geneticist put it in the early 1960s, "Human welfare, both with individuals and with society, is predicated on the health of the genetic endowment of human populations. . . . [An] appalling amount of human misery is due to defective heredity."[48]

Figure 10.4 An early twentieth-century demonstration of support for eugenics. The demonstrators did not necessarily agree with the statements on the signs they were holding; they were hired to participate in the demonstration.

Eugenics in the Twenty-First Century

Eugenics has ceased to be a social movement aimed at human betterment, but concerns about the genetic endowments of individuals are very much with us. The elimination of genetically borne defects is no longer a government policy. Instead, it is now an individual, personal matter, as physicians and parents-to-be make extensive use of a variety of tests to determine the presence of genetically linked disorders such as Tay-Sachs and Down syndrome. Through genetic counselling, prospective parents are helped to decide if a fetus with inborn defects should be aborted or carried to term.[49] For prospective parents using IVF, this decision is somewhat less painful because it is now possible to test a single extracted cell from an embryo consisting of only eight cells. Even so, the decision to terminate life at any stage of development is a difficult one. Yet for all the potential anguish they may cause, these choices will likely become more widespread as researchers find genetic links to a growing number of disorders.

Even labeling differences as disorders has become problematic. Consider autism or deafness; disability activists argue that the neurophysiological differences of autism or deafness are not disabilities, per se. It is the absence of accommodations and systems of discrimination that make differences into disabilities. For many in the Deaf community, not being able to hear is part of a cultural difference, not something to be cured with technology. Consider Down syndrome activism in the United States. Down syndrome children are cherished family and community members. As one of the oldest and most easily identified trisomy anomalies in gestation, it often leads to pre-implantation selection or abortion after amniocentesis. But one cannot know with certainty and in advance if a genetic anomaly discovered in utero will be incompatible with life or produce differences that are relatively benign.[50]

Although more medical problems may come to be associated with certain genetic malfunctions in the years to come, their influence should not be exaggerated. For one thing, a decade of research conducted after the initial sequencing of the human genome has shown that early expectations of tying certain diseases to specific gene abnormalities was far too optimistic. There is no question that genetic abnormalities can be implicated in many diseases, but how the two are connected is far more complicated than had been anticipated in 1990.[51] As we have seen earlier, making the transition from science to technology is often a long and difficult process. It is also important to understand that genes never determine physical and behavioral development by themselves. Genes function within individual organisms, and these organisms are always affected by their environments. As David S. Moore has noted, "It is a mistake to think that genes are necessarily the first link in the causal chains of biological events. From the moment of conception, environmental factors and genetic factors are in an ongoing "dialogue" with one another about building a person. Each of these sets of factors brings its own necessary information to this conversation."[52]

Focusing upon genetic endowments to the exclusion of everything else is a simplistic form of determinism that harks back to excesses of the eugenics movement. At the same time, it is more than likely that advances in biological and genetic research will serve as the basis for a plethora of potential medical applications. Taken to the extreme, expanding knowledge about genetics holds out the promise—or the threat—of what the eminent biologist Edward O. Wilson has called "volitional evolution," that is, directing human heredity to favor some traits while extinguishing others.[53]

The possibility of purposefully altering the course of human evolution has recently been boosted by intensive research into human genetics that followed the discovery of the structure of DNA and gave rise to the possibility of altering or "editing" genomes. During the 1970s gene editing became widely employed, but the process was slow, complicated, inaccurate, and expensive. This changed dramatically in the first decade of the twenty-first century with the use and development of CRISPR, a stretch of bacterial DNA characterized by *C*lustered *R*egularly *I*nterspaced *S*hort *P*alindromic *R*epeats.[54]

Before translating these elements of CRISPR, it is necessary to present an important term for understanding genetics: "nucleotide." This term refers to the key constituents of DNA and RNA. Nucleotides are organic molecules that consist of one of five bases: adenine, cytosine, guanine, and thymine, and RNA substitutes uracil for thymine. These molecules combine to form "base pairs" whereby adenine links with thymine and guanine links with cytosine. When connected to phosphate groups, a 5-carbon sugar, and a nitrogenous base, this aggregation forms a nucleotide.

The abbreviation CRISPR therefore refers to how nucleotide segments are situated within a gene's DNA. "Palindromic" indicates that the DNA sequence is the same forwards or backwards. When used for gene editing, CRISPR works in conjunction with the enzyme Cas9, the DNA of which is found near the CRISPR gene. Use of CRISPR-Cas9 for gene editing held great promise because so many human disorders are the result of errors in our genetic code. When it is recalled that the human genome's DNA contains 3 billion base pairs, the likelihood of such errors is all too apparent, and in fact, humans are subject to some 7,000 genetically based disorders.

CRISPR was discovered in the genome of E. coli in 1987, but 20 years passed before it was learned that it protected the bacterium from attacks by viruses called bacteriophages (literally "bacteria-eaters"). The use of CRISPR to edit genes was developed in 2011 by Jennifer Doudna and Jillian Banfield at the University of California, Berkeley. It was first applied to the human genome in January 2013.

Doudna then teamed up with Emmanuelle Charpentier to investigate how the Cas9 enzyme could be used in conjunction with CRISPR to edit genes. At about the same time, Feng Zhang and his colleagues were doing similar research and getting similar results at the Harvard/MIT Broad Institute. As might be expected, both groups filed patent claims that resulted in lengthy litigation that still remains unresolved at this writing.[55]

In clinical use, genetic intervention begins with the creation of a template based on a segment of CRISPR DNA that matches a segment of an organism's targeted DNA. This is used to mark the DNA sequence in the cell's nucleus. This "molecular guide" then directs an enzyme like Cas9 to cut the DNA sequence at this location. In some cases, removing this segment is enough to cure the malady. In other cases, a new segment of DNA is introduced to effect desired changes in an organism's genome.

As with most profound technological advances, the use of CRISPR has given rise to a host of ethical, social, and cultural issues and dilemmas. Most obviously, easy genetic modification returns us to the promises and perils of eugenics. Not much controversy surrounds the use of CRISPR to combat genetically based ailments, but what about using it to produce "designer babies" who will grow up to be superior to the general population? Would some authoritarian governments mandate this action? Would it be an appropriate policy? Could it properly be left to individual families? This could also be a perilous course of action for society as a whole. Not every family will have the resources for such an intervention, and under these circumstances, CRISPR has the potential to solidify and extend socioeconomic inequalities. As mentioned earlier with cloning, there is also the prospect of genetically engineering a future offspring in such a way as to prepare the child to serve as an organ donor to someone in the family with a similar genetic endowment who needs a new kidney or liver.

In the above examples, both the benign and problematic uses of CRISPR refer to "somatic" genetic interventions, i.e., changing the genetic blueprint for individual organs. A more troubling intervention is "germline" genetic editing, which involves changing the DNA of gametes—reproductive cells like sperm and eggs. In this way genetic alterations are not confined to a single organism but will be passed down from generation to generation through sexual reproduction.

Organisms and their progeny derived through germ line editing will forever have characteristics that set them apart from the original species. This would have unknown consequences. The "improved" organisms may out-compete other organisms in an ecosystem, or, conversely, have a flaw that results in their swift extinction. It should be emphasized that germline modification is a hazardous enterprise that eventually affects all members of a species, is virtually irreversible and has highly unpredictable consequences.

In assessing gene editing in general and CRISPR in particular, it must be recalled that the genetic endowments of organisms are spontaneously changing all of the time. This is, after all, the first phase of evolution through natural selection. However, until the past few years, conscious intervention into genetic modification has not been possible. But now that it can be done, how should it be overseen and guided? Governments worldwide have been struggling with this issue, but with no immediate resolution. In some cases, the control of government grants for scientific research has been the key means of doing so. Interventions involving human embryos and fetuses have been subject to sharp restrictions, especially in the United States.

These are reactive responses. More useful would be an approach that not only restricts certain actions but establishes protocols for weighing the positive and negative aspects of particular interventions. These could serve as the framework for determining if some interventions might be actively promoted or banned. But who should be involved in the setting and administration of such guidelines? Does it simply come down to scientific expertise, or should ethicists, religious leaders, government officials, medical doctors, and ordinary citizens also be on board? If so, how should they be chosen?

Should CRISPR-Cas9 and other twenty-first century genetic technologies live up to their potentials, we will be presented with "the most profound intellectual and ethical choices that humanity has ever faced."[56] We have yet to get to that point, but continued progress in understanding the structures and operation of the human genome will offer new ways of addressing a host of physical and psychological maladies and differences. The use of CRISPR and other genetic interventions will pose real and potential issues and dilemmas greater than any technology that we have ever created and used. Our success or failure in controlling and guiding it may serve as the final judgment of humanity's ability to make ethical, intelligent, and humane use of our technologies.

QUESTIONS FOR DISCUSSION

1. Where do you stand in the debate over using fetal tissue as a source of stem cells? Are embryos human beings? Even if they are, might it be justified to extract stem cells from them in order to cure an individual of a debilitating disease?
2. GM foods are sometimes derided as "Frankenfoods." Is this a fair way to characterize them? Do you have any qualms about consuming GM foods? Why?
3. Most specialists are of the opinion that there are no technical barriers to human cloning. Although no legitimate agency is likely to financially support human reproductive cloning, it could be funded by an individual seeking to perpetuate himself or herself. Should such a practice be forbidden with stiff penalties attached? If so, how would such a ban be enforced?
4. Gene-based therapies can now be used to successfully treat cystic fibrosis and a few other diseases, and more applications of this technology may be expected in the future. If they become available, should there be encouragement (financial and otherwise) for all gene therapies? Should they be used to treat differences like premature baldness, deafness, autism, or short stature? Who should be empowered to decide if these therapies should be developed and applied?
5. The field of epigenetics has challenged the utility of the long-standing "nature versus nurture" dichotomy. In its place, accumulating scientific evidence is demonstrating the effects of environmental factors on an organism's response to its genetic makeup. Are there any implications of this approach to heredity for both the individual and society?
6. What does Edward O. Wilson mean by "volitional evolution"? Is it something that should be pursued? If you were a prospective parent, would you consider using a gene-altering technology to affect the physical and mental attributes of your child? What attributes would you select? What might be the positive and negative consequences of this intervention?

Notes

1. The role of Rosalind Franklin in this discovery is often understated. See: www.nature.com/scitable/topicpage/rosalind-franklin-a-crucial-contribution-6538012.

2. Information on the project, both current and historical, can be found at Human Genome Project Information, "History of the Human Genome Project." Accessed from www.ornl.gov/sci/techresources/Human_Genome/project/hgp.shtml.
3. The majority ruling and the minority's dissent in the case, *Diamond v. Chakrabarty*, can be found at http://caselaw.lp.findlaw.com/cgi-bin/getcase.pl?court=us&vol=447&invol=303.
4. Human Genome Project Information, "Genetics and Patenting." Accessed from www.ornl.gov/sci/techresources/Human_Genome/elsi/patents.shtml. This site has many useful links pertaining to the patenting of genes.
5. Elliot Marshall and Michael Price, "Supreme Court Strikes Down Human Genome Patents," *Science Insider* (June 13, 2013). Accessed from December 21, 2015, at http://news.sciencemag.org/people-events/2013/06/u.s.-supreme-court-strikes-down-human-gene-patents.
6. Rohini Nott, "Moore v. Regents of the University of California (1990)." Accessed from https://embryo.asu.edu/pages/moore-v-regents-university-california-1990; and with regards to Henrietta Lacks, see: Rebecca Skloot, *The Immortal Life of Henrietta Lacks* (New York: Crown Publishing, 2010). For news of the Lacks family lawsuit see: www.wgbh.org/news/local/2022-05-17/thermo-fisher-seeks-dismissal-of-henrietta-lacks-familys-lawsuit-regarding-sale-of-her-cells.
7. Food and Agriculture Organization of the United Nations, "World Food Situation." Accessed from www.fao.org/worldfoodsituation/foodpricesindex%20/en. The 2022 war in Ukraine will stress global grain trade.
8. World Health Organization, "Twenty Questions on Genetically Modified Foods." Accessed from www.who.int/foodsafety/publications/biotech/20questions/en.
9. David Despain, "Farming the Future: GM Crops Recommended as Key Part of Obama's Evergreen Revolution" *Scientific American* (December 9, 2010). Accessed from www.scientificamerican.com/article.cfm?id=farming-the-future-gm-crops.
10. International Service for the Acquisition of Agri-Biotech Applications (ISAAA), "Global Status of Commercialized Biotech/GM Crops in 2014" (January 2015). Accessed from www.isaaa.org/resources/publications/pocketk/16.
11. International Service for the Acquisition of Agri-biotech Applications, "Biotech Crop Highlights in 2018," (2020). Accessed from www.isaaa.org/resources/publications/pocketk/16/
12. "Brazil GMO Planting to Increase by 6.8 Percent This Season" *Reuters* (December 18, 2013). Accessed from www.reuters.com/article/brazil-biotech-idUSL2N0JX0SF20131218.
13. Jorge Fernandez-Cornejo, Seth Wechsler, Mike Livingston, and Lorraine Mitchell, "Genetically Engineered Crops in the United States," *United States Department of Agriculture* (February 2014). Accessed from www.ers.usda.gov/media /1282246/err162.pdf.
14. Ibid. ISAAA, op. cit.

15. Matt Kallman, "Genetically Modified Crops and the Future of World Agriculture," *World Resources Institute* (June 17, 2008). Accessed from http://earthtrends.wri .org/updates/node/313, p. 3.
16. For the genetically modified crop market see: www.fortunebusinessinsights.com/industry-reports/genetically-modified-seeds-market-100389; and for seed saved year to year: "Concentration in Seed Markets," www .oecd-ilibrary.org/sites/9789264308367-5-en/index.html?itemId=/content/component/9789264308367-5-en.
17. For the four companies with more than 50 percent of the global seed market: https://www.foodandpower.net/gmos-seeds.
18. World Health Organization, "Twenty Questions," op. cit.
19. Dan Charles, "Top Five Myths of Genetically Modified Seeds, Busted," *NPR* (October 18, 2012). Accessed from www.npr.org /sections/thesalt/2012/10/18/163034053/top-five-myths-of-genetically-modified-seeds-busted.
20. David Ehrenfeld, "The Cow Tipping Point," *Harper's* (Oct 2002), pp. 13–20.
21. I. R. Dohoo, L. DesCôteaux, K. Leslie, et al. A Meta-Analysis Review of the Effects of Recombinant Bovine Somatotropin. 2. Effects on Animal Health, Reproductive Performance, and Culling. *Canadian Journal of Veterinary Research* 67, 4 (2003): 252–264.
22. Tony Hiss, "How Now Drugged Cow? Biotechnology Comes to Rural Vermont," *Harper's* (October 1994): 80–86, 90; and K. E. Jones, "Constructing rBST in Canada: Biotechnology, Instability and the Management of Nature. *The Canadian Journal of Sociology* 25, 3 (2000), 311–341.
23. David Suzuki and Peter Knudtson, *Genethics: The Clash between the New Genetics and Human Values* (Cambridge, MA: Harvard University Press, 1989), p. 205.
24. J. Stephen Lansing, "Balinese 'Water Temples' and the Management of Irrigation." *American Anthropologist* 89, 2 (1987): 326–341; and Alice S. Kirkham, "Balinese Rice Farming." Accessed from https://thesustainablenomad .blog/2019/02/25/balinese-rice-farming.
25. The Innocence Project, "DNA Exonerations Nationwide" (December 18, 2015). Accessed from www.innocenceproject.org/free-innocent/improve-the-law/fact-sheets/dna-exonerations-nationwide.
26. Mark A. Rothstein, "Keeping Your Genes Private," *Scientific American*, 299, 3 (September 2008): 64–6.
27. "GINA and You: Understanding GINA, The Genetic Information Nondiscrimination Act of 2008" (May 21, 2008). Accessed from http://ginahelp.org/GINA _you.pdf.
28. Robert G. McKinnell and Marie A. Di Berardino, "The Biology of Cloning: History and Rationale," *BioScience* 49, 11 (1999): 875–85.
29. See: Brown and Federoff, *Mendel in the Kitchen: A Scientists' View of Genetically Modified Foods* (Washington, DC: National Academies Press, 2004); and W. Popenoe, A. D. Shamel, and P. H. Dorsett, *The Navel Orange of Bahia: With Notes on Some Little-known Brazilian Fruits* (United States: U.S. Department of Agriculture, 1917).

30. Gina Kolata, *The Road to Dolly, and the Path Ahead* (New York: Morrow, 1998).
31. Rachel Dowty, "Clones and Cloning," in Sal Restivo (Ed.), *Science, Technology, and Society: An Encyclopedia* (Oxford and New York: Oxford University Press, 2005), p. 54.
32. David S. Moore, *The Developing Gene: An Introduction to Behavioral Epigenetics* (New York: Oxford University Press, 2015).
33. Rhordi Marsden, "Copy Cat: Why Cloning Dead Pets Doesn't Always have a Happy Ending," *The National News* (2019). Accessed from www.thenationalnews.com/arts/copy-cat-why-cloning-dead-pets-doesn-t-always-have-a-happy-ending-1.939673; and American Anti-Vivisection Society (AAVS) (2005), "Pet Cloning: Separating Facts from Fluff." Report at https://aavs.org/wp-content/uploads/2021/09/aavs_report_pet-cloning-facts-fluff.pdf
34. Stephen S. Hall, "Revolution Postponed," *Scientific American* 303, 4 (October 2010): 60–67.
35. Andrea L. Bonnicksen, "Embryonic Stem Cells," in Carl Mitcham (Ed.), *Encyclopedia of Science, Technology, and Ethics* (Detroit: Thomson Gale, 2005), p. 609.
36. Eugene C. Tarne "Trends Show More Federal Funds Awarded to Non-Embryonic Stem Cell Research," *Charlotte Lozier Institute* (February 8, 2018). Accessed from https://lozierinstitute.org/trends-show-more-federal-funds-awarded-to-non-embryonic-stem-cell-research.
37. Eryn Brown, "Stem Cells: Research Funding and the 2012 Elections," *Los Angeles Times* (November 21, 2011). Accessed from http://articles.latimes.com/2011/nov/21/news/la-heb-stem-cell-funding-2012-elections-20111121.
38. National Institutes of Health, "Stem Cell Information," (February 9, 2015). Accessed from http://stemcells.nih.gov/info/pages/faqs.aspx.
39. Adele E. Clarke, et al., *Biomedicalization: Technoscience, Health, and Illness in the U.S.* (Duke University Press, 2010).
40. The World Bank, "Poverty and Health," (August 25, 2014). Accessed from www.worldbank.org/en/topic/health/brief/poverty-health.
41. Suzuki and Knudtson, *Genethics,* op. cit., p. 205.
42. Ibid.
43. Ibid.
44. B. Weinhold, "Epigenetics: The Science of Change," *Environmental Health Perspectives* 114, 3 (2006): A160-7.
45. Daniel J. Kevles, *In the Name of Eugenics: Genetics and the Uses of Human Heredity* (Berkeley: University of California Press, 1985).
46. Philip Reilley, *The Surgical Solution: A History of Involuntary Sterilization in the U.S.* (Baltimore: Johns Hopkins University Press), pp. 94–95. Quoted in Joan Rothschild, *The Dream of the Perfect Child* (Bloomington and Indianapolis: Indiana University Press, 2005), p. 45.
47. Richard Weikart, "Eugenics," in *Encyclopedia of Science, Technology, and Ethics,* op. cit., pp. 708–09.
48. Theodosius Dobzhansky, *Heredity and the Nature of Man* (New York: Harcourt, Brace, and World, 1964), pp. 14–15.

49. Rothschild, *The Dream of the Perfect Child*, op. cit., pp. 52–131.
50. J. A. Rondal, "Down Syndrome: A Curative Prospect?" *AIMS Neuroscience* 7, 2 (2020): 168–193. And for the controversies about eradicating down syndrome: Sarah Zhang, "Down Syndrome: The Last Children of Down Syndrome," *The Atlantic* (December 2020). Accessed from www.theatlantic.com/magazine/archive/2020/12/the-last-children-of-down-syndrome/616928; and Conor Friedersdorf, "I Am a Man With Down Syndrome and My Life Is Worth Living," *The Atlantic* (October 30, 2017). Accessed from www.theatlantic.com/politics/archive/2017/10/i-am-a-man-with-down-syndrome-and-my-life-is-worth-living/544325.
51. Hall, "Revolution Postponed," op. cit.
52. David S. Moore, *The Dependent Gene: The Fallacy of "Nature" vs. "Nurture"* (New York: Henry Holt, 2001), p. 140.
53. Edward O. Wilson, *Consilience: The Unity of Knowledge* (New York: Vintage Books, 1999), p. 299.
54. Amy Maxman, "The Genesis Engine," *Wired*, August 2015.
55. Carolyn Shaffer, "Broad Defeats Berkely CRISPR Patent," *Nature* (March 14, 2022). Accessed from www.nature.com/articles/d41587-022-00004-2; and Heidi Ledford (2022) on how this will continue as new base editors come into play: www.nature.com/articles/d41586-022-00629-y.
56. *Consilience* Wilson, op. cit.

PART 4

Technology and the Transformation of Work

Few elements of human life have been as powerfully affected by technological change as the ways in which we spend our working lives. As is noted in Chapter 11, humans survived for most of their existence as a species through gathering and hunting. This mode of life was radically overturned by the adoption of agriculture, and then much later by industrialization. The introduction of new technologies has been a prominent feature in all of these transitions, but contrary to commonsense beliefs, technological advance did not always make life easier; in fact, sometimes the opposite has been true.

Although modern technologies have provided us with a plethora of goods and services with less physical effort, they also have brought with them the fear of widespread unemployment. Chapter 12 takes up this longstanding issue, suggesting when this fear is appropriate and when it is misplaced. Without downplaying the real pain caused by job losses due to technological change, this chapter provides some reasons to be optimistic about the future levels of employment.

Chapter 13 moves the discussion from the *quantity* of available work to the *quality* of working life. Industrialization radically changed the nature of work. New technologies altered many aspects of life on the job, as did the theories and practices of management that were developed at this time. In Chapter 14, we will examine the idea that much of the world is organized as a "post-industrial" economy and society. The distribution of skill and authority in the workplace is still a contentious issue, and will likely remain so, even as the distribution of jobs continues to change. We will discuss the intensification of ICTs (information and communication technologies) in production and work and the differential outcomes of their implementation.

Chapter 11

Work in Nonindustrial Societies

For most adults a large portion of their waking hours is spent doing some sort of work. Be it as an assembly line worker, a sales clerk, or a psychiatrist, we gain our livelihood and a good measure of our individual identity from the work that we do. Technological change has been a major influence on the way we go about our work. Today in the post-industrial economy, few of us are directly engaged in the actual production of things. Nor do many find it necessary to work from sunup to sundown in order to secure a livelihood. All in all, technological change appears to have lightened our physical workload substantially. Yet by now it should be apparent that the consequences of technological change are not always simple and straightforward. The development of technology over time has not always resulted in diminished workloads or an improved standard of living; in fact, the opposite has sometimes been the case. In this chapter we will examine the relationship between work and technology before the onset of industrialization. The facts and concepts developed here will provide perspectives that should be helpful when we consider subsequent changes in work and technology.

Working with the Earliest Tools

The history of work is virtually synonymous with the history of the human species. To trace the influence of technology on work we need to begin with our likely ancestors, the Australopithecines, who were shaping and using stone tools at least 2.6 million years ago. Paleoanthropologists are not in agreement as to which group of primates are entitled to be put in our genus, *Homo*, but there is no doubt that making and using tools was one of their distinguishing characteristics.[1] Although early stone tools have a crude appearance, this does not mean that the productive technologies early humans employed were primitive. From the outset, it is important to draw a distinction between the apparent crudity of the product and the skill that went into making it. Technologies in "primitive" societies are simple because only a few tasks are involved in the productive process, and a single person usually performs all of them.[2] In these societies the division of labor is limited; a worker does everything required to produce something, and the work process is not fragmented into separate tasks allocated to different individuals.[3] There may be some degree of specialization; a few jobs may be the exclusive work of women or men, or a worker may specialize in the production of a particular good, although again the task is his or hers from start to finish.[4] But most workers in nonindustrialized societies do not specialize in a single task, so they are not likely to be virtuosos in a particular area. Most people can do most jobs. They are, as the old description has it, jacks of all trades and masters of none.

A lack of specialized skills does not mean, however, that workers in nonindustrialized societies are lacking in technical ability. Although the overall work process is simple, the individual workers may be highly dexterous and ingenious as they go about their work. Take, for example, the production of stone tools. What could be more primitive than Stone Age technology? But in point of fact, the production of stone tools is a considerable achievement. In the first place, the idea of making an implement by chipping a stone is hardly obvious. Although sea otters commonly use stones for breaking clam shells, modifying a stone so that it can serve as a better tool is unknown in the animal world. Second, the fabrication of stone tools is a task that calls for considerable skill and practice. The effective chipping of a stone tool requires the preparation of a properly sited striking platform on the stone to be chipped, and the delivery of the striking blows at the proper angle. In some cases the stone is chipped by the use of a pressure chipping tool, such as a piece of antler, which forces a flake off as pressure is applied to it. Whatever the method employed, the production of stone tools is not a simple process; most of us would experience some difficulty in duplicating the work of early protohumans and humans.[5] At the same time, new generations of toolmakers did not have to develop these techniques by themselves. The regularity of the tools' shape and design suggests that the techniqucs of toolmaking were transmitted from person to person. An educational process was an integral part of Stone Age technologies.

The use of stone implements by prehumans and early humans was a crucial factor in the development of a species of animal that would forever be characterized by its ability to make and use tools. This ability was not simply the result of biological evolution; it was at the same time a cause of it. The process through which early hominids evolved into humans began with the assumption of upright posture and the anatomical changes that made the hand capable of performing an increasing variety of tasks. One of the most important of these was the production and use of tools. In turn, tool use stimulated brain development and gave an advantage to those individuals whose mental capabilities allowed them to be the most proficient makers and users of tools. In sum, manual dexterity, tool use, and brain development evolved together in a mutually reinforcing manner.[6] By the Upper Paleolithic era, this combination of evolved hands and brains had given rise to an impressive variety of equipment: pins, awls, needles with eyes, spoons, saws, axes, pestles for grinding, scoops, and shovels.[7] Some early hunters discovered that

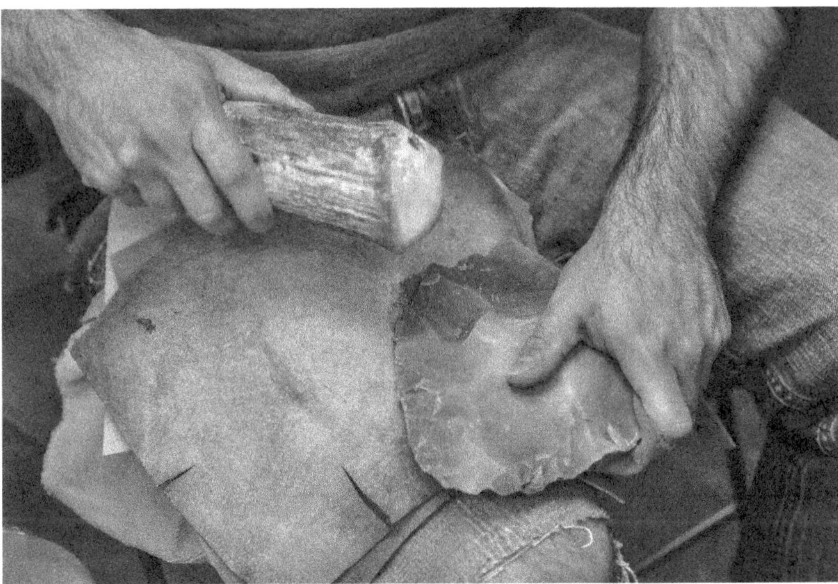

Figure 11.1 Contemporary flint knapping, Assistant Professor Metin Eren, Kent State University, circa 2017. (Source: U.S. National Science Foundation)

they could double the distance that a spear could be thrown through the use of a throwing stick (also known as an *atlatl*), long before Archimedes (c. 287–212 BC) explained the principle of leverage.

Work and Leisure in Pre-Industrial Societies

Significant as the first tools were for human development, the control over the environment that they offered was modest indeed. It is easy to visualize a life that, in the classic formulation of the seventeenth-century philosopher Thomas Hobbes, was "nasty, brutish, solitary, and short." So-called "primitive" technology seems to be associated with a precarious and insecure existence, one in which the daily struggle for mere subsistence occupied all of a person's time and energy. But once again we fail to give ancient peoples, and contemporary indigenous populations the respect they deserve. Research on the actual lives of these people indicates that Hobbes may have been wrong, and that a life of unremitting toil is not a necessary consequence of a basic level of technological development. Scholars such as Michael Adas have argued that technology was the means of colonization—weapons, ships, and interventions in production. But the *idea* of technology (remember its word origins as a concept in the early nineteenth century) was a justification for colonialism during the industrial revolution (mid-eighteenth to nineteenth century), as emerging ideas about "new" technology and productivity replaced religion as the justification for European "civilizing" missions and appropriating lands from people judged to be "primitive."[8]

Societies that sustain themselves through hunting and gathering are rare today. Still, there are a few parts of the world where agriculture and industry have not taken over and the ways of earning a living likely resemble the ones employed by our ancestors. One such place is the Kalahari Desert of southern Africa, the home of the !Kung Bushmen. It is a harsh, difficult environment, one that would seem the natural locale for the kind of life described by Hobbes. Since they possess only the simplest technologies, it might be expected that the life of the !Kung is one of unceasing hardship, and that the bare necessities of life are obtained only through continuous toil. Yet this is not the case. As the field research of Richard Lee has demonstrated, the !Kung do not have to work especially hard.[9] By gathering protein-rich mongongo nuts and over 80 other species of food plants, and supplementing these with the meat of game animals, the !Kung provide themselves with a nutritionally adequate diet. The work of gathering and hunting is not particularly arduous. According to Lee, adults put in a six-hour workday on average, about two and a half days a week. Earning a living thus requires only a total of 12 to 19 hours of labor each week. These numbers do not include times spent on food preparation and traveling to hunting grounds and areas suitable for gathering, however.[10] The remainder of the !Kung's hours are taken up with such activities as visiting and entertaining friends and relatives, and engaging in dances.

The modest work required to meet the needs of the !Kung has led one commentator to characterize the !Kung community as "the original affluent society."[11] Here, affluence is not the same thing as abundance, for the !Kung lack the material possessions common to more technologically saturated societies. A nomadic people, they have no permanent dwellings. Their clothing, as befits life in a warm climate, is minimal. They lack all but the most basic domestic utensils, and they have no way of conveying these except on their own backs.

Besides not producing many material goods, they have not produced many of their own kind. They have controlled their rate of reproduction and have avoided putting excessive pressure on their environment. The population density in their territory is only 41 persons per square mile. In contrast, agricultural societies have densities ranging from a few hundred to several thousand, while urban centers in industrial societies may pack tens of thousands of people into a square mile of space.

What the !Kung do have in great supply is leisure time. This leads to the seemingly paradoxical conclusion that technological progress may result not in the saving of labor but in its increase, or, as Marshall Sahlins puts it, "the amount of work per capita increases with the evolution of culture, and the amount of leisure per capita decreases."[12] Why is this so? The answer lies in the equivocal connection between technological change and the requirements of work. A technologically dynamic economy generates labor-saving devices, but at the same time it produces a steady stream of new goods that are eagerly sought after. This means that labor-saving technologies are generally used to increase income, not to reduce the hours of work. As workers in a technologically advanced society, we often find ourselves on a treadmill, working long and hard to obtain the material goods that we may have little time to enjoy.

Figure 11.2 A close up of Mongongo nuts. (Shutterstock/Fotografie-Kuhlmann)

The !Kung live lives of marked deprivation when measured against the material standards of most other societies. Their static economy and society have not produced the continually expanding and changing outflow of goods characteristic of technologically dynamic ones. Instead, they have maintained an equilibrium between their wants and their capacity to achieve them. Modest requirements and simple technologies have produced a way of living that lacks physical amenities, but leaves much more leisure time than can be found in our more "advanced" way of life.

Work and the Development of Agriculture

Although the !Kung way of life is a rarity today, it must be remembered that hunting and gathering was the dominant mode of existence over most of the history of humankind. This age-old pattern began to change about 12,000 years ago, when people first began to cultivate their food instead of foraging for it. After its likely beginning in the Tigris-Euphrates Valley of the Middle East, within a few thousand years farming independently appeared in other parts of the world, most notably in Egypt's Nile Valley, the Yellow River Valley in China, the Indus Valley in India, and parts of Central and South America. These were epochal developments, for in many ways the development of agriculture marked the most fundamental change in human life and the way it is sustained through work.

Why, after tens of thousands of years of successful foraging, did humans take up farming, a transition that has been dubbed "The Neolithic Agricultural Revolution"? Much debate still surrounds this question.[13] The movement from hunting and gathering to settled agriculture occurred over a long period of time in many different parts of the world, making it hard to generalize about its causes. Further complicating the matter is the lack of clearly defined boundaries separating the old ways of life from the new. Some hunting and gathering groups engaged in what might be termed proto-agriculture by encouraging the growth of certain plants and semi-domesticating animals. It also seems to be the case that in some regions hunting-and-gathering activities remained for hundreds if not thousands of years after the establishment of sedentary agriculture.[14] On a more positive note, despite a lack of a consensus on all of the causes of the Neolithic Agricultural Revolution, most scholars agree that changes in population size, the climate, and the natural environment strongly affected many hunting-and-gathering societies and, at least in some cases, impelled a turn to deliberate plant cultivation.

Whatever its causes, the result was a new symbiosis between plants and people. Domesticated grains could not reproduce themselves as they had done before humans intervened. People had to assiduously cultivate these descendants of wild grasses, and their lives increasingly depended on regular harvests of grain. Technological progress thus resulted in a greater regularity of work patterns. Also, people had to work harder and more consistently because there were more of them. Agriculture allowed considerably higher population densities than those found in hunting-and-gathering societies, but at the cost of a greater workload for each individual, and declines in human health due to pathogen exposure and sedentism in more crowded conditions, along with tooth wear and poor oral health.[15]

Farming Techniques and Patterns of Work

The interconnections between the development of agricultural technologies, population expansion, and increases in the amount of work performed can be seen through an examination of successive stages of agricultural technology. One of the earliest forms of farming, known as slash-and-burn (or swidden) cultivation, is based on cutting down indigenous vegetation and then burning it on the spot. This not only clears the land, but it also puts nutrients into the soil. A variety of crops are then planted on the cleared land. After harvesting, the plot is abandoned for a number of years, which allows the land to replenish itself before the cultivators return to begin the process once again. Slash-and-burn cultivation stands in an intermediary position between foraging and more intensive forms of farming, and it likely antedated the more settled forms of agriculture that were practiced in the aforementioned river valleys of India, China, the Middle East, and Central and South America.[16]

Although slash-and-burn cultivation alters the environment more than foraging does, on the whole it makes fewer demands on the land than other forms of farming. The soil is given ample time to recover, typically five years or more, and every effort is made to nurture the tree seedlings that will regenerate the forest. The natural variety of plant life is preserved, thus avoiding the radical simplification of the environment characteristic of settled forms of farming.[17] At the same time, slash-and-burn agriculture supports population densities greater than those found in hunting-and-gathering societies—up to 150 people per square mile—but with little increase in work effort. Slash-and-burn agriculture requires that an annual average of 500 to 1,000 person-hours of labor be expended on all phases of crop production.[18] This comes to a little more than 9.5 to 19 hours of work a week, figures that compare quite favorably with the work schedules of the !Kung. In some societies, such as the Bemba of Zimbabwe, even less work is required. Only three to four hours of work are performed each day even during the busiest agricultural seasons, and averaged over a year, only one to two hours per day are required.[19]

Slash-and-burn agriculture requires less effort than more "advanced" forms of farming, but its drawback is that it cannot support large populations. One study of three Mexican farm communities that employed different mixes of slash-and-burn and sedentary farming technologies showed that almost 14 times as many families could be supported on the same acreage when irrigated farming was used instead of slash-and-burn methods.[20] This, in fact, may be the primary reason for the development of settled, intensive agriculture. People began to practice this form of cultivation (perhaps while continuing to employ slash-and-burn techniques on other plots of land) in order to supply food to a growing population when no new land was available for cultivation. Settled forms of agriculture allowed more people to be fed because farm work was not dependent on the availability of vast tracts of infrequently cultivated land. The same farm plot could feed much larger numbers of people whose labor was in turn required by the much greater demands of sedentary farming. In extreme cases, such as that of early-twentieth-century China, with labor-intensive agricultural technologies the land could support enormous numbers of people—6,000 per square mile in some places.[21]

The paradox underlying the development of agricultural technologies until recently is that as the land becomes more productive, people have to work much harder, for the increased productivity of the land is the result of an increase in the number of workers and the amount of work that they do. Settled forms of agriculture require labor not just for the direct tasks of soil preparation, planting, and harvesting. If the land cannot be left fallow for long periods of time, it is imperative that farmers take active roles in maintaining the fertility of the soil; this requires gathering, storing, and applying organic wastes that range from animal droppings to the mud scraped from the bottom of ponds. Stubborn weeds and grasses must be removed at regular intervals. Unless rainfall is totally adequate, irrigation networks have to be built and maintained. Animals that provide power and manure must be cared for. Storage buildings have to be constructed. Farm implements need to be built and kept in repair. Plants require protection from insects, even if this means picking pests off them by hand, one at a time. The adoption of settled agriculture thus signifies a steep increase in the amount of work that has to be performed. Irrigated agriculture in particular requires sharp increases in the amount of work. Surveys of farms in India show that with the introduction of irrigation much more labor is required while the crops are growing; work is no longer largely confined to plowing, planting, and harvesting.[22] Twice as much labor may be required per unit of irrigated cropland than in the case of dry farming.[23] All in all, as sedentary agriculture develops, the amount of work required goes up sharply. In Southeast Asia, for example, an agricultural economy based on the transplantation of rice seedlings into irrigated fields requires an annual average of nearly 300 person-days of work. This situation changes only when industrial products such as farm machines and chemical fertilizers are introduced into the farm sector. This began to happen in Europe and North America a century and a half ago, and in many parts of the world it is yet to occur. In these places, the biblical injunction that "in the sweat of thy face shalt thou eat bread"[24] is a harsh reality.

The Ironies of Progress

At this point, one may question the benefits of an "advanced" technology. The three productive systems just examined represent successive stages in economic development, but it is by no means certain that they represent across-the-board progress, especially when the amount of time that has to be devoted to labor is considered. One cannot even say with certainty that they represent progressive increases in one critical component of technology, the skill exercised by individuals. The skills employed by foragers seem no less sophisticated than those employed by sedentary farmers. At the same time, however, the total skills found within the society as a whole may be greater in the latter case because the permanent communities that accompany sedentary agriculture allow specialization and a more extensive division of labor.

These comparisons of foraging, slash-and-burn, and sedentary agriculture give a strong indication that technological advance does not necessarily result in the saving of human labor. We are accustomed to thinking that new technologies result in the diminution of human drudgery, but the example of the epochal shift from hunting and gathering and slash-and-burn agriculture to settled crop cultivation shows that this need not be the case. We are left with the sobering realization that the development of agriculture, one of the greatest technological advances in human history, resulted in dramatic increases in the duration, pace, and extent of human labor, and decreases in human health, and increases in inequality.

Artisan and Craft Work

Not all of the consequences of sedentary agriculture were so grim. Settled farming was paralleled by the development of towns and cities, which were often characterized by monumental architecture and other artistic advances. Dense permanent settlements also gave rise to specialized religious practitioners, as well as more complex modes of civil and military organization. Occupational specialization flourished, allowing craftsmen to turn out higher-quality work and produce entirely new kinds of goods. Agrarian societies also made possible the development of roads and other elements of infrastructure that facilitated the work of merchants and artisans, giving a further stimulus to occupational specialization.[25]

In some cases, division of labor extended beyond workers specializing in a particular product; the work process was itself broken up. The Greek historian Xenophon (c. 430–354 BC) noted how on occasion shoe manufacture was subdivided: "One man earns a living by only stitching shoes, another by cutting them out, another by sewing the uppers together, while there is another who performs none of these operations but only assembles the parts."[26] A similar pattern can be observed in Rome during the days of the Republic, when the metalworking industry contained such occupational specialties as pattern makers, smelters, turners, metalchasers, and gilders.[27]

The change in economic relationships necessitated by specialization was even more important than the products themselves. Specialized producers need to exchange their goods and services with others. This is commonly done through some kind of market. Production for a market instead of for the direct use of family and community members implies a different kind of relationship with others. For most of human existence work was performed as part of an extensive set of mutual obligations shared by members of the same group. But when work is done in order to make goods for sale in a market, it ceases to be embedded in particular social relationships. Instead, it is done as a freestanding activity, governed by self-interest rather than a sense of obligation to members of one's group. The spirit of this kind of work was captured by a famous passage in Adam Smith's eighteenth-century classic, *The Wealth of Nations*: "It is not from the benevolence of the butcher, the brewer, or the baker, that we expect our dinner, but from their regard to their own interest. We address ourselves, not to their humanity, but to their self-love, and never talk to them of our necessities, but of their advantages."[28] A market-based exchange is one in which both parties participate in a transaction through which each expects to gain. When the transaction is completed, that is the end of the relationship. A car salesman may engage in some friendly conversation about your spouse and kids, but what he is really interested in is your money, just as your interest is getting a good price on the car. The relationship ends when they pocket your money and you drive off with the car.

An exchange system such as this, cold-blooded as it may be, has vast consequences for the way work is done. When, as often was the case in the past, work is firmly rooted in a larger set of social relationships, the way that work is organized will not be governed solely by the technical requirements of the job.[29] More will be at stake than efficiency, innovation, and the general development of production. It may be far more important to see to it that all members of the family or community are employed, or that jobs are parceled out not according to the abilities of the workers, but rather in accordance with their social position. For example, in India only members of the lowest castes engage in the tanning of leather, for this is considered an "unclean" occupation and therefore improper for a high-caste individual.

Although urbanization results in more occupational specialization and some disentangling of economic and social spheres, this separation may not be fully realized. Particular occupations may be the special province of specific social groups. In the African city of Timbuktu, until recently, butchers, barbers, slipper-makers, masons, tailors, and smiths were all hereditary tradesmen, and in the case of the smiths, almost all marriages were between members of families involved in the same occupation.[30] These alignments may be firmly buttressed by religious beliefs and practices that prevent the free movement of individuals into specific occupations. In Timbuktu it was widely believed that if someone other than a member of the Arna tribe tried to sew up a leather slipper, supernatural forces would cause the needle to jab his hand. In similar fashion, it was believed that one who was not a member of a mason family would suffer a fatal fall if he were so unwise as to attempt to build his own house.[31] In societies such as this, the division of labor and the market system have not completely taken over. Although the economy is fairly complex, economic relationships are still kept within the context of the established social order.

Figure 11.3 The caste of these leather workers in Bangalore, India, is a major determinant of their occupations and life chances. (Photo by Mike Prince/CC BY 2.0)

Guild Organization and Technological Change

Families and tribes are not the only social groupings that regulate economic activity in traditional societies. In medieval Europe, organizations known as guilds were of considerable importance. Guilds are groupings of people engaged in the same kind of occupation, be it manufacturing, trade, the provision of a service, or even begging and thievery.[32] One of the primary purposes of guilds is to restrict the practice of a particular craft to members of the guild, who also control recruitment into the guild. Close relationships between guild members are further reinforced by regular festivals and ceremonial functions, such as observance of the feast day of the guild's patron saint. These are not the only sources of solidarity; guilds also have a hierarchy of political offices that set rules, adjudicate disputes among members, and generally regulate the work activities of the members.

Guild organization had significant consequences for the pattern of technological change and for the work activities of guild members. On the one hand, the guild system facilitated the transfer of skills. Entry into a guild began with a long period of apprenticeship, which has been characterized as "the most widespread arrangement for transferring technical knowledge outside the family devised by human societies."[33] Apprentices spent much of their time on menial tasks, but they also had an extended opportunity to gain essential knowledge about their craft. Much of that knowledge involved the acquisition of tacit skills that could be acquired only through observing a master craftsman and by engaging in hands-on practice.[34] The relationship between master and apprentice was closely constrained; the apprentice agreed to work for his master and not leave or get married without the master's permission. In return, the master imparted craft skills and provided the apprentice with room, board, and clothing. Violation of these stipulations by either side could result in the payment of a cash penalty to the aggrieved party.[35]

While guild organization and practice helped to preserve and transfer skills, they also were a source of technological retardation. Although long apprenticeships facilitated skill acquisition, they also produced a kind of "trained incapacity" that locked guild members into set ways of doing things that inhibited innovation. Also, since the practice of a trade required membership in the appropriate guild, recruitment regulations and stipulated training programs set limits on who could practice a trade. In some instances, these regulations resulted in the exclusion of individuals who were not already members of a guildsman's extended family,[36] although this was not a universal feature of guild organization.[37] Many other guild regulations and procedures worked against technological change. In order to maintain their monopoly positions and keep prices high, guilds set exacting standards for workmanship and attempted to restrict output by individual members. Limits were placed on the number of apprentices and other workers that could be employed in a guildmember's shop, while deviations from normal work practices, such as working under artificial light, were prohibited. In general, technological innovations were discouraged, for the prime concern of the guild was the maintenance of the existing way of life.

The inhibitions created by guild organization meant that economic and technological progress often required the establishment of enterprises in areas remote from guild control. This occurred in the Roman Empire, where craft workshops were established on great estates in the outlying provinces, far from the cities and guild regulation. Here could be found such technological and economic

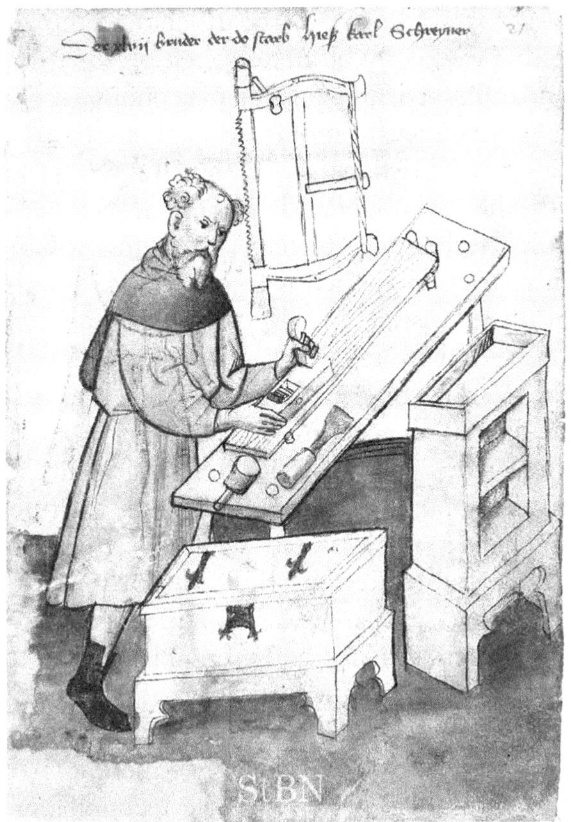

Figure 11.4 This fifteenth-century cabinetmaker practiced his trade as a member of a guild.

novelties as an establishment near Arles that boasted eight water mills, as well as glass and textile factories in eastern Gaul where work was done by wage laborers rather than independent craftsmen.[38] It probably is no coincidence that the rise of modern industry first occurred in England, where guild restrictions were weaker than they were in the rest of Europe.

Slavery and Innovation

The most extreme means of tying occupation to social status is through the operation of an institution that is unfortunately all too common in human history: slavery. The ownership of humans by other humans is a cruel and brutal system that persists today.

In times past it was a major source of labor in the empires of the ancient world; it would not be an exaggeration to say that slave labor was the foundation for a large portion of economic life. Slavery could be found in early medieval Europe, although its spread was inhibited by the belief that Christians should not wage wars against fellow Christians for the purpose of enslaving them. By the thirteenth century the church forbade the buying and selling of Christian slaves, but no such restriction applied to people of other faiths, especially Muslims. Scattered vestiges of slavery could be found in eighteenth-century Europe, but by that time the major center of slave labor was the New World, where enslaved Africans supplied much of the labor for the plantation economies of the Caribbean, the American South, and parts of South America. Its geographical limitation to these regions does not mean that slavery was of no importance to Europe and the parts of the United States where free labor prevailed. In these nations and regions a significant portion of the monetary capital produced by slave labor was an important source of funding for railroads, industries, banks, and even universities.[39]

Beyond its inherent evil, the institution of slavery shaped technological development, although not in a straightforward way. Chattel slavery had consequences for work, wage structures, worker motivation, and the development of workplace technologies. Slave labor was inefficient when compared to most forms of free labor because work had to be closely supervised with coercion as the main source of motivation. Slaves were charged with the performance of simple, if arduous tasks, but little was expected of them in the way of initiative, innovation, or attention to detail. Slaves produced great amounts of wealth for their owners, but the institution of slavery powerfully inhibited the development of a general skilled labor force, an essential component of long-term economic advance.

While slaves may have developed high skills in the trades, as cooks, millers, distillers, smiths, and in other occupations, that skill was not rewarded and suppressed the wages of free labor in similar trades.[40] It is sometimes asserted that slavery also inhibited the development and use of labor-saving technologies since slave owners paid very little for the labor supplied by their slaves, usually not much more than the cost of their subsistence. This is not altogether convincing; after all, slave owners could have increased their incomes by adopting technologies that enhanced the productivity of their slaves while still paying very little for their upkeep. When slaves themselves created innovations to improve their work, white slaveowners took credit and profited from their slaves' inventions.[41] More likely, slavery inhibited technological advance indirectly by engendering negative attitudes about work and its place in one's life.

The Greeks and Romans of the classical age often manifested a strong respect for work when it was part of a self-sufficient lifestyle. But work was despised when it was done for another—be he a master or simply a customer—because it signified that the worker was locked into a relationship of servitude that prevented him from being a free citizen, whose proper activities were directed toward political life and the pursuit of pure knowledge.[42] This attitude is exemplified by the assertion of the Greek philosopher Aristotle that "no man can practice virtue when he is living the life of a mechanic."[43] When slavery became widespread, as in the days of the Roman Empire, this attitude toward work hardened. Since more and more of it was being done by slaves, manual work was seen as an inherently degrading activity not fit for a free man. This attitude also characterized the American South before the Civil War. Although most White Southerners did not own slaves, they inhabited a culture in which manual work could easily be associated with slave labor. Due in part to the low status of manual work, the South had a weaker tradition of engineering and other technological endeavors. This was reflected in a much slower pace of industrialization than occurred in the North, and ultimately was one of the key reasons for the South's defeat in the Civil War.

The Measurement of Time and Changed Working Patterns

There are two ways to consider the relationship between work and time: in terms of the amount of time that is expended on work, and in the way that it is scheduled. Technological change altered both of these. As we have seen, in technologically simple societies where subsistence comes through foraging or slash-and-burn agriculture, relatively little time is expended on work. With the introduction of settled farming and the beginning of urbanization, the hours of work lengthened. Still, the sharp separation between work and other activities that we take for granted today was not nearly as pronounced.

Work was intertwined with social events, religious observances, and community social activities. Moreover, the scheduling of work activities was much more lax. The work week was punctuated with religious feast days and other excuses for the avoidance of work.

For the farmer, certain times of the year, such as sowing and harvesting, required long hours of work, but during other times of the year some leisure time could be enjoyed. Although a craftsman might spend long hours at his place of business, not all of his time was taken up with productive activities. Many hours were spent socializing with neighbors and customers. A good deal of time was spent on haggling over prices with prospective customers, an activity perhaps done more for the sheer pleasure of bargaining than for any real economic purpose. Concern about time did not pervade workers' consciousness. There was little of the frenetic desire to "get ahead" that is so typical of many inhabitants of the modern world. Jacques Le Goff's description of the Middle Ages can be taken as typical of the general attitude: "On the whole, labor time was still the time of an economy dominated by agrarian rhythms, free of haste, careless of exactitude, unconcerned by productivity—and of a society created in the image of the economy, sober and modest, without enormous appetites, undemanding, and incapable of quantitative efforts."[44]

Above all, the idea of keeping to a fixed work schedule rarely took hold. As Gideon Sjoberg describes work patterns in a preindustrial city, "Merchants and handicraft workers generally do not adhere to any fixed schedule. Shopkeepers open and close their shops as they see fit. They may open one morning at nine, the next at ten, and so on. The lunch hour is likely to be longer on some days than others. Ambulatory merchants, likewise, are apt to keep rather irregular schedules."[45]

These attitudes toward time and work seem to indicate that the producer was not concerned with the maximization of income. Work was not pursued with the kind of single-mindedness we seem to take for granted in the modern world. Nor was the time devoted to work considered a scarce commodity to be carefully utilized in order to bring the maximum economic return. These conceptions of work developed slowly and only in conjunction with major changes in the way people viewed themselves and their surroundings. In the Western world, a major impetus to these changes may have been religious. The rise of Protestantism, especially in its Calvinist form, gave a new centrality to work, for it made work into a quasi-religious "calling." According to Calvinist doctrine, only those predestined to salvation could hope to enter heaven, and material success was taken as a sign of being one of the chosen few. Equally important, this "calling" was manifested not just by a willingness to work hard in order to gain material success but also by a systematic and methodical approach to work.[46] This spirit was clearly congenial to the regulation of work activities according to the dictates of a precise schedule.

The Clock

Important as these changed attitudes might have been, "modern" attitudes toward work, especially the way it was organized and scheduled, also were influenced by a key invention that allowed the precise scheduling of work activities: the clock. To be sure, methods of telling time had existed throughout the ancient world. Sundials, candles with marked segments, and vessels that discharged water at a regular rate were all employed in antiquity. But each had drawbacks: sundials were useless at night and on cloudy days, candles could blow out, and water clocks froze in the wintertime. Equally important, most work activities in the ancient world required little in the way of precise timing and scheduling. These requirements arose only in conjunction with a new kind of social organization: the medieval monastery.[47] By the standards of their day, these monasteries were very large enterprises, many of them containing hundreds of monks and other workers. Their cooperative efforts required precise scheduling, as did their patterns of religious observance, which required regular times for prayers, masses, and other religious practices. This was most pronounced in the monasteries following the Rule of St. Benedict, which divided days and nights into intervals of 12 hours each and required prayers to be said at sunrise, at the third hour of daylight, at noon, at the ninth hour of daylight, and at sunset.[48]

The first clocks were built in the thirteenth century, their construction motivated by a growing awareness of the value of accurate timekeeping.[49] The scheduling of activities through the use of special timepieces had brought economic benefits to the monasteries and secular enterprises, and over time the idea of orderly routines and schedules had become an integral part of city life in late medieval Europe. There is the example of Philip VI of France giving to the city of Amiens in the fourteenth century a charter that allowed the issuance of an ordinance

> "concerning the time when the workers of the said city and its suburbs should go each morning to work, when they should eat and when to return to work after eating; and also, in the evening, when they should quit work for the day; and that by the issuance of this ordinance, they might ring a bell which has been installed in the Belfry of said city, which differs from the other bells."[50]

It was in these urban centers during the early fourteenth century that the mechanical clock began to exert its influence. The ringing of the town bell could be specifically tied to the needs of emerging occupations, as when the governor of the county of Artois granted to the government of the city of the same name the right to construct a special belfry because of the "cloth trade and other trades which require several workers each day to go and come to work at certain hours."[51]

In addition to laying the foundation for a more regularized pattern of work, the clock embodied all of the key characteristics of a machine. It used an external source of energy (a spring or a falling weight), unlike traditional tools and devices that required human or animal muscle power. The even distribution of this energy required some sort of regulator, as typified by the pendulum discovered by Galileo and first applied to timepieces by Christiaan Huygens in Holland. The clock's operation was automatic, requiring little human intervention. Finally, the clock put out a standard "product"—hours, minutes, and seconds.[52]

In producing this standard product, the clock had made time into a kind of substance. Without getting into deep philosophical waters, we can assert that time as we tend to think of it does not exist; when we think or speak about time what we are actually dealing with are *measures* of time: seconds, hours, decades, or centuries. These units of time are then taken to have real substance and are treated as valuable commodities. Hence, we use phrases like "saving time," "wasting time," "spending time," and "buying time." Note, too, the assertion that "time is money."

The conversion of time into a commodity reinforces the rational spirit that was briefly described in Chapter 1. It stimulates the conscious choice of specific means for the attainment of given ends

Figure 11.5 Based on a discovery by Galileo, this clock mechanism uses a pendulum to regulate its motion. (Catalogo collezioni/CC BY 4.0)

and provides a basis for judging the worth of a particular activity relative to others. Individuals in a society obsessed with time find it difficult to go about their work in an unplanned and irregular fashion.

The precise scheduling of work received further emphasis as new mechanical technologies began to make their appearance. Spinning jennies, power looms, flour mills, and refractory furnaces, as well as the water wheels and steam engines that ran them, called for a mechanically paced way of working that was profoundly different from the old ways of doing things. The process that began with medieval monasteries and thirteenth-century clocks moved inexorably toward the conversion of most work into routinized procedures governed by artificial time schedules. In Chapter 13 we will look at these historical developments and attempt to bring them up to date. But before analyzing the effects of technological development on the quality of work today, we will examine an issue even more fundamental: the prospect that technological advance might leave many people with no work to do.

QUESTIONS FOR DISCUSSION

1. After reading about the work habits of the !Kung, does trading places with one of them begin to seem attractive? How would a member of !Kung society feel about such a swap? All in all, who would be better off?
2. Which of the following two statements do you find more satisfactory: (1) "The development of agricultural technologies was beneficial to humankind because it allowed more people to be supported," or (2) "The development of agricultural technologies was harmful to humankind because it required people to work harder"?
3. The development of a market economy stimulated technological and economic development, but at the expense of a rather cold-blooded approach to human relationships. Is this a fair trade-off? Under what circumstances can marketplace exchanges coexist with emotionally satisfying interpersonal relationships?
4. Are there any modern occupations that retain characteristics of guild organization? Why have these persisted?
5. How often do you look at a clock or your watch (or cell phone) during the day? Is there anything unnatural about being governed by clocks? What would happen to modern society if all the timepieces stopped, never to run again?

Notes

1. Chris Stringer and Peter Andrews, *The Complete World of Human Evolution* (London: Thames and Hudson, 2005), pp. 130–139.
2. Manning Nash, "The Organization of Economic Life," in George Dalton (Ed.), *Tribal and Peasant Economies: Readings in Economic Anthropology* (Garden City, NY: The Natural His-tory Press, 1967), pp. 4–6.
3. Ibid, p. 4.
4. Melville J. Herskovits, *Economic Anthropology: The Economic Life of Primitive People* (New York: W.W. Norton, 1965), pp. 124–152.
5. For a video demonstrating the making of stone tools, see "Flint Knapping for Beginners, level up," HuntPrimitive, www.youtube.com/watch?v=ZzJWjlYwz2A.
6. Frank R. Wilson, *The Hand: How Its Use Shapes the Brain, Language, and Human Culture* (New York: Pantheon, 1998), pp. 15–34.

7. Gerhard Lenski and Patrick Nolan, *Human Societies: An Introduction to Macrosociology* (Boulder, CO: Paradigm, 2004), p. 83.
8. Michael Adas, *Machines as the Measure of Men: Science, Technology, and Ideologies of Western Dominance* (Ithaca, NY: Cornell University Press, 1989).
9. Richard B. Lee, "What Hunters Do for a Living, or How to Make Out on Scarce Resources," in Richard B. Lee and Irven DeVore (Eds.), *Man the Hunter* (Chicago: Aldine-Atherton, 1968).
10. "Hunter-Gatherers: Noble or Savage?" *The Economist* (December 19, 2007). Accessed from www.economist.com/node/10278703.
11. Marshall Sahlins, "Notes on the Original Affluent Society," in Lee, *What Hunters Do*, p. 85.
12. Ibid, p. 86.
13. For a comprehensive discussion of the theories, controversies, and empirical findings bearing on the transition to agriculture, see Graeme Barker, *The Agricultural Revolution in Prehistory: Why Did Foragers Become Farmers?* (Oxford: Oxford University Press, 2006).
14. T. Douglas Price and Anne Birgitte Gebauer (Eds.), *Last Hunters, First Farmers: New Perspectives on the Prehistoric Transition to Agriculture* (Santa Fe, NM: School of American Research Press, 1995).
15. See: Clark Spencer Larsen, "Biological Changes in Human Populations with Agriculture," *Annual Review of Anthropology* 24 (1995): 185–213; and Katherine J. Latham, "Human Health and the Neolithic Revolution: An Overview of the Impacts of the Agricultural Transition on Oral Health, Epidemiology, and the Human Body" *Nebraska Anthropologist* 28 (2013): 95–101.
16. Ester Boserup, *Population and Technological Change: A Study of Long-Term Trends* (Chicago: University of Chicago Press, 1981), p. 17.
17. Roy A. Rappaport, "The Flow of Energy in an Agricultural Society," in *Energy and Power* (San Francisco: W. H. Freeman, 1971).
18. Robert M. Netting, *Cultural Ecology*, 2nd ed. (Long Grove, IL: Waveland Press, 1986).
19. Boserup, *Population and Technological Change*, p. 46.
20. Netting, *Cultural Ecology*, 2nd ed., p. 68.
21. R. H. Tawney, *Land and Labor in China* (London: George Allen & Unwin, 1932), p. 24.
22. Boserup, *Population and Technological Change*, p. 52.
23. Ibid., p. 39.
24. Genesis 3:19.
25. Boserup, *Population and Technological Change*, pp. 71–72.
26. Quoted in Claude Mosse, *The Ancient World at Work* (New York: W.W. Norton, 1969), p. 79.
27. Ibid., p. 104.
28. Adam Smith, *An Inquiry into the Nature and Causes of the Wealth of Nations*, originally published in 1776, available at www.earlymoderntexts.com/assets/pdfs/smith1776_1.pdf.

29. Stanley Udy, Jr., *Work in Traditional and Modern Society* (Englewood Cliffs, NJ: Prentice-Hall, 1970), p. 66.
30. Horace Miner, *The Primitive City of Timbuctoo* (Garden City, NY: Doubleday, 1965), pp. 53–55.
31. Ibid, p. 57.
32. Gideon Sjoberg, *The Preindustrial City: Past and Present* (New York: The Free Press, 1960), p. 187.
33. Stephan R. Epstein, "Transferring Technical Knowledge and Innovating in Europe C.1200–C.1800," in Maarten Roy Prak and Jan Luiten van Zanden, *Technology, Skills and the Pre-Modern Economy in the East and the West: Essays Dedicated to the Memory of S. R. Epstein* (Leiden and Boston: Brill, 2013), p. 29.
34. Pamela O. Long, "Invention, Secrecy, Theft: Meaning and Context in Late Medieval Technical Transmission," *History and Technology* 16 (2000): 223–41.
35. Steven A. Epstein, *Wage Labor and Guilds in Medieval Europe* (Chapel Hill, NC: University of North Carolina Press, 1991), pp. 66–67.
36. Mack Walker, "Hometowns and Guilds in Early Modern Germany," in Edward Shorter (Ed.), *Work and Community in the West* (New York: Harper & Row, 1973), pp. 40–41.
37. Maarten Roy Prak and Jan Luiten van Zanden "Introduction," in Prak and van Zanden, *Technology, Skills and the Pre-modern Economy*, pp. 13–14.
38. Claude Mosse, *The Ancient World at Work*, trans. Janet Lloyd (London: Chatto and Windus, 1969) pp. 106–107.
39. Dina Gerdeman, "The Clear Connection between Slavery and American Capitalism," Forbes, May 3, 2017). Accessed from www.forbes.com/sites/hbsworkingknowledge/2017/05/03/the-clear-connection-between-slavery-and-american-capitalism/#695602217bd3.
40. Ta-Nehisi Coates, "A Species of Labor We Do Not Want," *The Atlantic* (June 27, 2016). Accessed from www.theatlantic.com/politics/archive/2016/06/a-species-of-labor-we-do-not-want/623679.
41. Shontavia Johnson, "With Patents or Without, Black Inventors Reshaped American Industry," *Smithsonian Magazine* (February 16, 2017). Accessed from www.smithsonianmag.com/innovation/with-patents-or-without-black-inventors-reshaped-american-industry-180962201.
42. Mosse, *The Ancient World at Work*, p. 45.
43. Politics 1278. Quoted in Alison Burford, *Craftsmen in Greek and Roman Society* (Ithaca, NY: Cornell University Press, 1972), p. 34.
44. Jacques Le Goff, *Time, Work, and Culture in the Middle Ages*, trans. Arthur Goldhammer (Chicago: University of Chicago Press, 1980), p. 44 (author's emphasis).
45. Sjoberg, *The Preindustrial City*, p. 209.
46. Max Weber, *The Protestant Ethic and the Spirit of Capitalism*, trans. Stephen Kalberg (Los Angeles: Roxbury, 2002). (Originally published in 1905.)
47. Lewis Mumford, *Technics and Civilization* (New York: Harcourt, Brace and World, 1934), pp. 12–18.

48. Arnold Pacey, *The Maze of Ingenuity: Ideas and Idealism in the Development of Technology* (Cambridge, MA: The MIT Press, 1976), p. 63.
49. David S. Landes, *Revolution in Time: Clocks and the Making of the Modern World* (Cambridge, MA: Harvard University Press, 1983), pp. 53–82.
50. Le Goff, *Time, Work, and Culture in the Middle Ages*, pp. 45–46.
51. Ibid., p. 46.
52. Mumford, *Technics and Civilization*, pp. 14–16.

CHAPTER 12

Technology and Jobs: More of One and Less of the Other?

Mention technological change to some workers, and a cold chill is likely to go down their backs. Lurking behind many an innovation is the threat of a job lost. This is not an irrational fear. Throughout history, and especially in the modern era, the application of many production technologies has been explicitly motivated by the desire to increase productivity. Since productivity is usually measured in terms of output per worker, the consequences of productivity increases seem clear: the reduction or elimination of labor. For example, if worker productivity increases at a rate of 2 percent a year—less than the post-World War II average in the United States—the size of the economy will double in about 35 years. On the other hand, productivity growth can be cast in negative terms by noting that the production of today's output of goods and services would require only half the number of workers 35 years from now.

While technologically induced productivity gains have made our material lives much richer, they also have raised the specter of mass unemployment. If a relatively few workers can perform all the necessary labor, large numbers of people will end up with no jobs and no work to do. The potential is certainly there. According to a widely publicized study conducted by Oxford University, about half of all work operations can be automated using existing technologies.[1]

Will that potential be realized, and will that realization lead to widespread unemployment? That is the fear, but fortunately the situation is more complicated. Technology's effects on employment are substantial, but they can also be contradictory. In this chapter, we will trace the complex connections between technological change and unemployment by looking at the historical record, and by considering where the productivity improvements wrought by technological change seem to be leading us. In addition, the chapter will briefly consider how technological change has affected the distribution of income, and will note some of the consequences of globalization, a process that has accompanied technological advance that also influences employment and worker remuneration.

The Technological Threat in Historical Perspective

A cursory examination of technological change reveals many examples of jobs lost because of technological advance. In 1920, American railroads employed 113,000 locomotive engineers and 91,000 firemen. In 2002, these railroads carried a much greater volume of freight but needed only 40,000 engineers and no firemen to do so.[2] During the early 1980s, studio musicians saw recording jobs erode by more than a third as a result of the growing use of musical synthesizers.[3] Over a five-year period beginning in 2002, steel production increased by almost 5 percent, yet employment fell 10 percent. During the same period, corn production went up by 30 percent while farm payrolls dropped by about the same percentage.[4] Technological change has even diminished employment prospects in as unlikely area as scouting baseball prospects, With the rise of computer-based analytics, many major league teams are now relying on sophisticated statistical analyses in the place of experienced scouts when accessing the value of potential major league players.[5]

Striking as these developments have been, they are not unique to our times. Beginning in the late eighteenth century, the power loom forced many handloom weavers out of their traditional craft, and those who stayed on did so only because they were willing to accept miserably low wages. Many similar examples could be extracted from the history of the Industrial Revolution. The introduction of machinery threatened the livelihood of large numbers of workers and produced a considerable amount of social unrest. Attempts to introduce the spinning jenny into the English woolen trade during the late eighteenth century resulted in numerous riots. By the early nineteenth century, worker resistance to new machinery was widespread in several regions of England, culminating, as we saw in Chapter 2, in the machine-smashing Luddite outbreaks.

Figure 12.1 Baseball scouting has been transformed, as an occupation, with the emergence of new data analytics. Some argue the games themselves are changing. (Intel Free Press/CC BY 2.0)

Opposition to new technologies because of their consequences for employment has not been confined to the workers whose jobs were threatened. The fear of the effects of new technology on employment has on occasion moved public officials to take drastic actions. In 1638, the British government banned the use of "engines for working of tape, lace, ribbon, and such, wherein one man doth more amongst them than seven English men can doe."[6] A more extreme (although perhaps apocryphal) example comes from the Polish city of Danzig, where in 1661, the municipal authorities destroyed a mechanical ribbon loom and drowned its inventor, for fear that the new device would put handweavers out of work.[7] Consignment to a watery oblivion was also advocated for a mechanical cotton picker by the Jackson, Mississippi, *Daily News* in the late 1930s: "It should be driven right out of the cotton fields and sunk into the Mississippi River, together with its plans and specifications. Nothing could be more devastating to labor conditions in the South than a cotton-picking machine."[8]

A Case for Optimism

Examples, however numerous, do not conclusively prove that technological advance is incompatible with high levels of employment. Although there can be no dispute that particular technological changes have resulted in job losses and attendant personal tragedies, this does not tell the whole story. The effects of technological change are complex, and it is easy to fix upon one particular result and ignore others. Rather than dwell exclusively on the fate of particular industries and occupations, it is essential to consider the larger picture.

When considering the effects of technological change on the overall level of employment, two things must always be kept in mind. First, the loss of *jobs* is not the same thing as loss of *work*. Throughout history particular jobs have diminished in number or disappeared altogether. Who is employed as a mule driver, linotype operator, or elevator operator today? Yet while these occupations have faded away, technological change has created many new jobs such as web designers and genetic counselors that did not exist less than a generation ago. Second, even if all available technologies were used to their utmost, there would still be plenty of work to do. In many parts of the world today, including many "developed" countries, there are still large numbers of people who lack the basic necessities and amenities of life. More and better food needs to be produced and distributed, more houses built, more clothing manufactured. People need to be educated, healed, and entertained. Even with high levels of affluence, needs and wants do not decrease. When it comes to the consumption of goods and services, it is likely that high levels of consumption do not diminish the desire for still higher levels of consumption; appetite may grow with eating. Although it may not be a particularly noble trait, the desire to acquire more and more is present in many human beings. As a result, progressively higher levels of economic development may generate the need for continued high (or perhaps higher) levels of production.

As economists have noted, scenarios of widespread unemployment induced by technological change are predicated on what has been called the "lump of labor fallacy," which postulates that there is only so much work on hand for a given workforce. But as we have just seen, there is no intrinsic reason that work cannot expand as more people enter the workforce or lose the jobs that they had held. At the same time, however, it has to be recognized that the realization of potential job opportunities will depend on having a labor force with the requisite skills (or the ability to develop them) and a political and economic system able and willing to pay for the work that needs to be done.

The seemingly endless expansion of needs and desires can be seen clearly in the field of medicine. As we have seen in Chapter 9, despite stupendous achievements in medical technology, we find that the monetary and human resources being devoted to health care have increased. Indeed, the rising cost of health care, which in 2020 took 19.7 percent of the gross national income of the United States, has become a major social and political issue.[9] New medical technologies have generated new possibilities and new hopes. Instead of accepting sickness and death as the inevitable working of fate or God's will, we expect that cures will be found and that medical personnel will be available to administer them. It should therefore come as no surprise to find that employment projections indicate that the demand for health care workers has increased year after year and in all likelihood will continue to do so in the future.[10] As this example indicates, technological advances will not lead to job losses if the demand for products and services increases at the same pace as increases in productivity. It is also important to bear in mind that the level of demand is affected by the price of a product or service. If productivity increases bring down the cost, more people can afford it, thereby increasing the size of the market for that product. This, in turn, may motivate an employer to take on more workers so that the increased demand can be met. Much of the history of mass production conforms to this pattern, as exemplified by the automobile industry and the interrelated increases in productivity, employment, production, and purchases that began when Henry Ford began to produce large numbers of Model T cars during the second decade of the twentieth century.

In other cases, lower prices may not result in significantly higher levels of demand, because people will buy only so much of a particular product irrespective of its price. In the jargon of economists, such a product is "price inelastic." Even so, since consumers pay a lower price for this product, they have more money left over to spend on other things. This increases the effective demand for other products and enhances employment prospects in the industries that make them. To be sure, many economists believe that there may be times when aggregate demand will be insufficient because consumers prefer to save the extra money rather than spend it on new purchases. According to many economic historians, this phenomenon of "underconsumption" may in fact have been the underlying cause of the Great Depression of the 1930s. Under such circumstances, governments may have to stimulate demand by increasing their own expenditures, even if this means running budgetary deficits.[11]

As a final point, it is also important to remember that technological change often generates problems that require a great deal of work be done. Animal and plant habitats need to be restored, bodies of water require purification, and victims of accidents and pollution need medical care. Many nuclear power plants are nearing the end of their useful lives, and safely decommissioning them will entail years of labor by thousands of workers. In general, the cost of environmental cleanups is hardly trivial; as was noted earlier, the United States will have to spend billions of dollars each year for many years just to deal with hazardous wastes and old refuse dumps. The funds spent on environmental cleanup and protection can be the basis of many jobs.

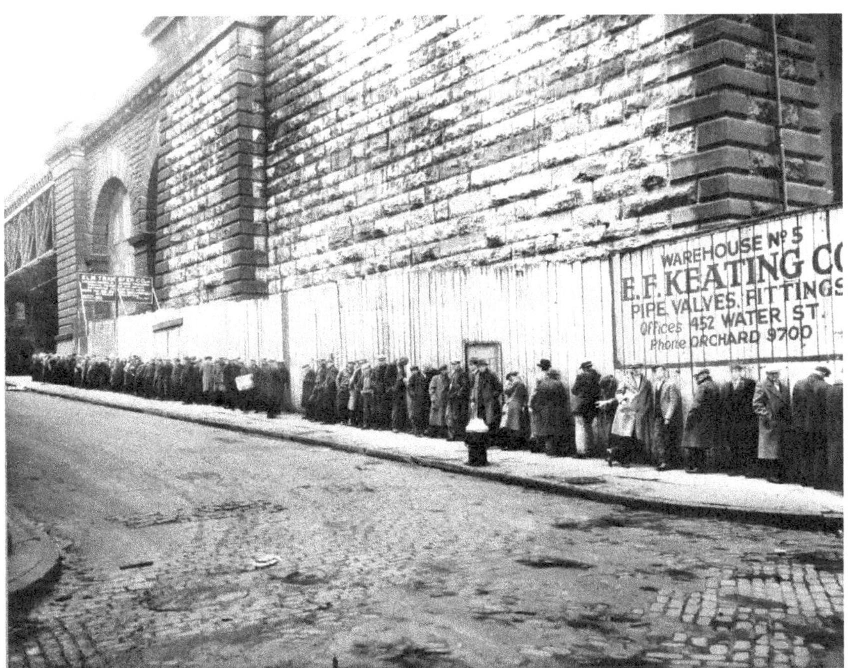

Figure 12.2 A breadline in the Great Depression.

How Technological Change Creates Jobs

Fixing upon particular examples of jobs lost to technological advance can lead to a kind of tunnel vision. While some technologies destroy existing human jobs, others produce entirely new ones. No one worked as a locomotive engineer before the coming of the railroad, and there was no need for X-ray technicians before the twentieth century. In fact, most of the occupations held today did not even exist one hundred years ago. This trend can be expected to continue. Not so long ago, occupations such as global positioning system technician, computer graphics animator, drone pilot, or social media strategist did not exist, but they are becoming increasingly significant as the technologies that gave rise to them are developed and deployed.

While the historical record shows that new technologies can give rise to whole new industries, it is often difficult to determine in advance the occupational consequences of a new invention. When the transistor was invented, it was at first thought that its primary application would be simple substitution: the replacement of vacuum tubes in amplifiers.[12] Because of its small size and low power requirements, the transistor's first practical application was to make hearing aids more compact and efficient. Today, transistors and other solid-state devices are the foundation of industries unknown a few decades ago—personal computers, smartphones, e-book readers, and MP3 players, to name but a few.

General technological advance has created a host of new products and the jobs necessary to make them. In this regard, too, actual events have run well ahead of many predictions. In 1963, one economist warned that American households were saturated with domestic appliances, resulting in a stagnant market for manufactured goods; the only significant new product was the electric can opener.[13] Today, if he is still alive, the man who made this prediction is likely to live in a house equipped with a smartphone, big-screen HD television, food processor, microwave oven, Blu-ray player, personal computer, and Wi-Fi router. He might even have embraced the economic and environmental benefits of solar energy by having solar panels installed on his roof. These products, all of them virtually nonexistent as consumer items five decades ago, have generated many new jobs in manufacturing, marketing, and servicing.

The Indirect Effects of New Technologies on Employment

In the past, a single new technology could be responsible for a large proportion of the new jobs that came into existence. Today's economy is much more complex, so no one technology is likely to have the same impact. It is unlikely that any recent technological innovation will equal the greatest job generator of the twentieth century: the automobile. Although direct employment in the automotive industry as a whole accounts for only about 5 percent of the workforce, the approximately 280 million automobiles[14] on U.S. roads provide employment for legions of mechanics, sales personnel, and insurance agents, as well as many workers whose jobs are indirectly tied to the automotive culture and are employed in everything from oil refineries to fast-food restaurants.

None of the new industries that have emerged as a consequence of recent technological change is likely to match this record. The jet airliner, as emblematic of our present transportation system as the automobile, has not led to the direct creation of many jobs; relatively few people are employed by America's airlines and passenger aircraft manufacturers. The computer, often hailed as the most significant invention of our age, has not added a great number of manufacturing and service jobs, and many of the former have been relocated to low-wage countries. The same can be said of the television industry, and just about any of the industries that owe their origin to recent technological changes.

At first glance, it may seem that the pessimists have got it right: technological development in established industries destroys jobs, and the occupations created by new technologies do little to offset the losses. But these conclusions only follow if one looks at only the specific industries without considering indirect effects on employment.

Consider the airline industry again. It is true that despite the industry's dramatic growth in recent decades, employment has gone up only slightly. But at the same time, air travel has stimulated a number of other sectors. Travel and tourism, powerfully stimulated by the growth of the airline industry, are now major sectors of the world economy, accounting for 9.2 percent of global GDP, 4.8 percent of world exports, and 9.2 percent of global investment according to one reckoning.[15] The growth of jet-age travel and tourism has brought with it a great many job opportunities in hotels, restaurants, travel agencies, tourist attractions, and the like. Television has also had a considerable, if unmeasurable, impact on economic growth by stimulating a demand for new products and services that are presented on programs and commercials. In similar fashion, the spread of computers into our economy has been essential to the development of new industries, with video games and social media being the most obvious. All in all, some jobs have been lost as computers have increased productivity in clerical, manufacturing, and other occupations; at the same time new jobs are being created as computers have stimulated the growth of new or existing industries.

The Robots Aren't Ready to Take Over

When people visualize technological change in the workplace, they often conjure up an image of legions of robots doing what people used to do. There is no denying that robots have been making an impact on factory production. By the end of 2020, more than 3 million robots were expected to be employed in the world's factories.[16] Moreover, robots are only the final step in a more general process of automating production. Computer-aided design (CAD) systems make it possible to draw a design on-screen, change its size and shape, call up stored shapes, and rotate the design to see how it looks from different perspectives. The design can then be tested through the use of computerized procedures. When the design work is completed, the new product can be built by a computer-aided manufacturing (CAM) system, in which computerized instructions run the robots and other machinery that fabricate and assemble the parts that have been designed with the help of computers. Computerized systems can also be used to organize the manufacturing process by scheduling production, directing the movement of parts, and generally keeping everything under control.

Industrial computers are also the basis of flexible manufacturing systems that alter the basic rules of mass production. Instead of using special-purpose machines that do only one thing, it is possible to use general-purpose machines that can be quickly reprogrammed to do a variety of things. For example, automobile manufacturers can now rapidly reprogram assembly-line robots to do different welding operations, depending on what is coming down the line, all of which happens in response to a barcode affixed to a rail running below the vehicle.[17] Flexibility of this sort is especially important for the manufacture of products that are made in small quantities or undergo rapid changes in specifications.

At present, these technologies have appeared piecemeal in most industrial settings; they are "islands of automation" on the shop floor. Over time, it is hoped, they will become part of an integrated system that governs the total manufacturing process from initial design to the testing of the completed product. When this occurs, a great deal of production will take place with very little direct human intervention.

As exciting as these developments are, the advance of industrial automation will be a slow process. Automated manufacturing systems are very expensive, with each workstation costing up to a million dollars. More important, many technical problems remain unsolved. There are still many difficulties surrounding the transport of materials from one station to another without human intervention. Robots

Figure 12.3 Robots in the automotive industry are expected to be a $13.6 billion dollar market. (Shutterstock/Rainer Plendl)

are particularly difficult to adapt to assembly operations, and a lot of hand labor, much of it done in low-wage countries, is still used for assembly operations. Robots and other elements of automated production technology have been eliminating some jobs, but there are limits to how far this process can go. Manufacturing systems that rely on automated production technologies also operate on the thin edge of disaster. If a robot or other computer-controlled process fails, production comes to a complete halt and everything stacks up behind the obstacle until human workers sort out the problem. Even worse, a small programming error can cause thousands of dollars of damage to machines and material in a few seconds. The challenges of programming, controlling, and maintaining computer-controlled industrial processes have put a high premium on skilled workers, who remain in short supply.[18]

Boeing discovered the limitations of fully automated assembly when it tried to rely on robots for the assembly of fuselage sections for its 777X jetliner. As it found to its sorrow, the robots were not up to the task, and flesh-and-blood workers had to take over the task of installing 60,000 rivets for each airplane.[19] Even more dramatic was Elon Musk's failed effort to fully automate the production of Tesla Model 3 sedans in Fremont, California. It soon became apparent that the complexity of the operation gave rise to small errors and unforeseen problems that cast the operation into "production hell." It soon became apparent that under these circumstances there is no substitute for human adaptability and ingenuity. Musk's initial failure to recognize this fact led to his observation that "excessive automation was a mistake" and that "humans are underrated."[20]

Problems like these have limited the advance of automated production. For all the talk about robots, panicked and otherwise, they constitute a relatively small segment of the total production of goods and services. This is reflected in modest financial investments being made in robots. In 2016, $11.3 billion was spent on robots in the United States, a tidy sum to be sure. The market is expected to be $15.7 Billion in 2022, and possibly $30.8 Billion by 2027,[21] but still only about one-half what Americans spend on their pets.[22] Yet purchases of robots reflect modest productivity gains; from 1947 to 1973 labor productivity increased by an average of 2.8 percent a year. In the years that followed it has grown by less than 3 percent a year—despite adding robots, productivity changes have been small and predictable. The pandemic years produced measures of higher productivity due to reduced employment.[23] If robot-based production is superior to an economy based on human workers, why has the economy not had higher productivity gains? Part of the answer lies not in the deficiencies of robots, but in the changing structure of the economy. Here in fact is another reason that technological advance is not producing widespread unemployment, an issue to which we will next turn.

Technology, Jobs, and the Changing Structure of the Economy

Although fully automated industrial processes have been only partially realized at best, there is no denying that they have made substantial inroads into manufacturing employment. Robots, computerized controls, and CAD-CAM systems have reduced the need for human labor in many industries, a process that will continue in the years to come. But alarming as this prospect may appear, it is nothing new. Over the past century, many of the jobs performed by manufacturing workers have been lost to mechanization, yet total employment has undergone a massive increase. This has occurred because the bulk of job creation, especially in recent decades, has not been in the production of things but in the supply of services, segments of the economy including, but not limited to, education, medicine, entertainment, law enforcement, wholesale and retail sales, and transportation. In 1950, 18.1 million workers were employed in manufacturing, construction, and raw materials extraction, while 26 million worked in the service sector, both public and private. By late 2021, with many millions of additional job holders, these figures were about 20 million and 126 million, respectively.[24]

In the 1950s, manufacturing, construction, and raw materials extraction were pillars of the economy. These occupations are still important, but they have steadily declined as sources of employment. The number of these jobs has fallen sharply both absolutely and relative to service occupations. From 1998 to 2008 manufacturing, construction, and resource extraction lost more than 600,000 jobs, rebounding through 2021, but with only modest change expected through 2031.[25] Even these figures may underestimate the extent to which services have been numerically dominant in modern economies. Many jobs in the manufacturing sector are actually service occupations. Independent consulting engineers provide a service, and are put in this category for statistical purposes, but if they perform similar duties as members of the engineering staff of a manufacturing enterprise, they appear as manufacturing employees. The number of workers performing service-type functions in the manufacturing sector is large and growing, another indication of the changing nature of work.

Many occupations fall into the service category, and not all of the jobs in this sector of the economy are equally desirable. Workers in car washes perform a service, just as neurosurgeons do. What many of these varied services have in common, however, is that they incorporate a personal element. This in fact is what makes them desired by consumers who have a surfeit of manufactured goods. For many people, the most sophisticated machine is no substitute for human contact. This desire will likely retard the replacement of human service providers by even the most sophisticated technologies. Computer-based expert systems can do an excellent job of diagnosing certain illnesses and even performing some surgical operations, but they fall short in providing a significant element in recovery from many illnesses—a patient's confidence in the services provided by a skilled medical practitioner.

The desire for services shows no signs of abating, and this sector will generate many future jobs. Still, relying on services to drive the economy and provide jobs seems to violate common sense. The basis of human survival is food, clothing, shelter, and the implements necessary to sustain life. How can the majority of the workforce be employed in occupations that directly contribute nothing to production? In fact, much economic progress can be interpreted as an ongoing reduction of the number of workers employed in producing the necessities of life. Two hundred years ago, more than half of the population of Europe earned their livelihood by growing food, as is the case in many nonindustrialized countries today. With the rise of industrialization and the provision of such things as irrigation pumps, mechanized farm implements, and chemical pesticides and fertilizers, the labor requirements of agriculture steadily dropped; today, fewer than 5 percent of the workforce in most industrial countries is engaged in farming. The same thing could happen in manufacturing in the years to come.

The displacement of agriculture by industry was closely tied to the rapid progress of technology during the nineteenth and twentieth centuries. In the case of the service sector, however, modern technologies are often of lesser importance. This is a major reason for the slow growth in productivity that has been characteristic of the American economy in recent years. While manufacturing employees have made much use of machines that embody increasingly sophisticated technologies, many service occupations have been only lightly touched by them. An attorney's office will make heavy use of information technology and a hardware store will use computers to track sales and inventory, but the main work of these enterprises entails face-to-face communication of the sort that only humans can provide.

Perhaps the clearest example of the labor-absorbing quality of services can be found in education. Despite efforts to use audiovisual aids, computer-based massive open online courses (MOOCs), and machine-scored exams, most teaching is conducted much as it was in the time of Socrates: through lectures and discussions, both of which require some degree of personal contact. Since much of the appeal of a service has to do with the human contact it provides, the more "productive" a service job is, the less satisfying it may be for the consumer. A teacher could greatly increase his or her productivity by lecturing to a class of 500 rather than 25, but most students would find so "productive" a class much less enjoyable, and the lack of personal attention would probably result in lowered levels of learning for many of them. It is also significant that one of the most successful online educational programs, Khan Academy, is most effective when students have the opportunity to interact on a face-to-face basis with human teachers who go over students' homework, monitor their progress, and help them with any difficulties they may have encountered.[26] As many of us experienced with the transition to online instruction due to the 2020–2022 COVID-19 pandemic, inequalities in online access, student disengagement, staffing losses, and decreases in instructional effectiveness led to significant learning losses, often exacerbating existing inequalities in schooling.[27] Add the fact that schools as physical places can be central to local communities, and the downsides to an entirely online educational universe become apparent.

To be sure, a good many services have experienced sizable productivity gains through the introduction of labor-saving technologies. The classic example of this can be found in the telephone industry.

If telephone calls were still manually switched as they were in the 1920s, today's level of telephone traffic would require no fewer than 50 million operators. Although the impact may not be as profound, it is likely that many of today's services will also experience technologically induced productivity gains and corresponding drops in employment. Email, word processing, automated banking, and the like have undoubtedly caused job losses, and will continue to do so. But does this mean that unemployment will necessarily increase? Here, again, the example of the telephone industry may be relevant. Some actual jobs, and millions of potential ones, were lost due to the introduction of automatic telephone switching, but total employment in the telephone industry continued to rise because increased levels of demand stimulated the industry's expansion. Equally important, the expansion of telephone communications has acted as a powerful stimulant to the economy as a whole, leading to the creation of many jobs that would not otherwise exist. The job-destroying and job-creating qualities of a new technology are usually hard to predict in advance, but they are there all the same.

Technology and the Distribution of Income

At this point, it can be fairly argued that the creation of new jobs through technological change is not the whole story; also in question is the nature of the jobs created, and in particular, how well they pay. In this regard, the record of recent years has been mixed. On the one hand, technological change has resulted in the emergence of many new, high-paying occupations and business opportunities. On the other hand, technological change has been implicated in diminished economic opportunities and an overall reduction in wages for significant portions of the labor force.

Technology and Income Inequality

The growth of income and wealth inequality has been one of the most profound economic and social changes over the past four decades. To cite one indication of growing inequality, in 1975, the top 20 percent of American households accounted for 40.7 percent of the income received in the United States. By 2018, their share had climbed to 52 percent.[28] By the second quarter of 2022, the top 10 percent of the U.S. population controlled 77% of the wealth.[29] Even more extreme has been the relative gain of households in the upper 1 percent, who earned 7.4 percent of the nation's total income in the 1979 but earned 20 percent of the country's total income in 2020. As of mid-year 2023, the top 1 percent hold 31.3 percent of all wealth and more than the entire middle class—those in the 50–90th percentiles holding 28.6 percent of the total wealth. The next 10 percent (90–99th percentile) holds 37.7 percent of the wealth while the bottom 50% holds a total of 2.4% of wealth in the United States.[30]

For many analysts, wide income and wealth gaps reflect the basic structure of twenty-first-century economies, where technical skill or the lack of it have been prime determinants of wages and salaries. This common belief is summarized in a publication of the International Monetary fund: "Technological progress . . . increases the premium on skills and tends to substitute away low-skill inputs. Technological progress thus increases the relative demand for higher skills, thereby exacerbating inequality in income."[31] According to one reckoning, from 2000 to 2009, the number of manufacturing employees with only a high school diploma declined by 38.7 percent, while the number of employees with college and postgraduate degrees increased by 2.4 percent and 44.4 percent, respectively.[32]

The growing importance of educational credentials has surely diminished incomes for certain groups of workers. For workers with only a high school education or less, automated production technologies have eliminated many jobs that a generation ago paid decent wages while requiring few skills. Undoubtedly, many workers have lost well-paying jobs because their limited skills made them vulnerable to being replaced by robots. For the economy as a whole, the loss of these jobs has not resulted in widespread unemployment, because these losses have been offset by large employment increases in the service sector. At the same time, however, service-sector jobs on average do not pay as well as manufacturing jobs, although there is wide variation in both.[33]

At the other end of the income scale, great fortunes have been made by the creators of search engines, social media sites, and widely used computer programs, while more modest fortunes are being earned by men and women with expertise in cybersecurity. A lot of money has also been amassed as an indirect result of technological change. For example, advances in communications technology make it possible for a musician or athlete to gain a worldwide following. This increases the stakes for competitors in a "winner-take-all" contest. A singer or soccer player may be only slightly better than his or her peers, but if he or she is considered "the best," their fans will number in the tens of millions, which in turn will be reflected in their salaries.[34]

Other students of economic and technological change have argued that some well-rewarded skills are not technical but are essential to make optimal use of advanced technologies: "realizing the potential of digital technologies often requires a fundamental restructuring of a firm's organizational structure and operations." As Erik Brynjolfsson and Andrew McAfee summarize, "The most productive firms reinvented and reorganized decision rights, incentive systems, information flows, hiring systems, and other aspects of organizational capital in order to get the most from the technology."[35] As a result, some high earners are not just the technically proficient members of the labor force but also include well-educated men and women with a broad range of aptitudes and skills.

Technology, Employment, and a Changing Workforce

It should be evident that technological change has not been kind to workers with only a high school education or less.[36] It can also make life very difficult for older workers, especially those with limited skills. Unemployed workers over 50 usually require more time to find new jobs, and when they do, remuneration is likely to be lower than what they had received in their previous job.[37] Many simply give up and drop out of the labor force (and no longer are counted as being unemployed). The fact that technological change may have increased economic opportunities overall is no comfort to a veteran autoworker who was displaced from a well-paying job by a robot welder. Although qualified workers will be needed to program and repair robots, such jobs are not likely to go to the former autoworker. They may go to his or her son or daughter, but for a mature worker with obsolete skills, the future is bleak unless training programs and some form of income supplement can help them through a difficult period of transition. For the nation as a whole, the job losses suffered by older workers will likely be a growing problem as the population of the United States ages; whereas there were 63 million men and women aged 55 or older in 2002; by 2025, there are projected to be 103 million.[38]

Robots and automated production processes will help to offsct a shrinking workforce, while many older workers who want to go on working will require re-training to prepare them for new jobs. Although older workers are sometimes stereotyped as being difficult to re-train, studies have shown that this need not be so.[39] The biggest impediment to equipping older workers with new skills may not inhere in the workers themselves, but rather stems from the erroneous belief that old dogs can't learn new tricks.

Other transformations of the workforce include the gender transformation. For example, in the United States in 1920, women were about 20 percent of the labor force, while at present today (2020), women are 47% percent of the labor force.[40] About twice that many African American women were in the labor force in 1920, as they more often continued working after getting married. The so-called "first wave" of the Women's Movement was primarily concerned with access to the right to vote, guaranteed by the 19th Amendment in 1920.[41] The "Second Wave" of the women's movement began in the 1960s, and besides sexual freedom, focused on economic equality and equality of opportunity. For many women, their experience in the labor force in World War II was part of their motivation for participating in the Second Wave—they resisted the "obsession" with a return to domesticity.[42] And while these social changes may have been somewhat independent of workplace technological changes, baby bottles and birth control technologies certainly facilitated workplace participation. Paradoxically, household technologies, such as appliances, have not actually freed women up for workforce participation. Contemporary western household technologies are designed with a full-time homemaker, or access to domestic servants, in mind,[43] leaving women with the infamous "second shift."[44]

The COVID-19 pandemic laid bare the sociotechnical systems and infrastructures that are part of family life, and the inequalities that may be hidden by such systems. U.S. women already faced wage inequalities, and even in high-tech fields they only earn 92 cents to the dollar relative to men.[45] But the collapse of schooling and childcare services outside the home disproportionately forced women out of the labor market, to the detriment of careers and productivity, and household earnings.[46] And as some families adapted to remote work, the return to the office in some sectors seems forced, as productivity, employee satisfaction, commuting, and environmental and health impacts, were all better for remote workers, although employers seem to be primarily concerned about control.[47] So it is not only technological infrastructures that shape opportunities and workplace outcomes, but their intersection with patterns and cultural attitudes about work, family, and who does what labor at home and for wages.

Recent decades have been the scene of breathtaking technological advances, but they have also been marked by economic globalization, altered family structures, reductions in government benefits for poor people, tax policies that have favored upper-income groups, increased immigration, stagnant minimum wages, and the erosion of union power, to name only the most prominent changes over the last few decades. Technological changes have surely contributed to both job losses and widening income economic inequalities, but all of the factors noted immediately above have also been highly relevant. When assessing the importance of technological change in promoting inequality, it is worth noting that all other industrialized nations have also been the scene of significant advances in technology, but few of them have exhibited income and wealth inequalities as large as those of the United States.[48]

One of the paradoxes, particularly in the United States, is that with increasing productivity, why do U.S. workers work *more* than their counterparts in other countries, and take on gigs and side hustles?[49] What are the policies which might distribute the benefits of technological change to workers themselves?[50] Iceland is moving to a four-day work week, finding stable productivity despite fewer hours at work.[51]

One would have to go back to the 1920s to find the same degree of income inequality in the United States as exists today. Substantial wage and salary differentials are likely to persist if employment trends continue. According to the U.S. Bureau of Labor Statistics, between 2018 and 2028 some of the largest projected additions to the U.S. workforce will come from the ranks of personal care aides, food preparation and serving workers, waiters and waitresses, retail salespersons, and janitors and cleaners.[52] These are for the most part poorly remunerated jobs that require little in the way of formal education. Does this mean that the men and women holding these jobs will be condemned to low levels of remuneration throughout their working lives? Wages and salaries supposedly are based on employees' contributions to the organizations they work for, but they also reflect factors external to individual workers such as minimum-wage laws, tax policies, and the extent to which federal, state, and local governments support or oppose unionization. Technology will continue to powerfully affect the distribution of income, but it is not the only influence. There is always the possibility that government policies can be used to move income distribution in the direction of greater equality.

Technology, Globalization, and Jobs

The loss of jobs as a result of technological change is an old story. A more recent source of job losses has been economic globalization—in particular, the offshoring of manufacturing jobs to low-wage countries. Much of this has been the work of multinational corporations (MNCs). The availability of low-wage labor in the poor countries of the world is often cited as the primary reason for moving jobs abroad, but it is not the only one; MNCs may have operations in other countries because they want to be close to important markets, insulate themselves from currency fluctuations, reduce transportation costs, and avoid tariffs and other trade barriers. Access to cheap labor is undoubtedly an important motivation for transplanting jobs to a poor country, but the advantages of a low-paid workforce may be offset by other costs. A successful offshore operation requires adequate communication and transportation infrastructures, trainable workers, and at least minimally effective and honest governments.

If cheap labor were all that mattered, all of the world's manufacturing would be done in sub-Saharan Africa and other places with large pools of unskilled workers. But the cost of transportation, the availability of expert labor, and the complementary firms that provide components and manufacturing tools and parts, become regionally specific as places develop infrastructures in support of manufacturing concerns. For example, Detroit symbolizes the height of automobile manufacturing in the United States in the 1950s and 1960s, with not just the automobile plants, but machine tool manufacturers, component manufacturers, and raw materials processing all within easy reach. The region known as the Rust Belt represents the decline of U.S. manufacturing that was integrated from raw materials (steel), machine parts, components manufacture, and final assembly. U.S. manufacturing was displaced by growing global commodity chain complexity, which focused on decreasing material and employee costs.[53] Other places, like Route 128 or Silicon Valley represent the specificity of place at a convergence of expertise and manufacturing capacity, for biotechnology and the computer and software industries, respectively.[54] There is no simple formula for repeating these kinds of concentrated regional successes, and these concatenations of people, skill, infrastructure, and resources make outsourcing difficult.

Although technological change and the offshoring of jobs are separate phenomena, they are closely related. Technological change in the form of shipping containers and intermodal transport systems has dramatically lowered the cost of transporting manufactured items from low-wage countries to consumers abroad.[55] A typical case of intermodal transportation begins with a manufacturing plant loading its products into shipping containers, which then travel by truck to a port where the containers are loaded onto a ship that may take the products halfway around the world. The containers are then put on railroad flat cars bound for a distribution point, where they are offloaded onto trucks that take them to their final destination. Through this process, the costs of loading, unloading, and reloading crates and boxes are eliminated, and losses due to the effects of weather, damage, and pilferage are greatly reduced.

While improved technologies have lowered the time and expense of the transportation of physical goods, the rapid advance of digital technologies has made it possible to transfer information at close to the speed of light—company memos, contracts, technical diagrams, legal briefs, market reports, and anything else that can be digitized. This has made it possible to offshore some activities requiring expert knowledge, such as interpreting lab tests and X-rays, as well as some requiring minimal skills, such as processing bill payments and staffing call centers. As of now, however, job losses due to moving white-collar work to low-wage countries have been less evident than the loss of manufacturing jobs to these countries. Many information-related occupations will likely remain in the industrially developed world because they require a fair amount of face-to-face interaction and relationships based on trust.[56] And even when some elements of the job can be reduced to digitized communications, as has been noted previously, it will still be necessary to employ creative, well-educated workers to initiate, guide, oversee, and evaluate the kinds of changes that are essential to the realization of the potential advantages of digital technologies.

In sum, economic globalization has brought benefits by lowering the price of goods, facilitating the diffusion of modern technologies, and in general opening up the world to new ideas and ways of doing things. But it has had its unfortunate consequences as well. Throughout the developed world, empty steel mills, abandoned factories, and depopulated communities bear witness to industries and jobs lost to foreign competition. As with technological change with which it is closely connected, economic globalization has created winners and losers. Manufacturing workers have been particularly hard hit as their jobs have been outsourced to countries with cheaper labor or other inducements. As we have seen, there is no inherent reason that technological advance should lead to long-term unemployment, and the same can be said of economic globalization. Both create new opportunities, but adjustments will not come spontaneously. As will be noted in the next section, matching job skills to these opportunities will require the implementation of appropriate policies in the private and public sectors.

Rebounding from Job Losses

A general optimism regarding the consequences of technological change for employment should not obscure the fact that adjustments will not be smooth and automatic. For example, high tech job growth seems to expand some low skill labor, although not to a large extent.[57] But overall, automation does drive income disparitics.[58] There will always be a need for purposive action if technological change is not to result in a great number of individual tragedies alongside a general economic advance.

In fact, there have been few effective programs to mitigate the inevitable destruction of some jobs by technological advance. When the necessary adjustments have been made, they have been done on an ad hoc basis. A few unionized workers have been able to nullify the job-destroying effects of technological advance by striking advantageous bargains with their employers. For example, locomotive firemen were left with little to do when the diesel-electric replaced the steam locomotive, but they were able to negotiate contracts that preserved their jobs for many years. Make-work schemes, however, do not address the real problem of technologically induced job loss. Most people seem to have a real need to work, and to do work that has some intrinsic meaning. At the same time, as has been noted, technological change has not eliminated the need for work. The problem, then, lies in putting the two together by preparing workers whose jobs have been eliminated by technological change to take on jobs for which a real need exists. In an era of rapid technological change there is a fundamental need for retraining programs that help displaced workers develop new skills and prepare them for new jobs. Unfortunately, current training programs in the United States have not been up to the task. Although federal, state, and local governments have a number of job training programs in place, they are still restricted in scope and scale. In part this is due to the chronic underfunding of such programs, resulting in their inability to meet demand for them. But even with adequate funding, formal training programs can do only so much.

The most effective training programs focus on skills learned on the job. But vocational and apprenticeship programs do have some risks: early advantages in hiring, such as good employment right out of high school, may face obsolescence more quickly in the face of technological change.[59] In general, the private sector has failed to take a leading role in improving workers' skills because employers fear that their workers will use their newly acquired skills to obtain better-paying positions elsewhere. The reluctance of employers to pay for retraining also results from tax policies. Some states offer a tax credit in return for incurring some training costs, and the federal government awards tax credits for the training of targeted individuals such as veterans, but tax credits and deductions are more generous for investments in physical capital than they are for human capital. At the same time, many workers fail to take advantage of the retraining programs that do exist because these programs do not take into account workers' skills and competencies, the need for continued work and wages (even if in low-wage jobs), or the need for childcare and related household responsibilities after a full day of work. Many of the jobs eliminated by technological advance have been held by unskilled workers with limited education. A basic remedial program may thus be an essential component of a successful retraining program. Workers operating at low levels of skill are especially likely to be victimized by technological change, and it will be a major challenge to help them develop the skills that allow them to adjust to a technologically dynamic economy.[60] For example Central Appalachia (United States) and South Wales (United Kingdom) were coal extraction regions, and with coal's decline, have faced challenges in revitalizing communities.[61]

As has just been noted, one of the best places to learn new skills is by working on a real job. For many workers, however, a new job is likely to pay less than the job that was lost to technological change (or globalization, for that matter). Under these circumstances, it would make sense for a government-sponsored insurance program to supplement the new wage up to a certain level. Such a program would likely motivate a worker to take an available job rather than hold out for a higher paying one that might never materialize. It would also lower resistance to job-related technological change. So far, no country has adopted such a program on a permanent basis, but it is worthy of serious consideration in this era of stagnant wages due in part to technological change.[62]

Benefits, but Disruption Too

All in all, the fear that technological advance will lead to widespread unemployment seems unwarranted. As noted at the beginning of this chapter, technological change does eliminate specific *jobs*, but it does not eliminate *work*. While some occupations are rendered redundant by new products and processes, others are being created. To be sure, this is cold comfort to mature workers who have been replaced by new machines. They may find work in America's expanding service sector or even in businesses that have been created by recent technological advances, but in all likelihood they will suffer a cut in pay. Even those who land good jobs may do so at the cost of considerable geographic, social, and psychological dislocation. Technological change often results in the rise of some geographical regions and the decline of others. Beginning more than a century ago, the rise of the automobile industry stimulated the rapid expansion of the Detroit area. In recent decades, the development of the semiconductor industry produced substantial growth in northern California's Silicon Valley. Taking advantage of new job opportunities may require pulling up stakes and moving to an unfamiliar part of the country, leaving friends, family, and community behind.

There is no getting around the fact that technological change can be a painful, disruptive process. Occupations and jobs rise and fall, and with them the fortunes of many individual people. There are spectacular winners and devastated losers. Special efforts must be made if the latter are not to be ruined by the advance of technology.

Technological change may destroy a career in which one has invested his or her whole life. It would be irresponsible and cruel to turn a blind eye to the needs of workers displaced by technological advance.

At the same time, it would be foolish to attempt to limit the development and spread of new technologies for fear of massive job losses. Far from being the product of technological advance, unemployment is much more likely to occur during periods of technological stagnation. One British labor historian has estimated that during the late seventeenth century, half the population of England were paupers because there was not enough work for them to do.[63] This situation began to change only when a series of technological changes helped to produce the Industrial Revolution. After two centuries of technological advance, significant numbers of people are still dogged by poverty and unemployment, but there is no denying that things are much better than they once were in large sectors of industrialized nations. This, however, is not the end of the matter. Although technological advance has expanded employment opportunities for millions of people, it can still be argued that it has led to the degradation of the work they do. People may have jobs, but many are dreary, alienating, and unrewarding. In the next chapter, we will try to determine if this is really the case.

Questions for Discussion

1. List a few jobs that have been rendered obsolete by technological change. Are some of these jobs so unpleasant that they are better done by machines? On the whole, have more "good" jobs or "bad" jobs been eliminated by technological advance?
2. What new products and services have been created by technological advance during the past 20 years or so? To what extent are they "essential" parts of contemporary life?
3. Is there anything problematic about an economy that centers on the production of services instead of the production of goods? Is a country with such an economy in an inherently dangerous position when it confronts international economic, political, and military competition?

4. In what ways is economic globalization similar to technological change in regard to job losses? Does it make sense to try to block either technological change or economic globalization in order to preserve jobs?
5. Should federal and local governments come to the aid of workers who have lost their jobs because of globalization and technological change? Is this a legitimate role of government? If so, what sort of policies and programs might be formulated to meet workers' needs?

Notes

1. James Manyika, et al., "Jobs Lost, Jobs Gained: What the Future of Work Will Mean for Jobs, Skills, and Wages," *McKinsey & Co.* (November 2017). Accessed from www.mckinsey.com/featured-insights/future-of-work/jobs-lost-jobs-gained-what-the-future-of-work-will-mean-for-jobs-skills-and-wages.
2. John H. White, Jr., "Oh, to Be a Locomotive Engineer," *Railroad History*, no. 190 (Spring/Summer 2004), p. 74.
3. Jalon, Allan, "High Tech Notes: Synthesizers: Sour Sound to Musicians. *Los Angeles Times* (December 8, 1985). Accessed from www.latimes.com/archives/la-xpm-1985-12-06-mn-554-story.html; see also J. P. Kraft, Musicians in Hollywood: Work and Technological Change in Entertainment Industries, 1926-1940," *Technology and Culture* 35, 2 (1994), 289–314.
4. Peter G. Gosselin, "What's Old Is New Again in the U.S. Economy," *Los Angeles Times*, May 25, 2008, p. A29.
5. Bill Shaiken, "Squeeze Play on Scouts, Minors," Los Angeles Times (January 10, 2020) E1, E3.
6. Withold Rybczynski, *Taming the Tiger: The Struggle to Control Technology* (New York: Viking/Penguin, 1985), p. 103
7. Ibid.
8. T. A. Heppenheimer, "The Machine That Killed King Cotton," *American Heritage of Invention and Technology* 20, 1 (Summer 2004): 39.
9. See the NHE Fact Sheet at www.cms.gov/Research-Statistics-Data-and-Systems/Statistics-Trends-and-Reports/NationalHealthExpendData/NHE-Fact-Sheet.
10. Bureau of Labor Statistics, "Industry Employment and Output Statistics to 2024," *Monthly Labor Review* (December 2015). Accessed from www.bls.gov/opub/mlr/2015/article/industry-employment-and-output-projections-to-2024.htm.

11. For a review of how economists have viewed the relationship between employment and technological change, see Gregory R. Woirol, *The Technological Unemployment and Structural Unemployment Debates* (Westport, CT: Greenwood Press, 1996).
12. Ernest Braun and Stuart McDonald, *Revolution in Miniature: The History and Impact of Semiconductor Electronics* (Cambridge: Cambridge University Press, 1978), p. 57.
13. James Fallows, "America's Changing Economic Landscape," *The Atlantic* 255, 3 (March 1985): 54.
14. See: Bureau of Transportation Statistics, "Number of U.S. Aircraft, Vehicles, Vessels, and Other Conveyances," from www.bts.gov/content/number-us-aircraft-vehicles-vessels-and-other-conveyances
15. Robert Greenhill, "Preface," in Jennifer Blank and Thea Chiesa (Eds.), *The Travel and Tourism Competitiveness Report 2011*, Geneva, Switzerland, World Economic Forum (2011). Accessed from www3.weforum.org/docs/WEF_TravelTourismCompetitiveness_Report_2011.pdf.
16. International Federation of Robotics, "Robots Double Worldwide by 2020, (May 30, 2018). Accessed from https://ifr.org/ifr-press-releases/news/robots-double-worldwide-by-2020.
17. Micheline Maynard, "Yes, Assembly Lines Can Mix Apples and Oranges," *The New York Times*, August 17, 2003, p. 5.
18. Peter Whoriskey, "U.S. Manufacturing Sees a Shortage of Skilled Factory Workers," *The Washington Post* (February 19, 2012). Accessed from www.washingtonpost.com/business/economy/us-manufacturing-sees-shortage-of-skilled-factory-workers/2012/02/17/gIQAo0MLOR_story.html
19. July Johnson and Paul Shukovsky, "Boeing Ditches Assembly Robots," *Los Angeles Times* (November 15, 2019).
20. Martin LaMonica, "Tesla's Problem: Overestimating Automation, Underestimating Humans," *The Conversation* (May 2, 2018). Accessed from http://theconversation.com/teslas-problem-overestimating-automation-underestimating-humans-95388.
21. Yahoo Finance, "The Industrial Robotics Market Is Projected to Grow From USD 15.7 Billion in 2022 to USD 30.8 Billion by 2027; It Is Expected to Grow at a CAGR of 14.3%," (October 17, 2022). Accessed form https://finance.yahoo.com/news/industrial-robotics-market-projected-grow-110900908.html
22. Pet Food Industry, "Pet Industry Spending Topped Us$66 Billion in 2016," (April 2, 2017). Accessed from www.petfoodindustry.com/articles/6370-pet-industry-spending-topped-us66-billion-in-2016.
23. See the U.S. Bureau of Labor Statistics: www.bls.gov/productivity.
24. U.S. Bureau of Labor Statistics, "Employment by Major Industry Sector." Accessed from www.bls.gov/emp/tables/employment-by-major-industry-sector.htm.
25. Ibid.
26. Clive Thompson, "How Khan Academy Is Changing the Rules of Education," *Wired* (July 15, 2011). Accessed from www.wired.com/2011/07/ff-khan.

27. Megan Kuhfeld, Jim Soland, Karyn Lewis, and Emily Morton, "The Pandemic Has Had Devastating Impacts on Learning. What Will It Take To Help Students Catch Up? Brookings (March 3, 2022). Accessed from www.brookings.edu/blog/brown-center-chalkboard/2022/03/03/the-pandemic-has-had-devastating-impacts-on-learning-what-will-it-take-to-help-students-catch-up; See also the UNC School of Education study at https://ed.unc.edu/2022/05/02/study-covid-19s-impact-on-learning-varied-widely.
28. United States Census Bureau, "Historical Income Tables: Households," (August 29, 2019). Accessed from www.census.gov/data/tables/time-series/demo/income-poverty/historical-income-households.html.
29. See "Wealth Distribution in the United States in the Second Quarter of 2023," at www.statista.com/statistics/203961/wealth-distribution-for-the-us.
30. See: "DFA: Distributional Financial Accounts," www.federalreserve.gov/releases/z1/dataviz/dfa/distribute/table/#quarter:134;series:Net%20worth;demographic:networth;population:all;units:shares.
31. Florence Jaumotte, Subir Lall, and Chris Papageorgiou, *Rising Income Inequality: Technology, or Trade and Financial Globalization?* (Washington, DC: International Monetary Fund, 2008), p. 15.
32. James Sherk, "Technology Explains Drop in Manufacturing Jobs" *The Heritage Foundation* (October 12, 2010).
33. David L. Kay and James E. Pratt, "Jobs, Good Paying Jobs, and Services," *Research and Policy Brief Series, Cornell University* (May 2009). And see: Peter Dizikes, "Study: Automation Drives Income Inequality," *MIT News* (November 21, 2022). Accessed from https://news.mit.edu/2022/automation-drives-income-inequality-1121
34. David Rotman, "Technology and Inequality," *MIT Technology Review* (October 21, 2014). Accessed from www.technologyreview.com/featured story/531726/technology-and-inequality/.
35. Eric Brynjolfsson and Andrew McAfee, *Race against the Machine: How the Digital Revolution Is Accelerating Innovation, Driving Productivity, and Irreversibly Transforming Employment and the Economy* (Lexington, MA: Digital Frontier Press, 2011), pp. 41–42.
36. David Card and John E. DeNardo, "Skill-Based Technological Change and Rising Inequality: Some Problems and Puzzles," *Journal of Labor Economics* 20, 4 (2004).
37. Lydia DePillis, "Losing a Job Is Always Terrible, for Workers over 50 It's Worse," *Washington Post* (March 30, 2015). Accessed from www.washingtonpost.com/news/wonk/wp/2015/03/30/losing-a-job-is-always-terrible-for-workers-over-50-its-worse/.
38. Sara J. Czaja and Joseph Sharit, *Aging and Work: Issues and Implications* (Baltimore: Johns Hopkins University Press, 2009), p. 259.
39. See, for example: T. F. Rizzuto and S. Mohammad, "Workplace Technology and the Myth about Older Workers," paper presented at the Annual Conference of the Society for Industrial and Organizational Psychology (2005), cited in Ron P. Githens, "Older Adults in E-Learning, Opportunities and Barriers,"; and Leora Friedberg, "The Impact of Technological Change on

Older Workers: Evidence from Data on Computers," Department of Economics, University of California, San Diego, 1999.

40. Janet L. Yellen, "The History of Women's Work and Wages and How It Has Created Success for Us All," *Brookings Institution* (May 2020). Accessed from www.brookings.edu/essay/the-history-of-womens-work-and-wages-and-how-it-has-created-success-for-us-all/.
41. See the 19th Amendment: www.archives.gov/milestone-documents/19th-amendment.
42. Dana Bisignani, "Feminism's Second Wave," *The Gender Press* (January 27, 2015). Accessed from https://genderpressing.wordpress.com/2015/01/27/feminisms-second-wave-2.
43. Ruth S. Cowan, *More Work for Mother: The Ironies of Household Technology from the Open Hearth to the Microwave* (NY: Basic Books, 1983).
44. Arlie Russell Hochschild, with Anne Machung, *The Second Shift: Working Families and the Revolution at Home* (NY: Penguin, 1989).
45. Valerie Bolden-Barrett and Kate Tornone, "All Things Equal, Women Still Earn Only 92 Cents on the Dollar," *HR Dive* (March 2, 2018). Accessed from www.hrdive.com/news/all-things-equal-women-still-earn-only-92-cents-on-the-dollar/518038.
46. Nicole Bateman and Martha Ross, "Why Has COVID-19 Been Especially Harmful for Working Women?" *Brookings Institution* (October 2020). Accessed from www.brookings.edu/essay/why-has-covid-19-been-especially-harmful-for-working-women; as well as McKinsey and Company, "Seven Charts That Show Covid-19's Impact on Women's Employment," (March 8, 2021). Accessed from www.mckinsey.com/featured-insights/diversity-and-inclusion/seven-charts-that-show-covid-19s-impact-on-womens-employment.
47. Susan Caminiti, "Flexibility Is Key to Finding, Keeping Workers Even as Companies Want People Back," CNBC (October 31, 2022). Accessed from www.cnbc.com/2022/10/31/flexibility-is-key-to-finding-keeping-workers-in-return-to-office-era.html; and paradoxical decline in productivity by going back to the office: Taylor Telford, "U.S. Workers Have Gotten Way Less Productive. No One Is Sure Why," *Washington Post* (October 31, 2022). Accessed from www.washingtonpost.com/business/2022/10/31/productivity-down-employers-worried-recession.
48. Colin Gordon, "The Computer Did It? Technology and Inequality," *Dissent* (Spring 2014). Accessed from www.dissentmagazine.org/article/the-computer-did-it-technology-and-inequality.
49. G.E. Miller, "The U.S. is the Most Overworked Developed Nation in the World," *20 Something Finance* (September 7, 2023). Accessed from https://20somethingfinance.com/american-hours-worked-productivity-vacation.
50. Josh Bivens and Lawrence Mishel, "Understanding the Historic Divergence Between Productivity and a Typical Worker's Pay," *Economic Policy Institute* (September 2, 2015). Accessed from www.epi.org/publication/understanding-the-historic-divergence-between-productivity-and-a-typical-workers-pay-why-it-matters-and-why-its-real.

51. Heejung Chung, "A Social Policy Case for a Four-Day Week," *Journal of Social Policy* 51, 3 (2022): 551–566.
52. United States Department of Labor, Bureau of Labor Statistics, "Occupations with the Most Job Growth, 2014–2024" (December 8, 2015). Accessed from www.bls.gov/news.release/ecopro.t06.htm.
53. Jeff Wallenfeldt," Rust Belt," *Encyclopædia Britannica*. Accessed From www.britannica.com/place/Rust-Belt.
54. Jarunee Wonglimpiyarat, "The Boston Route 128 Model of High-Tech Industry Development," *World Scientific Publishing*. Accessed from www.worldscientific.com/doi/abs/10.1142/S1363919606001399; and for the role of universities and new knowledge: Neil Bania, Randall W. Eberts, and Michael S. Fogarty, "Universities and the Startup of New Companies: Can We Generalize from Route 128 and Silicon Valley?" *The Review of Economics and Statistics* 75, 4 (1993): 761–766.
55. For a discussion of intermodal transport and its implications for labor, see Edna Bonacich and Jack B. Wilson, *Getting the Goods: Ports, Labor, and the Logistics Revolution* (Ithaca, NY: Cornell University Press, 2008).
56. Frank Levy and Richard J. Murnane, *The New Division of Labor: How Computers Are Creating the Next Job Market* (Princeton and Oxford: Princeton University Press, 2004), pp. 13–30.
57. Neil Lee and Stephen Clarke, "Do Low-Skilled Workers Gain from High-Tech Employment Growth? High-Technology Multipliers, Employment and Wages in Britain," *Research Policy* 48, 9 (2019): 103803.
58. Peter Dizikes, "Study Finds Stronger Links between Automation and Inequality," *MIT News* (May 5, 2020). Accessed from https://news.mit.edu/2020/study-inks-automation-inequality-0506.
59. Eric A. Hanushek and Ludger Woessmann, "Apprenticeship Programs in a Changing Economic World," *Brookings Institution* (June 28, 2017). Accessed from www.brookings.edu/blog/brown-center-chalkboard/2017/06/28/apprenticeship-programs-in-a-changing-economic-world.
60. Jeffrey Selingo, "The False Promises of Worker Retraining," *The Atlantic* (January 8, 2018). Accessed from www.theatlantic.com/education/archive/2018/01/the-false-promises-of-worker-retraining/549398.
61. Tom Hansell, *After Coal: Stories of Survival in Appalachia and Wales* (Morgantown: WVU Press, 2018). And also: Alana Semuels, "Imagining a Post-Coal Appalachia," *The Atlantic* (April 8, 2015). Accessed from www.theatlantic.com/business/archive/2015/04/imagining-a-post-coal-appalachia/38981.
62. "Creative Compensation," *The Economist* (January 23, 2016): 23–24.
63. Of course, one has to define poverty, and whether it is seasonal, permanent, whether people were housed, whether they had non-employment means to support themselves, whether they received "poor relief" or charity, or if they were recorded as paying or exempt from various kinds of taxes. Defining poverty is not simply a matter of income, but includes context and standards of the time. Tom Arkell, "The Incidence of Poverty in England in the Later Seventeenth Century," *Social History* 12, 1 (1987): 23–47.

CHAPTER 13

Technological Change and Life on the Job

The word *manufacture* literally means "the production of something by hand," and for most of human history, it was just that. Goods were produced in small numbers by individual craftsmen who relied on little more than their own skills and a few tools. This situation began to change markedly in the middle of the eighteenth century as industrialization began to transform the way that things were made. To be sure, this was a revolution that was a long time in the making; people did not pick up their newspapers one day in 1750 and read "Industrial Revolution Breaks Out in England!" Many of the changes that produced an industrial society were small and slow to develop. Others were more dramatic, but even these required numerous incremental changes before they were effective. But taken together, these changes utterly transformed the way things were made, and with them the way people worked. This process continues today, with technological advance constantly changing the nature of work. In this chapter we will consider the historical relationships between work and industrialization.

Industrial Production

An economy based on industrial production has a number of special characteristics. In the first place, it requires large amounts of energy. Before industrialization, almost all of the energy used came from organic sources: human and animal muscle power and the burning of wood and other plant products. Civilization rested on the physical labor of peasant farmers and artisans, and in many places, slaves, whose efforts produced the surplus that supported the activities of small numbers of artists, priests, and government officials. The use of new sources of energy allowed vast increases in production while at the same time lightening some of the physical burdens of work.

The first new energy source for the emerging industrial economy was water power. Water wheels date back to antiquity and were widely employed during the Middle Ages; over 5,000 of them could be found in England according to the eleventh-century Domesday Book. But these were simple devices that produced little power. With the mechanization of key industries such as spinning, weaving, and flour milling, larger, more elaborate wheels began to be used in great numbers. Of particular importance was the development of water turbines in the nineteenth century, which were far more powerful than the water wheels that had been used for centuries. To an increasing degree these sources of power were built according to systematic, empirically derived principles, making them significantly more efficient.[1]

Today, far from being obsolete in an age of fossil fuels, turbines in hydroelectric powerplants supply electrical power to many parts of the world.[2]

Moving water continued to be an important industrial power source well into the nineteenth century, at which time water wheels and turbines slowly began to be supplanted by steam engines. Originally built to pump water out of mines, during the late eighteenth century the steam engine was brought to a higher level of efficiency by James Watt's invention of the separate condenser. Of equal importance, Watt devised new linkages and gearing systems so that the up-and-down motion of a piston could be translated into rotary motion, which was essential for the powering of many types of industrial machinery.

During the Industrial Revolution steam and water animated a great number of new machines. Among the most important of these were machine tools that could produce large numbers of identical parts. The possibilities of this system were first indicated in the early nineteenth century when the Franco-British engineer Marc Brunel teamed up with Henry Maudslay, a pioneer designer and builder of machine tools, to produce large numbers of pulley blocks for the British Navy. After they installed 44 machines at the Portsmouth (England) Navy Yard, their staff of 10 workers was able to produce

160,000 blocks a year, a larger output than the one that had been achieved by 110 men using traditional techniques.[3]

The installation of power-driven machinery was complemented by the establishment of a new setting for productive work: the factory. Instead of the craftsman's shop with its half-dozen or so workers, the dominant manufacturing enterprise was the large factory in which hundreds and even thousands of men and women performed the tasks that large-scale production required. As industrialization took hold, the size of manufacturing establishments grew rapidly. In 1870, one of the largest industrial enterprises in the United States, the McCormick reaper plant in Chicago, employed no more than 500 workers. Thirty years later, more than a thousand factories had between 500 and 1,000 workers, and over 400 had more than 1,000.[4] Three steel plants and a locomotive works each had more than 8,000 workers.

With the spread of the factory system, most manufacturing workers no longer worked as self-employed artisans. They now worked for a business firm and were dependent on it for the tools they used, the facilities they worked in, and of course the money they took home. Independent workers ceased to be the foundation of the economy. The typical worker was now a hired member of a firm, a wage-earner rather than an independent producer. Industrialization had created a society of employees.

The supervision and coordination of large numbers of employees, in turn, required entirely new methods of organization. Not only were far more people working in a single enterprise, while on the job they were also socially and psychologically separate from their employers and detached from their other social roles and responsibilities. As a result, the face-to-face contact and personal attachments found in the craftsmen's shop gave way to bureaucratized managerial methods. Hierarchical command structures, written rules, strict job definitions, the precise scheduling of work, and rigid procedures became typical ways of organizing production.

Machine-Paced Labor

The spread of industrial technologies helped to make work a more regular and precisely scheduled process. New power technologies were of particular importance in making work follow strict temporal rhythms. Industrial processes that made use of steam engines and water wheels were obviously unsuited to irregularly paced methods of working. Since they were expensive pieces of capital equipment, these power sources and the machines they ran had to be operated constantly if they were to be used efficiently; sitting idle they produced no income. This meant that not only were the times of work more precisely scheduled, but work had to be performed at all hours. Shift work and labor at unsociable hours made their appearance, a situation that stood in marked contrast to the life of the craftsman working under guild regulations that usually prohibited working at night or under any form of artificial illumination.

Many of the new industries that emerged in the nineteenth century gave further impetus to precisely scheduled, clock-regulated work patterns. This was especially evident in the operation of one of the era's key industries, the railroad. The size of this new enterprise, the multiplicity of tasks required, and above all the need for adhering to precise schedules made the railroad a key example of the temporal regularization of work. The emphasis on scheduling and the efficient use of time also influenced the larger economy and society. As Henry David Thoreau said of the railroads of his era:

> They come and go with such regularity and precision, and their whistles can be heard so far, that farmers set their clocks by them, and thus one well-regulated institution regulates a whole country. Have not men improved somewhat in punctuality since the railroad was invented? Do they not talk and think faster in the depot than they did in the stage office?[5]

During the railroad's earliest years, railway operational workers were required to carry "good watches and correct them daily."[6] From that time onward, the conductor's pocket-watch, accurate to a few seconds a day, became emblematic of his occupation. On a larger scale, the demands of railroads led to the establishment in 1883 of Standard Time and the division of the United States into four time zones. No longer would each community observe a unique time that was based on a noontime determined by the highest point of the sun's passage. A new technological system had produced the artificial method of marking time that we take so much for granted today.

The railroad was not the only industry that required high standards of temporal regularity in the workplace. In factories of every

Figure 13.1 A railway engineer and a conductor check their watches. Time coordination was so important that by 1887 there was a Standard Code of railway rules that described approved watches.

sort, workers had to accommodate themselves to the demands of the machinery, and in many cases the machines were harsh taskmasters. And not only was the pace more frenetic and unvarying, the work itself was often highly unsatisfying. For many of the new industrial occupations, the required work often called for only the most rudimentary skills. Many machines were designed so that the skill required inhered in the machine and not in the worker who tended it. Although the traditional labors of artisans and farmers also had their periods of drudgery and monotony, the factories that sprang up in the nineteenth and twentieth centuries seemed to usher in a new era of deadening labor—repetitious, narrowly specialized, radically simplified, and paced by the machine.[7]

Even early on in the development of mechanized, water-powered, factory production, the severity of workplace dangers and the exploitation of already vulnerable workers was noted in the news and in literature, from Dickens to Melville. Dickens comment on the Victorian workhouse system, which provided indentured labor via the "Poor Laws" that incarcerated the poor and unhoused.[8] From the early nineteenth century through the early twentieth century, local workhouses provided labor by separating families and not even providing enough food to prevent starvation. Hence, Oliver Twist's request: "Please sir, I want some more." Some parishes in England legally forbade second helpings of food. Similarly, Melville's 1885 short story "The Tartarus of Maids," described the ghastly conditions of young, unmarried women compelled to provide for their families in the paper mills of New England,[9] while the profits from such ventures allowed for comfortable lives for mill owners and financiers in private clubs in London.[10] These gendered dimensions of labor, in terms of control and exploitation, repeat themselves time and again.

The employment of large numbers of workers, each performing specialized tasks that had to be closely scheduled and coordinated in accordance with the needs of machinery, necessarily resulted in the loss of worker autonomy. Even Friedrich Engels, Karl Marx's collaborator and a strong advocate of the rights of working people, was convinced that industrial technologies resulted in a regime that was "more despotic than the small capitalist who employs workers ever has been." Engels stressed that industrial technology required strict managerial authority:

> If man, by dint of his knowledge and inventive genius has subdued the forces of nature, the latter avenge themselves upon him by subjecting him, in so far as he employs them, to a veritable despotism independent of all social organization. Wanting to abolish authority in large-scale industry is tantamount to wanting to abolish industry itself, to destroy the power loom in order to return to [the technology of] the spinning wheel.[11]

Industrial technology thus stands indicted as the destroyer of long-established ways of working that had allowed workers to enjoy at least some measure of independence. Machinery and the accompanying factory system increased production, but only at the cost of

Figure 13.2 Poultry processing is dangerous, fast-paced work, often with exploitative working conditions and worker abuse. But the (dis-)assembly line process allows for tons of poultry to be processed daily, allowing for low-cost food supplies. (Shutterstock/vodograj)

a cold, regimented, on-the-job existence. Many would consider it a poor bargain. But the story is not yet complete; before we blame the advance of technology for the degradation of work, we must consider how these technologies came into being and whose interests they served.

Is Technology to Blame?

Industrialization resulted in rapid and sustained economic growth and a massive increase in consumer goods. But at the same time, for many people it meant a thoroughly unpleasant work environment. Were industrial technologies in themselves the cause of the long working hours, monotonous routines, and the general degradation of labor often found in industrial economies? Here we again meet a fundamental question about the nature of technology: Does technology determine a basic human process such as work, or is technology itself shaped by economic and social relationships?

The history of industrial technologies seems to indicate that similar technologies can support a variety of working arrangements. The use of steam power did not always require massive factories under centralized direction. In the Coventry ribbon industry, weaving was done on power looms situated in blocks of independent workshops that were arranged so that they could make use of a single steam engine.[12] By the mid-nineteenth century, these cottage factories contained over 1,000 power looms, as many as could be found in the conventional factories throughout the city.[13]

The Coventry ribbon-weaving industry is not an isolated example. In Sheffield and Birmingham, cities at the very heart of England's Industrial Revolution, local artisans could rent workspaces supplied with transmission belts connected to a centralized steam engine; "Power to Let" was a sign commonly seen in Birmingham during this period.[14] Fundamental mechanical innovations of the early nineteenth century, such as the Jacquard loom (which introduced the use of punch cards for the guidance of an industrial operation), were used not in large factories but in small workshops that were often part of the workman's home.

By themselves, these examples do not prove much. As we have seen, a fundamental feature of technological advance is that innovations begin to be used on a small scale and are steadily expanded as experience is gained with them. Still, the late-nineteenth- and early-twentieth-century deployment of machines in factories of ever-increasing size was not simply the result of increasing technological sophistication; putting workers into large factories may also have been motivated by a need to control and discipline them.

Employers had long contended with strong-willed, independent workers under a variety of working arrangements. Before the Industrial Revolution one such arrangement was the domestic (or "putting out") system of manufacture. An employer supplied workers with tools and raw materials that were to be used in the workers' own homes. The employer then picked up the finished product several weeks later. This system gave the workers a good deal of autonomy and flexibility—too much, as far as many employers were concerned. The lack of direct supervision often resulted in shoddy products and a slow pace of production. Even worse, some workers bilked their employers by selling the raw materials, buying an inferior substitute, and pocketing the difference.

Given these managerial problems, it is easy to see the appeal that the factory had for many employers. Within the walls of the factory, the workers were required to put in regular hours while the pace of their work was controlled by foremen who closely supervised their activities. Although perhaps overstated, one scholar's summation captures an essential reason for the development of the factory:

> It was purely for purposes of discipline, so that workers could be effectively controlled under the supervision of foremen. Under one roof, or within a narrow compass, they could be started to work at sunrise and kept going until sunset, barring periods for rest and refreshment. They could be kept working six days a week. And under the penalty of loss of all employment they could be kept going almost throughout the year.[15]

The development of the factory made close managerial control possible. Even so, the closest management by itself cannot ensure regular work efforts. Unless one supervisor can be assigned to each worker, the latter will always find ways to escape scrutiny and evade work if so inclined. In extreme cases, disgruntled workers can even resort to sabotage, a word that derives from the throwing of wooden shoes—sabots—into the machinery. In a setting where workers are employees and have no direct stake in the success of their enterprise, such inclinations are likely to be widespread. In working environments like these, machinery takes on an importance that goes beyond the improvement of productive processes. Machine-based technologies can ensure that work is steadily performed and that it is performed in accordance with the requirements of the management. As one critic of management-inspired industrial technologies put it:

> Machinery offers to management the opportunity to do by wholly mechanical means that which it had previously attempted to do by organizational and disciplinary means. The fact that many machines may be paced and controlled according to centralized decisions, and that these controls may thus be in the hands of management, removed from the site of production to the office—these technical possibilities are of just as great interest to management as the fact that the machine multiplies the productivity of labor.[16]

Machines can be complex and hard to manage, but people are even more difficult. Many employers eagerly sought machine technologies as a way of counteracting the vagaries of human motivation and performance. Machines provided a model of reliable performance, as well as a way of coaxing it out of their workers. The factory system that began to emerge in the late eighteenth century was not exclusively the consequence of the technologies employed; at least as important as the operational requirements of machine-based technologies were the desires and intentions of management.

Industrial Technology and the Division of Labor

A major characteristic of an industrial society is an extensive division of labor. Instead of a single craftsman performing all of the tasks involved in making a product, production is broken into small segments, each one performed by a different worker. This reduces production costs, as it is not necessary to pay top wages to a highly skilled worker to do everything; some of the work can be done by a person with lesser skills at lower wages.

Production based on division of labor requires managerial control, but equally important, it makes such control easier to exert. If tasks are kept simple so that workers can be readily hired and fired, there will be little need to depend on workers with all-around skills. This can benefit management because the special abilities of skilled workers generate an independent spirit and a fair amount of economic leverage, as John Delorean's description of the behavior of skilled toolmakers in the automobile industry illustrates:

> These fine old tool makers worked hard and were very proud of their craft. They kept their big tool boxes right underneath their work benches. If you looked at them the wrong way or dealt with them in any manner other than a man-to-man, professional fashion, they would simply reach under their workbenches without saying a word, throw their tools into the big box, lock it up and leave. That was it. Each guy figured, "What the hell. I don't have to take this from anybody. I'm a pro. I know this business and I am not depending on you or anyone else."[17]

The ultimate marriage of machine technology and the division of labor is the assembly line. Instead of using a variety of skills and working at their own pace, workers perform specialized and repetitious tasks that are dictated by the tempo of the moving line. The development of this system of production is attributed to Henry Ford, and rightfully so.[18] After taking his inspiration from the meat-packing industry (which, of course, used a "disassembly" line), Ford and his associates realized productive gains that had scarcely been imagined in the past. Before Ford's introduction of the moving line in 1913, automobiles were produced in small batches by workers with a multiplicity of skills. The work was undoubtedly more satisfying, but the product was high in price and necessarily limited to the rich. The assembly line allowed dramatic price cuts for two reasons: it drastically reduced the time needed for manufacture, and it allowed the substitution of low-wage, unskilled labor for the high-priced labor of all-around craftsmen. Ford's description of work in his factory captures the essence of this new form of manufacture:

> One man is now able to do somewhat more than four did only a comparatively few years ago. That line established the efficiency of the method and we now use it everywhere. The assembling of the motor, formerly done by one man, is now divided into eighty-four operations—those men do the work that three times their number formerly did. . . . In the chassis assembling [room] are forty-five separate operations or stations. The first men fasten four mudguard brackets to the chassis frame; the motor arrives on the tenth operation and so on in detail. Some men do only one or two small operations, others do more. The man who places a part does not fasten it—the part may not be fully in place until after several operations later. The man who puts in a bolt does not put on the nut; the man who puts on the nut does not tighten it.[19]

The transformation of the system of production from home or craft manufacture (for family use and for sale) impacted multiple industries in the nineteenth century. Mass-produced clothing emerged in the middle of the 1800s, providing low-wage and arduous employment for women of the lower classes, and expanding clothing options for the buying public.[20] What we now see as the origins of the global sweatshop system had its origins in the extension of the factory system into other domains of manufacturing, where standardized products could be made by low-skill and low-wage, and for women, presumably temporary workers who would ideally leave the workforce upon marriage. Their temporary nature kept them difficult to organize and enhanced their economic powerlessness. As global firms seek to maximize profits along complex commodity chains, the risks of direct physical harm or environmental degradation become redistributed to non-industrialized nations which gain jobs and industrial capacity, although the extraction of profit flows toward the financial centers of the industrialized global north.[21]

The extreme division of labor found on the assembly line was paralleled by an unrelenting tempo of work and the absence of any control over it by the workers. Until recent efforts at "job enlargement" began to change the nature of automobile production, manufacturing methods followed the Ford pattern closely. Employment changes in the present also demonstrate job enlargement: we see this in workplace reductions in staffing leading to expanding responsibilities, without compensation, for remaining staff. With the

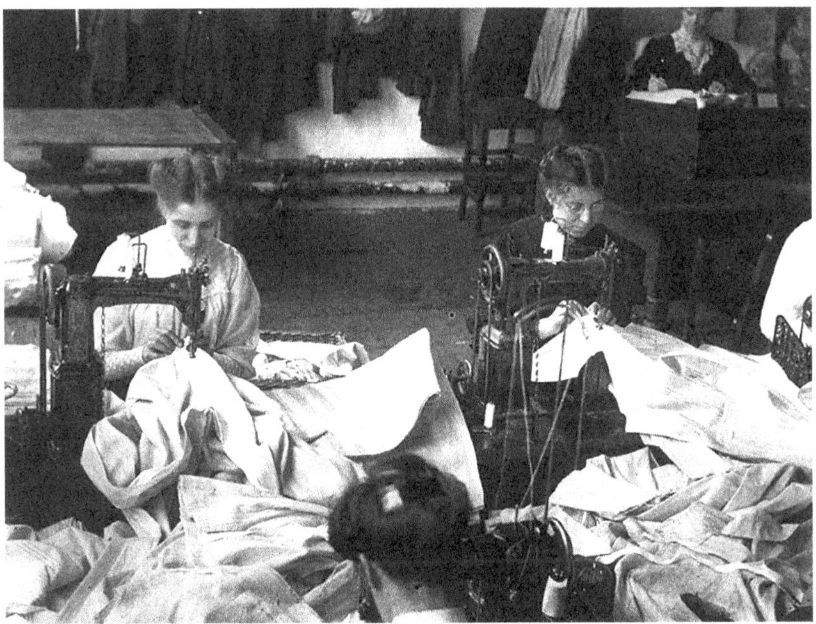

Figure 13.3 Factory girls in Manchester, England, making men's shirts in 1909.

Figure 13.4 The 1911 Triangle Shirtwaist Factory fire killed 146 women, with desperately trapped workers jumping from the building because they were imprisoned by locked doors. The tragedy inspired the first wave of factory safety reform and legislation in the United States.

typical automobile assembly line churning out 50 to 60 vehicles per hour, each worker had but a few seconds to complete an operation such as mounting a shock absorber or installing a windshield. When problems arose, the worker had to make the best of a bad situation, even if it meant installing something improperly. Under these circumstances work can be little more than a rather unpleasant way of earning a living. Extracting any sense of personal fulfillment from the work performed is just about impossible.[22]

Scientific Management Once Again

In addition to controlling the pace of work through the use of the assembly line, managers also attempted to regulate work by using elaborate systems that removed all vestiges of worker control over work processes. The most significant of these was Frederick Taylor's Scientific Management, which was briefly discussed in Chapter 2. The important thing to reiterate here is that Scientific Management stipulated the precise scheduling and organizing of work activities, and that these procedures were never to be left to the workers' discretion. As Taylor explained his system:

> Perhaps the most prominent single element in modern scientific management is the task idea. The work of every workman is fully planned out by the management at least one day in advance, and each man receives in most cases complete written instructions, describing in detail the task which he is to accomplish, as well as the means to be used in doing the work. . . . This task specifies not only what is to be done, but how it is to be done and the exact time allowed for doing it.[23]

The development of these instructions was the work of a small group of technical specialists who were not directly involved with the work being performed, for, as Taylor stipulated:

> All possible brain work should be removed from the shop and centered in the planning or laying-out department."[24] Efficient work required unfettered control by specially trained managers, who were to be the repository of all knowledge regarding the work being done: The managers assume . . . the burden of gathering together all of the traditional knowledge which in the past has been possessed by the workmen and then of classifying, tabulating, and reducing this knowledge to rules, laws, and formulae.[25]

Many would see Taylorism as a perversion of scientific means of inquiry and a travesty of engineering practice. But repellent as it may seem to us today, the spirit of Scientific Management still lives in technologically advanced societies. The explosion of knowledge that is the driving force of our sophisticated economy can fragment the population into groups of specialized experts. This may be a natural outcome of the growth of knowledge, but it takes on a pernicious quality when specialized knowledge is the exclusive possession of one group of people who attempt to use it as a means of dominating others. Under these circumstances science and technology do not simply supply detached knowledge that is used for universal human betterment. They are used to strengthen the position of a dominant group by removing the control over work processes from the hands of those doing the actual work.

Industrial Work and Recent Technological Developments

Many social critics have decried the human consequences of technologies and managerial strategies that take the skill out of human labor. Others have taken a different tack, arguing that an industrial system that "de-skills" workers is doomed to failure. A great deal of contemporary industrial work, according to this latter perspective, requires substantial amounts of independence if it is to be done effectively. A modern factory and its constituent departments cannot be expected to hum along indefinitely like a well-oiled machine, nor can all the jobs be precisely defined and completely subject to control from above. Breakdowns, disruptions, and normal deterioration are inevitable, and they all require prompt attention and often a fair amount of innovative ability on the part of the workers.[26]

Even routine tasks often require knowledge of special quirks in the machines and work processes. This knowledge can be gained only through intimate acquaintance with the work itself; engineers cannot plan everything in advance.[27] There are many occasions when shop-floor workers understand things better than their bosses, and their involvement in production processes may be essential to a firm's success. This point was strongly articulated by a General Electric executive when he noted, "All of the good ideas—all of them—come from hourly workers."[28] Workers who have been reduced to mindless operatives may not even be able to keep the machinery going when the inevitable problems emerge. Accordingly, effective managers realize that the destruction of worker skills through mechanization and rigid centralized controls is self-defeating, for an unskilled and hostile workforce results in production losses as well as high costs for maintenance and direct supervision.[29]

A precisely programmed production process that makes use of specialized machinery while sharply separating the tasks of those who plan and those who work is possible only when the product is a standardized one. This system generated a great outpouring of mass-produced goods, but it may not be appropriate to today's economy and society. With improvements in the overall standard of living, consumers are less willing to buy exactly the same things that their neighbors do. Instead, they have an appetite for goods and services with distinctive qualities. These cannot be produced by traditional manufacturing processes. As former Secretary of Labor Robert Reich has argued, rigid production technologies were appropriate only when business enterprises were "engaged in high-volume, standardized production, entailing a large number of repetitive tasks with clearly defined goals, in which almost all contingencies could be identified and covered."[30] In contrast, the production of nonstandardized items requires much more flexibility, making it difficult to replace human skills with machines and machinelike processes. At the same time, the continual modification and customization of products and procedures require the ongoing efforts of a variety of workers—designers, toolmakers, computer programmers, systems analysts, setup personnel, maintenance and repair staff, construction workers, and machinists, as well as personnel involved in marketing, distribution, sales, and service. All of these occupations require a fair degree of skill and on-the-job autonomy.

The production of standard products for a mass market creates a fertile ground for the extreme division of labor and the de-skilling of work that was typical of many businesses during the nineteenth century and a good part of the twentieth. But when consumers continually demand new, nonstandard products, existing ways of producing things no longer make sense from an economic standpoint. Firms that cling to the old routines not only oppress their workforce, they run the risk of eventual bankruptcy.

Technological Change and White-Collar Work

At this point it can be fairly argued that a concentration on industrial work is misplaced. As was noted in a previous chapter, the number of workers in manufacturing enterprises has steadily declined in relative terms, and these workers are now a small segment of the labor force. Work on the assembly line, often considered the epitome of industrial work, in reality occupies only a tiny fraction of workers in the manufacturing sector. Due to fundamental changes in the economy, the typical worker in the United States no longer toils on some factory floor. And even rapid expansion of manufacturing—on-shoring key industries in alternative energy production and electronics manufacturing—is only a drop in the historical bucket of well-paying industrial jobs.[31]

As the number of manufacturing jobs relative to the total labor force has declined, more and more people are now found in white-collar occupations. More workers are engaged in what the U.S. Bureau of Labor Statistics broadly classifies as "management, professional, and related occupations" than any other type of work, comprising about 41.2 percent of the nation's labor force in 2018.[32] Today, the typical workplace is not the factory but the office. Some analysts have seen this as an indication that the general occupational level has improved. Moreover, it has been argued that technological change has in many cases increased skill requirements and allowed clerical workers to take on broader responsibilities, an assertion that will be taken up in the last part of this chapter. As mentioned earlier, "job enlargement" or "job expansion" is taken by management as a way of motivating workers and rewarding them and increasing "buy in" without necessarily increasing compensation. It is not surprising many workers see through this.[33]

Although some technologies have the potential to empower workers, others are used to monitor them continuously. These technologies are especially appealing to managers like the one who stated, "I'm a great one for believing that people will really do what's *in*spected and not what's *ex*pected."[34] They now have at their disposal many ways to monitor their employees. As Ellen Alderman and Caroline Kennedy have noted: "Keystroke monitoring (where the computer counts the number of keystrokes per minute), telephone accounting monitoring (where the number of phone calls per hour and the length of each call are recorded), as well as service observation (where supervisors listen in on calls), and keeping track of unplugged time (measuring time spent away from the computer) are all becoming commonplace."[35] To these can be added the counting of mouse clicks and the monitoring of e-mail messages, sometimes with programs that detect potentially incriminating words like "boss" and "union."[36] Even executive and higher-level employees are not immune; laptop computers, cell phones, and voicemail all can keep employees tethered to an "electronic leash" that can make work a 24/7 obligation.

Industries and organizations, as well as the workers who are employed by them, vary immensely, making it impossible to come to a summary judgment of the consequences of technological change for white-collar work. Twenty-first-century technologies offer expanded opportunities for monitoring workers, but they also have the potential to empower these workers by making available vast troves of information that used to be available only to upper-echelon managers. It is also important to keep in mind that ascribed characteristics of workers, such as their gender and ethnicity, can affect the way in which technological changes have altered their jobs. There is some evidence that early examples of office automation were more beneficial to male rather than to female workers. In the insurance industry, for example, high-level positions in the management of electronic data processing systems became almost exclusively male preserves, while routine clerical work became an increasingly female occupation. A similar pattern was observed in the Internal Revenue Service, where computerization reduced the number of middle-level jobs, thereby blocking advancement opportunities for a largely female clerical staff.[37] These examples indicate that when a labor market is already segmented along gender or ethnic lines, new technologies can reflect or even reinforce existing divisions. One group may benefit from new ways of doing things, while another group may find themselves in the same, or possibly worse, circumstances.[38]

Telework

The differential effects of new workplace technologies can also be seen in the way new information and communications technologies (ICTs) have been used for work performed away from conventional workplaces. About 5 million workers—3.6 percent of the labor force—do at least half of their regular work at home. In 2014, five million workers or 3.6 percent of American workers were classified as teleworkers according to a definition that only encompassed employees who did at least half of their work at home. A more inclusive definition that included everyone who did at least some work at home, along with workers who do not have a fixed place of employment—plumbers, truck drivers, landscapers and the like—put 30 percent of employees in this category.[39] A survey conducted by the Gallup polling organization more defined a teleworker as someone who used a computer at home as part of his or her job. In this case the number of teleworkers was quite a bit higher at 37 percent, but on average, employees who worked at home did so for only two days a month. Only 9 percent of employees worked as telecommuters for at least half of the working week.[40]

Millions more joined the ranks of teleworkers when the COVID-19 pandemic forced the closure of many worksites, for many workers adding the burden of combining work with caring for children whose schools shut down during the contagion. Many workers came to prefer teleworking, with reduced commuting times, and the ability to self-manage work and family responsibilities. But as COVID-19 has been normalized, return-to-work expectations by employers have been rising, and despite stable productivity for many work-at-home employees. These expectations for presence at work are creating conflict, and in a seller's market for labor, employees have a great deal of power to select employers which accommodate their interests in hybrid and work-from-home options.[41]

Although it is based on modern technologies such as e-mail, fax machines, scanners, and teleconferencing, telework represents a return to a mode of work that prevailed before industrialization, suburbanization, and new transportation technologies combined to separate the workplace from the home. As we have seen, domestic production prior to industrialization was riddled with problems of motivation and discipline that were ultimately overcome by the development of the centralized factory and its regimen of close supervision. Modern communication technologies have allowed some reversal of this long-term trend. Through telework, workers can avoid expensive, stressful commutes and have more on-the-job flexibility.

In the past, homes adjoined the workshops of skilled craftsmen, but a home worksite also could be a place where miserably paid workers put in long hours in cramped cottages and fetid tenements. This dismal history caused an initial opposition to modern telework by organized labor. Labor unions also opposed work done in the home because dispersed and easily replaced workers are difficult to recruit and organize by labor unions.[42]

Union opposition to telework has softened in recent years with the realization that telework can be beneficial for many workers. Abundant possibilities for exploitation exist when workers are isolated in their individual homes, are dependent on their employers for their livelihoods, and can be easily replaced, but not all work in the "electronic cottage" conforms to this model. Electronic homework need not be confined to such things as routine word-processing tasks. Jobs requiring a high degree of skill and expertise can also be done in the home. It may even be possible for many specialized services to be supplied by homeworkers who are not employees of a firm but work on a contract basis, perhaps as a member of an autonomous teleworkers' association. At the same time, regular employees who work far from their firm's central office may not have to be electronically supervised as long as they have a personal stake in the success of the firm. This is more likely to happen when employees occupy upper-level positions within an organization. For these workers, new technologies may make their work more autonomous, convenient, and fulfilling.

Although telework can bring a number of benefits, workers should understand that a prolonged period of absence from a central workplace, even if it does not result in diminished output, may put these workers at a disadvantage when it comes to pay raises, promotions, and assignment to new projects. Teleworkers may be cut off from vital information, have difficulties forming job-related relationships with other workers, and may be excluded from essential job-related networks. As Joel Mokyr has noted of conventional working relationships, "Proximity in a plant or office created personal familiarity and thus conditions of trust and believability. There is always a role [for] body language, intonation, and general demeanor in human communications."[43] In sum, electronic communications technologies can be used for many work-related purposes, but there remains a need for at least occasional "face time" at a workplace.

Figure 13.5 Once adjustments were made, many employees found it advantageous to continue to work from home, at least in the absence of systemic child-care changes in the United States. (Shutterstock/Zivica Kerkez)

The "Gig Economy"

An intriguing change in the organization of work has been the recent rise of the "sharing" or "gig" economy. Its key characteristic is the replacement of traditional employees by people working as independent contractors. As briefly noted in Chapter 2, this has been exemplified by Uber, a firm whose drivers use their own vehicles and connect with customers using smartphones. Workers defined as contractors rather than employees can also work for firms by providing personal services, computer programs, and grocery deliveries, to name a few.

Although the expansion of the gig economy has been made possible by digital technologies, the basic idea is an old one. As we saw in the previous chapter, before the rise of the factory system, many employers did not directly hire workers; rather, they supplied tools and raw materials to workers who produced cloth and other goods in their homes. The workers were then paid after the finished goods had been collected by the employer.

Although the model is an old one, the development of the modern-day gig economy has generated some new issues about the worker-employer relationship. Whether a worker is an employee or a contractor will determine the applicability of laws governing wages, hours, and working conditions. This in turn will hinge on such issues as the extent of control exercised by the employer, the permanency of the employer-worker relationship, and whether the work being performed is an integral part of the employer's business.[44]

Other examples of the gig economy are direct connections between customers with suppliers of goods and services, thereby dispensing with an employer altogether. One of the best known of these is Airbnb, which allows owners of houses and apartments to temporarily rent all or part of their homes to short-term visitors. Once again, digital technologies facilitate making connections between buyers and sellers. At the same time, computer- and smartphone-based social media may also be important for the establishment of a modicum of trust between strangers.

Some observers see today's gig economy as the harbinger of dramatic changes in the nature of twenty-first-century employment. As evidence, they point to 30.2 million independent workers in the

U.S. economy, an increase of 12 percent from 2010 to 2015,[45] with that figure doubling again to an estimated 59 million workers in 2022.[46] On the other hand, skeptics note that the overwhelming majority of workers continue to be employees of business, governmental, and nonprofit organizations, and that the number of self-employed workers has actually fallen over the past decade.[47] All in all, it can be reasonably stated that the gig economy will expand in the years to come, but the rapidity and extent of that expansion is still in doubt.

Equally in doubt is the status of contract workers. Although many participants in the gig economy appreciate their independence in determining their working hours and other aspects of their job, they do not have the benefits and protections of regular employees. In recessionary times, people cut back on things like food deliveries, leading to more precarity in the gig economy.[48] Pension funds, health insurance, minimum wages, and employer contributions to Social Security have been nonexistent for Uber drivers, musicians, and for-hire computer programmers. California Proposition 22 (2020) was a ballot initiative to clarify "app-based drivers" as contractors, which when passed made it clear that labor law and wage policies would treat ride-share and delivery drivers as contractors, much to the satisfaction of Uber and related companies.[49] But then it was struck down as unconstitutional.[50] Some workers find great benefit in the flexibility of rideshare and delivery work,[51] although in the aggregate it seems that gig work depresses wages and promotes inequalities.[52] In March 2023, an appeals court upheld the ruling. As with much legislation, its mode of implementation will be as important as the wording of the law. Once again, technological advances like GPS and smartphones for Uber drivers, have created new possibilities, but their realization will require some significant changes in the relationships of workers and the people and organizations that pay for their services.

Smart Technologies and Dumb Jobs?

Optimistic students of technological change are fond of pointing out that as computerized processes continue to emerge, a growing segment of the labor force will be found in jobs that require high degrees of skill. Instead of monotonously tightening nuts and bolts or engaging in routine clerical operations, large numbers of workers will be engaged in tasks that cannot be performed by computers. Frank Levy and Richard Murnane have called attention to the many kinds of work activities that cannot be reduced to computer-friendly routines, the abilities and processes that they have dubbed "expert thinking" and "complex communication."[53] In the service sector, even though there is rapid proliferation of automated self-checkouts in grocery stores, or check-ins at hotels, regular human staffing is still required for complex transactions, often based on consumer preferences and need for interaction and problem-solving. And, as we saw in the previous chapter, even where computers have taken hold, a large number of employees will be required to continually modify organizational structures and operations in order to make optimal use of computer-based technologies. From these perspectives, occupational skill levels are going up, and the proportion of unskilled factory operatives and office clerks is declining in twenty-first-century enterprises.

It is undeniable that these trends can be found throughout the economy, but to less optimistic observers, predictions of the radical upgrading of workers' tasks are at best premature. Plenty of jobs still require only the most minimal levels of skill and training. The task of monitoring industrial processes can be just as dull as the most routine assembly operation, and in any event, many of these functions can be performed by sensors connected to computers. Even maintenance and repair jobs may represent only a slight improvement in required skill levels. Technological advances can lower maintenance requirements, and when repairs need to be done, they can often be accomplished by simply disconnecting a faulty module and replacing it with a new one. The task of diagnosis, which often requires the greatest amount of the repairperson's skills, has itself been simplified through the development of special test circuits and computer-based "expert systems." The need for repair skills will always be present, and it is unlikely that technological developments will allow the elimination of repair personnel. Still, any dilution of the repairperson's skill will have significant consequences for the overall skill levels of the labor force, for these workers comprise the largest single group of skilled workers.[54]

Moreover, even if the optimists are largely correct, their predictions only hold for industries that make extensive use of advanced technologies. High-tech factories and offices may require higher levels of employee skills, but these are not the workplaces of large numbers of workers today or in the near future. Nor will new high-technology industries generate great numbers of new jobs. To be sure, some high-tech occupations have grown rapidly in recent years and are projected to do so in the future. According to the U.S. Bureau of Labor Statistics, from 2014 to 2024 employment in "Computer and Mathematical Occupations" is expected to grow by 13.1 percent, for a net addition of 531,000 jobs.[55] This represents a healthy rate of growth, but it is outmatched by the 1,348,000,100 projected new jobs for "Healthcare Support Occupations" (about half of which are expected to be home health care aides), the 778,000 additional jobs in sales, and 812,900 more jobs in "Food Preparation and Serving Related Occupations" during the same period.[56] In fact, of the 10 occupations with the largest numerical employment increases projected from 2014 to 2024, only two—registered nurses and postsecondary teachers—require much in the way of advanced training. The remainder—occupations like personal care aide, truck driver, and customer service representative—hardly represent the high-tech future.[57] Advanced technologies have given rise to a number of occupations that have grown at a rapid rate from a low base; however, in absolute terms, most of the job growth in the near future will take place in occupations that have been around for a long time.

Although technological advance does not seem to be the source of large, across-the-board increases in workers' skills, it doesn't seem to be producing the opposite effect either. While the skill requirements in some occupations have decreased, they have increased in others, and it is likely that all of these changes have offset one another, leaving no net change one way or the other.[58] It also may be the case that there has been some degree of polarization as skill requirements have increased in some jobs and occupations and decreased for others.[59] For example, automobile manufacture has changed substantially as a result of the introduction of computer-controlled equipment. This, in turn, has led to changes in the skills exercised by the factory workforce; while skilled workers have experienced upgraded skill levels, the opposite has been the case for production workers.[60] This bifurcation was not the inevitable outcome of technological change, however, for other production workers. The distinction between skilled and unskilled workers has always been a prominent feature in the automobile industry, and in the absence of changes in organizational structure and culture, the introduction of new production technologies reinforced this division by giving skilled workers most of the responsibilities for the operation, maintenance, and repair of technologically sophisticated equipment while leaving ordinary production workers with jobs that continued to require little in the way of skill.

This is hardly an atypical pattern. As several researchers have noted, organizational patterns strongly influence the skills used by workers.[61] One aspect of organizational structure that may strongly affect workers' skill levels is the distribution of power. When there is a sharp division between those who manage and those who work, and when managers view their workers as hostile and unreliable, workers will be treated as replaceable parts. Their skill level will be low, and they will be controlled by centralized supervision and machine-paced work processes.[62] By contrast, as we will encounter in a later chapter, when a more democratic and egalitarian order prevails, technologies that require the initiative and commitment of workers are more likely to be selected, and workers will be given a chance to develop their skills and take their places as valued members of the organization. As Robert Thomas argued, workplace culture both shapes and is shaped by technological innovation in a firm, and attention to organization and culture is key to successful implementation, meaning one that enhances both employee and firm outputs.[63]

New technologies can be developed to increase levels of skill, or they can be used to diminish them. Like all matters regarding technology, choices must be made. These choices emerge in an environment where the intentions and relative power of employees, managers, and the owners of enterprises are usually at least as important as emerging technical capabilities in determining the course of technological change and its consequences for a firm's workers. Consider the French "right to disconnect" law, which allows employees of large companies to ignore emails after work hours.[64] Local organizational cultures, national cultures, and legal systems shape the development and adoption of new technologies.

Questions for Discussion

1. Do you agree that many early industrial technologies were used as ways of controlling labor? What sort of historical research could be done in order to test this proposition?
2. Do employers have a legitimate right to electronically monitor their employees? Would you object to monitoring by your employer? Should federal or state governments pass laws that limit electronic employee monitoring? If so, what would be reasonable limits to this practice?
3. Can you see any indications that consumers are showing a preference for nonstandardized products? How will changed consumer preferences affect workplace technologies?
4. What are some of the key skills used by practitioners of an occupation you are interested in? Which of these skills might be replaced by technological advances in the near future? Would a diminished need for these skills make this occupation more or less attractive to you?

Notes

1. Terry Reynolds, *Stronger Than a Hundred Men* (Baltimore: Johns Hopkins University Press, 1983).
2. B .J. Lewis, J. M. Cimbala, and A M Wouden, "Major Historical Developments in the Design of Waterwheels and Francis Hydroturbines," *27th IAHR Symposium on Hydraulic Machinery and Systems* (2014). Accessed from http://iopscience.iop.org/article/10.1088/1755-1315/22/1/012020/pdf.
3. Melvin Kranzberg and Joseph Gies, *By the Sweat of Thy Brow: Work in the Western World* (New York: G. P. Putnam's Sons, 1975), p. 112; and Carolyn C. Cooper, "The Portsmouth System of Manufacture." *Technology and Culture* 25, 2 (1984): 182–225.
4. Daniel Nelson, *Managers and Workers: Origins of the New Factory System in the United States, 1880–1920* (Madison: University of Wisconsin Press, 1975), pp. 4–7.
5. Quoted in Stewart Holbrook, *The Story of American Railroads* (New York: Crown, 1947), p. 15.
6. Walter Licht, *Working for the Railroad: The Organization of Work in the Nineteenth Century* (Princeton, NJ: Princeton University Press, 1983), p. 88.
7. Daniel T. Rodgers, *The Work Ethic in Industrial America* (Chicago, IL: University of Chicago Press, 1978), p. 67.

8. Jessica Brain, "The Victorian Workhouse," *Historic UK* (August 8, 2019). Accessed from www.historic-uk.com/HistoryUK/HistoryofBritain/Victorian-Workhouse. *Oliver Twist* was serialized from 1887–1839.
9. Herman Melville, *The Tartarus of Maids*. Accessed from https://42opus.com/v6n4/thetartarusofmaids.
10. Herman Melville, *The Paradise of Bachelors*. Accessed from https://42opus.com/v6n4/theparadiseofbachelors.
11. Frederick Engels, "On Authority," in Karl Marx and Frederick Engels, *Selected Works* (Moscow: Foreign Languages Publishing House, 1962), p. 637.
12. Malcolm I. Thomis, *The Town Labourer* (New York: Barnes and Noble, 1974), pp. 108–109.
13. Witold Rybczynski, *Taming the Tiger: The Struggle to Control Technology* (New York: Viking/Penguin, 1985), pp. 46–47.
14. Charles F. Sabel, *Work and Politics: The Division of Labor in Industry* (Cambridge: Cambridge University Press, 1982), p. 41.
15. N. S. B. Gras, *Industrial Evolution*, quoted in Sidney Pollard, *The Genesis of Modern Management: A Study of the Industrial Revolution in Great Britain* (Harmondsworth, England: Penguin, 1968), pp. 22–23.
16. Harry Braverman, *Labor and Monopoly Capital: The Degradation of Work in the Twentieth Century* (New York: Monthly Review Press, 1974), p. 195.
17. J. Patrick Wright, *On a Clear Day You Can See General Motors* (New York: Avon Books, 1979), p. 94.
18. David A. Hounshell, *From the American System to Mass Production, 1800–1932: The Development of Manufacturing Technology in the United States* (Baltimore: Johns Hopkins University Press, 1984), pp. 217–261.
19. Henry Ford, in collaboration with Samuel Crowther, *My Life and Work* (Garden City, NY: Doubleday, Page & Company, 1922), pp. 81, 82–83.
20. For background, see Sofi Thanhauser, "A Brief History of Mass-Manufactured Clothing." *Literary Hub* (January 27, 2022). Accessed from https://lithub.com/a-brief-history-of-mass-manufactured-clothing; and Dolores Mone, "Ready-to-Wear: A Short History of the Garment Industry," *Bellatory* (July 11, 2023). Accessed from tps://bellatory.com/fashion-industry/Ready-to-Wear-A-Short-History-of-the-Garment-Industry.
21. See: Astrid Becker "The Global South's Double Burden," *IPS* (December 11, 2021). Accessed from www.ips-journal.eu/topics/economy-and-ecology/the-global-souths-double-burden-5539; and Jason Hickel, Christian Dorninger, Hanspeter Wieland, Intan Suwandi, "Imperialist Appropriation in the World Economy: Drain from the Global South through Unequal Exchange, 1990–2015," *Global Environmental Change* 73 (2022).
22. See: Charles Rumford Walker and Robert H. Guest, *The Man on the Assembly Line* (Cambridge, MA: Harvard University Press, 1953).
23. Quoted in Braverman, *Labor and Monopoly Capital*, p. 118.
24. Ibid., p. 113.
25. Frederick Winslow Taylor, *The Principles of Scientific Management* (New York: Norton, 1967 [originally published in 1911]), p. 36.

26. Larry Hirschhorn, *Beyond Mechanization: Work and Technology in a Post-Industrial Age* (Cambridge, MA: MIT Press, 1984), pp. 61–86.
27. Ibid., p. 157.
28. "The Technology Payoff," *Business Week* (14 July 1993): 59.
29. Sabel, *Work and Politics*, op. cit., p. 74.
30. Robert B. Reich, *The Next American Frontier* (Harmondsworth, England: Penguin, 1984), p. 81.
31. Tim Mullaney, "Biden's climate change bill may produce your next job, and a half-million careers in all," *CNBC* (September 8, 2022). Accessed from www.cnbc.com/2022/09/08/biden-climate-act-may-create-your-next-job-and-a-half-million-careers.html.
32. U.S. Bureau of Labor Statistics, "Office and Administrative Support Occupations Make Up Nearly 16 Percent of U.S. Employment May 2013," (April 9, 2014). Accessed from www.bls.gov/opub/ted/2014/ted_20140409.htm; and Department for Professional Employees, "The Professional and Technical Workforce: By the Numbers," (August 16, 2019). Accessed from www.dpeaflcio.org/factsheets/the-professional-and-technical-workforce-by-the-numbers.
33. Indeed (www.indeed.com/career-advice/career-development/job-enlargement) sees this as a way of engaging worker satisfaction by rewarding employees with more expansive titles and responsibilities.
34. Michael W. Miller, "Computers Keep an Eye on Workers and See If They Perform Well," *The Wall Street Journal* (June 3, 1985): 1.
35. Ellen Alderman and Caroline Kennedy, *The Right to Privacy* (New York: Random House, 1995) p. 316.
36. Jill Andresky Fraser, *White Collar Sweatshop: The Deterioration of Work and Its Rewards in Corporate America* (New York and London: W.W. Norton, 2001), pp. 87–90.
37. Roslyn L. Feldberg and Evelyn Nakano Glenn, "Technology and Women's Degradation: Effects of Office Automation on Women Clerical Workers," in Joan Rothschild (Ed.), *Machina ex Dea: Feminist Perspectives on Technology* (New York: Pergamon Press, 1983), pp. 59–78.
38. Barbara A. Gutek and Tora K. Bikson, "Differential Experiences of Men and Women in Computerized Offices," *Sex Roles* 13, (1985): 3–4.
39. Alina Tugend, "It's Unclearly Defined, but Telecommuting Is Fast on the Rise," *The New York Times* (March 7, 2014). Accessed from www.nytimes.com/2014/03/08/your-money/when-working-in-your-pajamas-is-more-productive.html?_r=0; and Global Workplace Analytics, "Latest Work-at-Home Telecommuting/Mobile Work/Remote Work Statistics"(2020. Accessed from https://globalworkplaceanalytics.com/telecommuting-statistics.
40. Jeffrey M. Jones, "In U.S., Telecommuting for Work Climbs to 37%" *Gallup* (August 19, 2015). Accessed from www.gallup.com/poll/184649/telecommuting-work-climbs.aspx.
41. "Returning to Work Post-COVID: The Hybrid Work Model," *Kidder Mathews* (March 24, 2022). Accessed from https://kidder.com/trend-articles/returning-to-work-post-covid-the-hybrid-work-model/

42. See: Joe Swerdzewski, "Are Unions Relevant in the Age of Telework?" *FedSmith* (October 4, 2021). Accessed from www.fedsmith.com/2021/10/04/are-unions-relevant-with-telework/
43. Joel Mokyr, *The Gifts of Athena: Historical Origins of the Knowledge Economy* (Princeton, NJ: Princeton University Press, 2002) p. 160.
44. The Editorial Board, "Defining 'Employee' in the Gig Economy. " *The New York Times* (July 18, 2015). Accessed from www.nytimes.com/2015/07/19/opinion/sunday/defining-employee-in-the-gig-economy.html.
45. Patrick Gillespie and Sara Ashley O'Brien, "The Gig Economy: More People May Have Jobs Than You Think," *CNNMoney* (October 2, 2015). Accessed from http://money.cnn.com/2015/10/02/news/economy/jobs-report-gig-on-demand-economy.
46. Jacques Buffett, "Workers on the Gig Economy: 2022 Statistics," *Zety* (August 16, 2022). Accessed from https://zety.com/blog/workers-on-gig-economy.
47. Josh Zumbrun and Anna Louie Sussman, "Proof of "Gig' Economy Revolution Is Hard to Find," *The Wall Street Journal* (July 26, 2015). Accessed from www.wsj.com/articles/proof-of-a-gig-economy-revolution-is-hard-to-find-1437932539.
48. Jennifer Saba, "Gig worker rule comes at bad time for gig economy," *Reuters* (October 11, 2022). Accessed from www.reuters.com/breakingviews/gig-worker-rule-comes-bad-time-gig-economy-2022-10-11.
49. Ballotpedia, "California Proposition 22, App-Based Drivers as Contractors and Labor Policies Initiative (2020)." Accessed from https://ballotpedia.org/California_Proposition_22,_App-Based_Drivers_as_Contractors_and_Labor_Policies_Initiative_(2020).
50. Brian Chen and Laura Padin, "Prop 22 Was a Failure for California's App-Based Workers. Now, It's Also Unconstitutional," *NELP* (September 16, 2021). Accessed from www.nelp.org/blog/prop-22-unconstitutional.
51. Stephanie Whitfield, "Gig Workers and California Voters Deserve to Have Proposition 22 Upheld," *The Orange County Register* (November 29, 2022). Accessed from www.ocregister.com/2022/11/29/gig-workers-and-california-voters-deserve-to-have-proposition-22-upheld.
52. National Equity Atlas, "Prop 22 Depresses Wages and Deepens Inequities for California Workers" (September 21, 2022). Accessed from https://nationalequityatlas.org/prop22-paystudy
53. Frank Levy and Richard J. Murnane, *The New Division of Labor: How Computers Are Creating the Next Job Market* (Princeton and Oxford: Princeton University Press, 2004), pp. 47ff.
54. The changing skill requirements of auto mechanics are explored in Kevin Borg, *Auto Mechanics: Technology and Expertise in Twentieth Century America* (Baltimore and London: Johns Hopkins University Press, 2007).
55. Op. cit., p. 89. Ref: Bureau of Labor Statistics, "Occupational Employment Projections to 2024," *Monthly Labor* Review (December 2015). Accessed from www.bls.gov/opub/mlr/2015/article/occupational-employment-projections-to-2024.htm.

56. U.S. Bureau of Labor Statistics, "Occupations with the Greatest Job Growth." Accessed from www.bls.gov/emp/ep_table_104.htm.
57. Ibid.
58. Arne Kalleberg, *The Mismatched Worker* (New York and London: W.W. Norton, 2007), p. 50.
59. Ibid., p. 107.
60. Ruth Milkman and Cydney Pullman, "Technological Change in an Auto Assembly Plant: The Impact on Workers' Tasks and Skills," *Work and Occupations* 18, 2 (May 1991).
61. William Form, Robert L. Kaufman, Toby L. Parcel, and Michael Wallace, "The Impact of Technology on Work Organization and Work Outcomes: A Conceptual Framework and Research Agenda," in George Farkas and Paula England (Eds.), *Industries, Firms, and Jobs: Sociological and Economic Approaches* (New York: Plenum Press, 1988).
62. Louis E. Davis and James C. Taylor, "Technology, Organization, and Job Structure," in Robert Dubin (Ed.), *Handbook of Work, Organization, and Society* (Chicago: Rand-McNally, 1976), p. 412.
63. Robert J. Thomas, *What Machines Can't Do: Politics and Technology in the Industrial Enterprise* (Berkeley, CA University of California Press, 1994).
64. Library of Congress, "Telework and the French "Right to Disconnect," (August 21, 2020). Accessed from https://blogs.loc.gov/law/2020/08/telework-and-the-french-right-to-disconnect; and Vivienne Walt, "France's 'Right to Disconnect' Law Isn't All It's Cracked Up to Be," *Time* (January 4, 2017). Accessed from https://time.com/4622095/france-right-to-disconnect-email-work.

PART 5

Communication

The ability to express and understand complex thoughts is one of the most fundamental characteristics of the human species. For most of human existence, this was done primarily through the spoken word. The invention of writing greatly expanded the reach and permanency of communication, but for hundreds of years, reading and writing were confined to tiny segments of the population. As we will see in Chapter 14, the invention and diffusion of printed works marked a revolutionary stage in human communication, by making written works accessible to much larger numbers of people and at the same time contributing to profound social, political, cultural, and psychological changes.

For the next four centuries, communication technologies in Europe remained static. People waited weeks and even months for the arrival of a letter. But in the middle of the nineteenth century, a dramatic transformation occurred. Chapter 15 presents the transformative media technologies that began with the telegraph's bringing virtually instantaneous communication to every part of the world where wires had been strung. This was only the beginning of the next revolutionary phase; during the twentieth century, radio and then television vastly expanded the range and content of human communication.

Still, the revolution in human communications was far from over. While the consequences of radio and television are still being debated, recent years have seen the emergence and rapid diffusion of a host of new media technologies, carrying with them the potential for changes equal to those wrought by printing, the telegraph, radio, and television. In Chapter 16, we will consider information and communication technologies (ICTs), specifically the internet and mobile communication devices, focusing on questions about information, by looking at some recent case studies about censorship, surveillance, and (dis)information.

Today, many people view ICTs as virtually synonymous with technology in general. Chapter 17 looks at some additional case studies centered on ICTs, such as gaming, fads and social media, health and "Dr. Google", friendship and self-image and bullying, and artificial intelligence (AI) and algorithms, and finally schooling and student learning. ICTs occupy a special place in our everyday lives, which now incorporate heavy reliance on the internet and the frequent use of social media. Yet they echo some of the older issues with the technologies that have come before them, as the creation and use of these technologies highlight the complex issues presented by new ways of communicating.

CHAPTER 14

Printing, Literacy, and Media

We wrote a book. It could have been a magazine article. There are, of course, other ways of retaining and transmitting information and ideas. They can be stored in a person's memory and then communicated through the spoken word; indeed, that was the way things were done for most of human history and still are done in nonliterate societies. If present trends continue, at some point in the future, books, newspapers, and magazines will be largely supplanted by websites and e-books. But until very recently, nothing has equaled words printed on paper as a rapid, cheap, and portable means of storing and transmitting information. Moreover, the importance of printing goes well beyond the storage and communication of information. The social effects of printing are extensive, making printing one of the most significant technological innovations of all time.

The development of a written language is one of the hallmarks of civilization. While a spoken language is a basic element of all human societies, not all of them developed a written language. Of the approximately 5,000 to 8,000 languages that are spoken today, only a very small percentage appear in written form, and in many cases, these written languages have been recent inventions, often stimulated by missionaries or other contacts with the outside world. One interesting example of an effort to create a written language appeared in the 1820s, when a Cherokee Indian named Sequoyah single-handedly constructed a written language so that his people could more effectively resist the encroachments of white settlers. A written language was an essential technology, for as Sequoyah observed, "Much that red men know, they forget; they have no way to preserve it. White men make what they know fast on paper like catching a wild animal and taming it."[1] After much laborious effort,

Sequoyah succeeded in constructing a system of writing that could be mastered in a few weeks. Unfortunately for Sequoyah's people, the new written language by itself could not stem the advance of white Americans into Cherokee territory.

The writing systems of the major languages of the world were also constructed to serve practical purposes. In ancient Egypt, Babylonia, and China, writing was used to keep financial accounts and to facilitate communication between government officials. Writing also served a spiritual purpose in many cases, for it allowed the recording and diffusion of concepts that were central to the local religion. Since church and state were tightly interwoven in these ancient states, writing met the needs of the political elite in their dual role as priests and secular officials. As with many technologies, writing owed its development to the needs of one group of people and was used to maintain and extend their influence over others.[2] For example, the priests of Egypt during the days of the pharaohs were able to refer to written calendars that allowed them to predict the cyclical flooding of the Nile, thereby demonstrating their supposed indispensability.

The Printing Revolution

Other than the maintenance of a social and political hierarchy, the effects of writing were minimal when the written language was understood only by a small segment of the population. Few of the common people had the time and resources to become literate, and even if they were able to read, the great cost of producing written documents kept these works out of their reach. Written materials had to be laboriously drafted and copied by hand. Documents central to a civilization, such as the Bible of medieval Christianity, were scarce and inaccessible because monks and other scribes had to labor for many months to produce a single volume. Few individuals and institutions could amass even modest libraries. Two of the largest book repositories in England during the late Middle Ages (1300–1500), the abbey libraries of Canterbury and Bury, each held 2,000 books, while the library of Cambridge University had only 300.[3]

In the absence of widespread literacy, the ability to memorize was developed to a level scarcely imaginable today. People made use of elaborate mnemonic devices in order to recall important facts and concepts. This facility made it possible to preserve laws, ceremonies, and stories in the minds of individuals. It was not unusual to find people who could recite sacred texts and heroic sagas that might today occupy hundreds of pages of printed text. For example, the *Iliad* is about 16,000 lines of text, and the Odyssey about 11,000, and the structure and meter of the works facilitated its reproduction in recitations.[4]

Oral traditions began to change dramatically with the introduction of printing with movable type. As with many other technological innovations, printing originated in East Asia. In the fourth century CE the Chinese copied stone inscriptions through the use of paper rubbings. By the seventh century, Chinese printers were using carved wooden blocks to produce books on agriculture and medicine. Chinese accomplishments using this method were prodigious; during a 12-year period that began in 971, printers in the city of Chengdu published a 5,048-volume compilation of Buddhist scriptures that required the carving of 130,000 blocks. A significant advance came in the middle of the eleventh century when a man named Bi Sheng invented a system of printing that used individual pieces of clay type set in a wax matrix. Two hundred years later, another Chinese inventor, Wang Zhen, created a rotating type of frame that allowed the typesetter to easily pick out the required pieces. The Chinese also experimented with metal type, although it was their Korean neighbors who achieved success in this endeavor during the fourteenth century.[5]

Printing developed more slowly in Europe. In the early fifteenth century, playing cards and pictures of saints were printed from wooden blocks. It is likely that by 1440 woodblock printing was also used for the production of small books containing a few lines of text.[6] This method was not suitable for printing lengthy books, as the type for each page had to be laboriously carved from a single piece of wood. At about this time, some printers also made prints from copper plates that had words and images engraved into them.[7] As with woodblock printing, engraving was a very labor-intensive process that did little to lower the cost of the written word. What made printing a much more efficient process was forming each page from individual pieces of type that could be used over and over again. As we have just seen, this had already been done in China, but the basic idea had to be reinvented in Europe, stimulated perhaps by some knowledge of what the Chinese had been doing.[8]

Although there are other European claimants to this invention, Johann Gutenberg (ca. 1400–1468) is usually given credit for devising a workable system of printing with separate pieces of type. In order to put the idea into practice, Gutenberg also had to develop a number of new techniques, such as a method to uniformly cast pieces of type. This was done by punching the letter into a copper matrix and then using the resulting mold for casting the actual pieces of type. Setting these individual pieces of type in a straight line also presented a problem. Gutenberg solved it by designing a mold that kept each piece of type within acceptable tolerances, and by equipping each piece of type with a ridge on one side and a corresponding groove on the other. In this way, pieces of type could interlock with one another when they were set in a frame.[9]

Important as the invention of movable type was, the printing of books in large quantities required more than simply composing a page from individual letters. As is always the case with major technological changes, a whole system of interrelated elements had to be developed. Obviously, some sort of a press was required; here the ancient winepress provided a useful model. New kinds of ink, compounded from lampblack and linseed oil, were also necessary. The production of durable type required considerable experimentation with different alloys; a mixture of tin, lead, and antimony proved to be the winning combination. And to realize the economies offered by printing, costly parchment was replaced by paper, a Chinese invention that had passed through the Islamic world and began to be produced in Europe early in the fourteenth century.

In a world that produced its goods through individual craftsmanship, the first books printed from movable type were harbingers of the age of mass production. While the hand-copied and illuminated books of the past were works of art available to only a few, printed books, tracts, and pamphlets were standardized products that reached a large and growing audience. According to one estimate, more than 10 million copies of 40,000 different titles had been produced during the 50 years following Gutenberg's invention. Had they been copied by hand, the task would have taken a thousand years.[10] During the century that followed, as many as 200 million books were printed in Europe.[11]

Figure 14.1 A Gutenberg printing press at the International Printing Museum. (vlasta2/CC BY 2.0)

Printing and the Expansion of Knowledge

Printing and the consequent expansion of the number of books in circulation did much to transform European society. By systematizing knowledge and making it readily available, printing helped to promote the rational modes of thought that characterize a modern society. As happened five centuries later with the rise of the internet, information flowed much more readily, with everything from treatises on accounting to printed musical scores receiving widespread distribution. With so much information circulating, it was no longer necessary to simply accept the judgment of a few authorities, and a more critical spirit emerged. Printing also allowed the widespread distribution of maps and accounts of global exploration. As a result, new geographical discoveries could be widely publicized, and errors and inconsistencies could be made evident through the cross-checking of existing maps and printed descriptions.[12] The clearer definition of space made possible by printed maps gave a great impetus to voyages of discovery that opened the world to European civilization and vice versa. Better maps made journeys less hazardous, and the ongoing definition of the world's spatial contours motivated explorers and colonists to fill in the remaining blank spaces on the maps.[13]

While explorers were making fundamental geographical discoveries, others were gaining new insights into how the world worked. The great advances in scientific knowledge that began to emerge in the fifteenth and sixteenth centuries were pushed forward by printing. As with the progress of geographical exploration, the advancement of science was to a substantial degree the result of the spread of information through print media. Science requires accurate data, but before the age of printing, information was scattered, quirky, and often highly inaccurate. To be sure, much of the information disseminated by printing was bogus, but because information reached a much greater audience, it was subject to informed criticism, making for greater accuracy. In short, printing allowed the establishment of an intellectual feedback system that helped to drive out errors.[14]

More was involved than the transmission of factual information. Human knowledge was significantly advanced by a greater exposure to different theories, perceptions, and interpretations. As Elizabeth Eisenstein has summarized the process:

> Much as maps from different regions and epochs were brought into contact in the course of preparing editions of atlases, so too were technical texts brought together in certain physicians' and astronomers' libraries. Contradictions became more visible, divergent traditions more difficult to reconcile. . . . Not only was confidence in old theories weakened, but an enriched reading matter also encouraged the development of new intellectual combinations and permutations. . . . Once old texts came together within the same study, diverse systems of ideas and special disciplines could be combined. Increased output . . . in short, created conditions that favored new combinations of old ideas at first and then, later on, the creation of entirely new systems of thought.[15]

The spread of information through printing allowed major advances in scientific understanding. Everything from astronomical tables to accurate anatomical illustrations became grist for the scientific mill. Theories could be readily checked against a growing mass of information. Equally important, printing helped to change the nature of learning. Scientific investigators were freed from the drudgery of copying tables and texts. Also, as literacy spread and books were much more readily available, students of all ages did not need to spend vast amounts of time on rote memorization and the learning of elaborate mnemonic techniques. Freed from these tasks, advanced students and natural philosophers could devote their energies to "solving brain teasers, designing ingenious experiments and new instruments, or even chasing butterflies and collecting bugs if they wished."[16] The pursuit of knowledge could take on a playful quality, and play can be a key ingredient in the development of scientific understanding.

Printing and the Rise of Protestantism

Just as printing helped to alter people's views of the physical and natural world, the new medium played a key role in transforming conceptions of the spiritual world. It is certainly no coincidence that the Protestant Reformation began less than two generations after Gutenberg's invention. Before then, unorthodox interpretations of prevailing Christian religious beliefs had sprung up from time to time, but they remained confined to particular locations. Although several instances of religious dissent may have occurred simultaneously, they were not linked together into an integrated attack on the authority of the centralized Church. The circulation of printed religious books and tracts created a more unified challenge to the dominant Catholic order and helped to turn a number of separate revolts into the Protestant Reformation.

Literacy and Protestantism reinforced each other; while Protestantism benefited from the expansion of the printed word, the religious beliefs of Protestant reformers did much to encourage printing. One of the central tenets of Protestantism was that individuals should study the Bible and interpret its message unencumbered by the guidance of priests, bishops, and popes. The production of printed Bibles carried the word of God directly to the faithful, while at the same time a great outpouring of religious tracts spread the words of religious reformers throughout Christendom. For good reason, Martin Luther praised printing as "God's highest and extremist act of grace, whereby the business of the Gospel is driven forward."[17]

Luther later lost his enthusiasm for popular Bible reading when it became apparent that the unguided reading of the Bible could result in "dangerous" beliefs that contradicted his own.[18] But by then, individual reading of the Bible had become an integral part of Protestant life. Sixteenth-century English clergymen were told to "comfort, exhort, and admonish every man to read the Bible in Latin or English, as the very word of God and the spiritual food of man's soul, whereby they may better know their duties to God, to their sovereign lord the king, and their neighbor."[19]

The Catholic Church also used printed catechisms, tracts, and edicts as weapons in its battles against the Protestant "heresy." But the cultural order it attempted to maintain was fundamentally antagonistic to the use of printed sources and the spread of literacy. The Church was unwilling to allow laypeople to pursue Biblical scholarship, and it was deeply suspicious of popular Bible reading and interpretation. Catholicism was still rooted in the medieval past, and it was unprepared for the challenge presented by print technology.

If the Bible was to be made accessible to the Protestant faithful, it had to be printed in a language they understood, rather than in the Latin that had served as the universal language of the medieval clergy. While the Catholic Church strongly resisted the production of Bibles in modern European languages, Protestantism made it an imperative.[20] In Protestant lands, the translation and printing of Bibles in vernacular languages spread the Gospel, but equally important, stimulated nationalism. A Europe that had been united under the cultural domination of the Catholic Church gave way to a patchwork of separate states, each with its own established church and a growing literature printed in the local language. Printing helped to destroy the unity of the medieval world, while within the borders of discrete national entities, a considerable amount of homogenization took place. Printing was of great importance in standardizing the languages of these nations. Dialects unsupported by a printed literature slowly faded away and were supplanted by a common form of French, German, or English.

Printing, Literacy, and Social Change

Motivated by the religious requirement of Bible reading, Protestants became literate to an extent not previously found in European civilization. Throughout the Middle Ages, the great bulk of the population could not read. Even rich and powerful people were illiterate. Charlemagne, who reigned king of the Franks from 768 to 814 and after 800 was the emperor of much of present-day Western Europe, was an intelligent and learned man, but he needed to have books and documents read aloud to him, and he relied on scribes to write down his words.

The only people who found literacy to be a necessity were members of the Christian clergy. (The connection between literacy and a religious vocation can be seen in the etymological relationship between the words "clerk" and "clergy.") The ability to read was often taken as sufficient proof that a person was a member of the clergy, and therefore not subject to civil law. This could be a great advantage, for a person claiming "benefit of clergy" could escape corporal punishment or even execution if convicted of a crime. The 51st psalm of the Old Testament was known as the "neck verse," because a person able to read it would be spared the gallows.[21] As might be expected, many illiterates tried to escape punishment by memorizing the psalm and then pretending to read it at the appropriate moment. This resulted in some farcical scenes in which a person attempted to demonstrate his literacy by "reading" from an upside-down Bible!

Protestantism, with its emphasis on Bible reading, created strong incentives for laypeople to learn to read. This skill could, of course, be useful in other endeavors. A person who could read and write was able to keep accurate records and accounts, learn about new ways of doing things, understand the law, enter into contracts, communicate with large numbers of people, and make use of banks. Thus, the oft noted connection between Protestantism and rapid economic growth in early modern Europe was at least in part due to the superior levels of literacy in Protestant lands.[22]

As we have seen, the rise of Protestantism created a fertile environment for the development of the printing industry. At the same time, urbanization and the spread of capitalism also made for a society that was more open to—actually hungry for—the new ideas that books could disseminate. On a more utilitarian level, the development of commerce and industry generated a need for written records and accounts. In later centuries, the desire for literacy was further stimulated by the growing concerns with social equality. In particular, the greater accessibility of education that culminated with the introduction of mass schooling in the nineteenth century produced a literate public and expanded the market for the printed word.[23]

All of these historical forces produced an environment that was ripe for the large-scale production of printed works. Had they not been present, the invention of movable type would have been of much more limited significance. After all, printing had first emerged in East Asia, yet social conditions limited its influence in that region. Korean printers had independently invented movable type, but few books were printed by this means in Korea until the end of the nineteenth century. Korean scholars even invented a phonetic alphabet known as *Hangul* in the fifteenth century, which greatly facilitated the use of movable type. But Korea's traditional aristocracy clung to the use of Chinese characters for the written rendition of Korean words because their ability to understand the difficult-to-learn characters set them apart from the common folk. Although a mass readership literate in *Hangul* could have provided a large audience for printed works, no self-respecting writer was willing to use the "inferior" script. Consequently, printing with movable type had minimal effects on Korean society.[24]

In Europe, social and cultural conditions created a strong demand for the products of print technology. At the same time, the technological advances that made the mass production of printed works possible did much to create these conditions. As with all major technological advances, the influence of printing is not a simple matter of technological determinism. Instead, we find a reciprocal, reinforcing relationship between technological change on the one hand and social change on the other. Moreover, printing with movable type did not effect a sudden revolution in Western society; even as late as the middle of the eighteenth century, 40 percent of English men and more than 60 percent of English women were illiterate.[25] Still, working in conjunction with the great changes that were sweeping over Western civilization, the development of print technology produced a transformation that has only begun in some parts of the world today.

Psychological Effects of Printing

The effects of printing on religious, economic, and political life seem reasonably evident. A consideration of the consequences of reading for individual attitudes and ways of thinking, however, is a more speculative venture. The Protestant Reformation is an identifiable historical event, and its connections with printing seem clear. It is much more difficult to demonstrate that printing fundamentally changed basic ways of thinking, or that it altered how we understand our place in the world.

Still, we might reasonably speculate that printing resulted in a greater sense of one's separateness from the rest of society. Reading is usually a solitary activity. Novels, with their typical emphasis on one key figure, have been the central literary product of the past two centuries. It is therefore likely that a culture based on print is more individualistic than one that relies on collective means of communication, such as Sunday sermons or old-fashioned political oratory. It has even been suggested that the very concept of the "self" is foreign to people and cultures that have not been touched by literacy.[26]

Some fascinating possibilities of this sort have been suggested by Marshall McLuhan, for whom media such as print or television have had consequences that far outweigh the overt messages that they carry.[27] According to McLuhan, printed books fundamentally changed civilization not because of the information they transmitted; the greatest consequences of printing lay in the different modes of thought and perception that it fostered. In McLuhan's analysis, each medium of communication engenders a distinctive way of looking at the world; as his famous aphorism has it, "the medium is the message." The reading of the printed word makes readers think in sequence, just as a sentence is sequentially read from left to right in English, or right to left, or top-to-bottom. Reading also produces an egocentric view of the world, for the reader's involvement with the printed word is solitary and private. For McLuhan, the intellectual and social consequences of print were immense, encompassing such diverse things as the Newtonian view of the universe, the employment of fixed prices for business transactions, and the "detribalization" of society.

It is impossible to summarize McLuhan's ideas in a few sentences. The intrepid reader is advised to turn to McLuhan's writings after first being warned that their author can be cryptic, elliptical, hyperbolic, and at times simply maddening. Nonetheless, some taste of his thinking can be obtained by looking at how McLuhan has interpreted changed painting styles as a reflection of printing, or what he has termed "The Gutenberg Galaxy." Reading, as McLuhan points out, inculcates a single "point of view," the fixed position of the reader.[28] In similar fashion, the visual arts that developed after the invention of printing began to depict scenes as they might be viewed from a single point in space. Given this central vantage point, the scene appears to have a three-dimensional quality due to the artist's employment of the principles of perspective. The employment of perspective is not a natural or necessary part of pictorial art. As any art student knows, to paint or draw something in perspective requires the application of special procedures and a good deal of practice. These artistic techniques were not employed in medieval painting, which was characterized by a visual flatness. Figures were not presented according to the rules of perspective, but rather in accordance with the importance of the subjects; in a typical Medieval painting, Jesus, an archbishop, and the faithful appear in the same plane, but they are not all of the same size; their size is commensurate with their importance in the spiritual hierarchy. Until the advent of printing there was no reason to depict subjects according to rules of perspective because printing and widespread reading had not conditioned people to see the world as it might appear from a single vantage point. As McLuhan has argued, the rules of perspective were developed and used only after printing produced an individualistic perception of the world.

Newspapers

For several centuries after the invention of movable type, the most important product of the printing press was the book. Beginning in the eighteenth century, a new kind of publication appeared that was to have equally revolutionary consequences for society. That publication was the newspaper. The transmission of written news has a long history, originating with the *Acta diurna* that was posted in the Roman Forum, and continuing with the printed broadsheets that circulated from late medieval times onward. The first regularly appearing newspapers emerged in the seventeenth century, and by the beginning of the next century, newspapers were established institutions in a number of European cities. Colonial America followed suit, with the appearance of *The Boston News-Letter* in 1704 marking the beginning of newspaper publishing in what was to be the United States.

These early newspapers were unimpressive, usually consisting of no more than four pages of hard-to-read type. Their circulation was also limited, being confined to the small number of people who were both literate and able to pay for a newspaper. This situation changed dramatically during the nineteenth century, when the newspaper became a mass-produced item, a product of Industrial Revolution technology.

One of the distinguishing characteristics of the Industrial Revolution was the use of new sources of energy, with steam power taking on a growing importance. The first steam-powered printing press was invented by Friedrich Koenig and Andreas Bauer in 1812 and was put to use by the *Times* of London in 1814. The use of steam power for this purpose necessitated a fundamental redesign of the printing process. Instead of manually pressing a sheet of paper against a flatbed containing the type, the job was done by a steam-powered rotating cylinder, while another cylinder did the inking. This process ran off 1,000 copies of the newspaper each hour, one-third the time that the handpress required.[29]

Other refinements ensued. Printing on a continuously running roll of paper (instead of feeding single sheets of paper) greatly speeded up the process, as did the development of presses that printed on both sides of a sheet and automatically cut individual pages. Improved inking processes contributed to the speed of production, while the use of wood pulp for paper also lowered production costs.

Improvements to the rotary press concept followed, most notably Richard Hoe's use of another rotating cylinder instead of a flat horizontal bed to hold the type. First employed in 1846, by the beginning of the Civil War improved variants of the rotary press could produce as many as 20,000 impressions per hour.[30] Printing was further accelerated by the invention of the stereotype plate, which allowed the easy production of duplicate plates for use on multiple presses. The stereotype plate was made by pressing a paper mat onto a frame of set type. After the mat was dried and shaped into a half of a cylinder, molten metal was forced into it, forming the actual printing plate. The other side of the rotating cylinder held type for another page. The first of these presses contained two cylinders, each with four plates, so an eight-page newspaper could be printed with each turn of the cylinders.[31] By 1890, improved rotary presses were capable of printing 48,000 twelve-page papers each hour.[32]

While the presses turned out newspapers with increasing rapidity, the need to set type by hand restricted the rate of production. A great deal of inventive energy therefore was expended on the development of automatic typesetting. Even Mark Twain was involved as a financial backer for such a venture; it unfortunately failed, dashing the great author's dreams of making a vast fortune.[33] Success instead came to a German immigrant, Ottmar Mergenthaler, whose Linotype machine was the first effective typesetting device. First used in 1886, by the end of the century 3,000 of these machines were in use around the world.[34]

The development of the mass-circulation newspaper was further stimulated by a host of nineteenth-century inventions that allowed the rapid gathering and dissemination of news. Railroads and steamships conveyed reporters to scenes of newsworthy events. The telegraph quickly relayed their stories back to the newsroom and had the incidental effect of making stories less verbose because the senders of telegraphic messages were charged by the word.[35] The news-gathering power of the telegraph was enhanced by the laying of transatlantic cables that greatly sped up communications between Europe and the United States. The telephone and, by the end of the century, the radio further assisted reporters and the newspapers for which they worked.

Circulation Wars and the Shaping of Public Opinion

As with other mass-production technologies, the improvements in newspaper printing dramatically lowered production costs. Before the 1830s, newspapers sold for 6 cents a copy, which in those days could buy a quarter pound of bacon or a pint of whiskey.[36] As production costs fell, the first American penny paper, the *New York Sun*, made its appearance in 1837. Aiming for a mass audience, the *Sun* offered sensational stories of dubious veracity, such as a report on the observance of life on the moon through the use of a powerful telescope located in South America. Other newspapers followed suit, and by the second half of the nineteenth century the cheap mass-circulation daily newspaper was a fixture in every city and town of Europe and North America. The expansion of newspaper readership occurred at a rapid rate. During the early decades of the nineteenth century, the *Times* had led all English newspapers with 10,000 readers,[37] while the largest New York newspapers usually printed editions of no more than 2,500.[38] By 1870, 2.6 million newspapers were printed each day in the United States, and by the end of the century, that figure had risen to 15 million.[39]

Unfortunately, the spread of these newspapers was often stimulated by journalism of the worst sort. Newspapers built up circulation by running stories that stressed scandal, crime, and bizarre occurrences. The debasement of news reporting accompanied the increase in newspaper readership; the "yellow journalism" of that era, in the words of Edwin and Michael Emery, "turned the high drama of life into a cheap melodrama and led to stories being twisted into the form best suited for sales by the howling newsboy."[40] Sensationalist newspapers had a long life. Until the internet diminished their readership, the supermarket tabloids enjoyed a wide circulation by running stories about miraculous medical cures, sex scandals, and UFO sightings.

Both in the past and today, reportage of this sort demonstrated many newspapers' low regard for the taste and intelligence of their readers. Beyond this, the mass-produced newspaper became a force in its own right, and its editorial policies did much to shape public opinion. This can be seen in the way the popular press helped to push America into a war with Spain in 1898. To be sure, other forces were at work; having conquered the frontier of its own territory, the United States was seeking new triumphs of Manifest Destiny. Still, without the galvanization of public opinion by newspapers such as William Randolph Hearst's *New York Journal*, it is possible that the war could have been averted.

Hearst was engaged in a fierce struggle with Joseph Pulitzer's *New York World*, and inflamed coverage of a rebellion against Spanish rule in Cuba was a surefire way to sell papers. Through the use of sensational reporting, both papers enlarged their circulations to over a million each, a record at that time.[41] Reports of Spanish atrocities in putting down the rebellion, while having some foundation, were wildly exaggerated and given extensive coverage. Stories in the

World left little to the imagination: "Blood on the roadsides, blood in the fields, blood on the doorsteps, blood, blood, blood!"[42] Not to be outdone, Hearst's *New York Journal* played a prominent role in whipping up public enthusiasm for a war with Spain to "liberate" Cuba. When Hearst's artist-reporter in Cuba notified the boss that war was unlikely, Hearst reportedly sent him a telegram with the command "You furnish the pictures and I'll furnish the war."[43]

From the beginning of the revolt to America's going to war with Spain, hardly a day went by without a story about Cuba in one of the New York newspapers.[44] And when the American battleship *Maine* blew up in Havana harbor, many newspapers were quick to attribute it to a Spanish mine. (To this day, the cause of the ship's sinking remains unknown, although it was likely due to internal causes.) Even though disputes with Spain over a variety of issues were almost fully resolved by this time, a significant portion of the popular press had whipped up a war hysteria that could not be extinguished. The brief war that ensued concluded with the United States assuming the role of an imperial power, with colonial holdings that stretched from the Philippines to Puerto Rico.

It would be an exaggeration to say that the Spanish–American War, or any other major historical event, was directly caused by the reportage of circulation-hungry newspapers. The remnants of Spain's American empire were an inviting target for turn-of-the-century America's expansionist ambitions. Under these circumstances, overheated news reporting reinforced opinions and ambitions that were already evident. Still, there can be no denying that the journalistic policies of Hearst and Pulitzer made a peaceful resolution of disputes between the United States and Spain far more difficult.

It would be even farther from the truth to attribute such events as the Spanish–American War to technological advances in printing. To be sure, the mass-circulation dailies depended on improved methods of newspaper production, but much more was involved than such innovations as the steam-powered rotary press. The spread of education and literacy, spurred by the expansion of public schooling, created a large potential readership for these publications. The expansion of industry and commerce was accompanied by the emergence of a growing advertising industry that provided vast revenues for the newspapers. Urbanization and immigration produced large concentrations of population from which a mass readership could be drawn. Finally, a more democratic social order generated an environment in which the "common man" gained in political and economic importance; as the first of the mass media, newspapers were a natural outgrowth of mass society.

All of these trends reinforced one another. New printing technologies contributed to literacy and the rise of a mass readership, commercial expansion, and perhaps even the expansion of democracy. At the same time, the emergence of new printing technologies was stimulated by all of these economic, social, political, and cultural factors. As we have already seen, Gutenberg's invention of movable type was accompanied by changes on many fronts. The same thing happened during the second half of the nineteenth century as a series of technological innovations made possible the publication of mass-circulation newspapers. In neither case did technological developments by themselves produce the epochal changes with which they are associated. The development of these technologies was part of a larger process of change. Yet at the same time, these technologies were major contributors to this process. From movable type to the steam-powered rotary press, the technology of printing exemplifies the complex, mutually reinforcing interactions between technological and social change.

Figure 14.2 Linotype machines were the first effective typesetting device and were still being used up until the late 1970s and early 1980s.

Digitizing the News

A similar process can be seen today, as digital technologies now allow the transmission and reception of the printed word without the need for ink and paper. With the advance of digital technologies, readers can access vast amounts of material through the use of computers, smartphones, tablets, and electronic books. In recent years, e-books and tablets have become especially important sources of reading material. According to a survey conducted in 2015, 72 percent of adults had read a book during the past twelve months; of these, 27 percent had read at least one e-book.[45] In 2022, book readership had dropped to 64 percent, with less than half of adults reading literature, as opposed to non-fiction.[46]

So far, e-books have not threatened to replace printed books. The same cannot be said about printed newspapers and magazines. According to the Pew Research Center, print newspaper circulation peaked in the mid-1980s and then declined in every subsequent year as increasing numbers of readers turned to digital sources of information.[47] During that period hundreds of newspapers, many of them in small communities, went out of business altogether. Newspaper employment has reflected this downward trend. In 2008, newspapers employed 71,000 workers; ten years later their number had dropped to 38,000, a 47 percent decline.[48] Magazines have fared no better; circulation losses, especially for newsstand purchases, and declining revenues have for the most part paralleled the experience of the newspaper industry.[49]

The overall decline of newspaper circulation has resulted in significant financial losses. Newspapers earn more of their revenues from advertising than they do from subscription payments and newsstand purchases, so lower circulation means lower advertising rates. Once again, the historical trend has been grim; advertising revenue from print editions fell from $48.6 billion in 2000 to $9.6 billion in 2020.[50] This situation will not likely improve in the years to come. In 2009, 60 percent of teenagers read newspapers; by 2018 the number had fallen to 20 percent.[51] Some of these losses have been offset by increased readership and income from digital editions. It would be helpful to note the size of the digital readership, but determining the number of readers of digital newspapers is difficult.

Some readers peruse digital versions of print newspapers, but many more read a single article after searching the internet for specific information, or clicking on a link to an article that appeared in e-mail, social media, or other sources.

Traditional newspapers now receive about 35 percent of their revenue from digital editions. This, however, does not so much reflect the growth of digital income as it does the decline of traditional sources of revenue. In any event, digital earnings have not compensated for the losses caused by the switch from paper to digital media. Some newspapers have attempted to increase revenue from digital editions by requiring a payment each time an article is accessed. In 2016, among the 98 U.S. newspapers with circulations over 50,000, the American Press Institute found that 77 of them use a digital subscription model. Some charge a flat fee for access, but much more common is a "metered" model, that allows a certain number of articles to be accessed (typically ten per month) before the payment of a fee is required. It still remains an open question as to whether readers will be longtime subscribers willing to consistently pay for the digital edition; crucial to the success of paid digital subscriptions will be convincing more people to sign up in years to come.[52] In 2022, it is clear that purely local newspapers have declined, with larger news conglomerates buying up local newspapers, and national publications solidifying their centrality to the industry in print and digital forms.[53]

Although some readers access digital versions of established newspapers, online news sources have become important players in the news business. Moreover, many readers regularly get their news via social media platforms like Facebook (31 percent), YouTube (25 percent), and X (formerly known as Twitter; 17 percent).As might be expected, there are age-group differences in the use of social media as a source of news:

> [A]dults under 30 make up the largest share of those who regularly get news on the site. For example, half or more of regular news consumers on Snapchat (67%), TikTok (52%) and Reddit (50%) are ages 18 to 29. Additionally, women make up a greater portion of regular news consumers on Facebook, while the opposite is true for sites like Twitter and Reddit.[54]

While social media has become an important means of accessing the news, even those who use it bring some skepticism to these sources. Perceived importance is not the same thing as trust. This has transformed rapidly, from social media being a disreputable source, to U.S. adults under 30 who trust social media almost as much as regular national news outlets.[55]

Trustworthy or not, digital news sources have overtaken newspapers as sources of news and information. A recent Pew Research Center survey found that 36 percent of U.S. adults learned something about the 2016 election in the week prior from a print newspaper. This was lower than the portion that learned from radio (44 percent), digital sources (65 percent), or television (78 percent).[56] This is a substantial change from only a decade before, when newspapers outranked radio and the internet as the public's main source of news.[57] Another indication of the declining influence of newspapers came from the Pew Research Center survey, which found that a mere 5 percent of U.S. adults named print newspapers as their "most helpful" source for learning about the election, a percentage that trailed nearly every other media category (cable, local and national TV, radio, social media, and news websites) by wide margins.[58] In 2022, candidates issued approximately 3.4 million tweets in total from the start of the year.[59]

Reading Digitally

The decline of printed newspapers may not be terribly troubling unless one is somehow connected to that industry. After all, the medium, be it a paper newspaper or a social media site, that conveys fact and ideas should not alter their substance. But remember Marshall McLuhan's dictum that "the medium is the message." We won't concern ourselves here with the possible psychological and cultural effects of digital media. Instead, we will focus on a narrower issue: the extent to which digital media is comprehended and retained when compared with conventional print media.

Due to the relatively recent arrival of digital reading materials, only a few empirical studies have been conducted on its consequences, while long-term studies are notably absent. Even so, some aspects of reading on-screen seem to be evident. For one thing, reading from a screen is likely to make concentration difficult. Embedded videos, pop-up advertisements, and links to other materials can be powerfully distracting. It also appears to be the case that on-screen readers are more likely to engage in multitasking than readers of paper materials.[60]

These attributes create obstacles to what has been called "deep reading."[61] Readers of on-screen materials tend to skim the text; as one researcher has noted, "When reading on screens, for example, people seem to reflexively skim the surface of texts in search of specific information, rather than dive in deeply in order to draw inferences, construct complex arguments, or make connections to their own experiences. Research has also found that students, when reading digitally, tend to discard familiar print-based strategies for boosting comprehension.[62] In the words of another researcher, "Skimming, had become the new reading: the more we read online, the more likely we were to move quickly, without stopping to ponder any one thought."[63] This tendency can be seen in the way people access online news articles. According to one survey, a typical reader quickly scans the headlines and spends fifteen seconds or less on individual articles.[64]

In assessing the effects of on-screen reading, a lot depends on what is being measured. This is exemplified by one study that took a college-level course and divided the students into two groups, one reading an assignment on a tablet, the other as printed matter. They were then given two tests, one a set of multiple-choice questions, the other a few short answers. When the scores of both groups were compared, there was no statistically significant difference in their mean scores. But according to the author of the study, when the range of score distributions was examined, the group who read the assignment on paper had greater frequencies of higher scores for both multiple-choice recall and the short answers that measured comprehension.[65] Similarly, while multitasking can happen offline, it is more significant in online environments, although both correlate poorly with academic performance.[66]

Why this was so is unclear; quite a lot of future research is needed. It is far too late to argue that school districts, colleges, and other

Figure 14.3 There remains a great deal of optimism that technology can be effectively harnessed for early childhood, indeed any level of education. (Shutterstock/Monkey Business Images)

institutions would be wise to postpone massive purchases of tablets, e-readers, and other devices for the presentation of digital materials. The rapid transformation of schooling online to accommodate COVID-19 solidified the centrality of technology in education at nearly all levels. But particularly due to COVID-19, at least in the United States, students learning using information and communication technology has been associated with significant learning loss, although teasing out the effects of the technology from issues of isolation and differential access to and quality of online instruction will be difficult.[67]

The rise of digital texts is only one component, and a small one at that, of the revolutionary changes that have been transforming media from the middle of the nineteenth century onward. These epochal changes are the topic of the next three chapters.

Questions for Discussion

1. Do statements that appear in print seem more believable than those that are presented through some other medium? Why?
2. One of the major trends in the world today is the spread of literacy. In many parts of the world, only a small percentage of the

population could read a generation ago. Today, the majority of the population is literate. What sort of social, political, and economic effects will this produce? Will all of them be good?

3. Some social critics are of the opinion that the spread of digital media is destroying literacy. According to some studies, standardized test scores of reading and writing ability have gone down in recent years. Are the new media the main cause? If so, is the slow displacement of the printed media by the electronic media necessarily a bad thing?
4. Lurid newspaper stories helped to fan a conflict between Spain and the United States that ultimately led to war. The Vietnam War, another overseas conflict involving the United States, met with a great deal of opposition, due in part to the continual presentation of that war on television. Marshall McLuhan would have thought this significant. Can you imagine why? Can you think of any fundamental differences between newspaper accounts of a war and those presented on television?
5. Do you think that digital technologies in the form of e-books, tablets, and smartphones will soon replace newspapers and magazines? Do conventional printed materials have a future? What, if any, advantages do they have over digital media?

Notes

1. Quoted in Maurice N. Richter, Jr., *Technology and Social Complexity* (Albany: State University of New York Press, 1982), pp. 16–17.
2. Jared Diamond, *Guns, Germs, and Steel: The Fate of Human Societies* (New York and London: W. W. Norton, 1997), pp. 215–38.
3. James Thorpe, *The Gutenberg Bible: Landmark in Learning* (San Marino, CA: Huntington Library, 1975), p. 3.
4. See: J. M. Foley, "Reading" Homer through Oral Tradition. *College Literature*, 34, 2 (2007): 1–28.
5. Xing Runchuan, "The Invention and Development of Printing and Its Dissemination Abroad," in *Ancient China's Technology and Science* (Beijing: Foreign Languages Press, 1983), pp. 383–91.
6. Albert Kapr, *Johann Gutenberg: The Man and His Invention*, translated by Douglas Martin (Aldershot, England: Scolar Press, 1996).
7. Frances and Joseph Gies, *Cathedral, Forge, and Waterwheel: Technology and Invention in the Middle Ages* (New York: HarperCollins, 1994).
8. Kapr, *Johan Gutenberg*, pp. 109–22.

9. D. S. L. Cardwell, *Turning Points in Western Technology: A Study of Technology, Science and History* (New York: Science History Publications, 1972), pp. 20–24.
10. Thorpe, *The Gutenberg Bible*, p. 4.
11. Ferdinand Braudel, *Capitalism and Material Life, 1400–1800* (New York: Harper & Row, 1975), p. 298.
12. Elizabeth Eisenstein, *The Printing Revolution in Early Modern Europe* (New York: Cambridge University Press, 1984), pp. 195–99.
13. Lewis Mumford, *Technics and Civilization* (New York: Harcourt, Brace and World, 1934), pp. 21–22.
14. Eisenstein, *The Printing Revolution*, p. 75.
15. Ibid., pp. 43–44.
16. Ibid., p. 239.
17. Ibid., p. 147.
18. Richard Gawthorp and Gerald Strauss, "Protestantism and Literacy in Early Modern Germany," *Past and Present* 104 (August 1984): 6.
19. Quoted in David Cressy, *Literacy and the Social Order: Reading and Writing in Tudor and Stuart England* (New York: Cambridge University Press, 1980), p. 3.
20. Eisenstein, *The Printing Revolution*, p. 170.
21. Cressy, *Literacy and the Social Order*, p. 16.
22. A. L. Stinchcombe, "Social Structure and the Invention of Organizational Forms," in Tom Burns (Ed.), *Industrial Man* (Harmondsworth, England: Penguin, 1969), pp. 163–65.
23. Scott Cook, "Technology and Social Change: The Myth of Movable Type," *The Weaver* 4, 1 (Fall 1985): 5.
24. Terry Forsyth, "A Brief History of the Korean Language," Department of Linguistics, Brigham Young University (1999). Accessed from http://linguistics.byu.edu/classes/ling450ch/reports/korean2.html. See also "Superscript," *The Economist*, October 10, 2015, p. 40.
25. Cressy, *Literacy and the Social Order*, p. 176.
26. Barry Sanders, *A Is for Ox: The Collapse of Literacy and the Rise of Violence in an Electronic Age* (New York: Random House, 1995), pp. 29–32.
27. See Herbert Marshall McLuhan, *The Gutenberg Galaxy: The Making of Typographic Man* (Toronto: University of Toronto Press, 1962); and *Understanding Media: The Extensions of Man* (New York: New American Library, 1964).
28. McLuhan, *The Gutenberg Galaxy*, p. 111.
29. Robert W. Desmond, *The Information Process: World News Reporting to the Twentieth Century* (Iowa City: University of Iowa Press, 1978), pp. 62–63.
30. Edwin Emery and Michael Emery, *The Press and America: An Interpretive History of the Mass Media*, 7th ed. (Englewood Cliffs, NJ: Prentice-Hall, 1992), p. 115.
31. Desmond, *The Information Process*, p. 295.
32. Emery and Emery, *The Press and America*, p. 188.

33. John F. Kasson, *Civilizing the Machine: Technology and Republican Values in America, 1776–1900* (New York: Penguin Books, 1977), pp. 203–04.
34. Desmond, *The Information Process*, p. 298.
35. John Tebbel, *The Compact History of the American Newspaper* (New York: Hawthorne Books, 1969), p. 121.
36. Emery and Emery, *The Press and America*, p. 93.
37. Desmond, *The Information Process*, p. 77.
38. Emery and Emery, *The Press and America*, p. 85.
39. Edwin Emery, *The Press and America*, 3rd ed. (Englewood Cliffs, NJ: Prentice-Hall, 1972), p. 285.
40. Emery and Emery, *The Press and America*, p. 191.
41. Desmond, *The Information Process*, p. 388.
42. Edwin Emery, *The Press and America*, 3rd ed. (Englewood Cliffs, NJ: Prentice-Hall, 1972), p. 365.
43. See https://historynewsnetwork.org/article/173692 and other sites that discuss whether or not this statement is apocryphal.
44. George H. Douglas, *The Golden Age of the Newspaper* (Westport, CT, and London: Greenwood, 1999), p. 111.
45. Lee Rainie and Andrew Perrin, "Slightly Fewer Americans Are Reading Print Books, New Survey Finds," *Pew Research Center* (October 19, 2015). Accessed from www.pewresearch.org/fact-tank/2015/10/19/slightly-fewer-americans-are-reading-print-books-new-survey-finds.
46. Riley Stoltenburg, "US Book Reading Statistics (National Survey 2022)," *Test Prep Insight* (January 14, 2023). Accessed from https://testprepinsight.com/resources/us-book-reading-statistics.
47. Pew Research Center, "Newspapers Fact Sheet," (June 29, 2021). Accessed from www.pewresearch.org/journalism/fact-sheet/newspapers.
48. Elizabeth Orieco, "Despite Subscription Surges for Largest Newspapers, Circulation and Revenue Fall for Industry Overall," *Pew Research Center* (July 9, 1019). Accessed from www.pewresearch.org/fact-tank/2019/07/09/u-s-newsroom-employment-has-dropped-by-a-quarter-since-2008.
49. Tess Stynes, "Print Magazine Sales Decline in 1st Half of 2014," *The Wall Street Journal* (August 7, 2014). Accessed from www.wsj.com/articles/print-magazine-sales-decline-in-1st-half-of-2014-1407423959.
50. Pew Research Center, "Newspapers Fact Sheet," (June 29, 2021). Accessed from www.pewresearch.org/journalism/fact-sheet/newspapers.
51. "Seize the Memes," *The Economist* (December 31, 2019–January 3, 2020), p. 85.
52. Alex T. Williams, "Paying for Digital News: The Rapid Adoption and Current Landscape of Digital Subscriptions at U.S." *American Press Institute* (February 29, 2016). Accessed from www.americanpressinstitute.org/publications/reports/digital-subscriptions/single-page.
53. See: Katerina Eva Matsa and Kirsten Worden, "Local Newspapers Fact Sheet," *Pew Research Center* (May 26, 2022). Accessed from www.pewresearch.org/

journalism/fact-sheet/local-newspapers; and www.pewresearch.org/journalism/fact-sheet/newspapers.

54. Pew Research Center, "Social Media and News Fact Sheet," (September 20, 2022). Accessed from www.pewresearch.org/journalism/fact-sheet/social-media-and-news-fact-sheet. www.pewresearch.org/fact-tank/2022/10/27/u-s-adults-under-30-now-trust-information-from-social-media-almost-as-much-as-from-national-news-outlets/
55. Jeffrey Gottfried, Michael Barthel, Elisa Shearer, and Amy Mitchell, "The 2016 Presidential Campaign—A News Event That's Hard to Miss," *Pew Research Center* (February 4, 2016). Accessed from www.pewresearch.org/journalism/2016/02/04/the-2016-presidential-campaign-a-news-event-thats-hard-to-miss.
56. American Press Institute, "How Americans Get Their News" (March 17, 2014), www.americanpressinstitute.org/publications/reports/survey-research/how-americans-get-news/
57. Gottfried, et al., "The 2016 Presidential Campaign."
58. Sono Shah and Samuel Bestvater, "As the 2022 Campaign Draws to a Close, Here's How Federal, State and Local Candidates Have Used Twitter," *Pew Research Center* (November 2, 2022). Accessed from www.pewresearch.org/fact-tank/2022/11/02/as-the-2022-campaign-draws-to-a-close-heres-how-federal-state-and-local-candidates-have-used-twitter/
59. Lisa Allcott, "Reading On-Screen vs Reading in Print: What's the Difference for Learning?" *National Library* (October 11th, 2021). Accessed from https://natlib.govt.nz/blog/posts/reading-on-screen-vs-reading-in-print-whats-the-difference-for-learning.
60. Ferris Jabr, "The Reading Brain in the Digital Age: The Science of Paper versus Screens," *Scientific American* (April 11, 2013). Accessed from www.scientificamerican.com/article/reading-paper-screens.
61. Benjamin Herold, "Digital Reading Poses Learning Challenges for Students," *Education Week* (May 6, 2014). Accessed from www.edweek.org/teaching-learning/digital-reading-poses-learning-challenges-for-students/2014/05?cmp=ENL-EU-MOSTPOP.
62. Maria Konnikova, "Being a Better Online Reader," *The New Yorker* (July 16, 2014). Accessed from www.newyorker.com/science/maria-konnikova/being-a-better-online-reader.
63. Nicole Martin, "How Social Media Has Changed How We Consume News," *Forbes* (November 30, 2018). Accessed from www.forbes.com/sites/nicolemartin1/2018/11/30/how-social-media-has-changed-how-we-consume-news/#5ef1d87b3c3c.
64. Anne Niccoli, "Paper or Tablet? Reading Recall and Comprehension," *EDUCAUSE Review* (September 28, 2015). Accessed from https://er.educause.edu/articles/2015/9/paper-or-tablet-reading-recall-and-comprehension.
65. Ahlam Alghamdi, et al., "Online and Face-to-Face Classroom Multitasking and Academic Performance: Moderated Mediation with Self-Efficacy for Self-Regulated Learning and Gender," *Computers in Human Behavior* 102 (2020):

214–222; and Andrew Lepp Aet al., "College Students' Multitasking Behavior in Online Versus Face-to-Face Courses." *SAGE Open* 9 (2019).

66. Bianca Vázquez Toness and Jocelyn Gecker, "Online School Put Us Kids Behind. Some Adults Have Regrets," *Associated Press* (October 21, 2022). Accessed from https://apnews.com/article/online-school-covid-learning-loss-7c162ec1b4ce4d5219d5210aaac8f1ae; and Megan Kuhfeld, et al., "The Pandemic Has Had Devastating Impacts on Learning. What Will It Take To Help Students Catch Up?" *Brookings Institution* (March 3, 2022). Accessed from www.brookings.edu/blog/brown-center-chalkboard/2022/03/03/the-pandemic-has-had-devastating-impacts-on-learning-what-will-it-take-to-help-students-catch-up.

Chapter 15

Electronic Media: From the Telegraph to the Television

During the 450 years that followed the invention of movable type in the West, the printed word was the dominant mode of mass communication. By the middle of the nineteenth century, words could be sent and received at unprecedented speeds through the newly invented telegraph. Although it was not identified as such when it was invented, the telegraph is a digital device, one that uses electrical pulses to transmit combinations of dots and dashes representing individual letters—the Morse Code. Few people send or receive telegrams today, but in its time the telegraph was as revolutionary as the internet is in ours. It lifted the isolation of remote communities, knit commercial enterprises together, greatly facilitated the operation of vast railroad networks, and brought news from all around the world.[1] Still, the telegraph had its limitations. Not only were trained operators required to encode and decode messages, but large amounts of capital were also needed to string up telegraph wires between communication points. Moreover, telegraphic communication was thwarted when the installation of wires was physically impossible or costs were prohibitive, as in the case of ships at sea or geographically remote places.

These limitations began to be overcome toward the end of the century as scientists and engineers in several countries learned how to send and receive signals through empty space. The first product of their efforts was radio. The experience gained with radio subsequently supplied the foundation for an even more far-reaching medium: television. The social, cultural, political, and economic effects of these media have been at least as great as those of printing.

We are still attempting to comprehend how radio and television have affected our lives, even as we now find ourselves facing the opportunities and challenges of digital media. The latter will be explored in the next chapters, but first we will survey the causes and consequences of advances in radio and television technology.

The Invention of Radio

It might be thought that the limitations of telegraphic communication created the motivation to develop a method of electrical communication that did not require wires. And it is certainly true that much of radio's development was "pulled" along by the demand for better methods of communication. But that came later; the first successful attempt at sending and receiving wireless communications had nothing to do with practical concerns. As we saw earlier, radio transmission and reception were first used by Heinrich Hertz as a way of empirically testing the theoretical formulations of James Clerk Maxwell. Hertz produced radio waves with an oscillator, a device that rapidly generated electrical pulses. To receive or "detect" the radio waves produced by these pulses, Hertz employed a piece of wire with small balls at each end, bent into a near circle. When this was placed in the vicinity of the oscillator, a spark jumped between the two balls in conjunction with the sparking of the oscillator. Through a series of experiments using these devices, he was able to confirm Maxwell's calculations that electrical waves traveled at the speed of light. Additionally, he showed that these waves propagated themselves in accordance with the laws of optics that had first been formulated for visible light.

Hertz's concerns were entirely in the realm of scientific inquiry, but others saw in his apparatus an embryonic technology that might realize the dream of wireless communication. Many experimenters tackled the numerous technical problems that kept radio from being a practical possibility. Although Guglielmo Marconi is often hailed as the inventor of radio, as with most significant inventions, there are many others whose contributions were equally important, such as Oliver Lodge, who developed the method of tuning a receiver so it could receive signals of a specific wavelength.

At this point, radio was simply wireless telegraphy, useful only for sending and receiving dots and dashes. The transmission and reception of actual sounds was made possible by using continuous waves to carry a signal. This, in turn, had been made possible by Ronald Fessenden's incorporation of the heterodyne principle into radio technology. Heterodyne transmission and reception used two separate alternating currents, one generated by the transmitter and the other by the receiver. Oscillating waves transmitted at a certain frequency mixed with the receiver's lower frequency waves to produce a wave that could be easily received.

Important as the heterodyne principle was, an even greater breakthrough came with the invention of better equipment for the transmission and reception of radio waves. These devices were based on the principle of using a small flow of current to govern a larger current. In this application the small current produced by a radio wave could be used to regulate the stronger current that drove the speaker of a radio. The first of these devices was the diode, invented in 1904 by John Ambrose Fleming, whom we met briefly in Chapter 4. It established the principle of controlling current electronically, but it had its limitations. A much more workable device was the triode, an invention of Lee de Forest. By using a separate electrode to control the current, the triode served as an excellent basis for a radio receiver, as well as for its amplifier.[2]

Figure 15.1 Although Guglielmo Marconi did not invent radio, he played an important role in its commercial application. Here, he is with some early devices for transmitting and receiving wireless telegraphic messages.

The Origins of Commercial Radio

While technical development proceeded at a good clip during these early years, few thought of radio as a means of providing entertainment and information for a large audience. Radio was seen simply as wireless telegraphy, and for many years its main use was for ship-to-shore communications. Radio was also used to transmit messages across oceans, thereby obviating the need for undersea cables, but high cost and unreliability plagued these efforts. Military forces also made extensive use of radio communications, and World War I stimulated a good deal of technical progress, especially in the development of mass-production techniques for the manufacture of radio components.

During the 1920s, interest in radio mounted as large numbers of amateur radio operators constructed their own equipment, sent out messages, and attempted to pick up the signals of others. Their chatter cluttered up the airwaves, and with a mischievous intent like that of today's hackers, a few even sent out bogus orders to naval vessels. But others had more serious aims. Throughout the United States, amateur radio operators, many of whom had developed their skills during World War I, began to broadcast news, weather bulletins, musical recordings, and even live musical performances. In a manner resembling the early days of the personal computer, enthusiastic amateurs made significant contributions to the new technology by constructing their own apparatus, finding new applications for radio communication, and in general exploring the potentials of the new medium.

At this time, business enterprises began to take notice of radio's commercial possibilities. Realizing that regular programming could stimulate a sizable market for its radio sets, in 1920, the Westinghouse Electric and Manufacturing Company engaged one of its employees, himself a dedicated radio amateur, to construct a broadcasting station atop one of the buildings of its Pittsburgh headquarters. On November 2, 1920, radio station KDKA went "on the air," giving up-to-the-minute results of the 1920 presidential election to an audience of several hundred. Within a matter of months Westinghouse and General Electric had set up radio stations in several cities. Other entrepreneurs followed suit, and so by 1922 more than 500 stations were transmitting music, sporting events, speeches, and news programs. Commercial broadcasting took on a growing importance as radio began to reach the public and not just a relatively small group of technically minded amateurs.

At first, many of these programs could be low-budget endeavors; phonograph records required only small initial costs, and most live performers were willing to appear free of charge. This did not last for long; the American Society of Composers, Authors, and Publishers (ASCAP) began to demand payment for the broadcast of recorded music, and in 1923, it backed this demand with a successful lawsuit. In similar fashion, live performers started to expect payment for their services. Other countries, most notably Great Britain, had by this time established a national radio system through which the expenses of broadcasting were defrayed by licensing fees paid by owners of radio sets. But in the United States, free enterprise was the order of the day, and radio broadcasting was sustained by companies willing to sponsor programs in return for the airing of their commercial messages. In the years that followed, listening to radio programs became an integral part of American life as families gathered to listen together, much as was to happen during the early days of television. By this time, several decades had passed before radio technology had made the passage from wireless telegraphy to commercial broadcasting. Along the way, various entrepreneurs looked for ways to use radio technology as a basis for making money, just as has been happening with the internet today.

Figure 15.2 Radcliffe College students operating an early radio station.

Figure 15.3 In pre-television days, radio was a chief source of family entertainment.

The Rise of Television

For all of radio's accomplishments, only a portion of the electronic media's potential had been demonstrated. Radio could only reach one of the senses; if sight could be added to sound, the appeal of electronic communications could be greatly enhanced. The dream had long been there; the idea of electrically transmitting and receiving visual images existed even before radio was a practical reality. In 1879, readers of the English magazine *Punch* were presented with an illustration of a couple watching a tennis match being shown on a large screen over their fireplace (Figure 15.4). Three years later, a French artist drew equally prescient illustrations that depicted people viewing television screens filled with pictures of distant lectures, wars, and even girlie shows.[3] And in 1907, in the pages of *Scientific American,* the term "television" appeared for the first time in print.

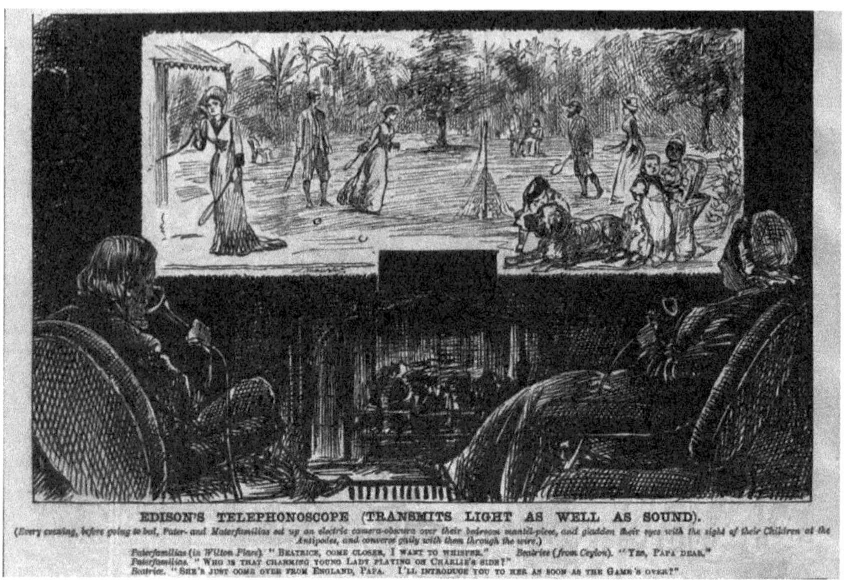

Figure 15.4 A *Punch* magazine illustration from 1879.

A substantial amount of inventive activity took place in the 1920s on both sides of the Atlantic as inventors, government agencies, and private firms sought to make television a practical reality. The British Broadcasting Corporation (BBC) took an early lead in 1929 when it began to transmit half-hour television programs five days a week. The technology employed by the BBC used an electromechanical system in which a rotating disc was used to convert electrical signals into visible images. This technology had many drawbacks, not the least of which was that it required extremely high levels of illumination when attempting to capture the image of a person or object.

By 1937 the BBC abandoned electro-mechanical television in favor of an all-electronic system. The genesis of this technology went all the way back to a series of fundamental discoveries and inventions that began at the end of the nineteenth century. Most important of these was the cathode ray tube. This device was based on a principle, discovered by Ferdinand Braun in 1897, that a beam of electrons moving along a closed tube could be deflected by a magnet. In Russia, Boris Rosing used this tube for television reception by placing a screen of photoelectric cells in the tube and activating them with moving electrons.

This was only a laboratory demonstration and not a complete system of sending and receiving television images. An important step was taken by Vladimir Zworykin, a Russian émigré who had served as an assistant to Rosing while a student at the Institute of Technology in St. Petersburg. After coming to the United States, he was employed by Westinghouse and then by the Radio Corporation of America (RCA), where he worked on television in addition to several other projects. In 1928, he produced the first workable electronic television camera, which he called the "iconoscope." It employed a screen made up of thousands of individual elements that took on an electrical charge when they were struck by light reflected from the object whose image was to be transmitted. The screen was rapidly scanned by a beam from an electron gun, which resulted in an electrical discharge that was amplified and transmitted to the receiver. The receiver then reconstructed these electrical discharges into points of light on the television screen.[4]

Zworykin's iconoscope was a major accomplishment, but as often happens in the history of technology, parallel inventions were being made elsewhere. While Zworykin was engaged in the early phases of his research, an Idaho farm boy named Philo Farnsworth astonished his high school science teacher by presenting him with plans for an apparently workable television system. A few years later, Farnsworth succeeded in getting financial backing for his research, and by the late 1920s, he had produced a viable, if crude, system of electronic television.

Both Farnsworth and Zworykin can be credited with the invention of television. Zworykin and his team did important work, while years of litigation culminated in Farnsworth's patents being upheld. RCA paid a substantial sum in royalties for these patents, but it succeeded in putting together the first television network, the National Broadcasting Corporation (NBC), which began operations in New York City in 1939. Several drawbacks were still evident; most of the early receiving sets had five-inch screens, and their dim pictures meant that they had to be viewed in complete darkness. Many technical problems remained to be solved in such areas as the design of antennas and the relaying of signals over large distances. Still, television had become a practical possibility.

The Federal Government Steps In

The development of radio and television in the United States was largely the work of private enterprise. Even so, the activities of the federal government gave a substantial boost to the electronic media. Military contracts stimulated a great deal of useful research. National security was also invoked when the government played the leading role in the creation of a unified radio industry. In the years after World War I, government officials, especially those in the Navy, were increasingly concerned about the monopoly that the British Marconi Company had over international radio communication. After some prodding by the Navy, the General Electric Company bought a controlling interest in Marconi's American subsidiary and transferred it to the newly formed Radio Corporation of America. There remained the problem of RCA's gaining access to the numerous patents that covered various aspects of radio production and broadcasting. This was ultimately resolved by giving the two other major patent holders, Westinghouse and American Telephone and Telegraph, a financial stake in RCA in return for the use of their patents. The creation of a radio monopoly was thus the result of an action undertaken at the behest of the U.S. government.

In addition to these activities, the federal government used its regulatory power to ensure an orderly environment for broadcasting. For clear reception it is necessary for each broadcaster to remain on a single frequency. (When you "tune in" to a radio station, you are selecting a specific frequency, such as 790 kHz or 89.3 MHz). Should several radio stations broadcast their programs on the same frequency, the result is a chaos of conflicting sounds. To address this problem, in 1912, the Department of Commerce and Labor began to require the licensing of all broadcasters. When a license was awarded, it stipulated the exact frequency on which the station could broadcast, as well as its times of operation. But this authority was successfully challenged in court, and by the mid-1920s, many broadcasters were transmitting all over the frequency spectrum. The airwaves had become a cacophony of interfering signals.

To remedy this situation, Congress created the Federal Radio Commission in 1927, giving it broad powers to issue licenses and assign frequencies. In 1934, government oversight of communications media was consolidated through the creation of the Federal

Communications Commission (FCC). The licensing and frequency-allocation powers of the FCC worked to the detriment of the very few stations whose licenses were revoked because their broadcasts were deemed not to be in the public interest. The strict assignation of frequencies also frustrated would-be broadcasters who were unable to gain licenses. But for the industry as a whole, the Federal Radio Commission and its successor, the FCC, were essential elements in the system of radio broadcasting. Without federal regulation, broadcasters could not be assured of their own special frequencies, and the air would be filled with signals that constantly intruded on one another. Most radio entrepreneurs probably believed in free enterprise, but they realized that in the absence of regulation their pursuit of individual self-interest would result in the destruction of their industry.

The issue of federal regulation once again came into prominence when television became ripe for commercial development in the late 1930s. One of the major issues confronting the industry was the setting of technical standards. Each one of the major companies involved in television research employed a different system for transmitting and receiving televised images. At the same time, one of these companies, RCA, had made the most progress in putting together a total television system. This meant that the standards employed by RCA would have to be used if the FCC decided to permit the immediate broadcast of commercial television programs. The other manufacturers knew that they lagged RCA, but they felt that their own technologies had a great deal of promise. They certainly were not eager to adopt RCA's standards and pay royalties for its patents.

Faced with these contentious issues, the Commission gave the go-ahead for commercial television broadcasting in 1940, only to rescind its decision less than a year later due to America's entry into World War II. A special engineering committee was then set up to consider the issue of standards. By the middle of 1941, these had been successfully resolved. When television went into its period of rapid growth in the late 1940s and early 1950s, some of its success could be attributed to the uniform technical standards that were developed under FCC sponsorship.[5] Unlike the early years of the personal computer, the television market was not fragmented by incompatible standards and equipment.

Problems of Regulation

The efforts of the FCC to promote uniform technical standards were essential to the growth of the electronic media industry. The consequences of the FCC's other activities, however, are more controversial. According to the 1934 Communications Act, the FCC is supposed to grant and renew broadcasting licenses based on "the public interest, convenience, and necessity." In fact, the FCC has been quite reluctant to influence the radio and television industries through the exercise of its licensing powers. Very few licenses have not been renewed; as long as a station has met some minimal level of community service it is deemed to have met the letter, if not the spirit, of the regulations. An FCC chairman might indict American television as "a vast wasteland," as Newton Minow did in 1961, but other than sponsoring the development of noncommercial television, his agency had scant influence over the quality of television programming.

It can be fairly argued that the FCC's active involvement in the shaping of television programming amounted to unwarranted government interference with the private sector. There have been times when FCC actions have had significant consequences for the broadcasting industry, and its decisions have on occasion benefited some firms at the expense of other firms. One particularly notorious example was the FCC's role in reallocating established FM frequencies in the 1930s, a move that was advantageous to RCA, the dominant radio corporation, but detrimental to the interests of FM's inventor, Edwin Armstrong.[6] RCA also reaped considerable benefits from the FCC's decision in 1953 to support its color television technology rather than the one developed by CBS. Seven months after this decision, the chairman of the FCC resigned his post and took a high executive position with NBC, at that time an affiliate of RCA.[7] Much the same had happened during the controversy over the shifting of FM frequencies, when the FCC's chief engineer left his post to become the head of RCA's engineering department.

More recently, Michael K. Powell, a former chair of the FCC, left that position to head the National Cable & Telecommunications Association, the industry's chief lobbying group.[8] In similar fashion, Meredith Attwell Baker left her position as FCC commissioner to become a lobbyist for Comcast's NBCUniversal, a few months after favorably ruling on a proposed acquisition of NBCUniversal by

Comcast.[9] Examples of regulators passing through the "revolving door" to take positions with enterprises they had previously regulated are not confined to the FCC nor are they illegal. But, they are cited as evidence that regulatory agencies such as the FCC are vulnerable to "capture" by the industry they are regulating. It should not be too surprising that this occurs; "regulation" has become a dirty word in some political circles, and the lack of support for regulatory agencies forces them to find it in the very industries that they are regulating. Moreover, regulatory agencies depend on the regulated industries for expertise and information. Compounding a potential conflict of interest, there is often close personal contact between the personnel of these agencies and representatives of the industry that they are supposed to be regulating. And, as in the case of the FCC officials who took positions in media organizations, they are often alert to the possibility of going over to the other side.

Occasionally, a regulatory issue will leap into political prominence and disrupt the comfortable relationship between a regulatory agency and the industry that it is regulating. This is what happened when a major accident at the Three Mile Island nuclear power plant in Pennsylvania put the Atomic Energy Commission's role in regulating nuclear power into the political spotlight. But this is rare; for the most part, regulatory agencies view the industry that they regulate as a client to be served, sometimes to the detriment of the public interest.

The FCC's proclivity for maintaining existing arrangements and avoiding conflict may have slowed technological innovation because of the agency's weak support for innovations such as FM broadcasting, UHF, and pay and cable TV.[10] But to reiterate a key point, political realities sharply limit the ability of regulatory agencies like the FCC to act as independent and aggressive foes of the industries they oversee.[11]

The Television-Viewing Public

Television ownership has been virtually universal for decades. Today it no longer requires a conventional television set, as increasing numbers of viewers are accessing programs on their computers, tablets, and smartphones. Although the time spent watching television has declined slightly in recent years, it remains the number one leisure activity in the United States, accounting for an average of 2.8 hours of viewing each day in 2014.[12] Only work and sleep take up more time than watching television, and on an annual basis children spend more hours in front of a television set than they do in school.

In recent years, other screen-based activities like surfing the internet and playing electronic games have taken some time away from watching conventional programming. There have always been age-related differences in television-viewing time, and these have been magnified by the greater use of alternative forms of digital media by younger age cohorts. Children and teenagers watch less television than adults, while individuals over the age of 65 spend about 13 more hours weekly watching television than the statistically average viewer.[13] In regard to other social variables, the amount of time spent watching television is negatively correlated with educational attainment[14] and it also declines at higher income levels.[15] In both cases, the correlations are apparent, but the direction of causality is not self-evident; after all, it is possible that watching television takes away time that could have been used for educational purposes. One thing, however, is clear, although the extent of television viewing is affected by socioeconomic variables like education, income, and age, watching television is embedded in the daily routines of people in every station of life.

Violence on Television and Its Consequences

One of the most pervasive concerns about television's influence centers on the vast amount of violent and aggressive behavior depicted on the screen. Violence is a programming staple. According to one account, before reaching the age of 18, the average child in the United States will have viewed more than 200,000 televised acts of violence, including 16,000 acts of murder.[16] With all of this mayhem appearing on television, it is understandable that critics have indicted television for contributing to aggressive and violent behavior of children.

Many studies have been devoted to this topic. Some of these studies have been based on laboratory experiments that allow the researcher to eliminate extraneous factors to determine the relationship between exposure to a filmed act of aggression and subsequent behavior. In a typical experiment, one group of children is exposed to a film depicting violence while a control group is not. The subsequent behavior of the two groups is then compared. A classic experiment of this sort was conducted in the 1960s. One group of preschool children witnessed an actor physically and verbally assaulting Bobo the Clown, a large, inflated plastic doll. The control group saw nothing. The children were then allowed to play in a room containing a Bobo doll and other toys. The results were unequivocal; the children who had seen the simulated aggression displayed higher levels of aggressive behavior toward the hapless toy, in some instances even directly imitating the acts they had viewed, such as yelling out "Lickit! Stickit!" while pounding on the toy with the handle of a mallet.[17]

Many studies of this sort have been conducted. Some have found indirect as well as direct consequences of exposure to filmed violence. For example, children are more likely to play with toy guns after seeing filmed aggressive acts, even if gunplay did not appear in the film. All in all, these experimental studies are nearly unanimous in finding a connection between seeing acts of violence and subsequently engaging in aggressive acts.[18]

Laboratory experiments such as these can always be criticized because of their artificiality. In their attempt to reduce or eliminate extraneous influences, experimenters may create an environment that bears little resemblance to reality. A film clip presenting violent behavior is not a television program, in which violent acts are presented in conjunction with a variety of other messages. Experiments also deal with only immediate cause-and-effect relationships. They therefore do not take long-term, cumulative effects into account. Finally, in these experiments the subjects may think that aggressive acts are permitted and even expected. For all these reasons, the real-world applicability of these experiments can be questioned.

Other researchers have attempted to circumvent the inherent limitations of laboratory experiments by conducting more "naturalistic" studies. In these, one group of children views "normal" violent programming, while another group does not. Their subsequent behavior in everyday settings is then directly observed or is reported by parents or teachers. Although they have a somewhat contrived quality, these studies can be taken as a reasonable reflection of real-world situations. And as with the experiments previously noted, viewing violent programming was associated with subsequent aggressive behavior, while neutral programming had no such effect.[19]

A third type of research examines actual viewing habits to see if there is an association (or "correlation") between viewing violence on television and aggressive or violent behavior. These studies generally use interviews and self-reports as their primary source of information about viewing habits and behavior. For the most part, research of this sort does show a positive correlation between watching violent television programs and aggressive behavior. Children and adolescents who prefer violent television programs tend to be more aggressive in their behavior.

There is a basic problem with correlational studies, however, for correlation is not the same thing as causation. The correlation of A with B does not necessarily mean that A causes B. It is also possible that B causes A, or that a third factor or factors could be the cause of both. Accordingly, a child who was already violence-prone might seek out violent programs. It is also possible that an impoverished, culturally deprived background may be the cause of both the viewing of violent programming and aggressive behavior. These are reasonable objections, but several research programs have been able to statistically isolate the viewing of violent television programming from other factors that may cause violent behavior. When this is done, the viewing of violent television shows is shown to be an independent source of violent and aggressive behavior. It may most strongly affect those who are already predisposed to this kind of behavior, but it is not confined to them.[20]

It must be stressed that the conclusions drawn from all these different studies cannot be applied to the behavior of specific individuals. For most children, watching violent acts on television probably will not lead to aggressive or violent behavior, but for some children it may. Television affects different people in different ways. Hard-and-fast generalizations are difficult to come by, but what Wilbur Schramm noted many years ago still holds today: "For some children under some conditions, some television is harmful. For other children under the same conditions, or for the same child under other conditions, it may be beneficial. For most children, under most conditions, most television is neither particularly harmful nor particularly beneficial."[21]

Like any voluntary act, violent behavior is a product of a complex set of motivations and inhibitions. All of us probably have had moments when we wanted to commit a violent act. We usually have not done so for three basic reasons: (1) we have learned that such actions are likely to result in retaliation, (2) we know that they usually do not solve the problem, and (3) we have internalized a code of behavior that discourages such acts. Most likely, the operation of all of these prevents us from giving vent to our violent urges.

Television has the capacity to alter all these inhibiting factors. In some televised depictions of violence, the retaliation for the violent act is delayed and muted, if it appears at all. For nearly 75 percent of televised violent scenes, there is no punishment or even condemnation of the violence that had been depicted.[22]

To make things worse, the second inhibiting factor, a belief that violence does not solve most problems, is not always supported by television programming. Illegal and violent acts are sometimes used to attain socially approved goals.[23] Finally, there is the problem of television's contribution toward desensitizing people, especially children, to violence, thereby attenuating the development of a personal code of ethics that discourages violent acts. Such long-term effects of televised violence are difficult to prove, but there is a fair amount of evidence that exposure to televised violence dampens a child's emotional response to the witnessing of violent acts and diminishes his or her concern about such acts.[24]

Television can be a powerful influence on thought and behavior, although it does not necessarily override the influence of family, school, church, and peers. Research into the consequences of television viewing faces formidable challenges in separating television's effects from other influences. Even so, the majority of the many studies that have been done have found a connection between on-screen violence and real-world violence. Based on these studies, the American Psychological Association, in 1985 and again in 2016, indicted media violence for contributing to violent and aggressive actions, desensitizing individuals to the consequences of violence, and increasing fears of being a victim of violence.[25]

Television violence has been identified as a significant social problem, but it is not one with an easy solution. The depiction of violence may be essential to a story, even one with great artistic merit; some of Shakespeare's works include scenes of appalling brutality. Occasional violence also accounts for some of the appeal of sports like football and hockey. In the absence of widespread popular support for government censorship, which would likely be overturned on constitutional grounds, about all that can be hoped for is a greater degree of responsibility on the part of the television industry. A technological fix of sorts is available in the form of the V-chip, which can be used to block programming according to an age-based rating system.[26] But the most effective way of avoiding violent programming is for parents to take an active role in monitoring their children's consumption of television programs and to use the most effective way of regulating viewing: the on-off control.

Delivering the News

Until recent times, most people were blissfully ignorant of the world around them. Travelers might bring stories of distant places, and the literate few could read of them. Information traveled very slowly; the Battle of New Orleans was fought two weeks after a treaty formally ended the War of 1812, as the combatants were unaware of its signing. During the nineteenth century, the railroad, the telegraph, and the appearance of popular newspapers brought the world closer, but coverage was still slow, spotty, and often inaccurate. Past the borders of one's own community was a world dimly perceived by most people. All of that has changed today. Instantaneous communications tell us of uprisings in the Middle East, wars in Africa, and financial crises in Europe. Although we do not quite live in a "global village," modern communications technologies have extended our vision to a degree scarcely imaginable not long ago.

As we saw in the previous chapter, the rise of digital sources has brought important changes to the way that news is presented and consumed. This trend can be seen as a continuation of a process that began in 1960, when survey respondents indicated for the first time that television had replaced newspapers as their prime source of news.[27]

For decades, television has been a prime source of news about the nation and the world; in 2015 the evening news programs on ABC, CBS, and NBC had a total viewership of 24 million.[28] But lying behind this figure is a steady erosion of viewership as the internet and social media have become a major source of news. This trend is especially notable when the viewing audience is broken down into age categories, as younger cohorts are much more likely to use digital news sources. According to one study, 71 percent of young adults

Figure 15.5 Television provided an "electronic hearth" with shared family viewing of important events and entertainment. (Source: Evert F. Baumgardner)

aged 18 to 29 cited the internet as their main source of news, while only 55 percent cited television as their prime source.[29]

Although television has a great potential to bring the outside world into our homes, the effects of watching television news have been negligible for large segments of the viewing public. According to a survey conducted by the Pew Research Center in 2007, only 38 percent of the respondents who got most of their information from evening television news programs correctly answered at least 15 out of 23 basic questions about current events. In contrast, 51 percent of respondents who were regular listeners to National Public Radio were able to do so. But NPR enthusiasts should not feel too smug; 54 percent of the surveyed population who listed *The Daily Show* and *The Colbert Report* as sources of their knowledge of current events also answered at least 15 out of 23 questions.[30]

The way the news is covered in print differs significantly from televised coverage. Newspaper reporting tends to be detached and impersonal, whereas television reporting is more like storytelling. A news report is presented in narrative form, with a theme being developed using sound and pictures as well as verbiage. A great deal of television news is presented as a kind of visually oriented entertainment, narrated by newscasters who are selected, at least in part, based on their physical attractiveness. The stories are short and fragmentary, with very little contextual information. This is inevitable, since the average network news story runs for only 1 minute and 20 seconds.[31] News of this sort is quickly forgotten; one study found that more than 20 percent of the viewers of television news could not recall a single item of news an hour after having seen a news broadcast. Another study found that the average viewer retained only 20 percent of the information presented in a simulated news story.[32]

In contrast, newspaper stories are written so that the key item can be quickly grasped. Background and analysis then follow for the reader who wants to continue with the story.[33] Viewers of television news and casual newspaper readers may have equally superficial understandings of the events in the news, but at least the latter can extend their knowledge. Television news gives few such opportunities.

Television and Politics

Numerous analysts have charged that television has fundamentally altered the political process. Television, they assert, has reduced political campaigns to trivial "media events," selected the sensational over the substantive, undermined political parties, greatly inflated the cost of political campaigns, and made "image" the primary criterion of a candidate's appeal. Are these accusations true?

There can be little question that the politics of the television age is different from the politics of earlier times. In 1948, Harry Truman could wage a successful presidential campaign from the back of a railroad passenger car. Today, televised news and political commercials have taken the place of the whistle-stop tour. On occasion, candidates have successfully contended for office without the aid of television advertising; in 1974, a Democrat won Gerald Ford's vacant congressional seat in a special election despite being too impecunious to pay for television commercials. But incidents such as this are highly atypical. Televised political advertising has become an inescapable part of electioneering. Presidential and congressional campaigns devour enormous amounts of money, much of it used for televised appeals. A typical congressional campaign budgets at least one-third of its funds for television advertising. During the 2016 presidential campaign, $4.4 billion was spent on television ads, and a record $14.4 billion in 2020.[34]

As television has greatly increased the costs of conducting campaigns, the need for large campaign contributions has mounted. This, in turn, has increased the danger of successful candidates being beholden to the special interests that supply a significant portion of these contributions, especially now that there are no limitations on the amount that can be spent by corporations and political action committees. As cynics (or maybe realists) have suggested, we may end up with "the best elected officials that money can buy." Corruption has always been a part of political life, but the demands of television campaigning have created something new, a scramble for campaign funds that is perfectly legal but nonetheless a threat to democratic political institutions.

Figure 15.6 For a presidential candidate, intense scrutiny by the media is an inescapable part of campaigning.

Although television has become a major force in American political life, there are limits to its influence. Televised political advertising is not always decisive, and money does not always win elections. This became painfully evident to Michael Bloomberg, who during the contest for the 2020 Democratic presidential nomination spent $552 million of his own money on television, radio, and digital advertising.[35] In most elections, most voters have made up their minds about who will get their vote well before the campaign gets under way. For these voters, televised advertisements help to crystallize their choices and reinforce their existing beliefs but do not produce any fundamental alterations. At the same time, however, there are many elections where the outcome hinges on the choices made by voters who make their decisions late in the political campaign. Although the 2016 election offered a clear-cut choice of candidates and platforms, many voters made their decisions very late in the campaign. In the key states of Michigan and Wisconsin, 20 percent of voters chose their candidate during the last week before the election.[36] Under these circumstances, even if they swayed only a few voters, television ads may have been a sufficient basis for Donald Trump's upset victory.

In most cases, the voters who have made up their minds long before the election takes place are the ones most strongly influenced by party loyalties. This makes them impervious to televised appeals to vote for another party's candidate. Yet, identification with a party has been eroding in recent years, and many people who call themselves Democrats or Republicans do not have a strong sense of commitment to their party. Television has been responsible for a good deal of the erosion of party loyalties because a candidate can use the medium to appeal directly to the electorate.[37] Also, presidential and other candidates for national office are no longer selected by party bosses in smoke-filled rooms. They usually are the products of state primaries, in which televised appearances and political commercials are usually the most important part of the campaign.

At first glance, the ability of candidates to use television to directly appeal to the electorate seems like a good thing, an improvement over the days of party bosses and political machines. In reality, electronically transmitted communications from candidates to voters are usually devoid of substance. Numerous critics of television have pointed out that the effectiveness of a television presentation hinges on visual appeal and memorable sound bites, with content a secondary consideration. Recent trends have exacerbated this tendency. One study found that from 1968 to 1992, network news programs reduced the average length of presidential candidates' spoken comments from 1 minute to 10 seconds. Since then, coverage has dropped to an average of 8 seconds, and only a third of these clips addressed substantive policy issues.[38] Televised political appeals make little room for ideas, policies, and well-articulated political positions; what really matters is the kind of personality and presence that a candidate projects on the screen. At the same time, television often conveys the idea that problems can be quickly resolved—requiring less than 30 minutes in the case of a show, and 30 seconds for individual commercials. It is no wonder, then, that the electorate is often attracted to candidates who offer simple and quick solutions. And these "solutions" are, of course, neatly packaged for television; instead of lengthy discussions of a candidate's stand on particular issues, the typical commercial consists of 30 seconds of visual images and a few empty slogans.

The pervasive influence of television on political life should worry anyone concerned about the fate of our political institutions. As some critics have argued, the greatest threat to democracy may come not from the assaults of hostile nations but from the trivialization and even corruption of the political process that occurs when television dictates the general mode of discourse and comprehension.[39]

Television and Thought

For some students of the media, the effects of television go well beyond the ones just described. The most notable of these has been Marshall McLuhan, whom we met briefly in the preceding chapter. In McLuhan's schema, television is a "cool" medium: The televised image is indistinct and requires that viewers "fill in the blanks" in a way that involves all the senses. Viewing television is not a single-minded, linear process, as reading is. Exposure to television in turn leads to a mode of perception that affects many other aspects of life. For McLuhan, a television-based culture is less concerned with sequence than with complete, all-at-once involvement. In his view, these changes in perception are transforming every aspect of our lives; McLuhan attributed everything from consumers' preference for small cars to changes in church liturgy to the influence of television. Examples such as these seem far-fetched, and some of his predictions have not been borne out, such as his prophecy that baseball will necessarily decline because it is a "linear" game. McLuhan's focus on the influence of different types of media opens all kinds of fascinating possibilities, but his conclusions lack clear proof, and in fact the basic thesis is virtually unprovable.

More concretely, it has been claimed that television helps young children to develop visual perception, such as learning that an object viewed on a two-dimensional television screen actually has three dimensions, and that an object viewed from different perspectives is still the same object.[40] This may be true for programming that has been consciously designed to develop these skills, but most television shows have little to offer in this regard. Moreover, when they occur, advances in perceptual skills are limited and temporary. Even these gains are largely restricted to children with low language ability and visual skills; television viewing may actually be detrimental to the perceptual development of children not deficient in these areas.[41]

Other deleterious effects of television on children can be noted. Experimental studies have shown that children supply more imaginative continuations to interrupted stories that they read than they do to televised stories. In similar fashion, young children who are heavy watchers of television engage in less imaginative play than other children.[42]

There is also a fair amount of evidence that television watching may contribute to underdeveloped reading skills in children. The relationship is not a perfectly direct one; it may emerge only when a certain threshold of television watching is reached, and it seems to be affected by the kind of programming selected.[43] Still, it seems evident that for many children television is a negative influence in this regard. Finally, one might also speculate that the lack of concentration and the short attention spans that seem so common today are at least partially the result of growing up with television programs that are constantly punctuated by commercial messages and other distractions.

Television was the most pervasive and significant medium for most of the second half of the twentieth century. But just as it brought about a revolution in how we are informed and entertained, it too is being reshaped by technological change. A television set is no longer the only way of viewing televised programming. In recent years, growing numbers of viewers have been able to watch television—everything from network shows to the latest YouTube clip that has gone viral—on laptops, tablets, game consoles, and smartphones. At the same time, television viewers are no longer limited to programming supplied by cable and over-the-air networks; increasing numbers of them are subscribing to online video services like Netflix and Amazon Prime Video.[44] Although long-established television networks are providing much of the content being viewed today, many media analysts foresee a time when most programming will be supplied by streaming services operating independently of these networks. Conventional television viewing has also been affected by digital video recorders (DVRs) and digital video discs (DVDs) that allow prerecorded programs to be accessed at a viewer's convenience. This often brings the added benefit of eliminating commercial interruptions, a boon to viewers but a source of considerable consternation for advertisers. Meanwhile, the integration of television and the internet is proceeding apace, as a new generation of sets allows viewers to simultaneously watch television, access the internet, access and apply various apps, and communicate through social media sites.

While television is being reshaped by new technologies, other forms of electronic media have been taking on growing importance for communication and entertainment. In the next chapters we will take a look at video games, social networking, and the many components of the internet. In so doing, we will consider the economic, political, cultural, and social contexts that have shaped their development. We will then bring radio and television back into the picture for an overall assessment of the electronic media's place in social life.

QUESTIONS FOR DISCUSSION

1. In many countries, a considerable amount of television and radio broadcasting is done by government organizations such as the BBC. In the United States, most broadcasting is done by commercial stations. What are the advantages and disadvantages of the two systems?
2. Political scientists have noted that government regulatory agencies tend to be "captured" by the industry that they are regulating. Is this inevitable? Is it possible for regulatory agencies to prevent this from happening?

3. Do you agree with this chapter's general conclusions about television's influence on violent behaviour? Do you think that there is too much violence on television? Should television be more tightly regulated so less violence appears on the screen?
4. Television commercials are a major part of campaigns for the presidency and congressional offices. In what ways has television affected the way that campaigns are run? On balance, has television been a positive or negative influence on political campaigns and campaigners?
5. Should parents try to control the television-viewing habits of their children? To what extent is it even possible to do so? How do you watch televised material? Do you use a conventional television set or a new media platform like a smartphone? What are the advantages or disadvantages of each?
6. Although many thought that radio would be completely replaced by television, this never happened, and, in the aggregate, radio stations are quite profitable today. How is radio programming today different from the "golden age of radio" in the 1930s and 1940s? Might network television also have to undergo major changes if it is to survive?

Notes

1. Tom Standage, *The Victorian Internet: The Remarkable Story of the Telegraph and the Nineteenth Century's Online Pioneers* (New York: Berkley Publishing Group, 1999).
2. For extensive coverage of the early history of radio, see: Hugh G. J. Aitken, *Syntony and Spark: The Origins of Radio* (Princeton, NJ: Princeton University Press, 1985); and *The Continuous Wave: Technology and American Radio, 1900–1932* (Princeton, NJ: Princeton University Press, 1985).
3. Eric Barnouw, *Tube of Plenty: The Evolution of American Television* (New York: Oxford University Press, 1975), pp. 4–5.
4. Albert Abramson, Zworykin, *Pioneer of Television* (Urbana and Chicago: University of Illinois Press, 1995), pp. 87–113.
5. W. Rupert Maclaurin, *Invention and Innovation in the Radio Industry* (New York: Macmillan, 1949), pp. 225–240.
6. Tom Lewis, *Empire of the Air: The Men Who Made Radio* (New York: HarperCollins, 1991), pp. 268–69, 300–307.
7. Barnouw, *Tube of Plenty*, p. 100.
8. National Cable & Telecommunications Association, "Michael Powell." Accessed from www.ncta.com/who-we-are/leadership/bio/169.

9. Preeti Vissa, "The Revolving Door Continues," *HuffPost* (July 17, 2011). Accessed from www.huffpost.com/entry/the-revolving-door-contin_b_862784.
10. Barry Cole and Mel Oettinger, *Reluctant Regulators: The FCC and the Broadcast Audience* (Reading, MA: Addison-Wesley, 1978).
11. Megan McArdle, "It's Normal for Regulators to Get Captured," *Bloomberg View* (October 1, 2014). Accessed from www.bloombergview.com/articles/2014-09-30/it-s-normal-for-regulators-toget-captured.
12. Nielson, "Connectivity is driving how Americans are engaging with TV," www.nielsen.com/insights/2023/connectivity-is-driving-how-americans-are-engaging-with-tv/; and Ben Leubsdorf, "We're Working More Hours—and Watching More TV," *The Wall Street Journal* (June 24, 2015). Accessed from www.wsj.com/articles/were-working-more-hoursand-watching-more-tv1435187603.
13. "Time Spent Watching TV," *AdAge Mediaworks* (April 18, 2011). Accessed from http://adage.com/article/mediaworks/time-spent-watchingtv/227022/
14. Wayne Friedman, "Income, Education Levels Impact Television Viewing," *MediaPost* (April 2, 2013). Accessed from www.mediapost.com/publications/article/197108/income-education-levelsimpact-tv-viewing.html.
15. Mary Dooe, "Beyond the Data: TV Viewing and Income." *Marketplace* (November 4, 2013). Accessed from www.marketplace.org/2013/11/04/wealth-poverty/income-upshot/behind-data- tv-viewing-and-income.
16. Eugene V. Beresin, "The Impact of Media Violence on Children and Adolescents: Opportunities for Clinical Interventions," *American Academy of Child and Adolescent Psychiatry* (2015). Accessed from www.aacap.org/aacap/Medical_Students_and_Residents/Mentorship_Matters/De.
17. Albert Bandura, Dorothea Ross, and Sheila A. Ross, "Transmission of Aggression through Imitation of Aggressive Models," *Journal of Abnormal and Social Psychology 63*, 3 (1961): 575–82.
18. George A. Comstock and Erica Scharrer, *Television: What's On, Who Is Watching, and What It Means* (San Diego and London: Academic Press, 1999), p. 278.
19. George Comstock and Erica Scharrer, *Media and the American Child* (Amsterdam: Elsevier, 2007), pp. 204–7.
20. Ibid., pp. 210–19.
21. Wilbur Schramm, Jack Lyle, and Edwin B. Parker, *Television in the Lives of Our Children* (Stanford, CA: Stanford University Press, 1961), p. 1.
22. "Youth Violence: A Report of the Surgeon General," *Office of the Surgeon General* (November and December 2001). Accessed from www.surgeongeneral.gov/library/youthviolence/default.html.
23. George Comstock, *Television in America* (Beverly Hills, CA: Sage, 1980), p. 83
24. Comstock and Scharrer, *Media and the American Child*, pp. 239–41.
25. American Psychological Association, "Violence in Mass Media." Accessed from www.apa.org/about/policy/media.aspx
26. Federal Communications Commission, "The V-Chip: Putting Restrictions on What Your Children Watch" (November 4, 2015). Accessed from www.fcc.gov/consumers/guides/v-chip-putting-restrictions-what-yourchildren-watch.
27. Barnouw, *Tube of Plenty*, p. 314.

28. Pew Research Center, "State of the News Media 2015," https://assets.pewresearch.org/wp-content/uploads/sites/13/2017/05/30142603/state-of-the-news-media-report-2015-final.pdf, p. 38.
29. Andrea Caumont, "12 Trends Shaping Digital News," Pew Research Center (October 16, 2013), accessed on August 4, 2016, atwww.pewresearch.org/facttank/2013/10/16/12-trends-shaping-digital-news/#twt2.
30. Pew Research Center for People and the Press, "Public Knowledge of Current Affairs Little Changed by News and Information Revolutions" (April 15, 2007), accessed on April 3, 2012, at www.people-press.org/2007/04/15/publicknowledge-of-current-affairs-little-changed-by-news-and-informationrevolutions/.
31. Comstock and Scharrer, *Television: What's On*, p. 123.
32. Neil Postman, *Amusing Ourselves to Death: Public Discourse in the Age of Show Business* (New York: Viking/Penguin, 1985), p. 152.
33. Melvin L. DeFleur and Everette E. Dennis, *Understanding Mass Communication*, 2nd ed. (Boston: Houghton Mifflin, 1985), p. 450.
34. Danielle Kurtzleben, "Campaigns Will Spend $4.4 billion on TV Ads, But Why," *NPR* (August 19, 2015). Accessed from www.npr.org/sections/itsallpolitics/2015/08/19/432759311/2016-campaign-tv-ad-spending; 2020 figures are summarized here: www.cnbc.com/2020/10/28/2020-election-spending-to-hit-nearly-14-billion-a-record.html.
35. Seema Mehta, "Bloomberg Exits the Race and Quickly Endorses Biden," *Los Angeles Times* (March 5, 2020).
36. Aaron Blake, "How America Decided at the Last Minute to Elect Donald Trump," *The Washington Post* (November 17, 2016). Accessed from www.washingtonpost.com/news/the-fix/wp/2016/11/17/how-america-decided-at-the-very-last-moment-to-elect-donald-trump.
37. D. Garzia, Ferreira da Silva, & A. De Angelis, "Image that Matters: News Media Consumption and Party Leader Effects on Voting Behavior. *The International Journal of Press/Politics*, 25, 2 (2020): 238–259. Accessed from https://doi.org/10.1177/1940161219894979.
38. Ezra Klein, "Your Slip is Showing and Reshowing," *Los Angeles Times* (May 18, 2008), p. M5.
39. Postman, *Amusing Ourselves*, pp. 125–141; *Jerry Mander, Four Arguments for the Elimination of Television* (New York: Quill, 1978).
40. Patricia Marks Greenfield, *Mind and Media: The Effects of Television, Video Games, and Computers* (Cambridge, MA: Harvard University Press, 1984), pp. 17, 33.
41. Comstock and Scharrer, *Television: What's On*, pp. 231–32.
42. Greenfield, *Mind and Media*, pp. 88–89.
43. Johannes W. J. Beentjes and Tom H. A. Van der Voort, "Television's Impact on Children's Reading Skills: A Review of Research," *Reading Research Quarterly* 23, 4 (Fall 1988).
44. Shalini Ramachandran, "TV Viewing Slips as Streaming Booms, Nielsen Report Shows," *The Wall Street Journal* (December 3, 2014). Accessed from www.wsj.com/articles/tv-viewing-slips-as-streaming-booms-nielsen-report-shows-1417604401.

CHAPTER 16

The Internet Age

As we saw in the last two chapters, a succession of new technologies has vastly enhanced the ability of people to send and receive messages. Today, we are experiencing the latest phase of this ongoing revolution as the internet, mobile phones, and social media are reshaping the ways in which we communicate with one another. But to repeat what by now should be a familiar theme, these advances have been accompanied by a number of new issues and problems. The same free flow of information that enhances our ability to communicate also gives rise to a number of concerns: the protection of information that we do not want to move so freely, such as intellectual property and personal data; unequal access to information exacerbating the gap between the haves and have-nots; and the difficulty and expense of filtering, managing, and using so much information. This chapter traces the development of digital communication technologies and the challenges they pose, and also shows how technological advances can move along unanticipated paths.

The Birth and Growth of the Internet

The internet can be described as a giant network of smaller computer networks that allow users to access files located anywhere within these individual networks. These files can be email messages, web pages, YouTube videos, and social networking updates, to name the most prominent.

Constructing the internet has been a massive undertaking, one that required the active participation of the federal government. As with many other far-reaching technologies, the invention of the digital network that eventually evolved into today's internet began with the needs of the military. The sponsoring agency for the development of the first computer network was the U.S. Department of Defense's Advanced Research Projects Agency (DARPA), hence the name of the first computer network, ARPANET. Organized in 1969, its initial purpose was to tie together a small number of mainframe computers so that data could be uploaded and downloaded to and from any computer in the network.[1]

An important aspect of ARPANET was the division of each block of data into a number of smaller "packets." In this way the capacity of the network could be maximized by sending each packet over the route that was least congested at the time. This technique had the additional virtue of ensuring that messages would go through even if some routes were destroyed by an enemy attack, although this was not its primary purpose.[2] This was not a unique occurrence; as we have seen before, technologies created for one purpose can end up being used for something quite different, as when radio mutated from wireless telegraphy into commercial broadcasting.

Computer networking through ARPANET remained largely in the military realm in the 1970s; but the appeal of networked computing stimulated the development of alternative systems. Responding to growing demand, another federal agency, the National Science Foundation (NSF), provided financial sponsorship for the Computer Science Network (CSNET), which went into operation in 1982. Computer networks expanded throughout the 1980s, facilitated by the development of programs for the efficient routing of digitized information, notably TCP (Transmission Control Program). It was during this period that the power and utility of the internet was given a massive boost when programmers began to build systems that allowed users to access other network sites by clicking on highlighted text or an icon that appeared on the on-screen document. By clicking on these links, users could easily move all over the internet as they accessed related text, image, and sound files.

The first of these networks were proprietary, and to be included required the payment of a fee. A different approach was taken by Tim Berners-Lee at another government endeavor, the European Organization for Nuclear Research (known by its French acronym, CERN). Berners-Lee was the primary developer of another software system for accessing files within computer networks. Known as hypertext transfer protocol (http), it was released in 1991 and formed the basis of a network that Berners-Lee called the World Wide Web (WWW). Two years later, the WWW team announced that they had made a crucial decision; CERN would not attempt to patent or claim any copyright fees for the technology underlying the web. This meant that there would be no restrictions on becoming part of the web, and as a result it became a virtually universal repository of computer-based information and entertainment.[3]

The web underwent explosive growth in the years that followed. In December 1993 there were 623 websites. Today there are over a billion, double the number of only five years earlier.[4] The growth of the web has been paralleled by the unprecedented growth of the internet as a whole. Its expansion constitutes one of history's fastest rates of technological diffusion. According to a United Nations study, it took radio broadcasters 38 years to reach an audience of 50 million, while television took 13 years. For the internet, only four years were required to assemble 50 million users.[5] By 2015 more than 3.4 billion people, about 40 percent of the world's population, were "netizens."[6] Just over 5 billion people use the internet as of 2023.[7] But as we shall see, worldwide internet usage is very unevenly distributed, with high rates of usage in some places offset by low rates in others.

The revolutionary changes wrought by the internet reflect rapid advances in the development of computer hardware and software. As has been noted earlier, large-scale changes are not just the product of solitary inventions. Rather, they entail the prior existence or creation of complementary elements. This is particularly evident when we consider the use of the internet for the buying and selling of goods—e-commerce, as it has come to be called. But the internet and means to access it are not enough; online buying also requires credit or debit cards and the security measures employed to protect the customers using them. For sellers, successful e-commerce depends on the use of barcode readers for selecting and keeping track of items, forklifts for retrieving them from high shelves, and conveyor belts to move them to the packing room.[8] The transport of these items to the customers who ordered them is usually performed by nothing more avant-garde than a delivery truck, often in the service of the venerable U.S. Postal Service. The internet has brought substantial changes to retailing, but much more has been required than computers and software.

As with all significant technological innovations, the internet has been beset by challenges and conflicts. One major issue is "net neutrality," the expectation that every internet site will be accessible, irrespective of which internet service provider (ISP) is providing the access. Without net neutrality, an ISP could enable access to its chosen sites while restricting access to those of its competitors. This is particularly blatant when it is practiced by countries that attempt to block access to sites that their governments view as threats to their power and authority. It also has emerged as an issue in the United States where private-sector ISPs have been accused of giving preferential treatment to allied TV services while limiting access to other providers.[9]

Another restriction on complete access to the internet occurs when users confine most of their usage to what have been called "walled gardens." Selective access began when providers like America Online offered an easy route to websites. Today, Apple, Facebook, and Google serve as portals to the internet for many users, while at the same time retaining some degree of control over access.[10] This is particularly evident with Google searches that provide links in accordance with the user's search history. Although perhaps convenient, this practice may result in someone with a perceived political orientation being initially presented with links to sites that are in accord with this orientation. This is analogous to what has been termed "narrowcasting," the tailoring of radio and television programming to fit the interests and beliefs of specific audience segments.

Social Media and the Network Effect

People have always sought to connect with one another; it is one of the defining characteristics of our species. But for most of human existence the ability to interact with one another and to create, join, and participate in social networks did not go beyond face-to-face interactions. As we have seen in Chapter 12, written language and printing greatly expanded the boundaries of communication, and during the last two centuries the telegraph, telephone, radio, and television stretched these boundaries even further.

What has differed from old patterns is the speed at which digital communications like social media have gained large audiences.

Facebook, the most commonly used site, began with a few dozen users at Harvard University in 2004. It had one billion users in 2012, and by 2016, 2.38 billion users had Facebook pages.[11] Other, more specialized, social networking sites also exhibited impressive rates of growth. LinkedIn, a site oriented toward work and careers, reached more than 575 million members by 2020, seventeen years after its initial launch.[12] X (formerly known as Twitter) has also grown at an impressive rate. Founded in 2006, sixteen years later it had 368 million monthly users[13] who collectively accounted for an average of 500 million tweets every day.[14]

Facebook was not exactly a unique service; other general social media sites, such as Friendster and Myspace, were created at about the same time. But having a multitude of sites is like having a multitude of telephone companies, each issuing their own telephone numbers. Communications systems like telephones and social media sites are governed by the "network effect." This simply means that an interconnected system becomes increasingly valuable as more members are connected to it. A telephone, computer, or any kind of communication device would be of limited use if it were connected to only a few dozen similar devices. But anyone with a Facebook account and a device to access it has the potential to connect with hundreds of millions of people. The growth of a network also illustrates the power of positive feedback; as a network expands it becomes of greater potential value to people and organizations who are still outside the network. This motivates them to join the network, which makes the network even more valuable, which induces more people and organizations to join it, and so it goes until a saturation point is eventually reached.

There is no upper limit to the number of individuals who can be counted as one's Facebook "friends," although in many cases the term "friend" has been stretched almost beyond recognition. When a study was conducted in 2014, 39 percent of adult Facebook users had 1 to 100 Facebook friends, while 15 percent had more than 500. As with many aspects of digital technologies, sharp age differences are apparent in Facebook use. Users between the ages of 18 and 29 have an average of 300 friends, but those between 50 and 64 average only 75, and seniors aged 65 and above have an average of 30.[15]

Unlike the number of friends, the amount of time that can be spent on Facebook necessarily has some limit, as there are only

24 hours in a day. Facebook users in the United States spend an average of 35 minutes a day on the site, less than the time spent watching television (2.8 hours), but considerably more than the time spent reading (17 minutes).[16] Not all of that time is necessarily well-spent. Many Facebook pages contain messages and pictures of what a friend had for dinner or a picture of their dog wearing a Spiderman costume, and so on. Facebook users are also exposed to a lot of advertising. With such a large number of users, Facebook has been very successful as a vehicle for advertisers. Its pages feature a plethora of ads along with "clickbait" that encourages a reader to view sites featuring such attractions as photographs of movie stars who have not aged gracefully, all the while presenting still more advertisements.

Sometimes damage due to misuse of social media is self-inflicted. Prospective employers may look at the sites of job seekers and discover pictures or information that are anything but beneficial to a job applicant. Another potential pitfall is "sexting"—sending sexually explicit pictures and texts over social media and other digital sites. According to a study conducted at one university, 54 percent of respondents said they had done so before the age of 18.[17] Although these were often sent to boyfriends, girlfriends, or close friends, there is no guarantee that they would not be sent elsewhere. Would-be "sexters" should also be aware that sending or receiving sexually explicit materials may result in prosecution under laws covering child pornography.

For individual users, Facebook and other social networking sites bring evident benefits and potential liabilities to individual users. Less clear, however, is their effects on society and culture in general, and on communication in particular. In contrast to firm statistics on the number of users, an accounting of social media's influence on society and culture is necessarily speculative. The relative newness of social media means that definitive statements are premature and are subject to later review, but some interesting ideas and findings have emerged. On the one hand, social media and other internet-related media such as email have moved us closer to Marshall McLuhan's "Global Village" by connecting people on the basis of interests, work relationships, and political causes rather than geographical proximity. At the same time, however, digitally mediated connections may be undermining person-to-person relationships by making it all too easy to substitute the latter for the former. As Sherry Turkle has observed:

> We expect more from technology and less from one another and seem increasingly drawn to technologies that provide the illusion of companionship without the demands of relationship. Always-on/always-on-you devices provide three powerful fantasies: that we will always be heard; that we can put our attention wherever we want it to be; and that we never have to be alone. Indeed, our new devices have turned being alone into a problem that can be solved.[18]

This is a powerful indictment of digitally based relationships, but other studies have found that an expansion of online relationships has not in fact led to a decline of traditional, face-to-face interactions or displaced traditional, nondigitized social relationships.[19] Moreover, while involvement with social media sites absorbs a fair amount of time, it does not come at the expense of traditional face-to-face contacts. If anything is displaced, it is the time spent eating, sleeping, and watching television.[20] All in all, the effects of digitally based social media seems remarkably similar to those in the early days of the telephone; despite widespread concerns at the time, chatting on the phone did not displace traditional modes of communication and socialization. If anything, telephone communications appear to have extended and intensified existing social relationships.[21]

Mobile Communications

Many social media sites are now accessed on smartphones. The rapidly expanding capabilities of these devices have led to the point where their ability to send and receive telephone calls seems almost incidental to their other functions. In parallel with the spread of smartphones, the introduction of tablet computers, such as Apple's iPad, has given their owners a larger screen and new functions. The expanded capabilities of smartphones and tablets have made them the favorite devices for social networking, game playing, email, and other internet-based activities. In addition, their portability allows their users to do a variety of tasks that would be impossible or impractical with a personal computer, such as GPS navigation and paying for purchases. Their specific features aside, smartphones and tablets have ushered in a new phase of human existence, what has been called "The Age of Continuous Connection," an era in which anyone with a telephone or internet access can potentially reach anyone similarly equipped almost instantaneously.[22]

As with the other elements of twenty-first-century information and communication technology, the number of smartphones has grown dramatically. About 85 percent of U.S. adults own a smartphone,[23] and in 2022, global smartphone sales were about 1.4 billion units.[24]

Much of the attractiveness of smartphones and tablets lies in the applications ("apps") that have been developed for them, allowing their users to track their investments, receive sports scores, get turn-by-turn driving instructions, find their cars in a parking lot, and much more. Between the Apple App Store and Google Play, more than 100 billion apps are downloaded each year.[25] The development of these apps has become an industry in itself that collectively produces hundreds of thousands of specialized apps. Although vast numbers of them are available for downloading, only a few account for the majority of actual downloads, and only a few of these are widely used. According to one study, on average one smartphone app accounted for half the time spent with all apps. The second most used took up 18 percent of app time and the third accounted for ten percent. Everything else accounted for only about 20 percent of app use.[26]

Although we tend to associate twenty-first century technologies like mobile phones with the industrially developed countries, these phones have become widespread in the poorer parts of the world. Most of them lack the features of smartphones, but they have facilitated communications while circumventing the need for a wired infrastructure. Mobile phones are used to maintain family ties for migratory workers, to apprise farmers of crop prices, and even to provide basic medical diagnoses.[27] Some mobile phone users in Africa are able to use their phones as mobile banks that allow them to electronically store money, transfer funds, and pay bills.[28]

Social Media and Social Movements

Social networks can promote stability within a society by forging and strengthening linkages between individuals. At the same time, however, they can be powerful tools for undermining an existing social order. Turning individual grievances into a mass movement requires communication links to create a sense of common purpose, to forge a group identity, to recruit new members, to inform participants of intended actions, and to schedule these actions. The power of social networks to do all of these was demonstrated in August 2011 after the death of a man who had been killed by London policemen in the course of a bungled arrest. After a day of rioting, some individuals began to use a smartphone feature that made it possible to send the same message to everyone in their phone's directory. The message was simple: it told them to congregate at a particular time and place so they could "linkup and cause havic (sic), just rob everything. Police can't stop it."[29]

The looting conducted by the resultant flash mob was only one of many recent examples of collective action by individuals who coalesced and took action in accordance with mobile phone messages. Other mass movements energized by social media have been oriented toward political transformation rather than illegal personal enrichment. Up to now, the most far-reaching of these has been the use of social media to bring down the dictatorial regime of Hosni Mubarak in Egypt during the "Arab Spring" of 2011. At that time, hundreds of thousands of disaffected Egyptians had been regular readers of and contributors to Facebook pages that aired their grievances and motivated them to demonstrate against the government at particular times and places. Twitter feeds provided up-to-the-minute information, and mobile phones served as crucial communication links during this turbulent period. For some of the insurgents, these new communication technologies were viewed as powerful but nonviolent weapons for successfully challenging oppressive regimes and effecting political transformation. As the creator and administrator of the most influential Egyptian Facebook site put it:

> Now that so many people can easily connect with one another, the world is less hospitable to authoritarian regimes. Humanity will always be cursed with power-hungry people, and the rule of law and justice will not automatically flourish in all places at all times. But thanks to modern technology, participatory democracy is becoming a reality. Governments are finding it harder and harder to keep the people isolated from

> one another, to censor information, and to hide corruption and issue propaganda that goes unchallenged. Slowly but surely, the weapons of mass oppression are becoming extinct.[30]

Might this evaluation of new communications technologies be excessively optimistic? Social networking sites and mobile phones helped to create the collective identity and actions that rocked Egypt and other parts of the world, but it is important to note that they were an enabling factor, not an independent force for change. As ever, we have to avoid invoking implicit technological determinism that would make digital media autonomous sources of political transformation. In Egypt and other places where users of social media challenged the existing order, grievances had long festered within a significant portion of the population. The Mubarak regime was associated with decades of corruption, political repression, a stagnant economy, and severely limited opportunities for young people. It was a tinderbox waiting for a spark; this was supplied by the self-immolation of a young vegetable seller in Tunisia. New communication technologies helped to organize and energize the movement, but they did not start or sustain it.

No less important, digital communication is a two-edged sword that can be a force for oppression as well as of liberation. Technologically sophisticated authoritarian regimes can tap into email accounts and mine social networking sites in order to locate associates of known dissidents and to track their activities. They can also plant bogus webpages, blogs, and videos to discredit their opponents. And they can censor internal and external news sources and communications. Also, it cannot be assumed that widespread connection to digital media will impart unity. Tweets, blogs, and instant messages are inherently decentralized communication modes, and their fragmented nature may prevent disaffected citizens from coalescing into a unified opposition movement.[31]

Even when the effective use of social media helps to foment a large-scale uprising, it is less useful for generating long-term purposive action, especially when this requires discipline, organization, and sacrifice. As Malcolm Gladwell has argued, Facebook and similar sites excel at constructing networks, but these are not permanent structures with well-defined individual responsibilities and unambiguous centers of authority. As seen by Gladwell, social networks are deficient when it comes to setting long-term goals and creating effective strategies for achieving them.[32] The case of Egypt seems to bear out Gladwell's critique. Mubarak and his henchmen were swept aside by a movement that involved millions of Egyptians, but after the dust had settled, the army was in control and democracy faced an uncertain future.

In the broadest terms, the role of digital media in the Egypt revolution is reminiscent of the connection between printing and the Protestant reformation nearly five centuries earlier. Neither movement would have unfolded the way it did in the absence of new media, but much more has been involved than enhanced communications capabilities. Without denying their unquestioned importance in promoting social and political change, printing, social media, and mobile phones by themselves do not a revolution make.

The Digital Divide

A computer, tablet, or smartphone and an internet connection are all that is needed for buying things online, staying in touch with friends through email, or using social media to foment revolution. But for many people, access to cyberspace remains limited at best. The internet has been characterized as a highly democratic system of communication because it is not under centralized control, yet it continues to reflect inequalities within and between nations. According to statistics compiled toward the end of 2022, although 67.9 percent of the world's population used the internet to some extent, vast disparities existed within different regions of the world. To take one of the starkest contrasts, 93.4 percent of North America's residents were internet users, but only 43.2 percent of Africa's population had this capability.[33] To take another example of a significant disparity in internet connectivity, 99 percent of the population of Iceland, a country with one of the greatest degrees of internet penetration, used the internet, while only 22.9 percent of Afghans could connect to it in 2022.[34]

As might be expected, all of the countries with high rates of internet usage lie within the developed world, but substantial disparities in usage can be found within these countries. Numerous critics have pointed to the "digital divide" that separates people with easy access to the internet from those with no access. In the United States, access in schools is virtually universal, but the ability to go online at home has differed according to race, ethnicity, and especially social class. According to the Pew Research Center, 93 percent of U.S. adults use the internet. Differences of race and ethnicity are evident but not massive; 80 percent of white households had a broadband internet connection, while 71 percent of African American and 65 percent of Latino households had one.[35]

COVID-19 laid bare the digital divide in two important ways. News reports in many towns illustrated the problem of the move to remote schooling, with kids taking to parking lots outside libraries or businesses trying to connect to Wi-Fi.[36] Globally, about two-thirds of school-aged children do not have home internet.[37] Even as the pandemic rolled on, it was difficult to address inequalities in digital access.[38] More vulnerable learners do poorly in online-only environments, leading to lasting deficiencies in learning, a phenomenon not limited to first-world learners.[39] And significantly, the privilege of working remotely fell unequally across employment sectors.[40] "Essential workers" in the United States were disproportionately minorities such as female African American workers in service occupations such as nursing and personal care work, and a vast majority are blue-collar workers without advanced degrees.[41] In such work , people faced higher risks of continued exposure to COVID-19 by remaining at necessary in-person jobs, with predictably higher rates of infection and poor outcomes.[42] As with other technologies, the risks and benefits of technological systems do not fall equally on all people.

Income remains one of the greatest markers of internet access, with 97 percent of U.S. households with incomes over $75,000 connected to the internet, considerably more than the 74 percent of households with incomes under $30,000.[43] Geographical location is also significant; 85 percent of urban households have an internet connection, whereas 78 percent of rural households are connected.[44] The way that different groups log onto the internet also shows some interesting differences. The growth of smartphone ownership has created a new means of access. African Americans and Latinos own smartphones at about the same rate as Whites, but they are more likely to use these devices for online access.[45] This may result in the "walled garden" phenomenon whereby control by mobile carriers, which are not governed by net neutrality laws, results in more limited access to the internet, as when content deemed "controversial or unsavory" has on occasion been blocked by mobile carriers that serve as ISPs.[46]

In a world where the ability to acquire and make use of information has taken on increasing importance, the internet has the potential to empower people by making information more accessible than it ever has been. But the internet is not a technological fix. As we have just seen, while the internet has been a revolutionary technology, access to it also reflects prevailing social and economic conditions. Perhaps of equal importance, even with widespread online access there remain significant disparities in the ability to make effective use of it. In the absence of parallel social changes, the internet may intensify rather than diminish global and national social inequalities.

Intellectual Property

One of the most contentious issues surrounding the use of computer networks has been the innocuous-sounding practice of file sharing. In this context, a file can be anything in a digital format—text, sound, or pictures. No one objects when people use the internet to send a Christmas letter or a photograph of a new grandchild to friends and family, but a host of legal and commercial issues arise when a digitized movie or song is sent to members of a file-sharing service. At the heart of the matter is intellectual property, the ownership of a product of one's creativity, such as a novel or a musical performance. Written works, music, and movies intended for commercial use are protected by copyrights, which prevent the unauthorized use of the material and usually require the payment of fees when they are legally used. Governments grant copyrights for the same reason that they issue patents: both are thought to stimulate creative efforts by giving individuals and firms monopoly control over their creations, and with it the chance to reap financial rewards.

In previous decades, the development of radio, television, movies, and recorded music greatly expanded the commercial value of creative efforts. Generations of actors, artists, musicians, and others have owed their livelihoods to the media in its various forms, but more recent technologies have been more problematic, and their use has exemplified the ways in which new technologies may create winners and losers, as noted in Chapter 1. One early example is the videocassette recorder (VCR). When the VCR first came out, it was thought that its primary function would be the playing of prerecorded tapes of movies rented from video stores. But many VCR owners also used their machines to record programs that had been broadcast on commercial television. The movie industry sued the manufacturers of VCRs and videotapes, claiming copyright infringement. In a case that eventually went all the way to the U.S. Supreme Court, the Court ruled in 1984 by a 5 to 4 margin that home recording constituted "fair use" under existing copyright laws, allowing individuals to videotape movies and TV shows with impunity.[47]

In the late 1990s, the rapid spread of computer networks presented a new challenge to existing conceptions of intellectual property, because after being compressed and digitally encoded, a musical selection could be sent to anyone with a networked device and the requisite software. By the start of the new century, 60 million persons were sharing music that was being routed through Napster, the most prominent file-sharing service at the time. This was done by connecting one's computer to Napster's network servers, which kept a list of all the music that was available on clients' computers in the MP3 format. Napster and its clients claimed that all that was going on was the sharing of files by private parties, which constituted "fair use" under existing copyright laws. But the courts did not agree, ruling that Napster's central database of music titles gave it primary responsibility for what was deemed illegal downloading.[48] Napster tried to reconstitute itself as a commercial distributor of digitized music, but it failed to survive in this guise. Napster's demise did not put an end to file sharing. New services allowed individuals to access music residing in millions of individual sites but avoided the legal problems caused by a maintaining a central database that used the internet to channel requests and to inform users of available music, movies, and other digitized files.

While the music industry was trying to stop "piracy," growing numbers of individuals were subscribing to fee-based services that allowed them to stream or download music and movies without fear of legal reprisals. The services provided by iTunes, Rhapsody, Amazon, and now Spotify and Pandora have been quite successful in getting people to pay for downloaded music. Apple, which pioneered the sale of downloaded music for its line of iPods, tallied its 10 billionth iTunes download in early 2010.[49] The growth of these enterprises made major inroads into the practice of unauthorized file sharing.[50] After all, it didn't require much of a financial commitment, usually not much more than a dollar, to obtain a hit song. Legal downloading also retained a key advantage of illegal downloading—the convenience of instantly acquiring a song without having to go to a store or wait for a mail delivery.

The rise of file sharing, both legal and illegal, coincided with a significant drop in CD sales, but there is no agreement that this reflected a cause-and-effect relationship. As might be expected, the music industry attributed falling sales to widespread "piracy" and claimed that the very survival of the music industry depended on the suppression of unauthorized file sharing. Although the person-to-person transfer of files was a common occurrence, the industry may have overstated its impact; one study concluded that unauthorized file sharing had only a minimal effect on CD sales.[51] Falling CD sales, it was argued, were the result of high prices, the lack of new and interesting music, and the appeal of alternative forms of entertainment such as video games, smartphones, tablet computers, and movies on demand. These claims were of course challenged by the music industry, which pointed to other studies that purportedly proved that file sharing definitely damaged CD sales.

Neither the industry nor file-sharing services were right. The real culprit in the decline of CD sales has been the rapid growth of music streaming through subscription services like Spotify, Apple Music, and Amazon Music. These sites now account for over $10 billion, or almost 70 percent of music industry revenue, vastly more than the revenues earned though digital downloads. In 2006 (well into the era of illicit file sharing) the music industry earned $9.4 billion from CD sales in the United States (79.7 percent of total revenue). By 2022, CD sales stood at just $483 million (3 percent of total revenue), Ironically, the only bright spot in this dismal financial picture has been a resurgence in the sale of vinyl records, which earned $1.2 billion for the industry.[52]

The Electronic Media in Modern Society

In considering the consequences of electronic media, beginning with radio and television, and continuing with the numerous forms of digital media available today, we are left with many unresolved issues. As with other major technological changes, the electronic media have advanced in conjunction with other historical changes, making it very difficult to come to an exact reckoning of their unique contributions to culture and society. To put it slightly differently, to understand the influence of the electronic media, it is necessary to consider the larger contexts within which they emerged and diffused.

The electronic media are only the latest in a series of technologies that have expanded our ability to communicate with one another. As human societies have expanded in size and complexity, and facilitated mobility, the need to tie them together through improved communications media has increased apace. As we have seen, visionaries described technologies that would transmit sounds and pictures over great distances long before the realization of these dreams was technically possible. Even so, the need for something does not ensure that it eventually will be produced. The technical capacity has to be created, and individuals and organizations have to come forward with money and other resources to support it. One of the major reasons for the success of radio, television, and the internet can be found in the solid financial support that they have attracted. During its early years, radio advanced because military, governmental, and business organizations perceived its relevance to their needs. Later, the profit-making potential of radio and then television was eagerly exploited, and the market created by commercial broadcasting further stimulated technological development. More recently, the internet began as a creation of the military, was further developed as a means of business and personal communication, and toward the end of the twentieth century emerged as a popular medium with seemingly endless commercial possibilities. In the opening years of the twenty-first century there has been a massive expansion in the products and services tied to the internet, and we surely have not seen the last of it.

The electronic media have become essential features of modern life because they have helped to mitigate the disruption of stable community ties that has been a prominent feature of the modern era. Economic and social changes have produced a great deal of social and geographic mobility, as well as corresponding feelings of rootlessness. In response, first radio and then television contributed to the construction of a common culture that served as at least a partial substitute for attenuated local ties. The effects of the internet, social media, and other digital media are less easily characterized. Whereas television and radio formerly attracted large audiences through the broadcasting of a small range of programs, cable television and some of the key products of the internet—websites, blogs, tweets, and social networks—are engaged in "narrowcasting" that plays to a much more limited audience. The consequences of audience segmentation are still somewhat speculative, but there can be little doubt that they have contributed to the political and cultural polarization that is characteristic of our times.

Although participants in social media sites may list hundreds of "friends," a virtual community is not the same as one based on long-lasting face-to-face relationships. As we have seen, person-to-person media such as the telephone and Facebook can strengthen existing social connections, but they should not be seen as substitutes for them. The loosening of local ties has not been reversed by the spread of new media, while according to some social scientists we have become an increasingly individualized society.[53] For many people the electronic media have filled a void by bringing information, entertainment, and new means of communication that have helped to mitigate weakened connections to family, friends, and community.[54] In the next chapter we will look in a bit more detail about gaming, internet culture and mental health, surveillance and privacy, and information and disinformation dynamics. Both old and new media can be characterized as "technological fixes," and as we have seen, technology cannot always fix problems that are social and cultural rather than technical in nature.

QUESTIONS FOR DISCUSSION

1. For long-range communication, do you prefer email, video messaging, texting, or the telephone? Are some communications more suitable for email and others for the telephone? Why? Are there times when a barrage of email messages leaves you feeling overwhelmed and frustrated?

2. Is the digital divide still relevant? Have you had any experiences with ICTs that have left you feeling left out or left behind?
3. The rapid expansion of the internet has put vast amounts of information and entertainment within easy access. The web also contains many sites that dispense pornography, gambling opportunities, and the ravings of hate groups. Although these sites can be blocked through the installation of filters, these are not completely effective and may limit access to important information such as AIDS awareness. Should efforts be made to limit the spread of "objectionable" material? How might this be done? What could be lost as a result?

Notes

1. Brian Winston, *Media Technology and Society: A History: From the Telegraph to the Internet* (London and New York: Routledge, 1998), p. 324.
2. Alex Roland, "Will Nuclear War Interrupt My Email?" Address at the annual meeting of the Society for the History of Technology, Henry Ford Museum, Dearborn, Michigan, October 14, 1999.
3. World Wide Web Foundation, "History of the Web." Accessed from http://webfoundation.org/about/vision/history-of-the-web.
4. Adrienne LaFrance, "How Many Websites Are There?" *The Atlantic* (September 30, 2015). Accessed from www.theatlantic.com/technology/archive/2015/09/how-many-websites-arethere/408151.
5. "50 million Users: How Long Does It Take Tech to Reach This Milestone?" Interactive Schools (February 8, 2018). Accessed from https://blog.interactiveschools.com/blog/50-million-users-how-long-does-it-take-tech-to-reach-this-milestone.
6. Internet Live Stats, "Total Number of Websites." Accessed from www.internetlivestats.com/total-number-of-websites.
7. See: The World Bank, "Individuals Using the Internet (% of Population)." Accessed from https://data.worldbank.org/indicator/IT.NET.USER.ZS; and Data Reportal, "Digital Around the World." Accessed from https://datareportal.com/global-digital-overview for estimates.
8. Malcolm Gladwell, "Clicks and Mortar," *The New Yorker* (December 6, 1999): 106–115.
9. Eduardo Porter, "Keeping the Internet Neutral," *The New York Times* (May 8, 2012). Accessed from www.newyorktimes.com/2012/05/09/business/economy/net-neutrality.
10. Matthew Ingram, "Open vs. Closed: What Kind of Internet Do We Want?" *GigaOM* (March 23, 2012). Accessed from http://gigaom.com/2012/03/23/open-vs-closed-what-kind-of-internet-do-we-want.

11. Seth Fiegerman, "Facebook is Unstoppable," CNN Money (July 27, 2016). Accessed from http://money.cnn.com/2016/07/27/technology/facebook-earnings-highexpectations.
12. Statista, "Numbers of LinkedIn Members from 1st quarter 2009 to 3rd quarter 2015," Accessed from www.statista.com/statistics/274050/quarterly-numbers-of-linkedin-members; and Maddy Osman, "Mind-Blowing LinkedIn Statistics and Facts, 2020," *Kinsta* (September 22, 2023). Accessed from https://kinsta.com/blog/linkedin-statistics.
13. Statista, "Number of Twitter users worldwide from 2019 to 2024." Accessed from www.statista.com/statistics/303681/twitter-users-worldwide.
14. Internet Live Stats, "Twitter Usage Statistics." Accessed from www.internetlivestats.com/twitter-statistics.
15. Aaron Smith, "6 New Facts About Facebook," *Pew Research Center* (February 3, 2014). Accessed from www.pewresearch.org/fact-tank/2014/02/03/6-new-facts-about-facebook.
16. James B. Stuart, "Facebook Has 50 Minutes of Your Time Each Day. It Wants More." *The New York Times* (May 5, 2016). Accessed from www.nytimes.com/2016/05/06/business/facebook-bends-the-rules-of-audience-engagement-to-its-advantage.html?_r=0.
17. Randye Hoder, "Study Finds Most Teens Sext before They're 18," *Time* (July 3, 2014). Accessed from http://time.com/2948467/chances-areyour-teen-is-sexting.
18. Sherry Turkle, "The Flight from Conversation," *The New York Times* (April 21, 2012), p. 2. Accessed from www.nytimes.com/2012/04/22/opinion/sunday/the-flight-from-conversation.html?pagewanted=1&_r=2.
19. Vincent Chua, Julia Madej, and Barry Wellman, "Personal Communities: The World According to Me," in Peter Carrington and John Scott (Eds.), *Handbook of Social Network Analysis* (Thousand Oaks, CA: Sage, 2011) p. 106.
20. Ibid., p. 105.
21. Claude S. Fischer, *America Calling: A Social History of the Telephone to 1940* (Berkeley, CA: University of California Press, 1992).
22. Niclaj Siggelkow and Christian Terwiesch, "The Age of Continuous Connection" *Harvard Business Review* 2019: 64–73.
23. Pew Research Center, "Mobile Fact Sheet." Accessed from www.pewresearch.org/internet/fact-sheet/mobile.
24. Statista, "Number of Smartphones Sold to End Users Worldwide from 2007 to 2022." Accessed from www.statista.com/statistics/263437/global-smartphone-sales-to-end-users-since-2007.
25. Statista, "Number of Apple App Store and Google Play Mobile App Downloads Worldwide from 3rd Quarter 2016 to 1st Quarter 2023." Accessed from www.statista.com/statistics/695094/quarterly-number-of-mobile-app-downloads-store.
26. Victor Luckerson, "This Is How Many Apps You're Really Using on Your Smartphone," *Fortune* (September 24, 2015). Accessed from http://fortune.com/2015/09/24/apps-smartphone-facebook.

27. Antonio Regalado, "How Mobile Phones Jump-Start Developing Economies," *Technology Review* (November 22, 2010). Accessed from www.technologyreview.com/news/421769/how-mobile-phones-jump-startdeveloping-economies.
28. Grace Dobush, "How Mobile Phones Are Changing the Developing World," *Consumer Technology Association* (July 27, 2015). Accessed from www.cta.tech/Blog/Articles/2015/July/How-Mobile-Phones-Are-Changing-the-Developing-Worl.aspx.
29. Bill Wasik, "Crowd Control," *Wired* (January 2012): 78.
30. Wael Ghonim, *Revolution 2.0: The Power of the People Is Greater Than the People in Power* (New York and Boston: Houghton Mifflin Harcourt, 2012), pp. 292–293.
31. Evgeny Morozov, "The Digital Dictatorship," *The Wall Street Journal* (Feb. 20, 2010). Accessed from http://online.wsj.com/article/SB10001424052748703983004575073911147404540.html.
32. Malcolm Gladwell, "Small Change: Why the Revolution Will Not Be Tweeted," *The New Yorker* (October 4, 2010). Accessed from www.newyorker.com/reporting/2010/10/04/101004fa_fact_gladwell?currentPage=4.
33. Internet World Stats, "Internet Users in the World by Regions." Accessed from www.internetworldstats.com/stats.htm.
34. Ibid.
35. Pew Research Center, "Internet/Broadband Fact Sheet." Accessed from www.pewresearch.org/internet/fact-sheet/internet-broadband/?tabId=tab-d5edf003-5858-4269-89c5-f2889ecf7951.
36. N'dea Yancey-Bragg, "More than $130K Raised for California Family after Girls Seen Using Taco Bell WiFi for School Work." Accessed from www.usatoday.com/story/news/nation/2020/09/01/thousands-raised-girls-who-had-use-taco-bell-wifi-school/5680992002.
37. UNICEF, "Two thirds of the world's school-age children have no internet access at home, new UNICEF-ITU report says," (November 30, 2020). Accessed from www.unicef.org/press-releases/two-thirds-worlds-school-age-children-have-no-internet-access-home-new-unicef-itu.
38. Erin Richards, Elinor Aspegren, and Erin Mansfield, "A Year into the Pandemic, Thousands of Students Still Can't Get Reliable WiFi for School. The Digital Divide Remains Worse than Ever," *USA Today* (February 4, 2021). Accessed from www.usatoday.com/story/news/education/2021/02/04/covid-online-school-broadband-internet-laptops/3930744001.
39. See: Megan Kuhfeld, Jim Soland, Karyn Lewis, and Emily Morton, "The Pandemic Has Had Devastating Impacts on Learning. What Will It Take To Help Students Catch Up?" *The Brookings Institution* (March 3, 2022). Accessed from www.brookings.edu/blog/brown-center-chalkboard/2022/03/03/the-pandemic-has-had-devastating-impacts-on-learning-what-will-it-take-to-help-students-catch-up/. For global learning loss see: "COVID Learning Loss Has Been a Global Disaster," *The Economist*

(July 7, 2022). Accessed from www.economist.com/international/2022/07/07/covid-learning-loss-has-been-a-global-disaster.

40. Colleen Mcclain, Emily A. Vogels, Andrew Perrin, Stella Sechopoulos, and Lee Rainie, "The Internet and the Pandemic," *Pew Research Center* (September 1, 2021). Accessed from www.pewresearch.org/internet/2021/09/01/the-internet-and-the-pandemic.
41. Celine McNicholas and Margaret Poydock, "Who Are Essential Workers?" *Economic Policy Institute* (May 19, 2020). Accessed from www.epi.org/blog/who-are-essential-workers-a-comprehensive-look-at-their-wages-demographics-and-unionization-rate; and National Conference of State Legislatures, "COVID-19: Essential Workers in the States," (January 11, 2021). Accessed from www.ncsl.org/research/labor-and-employment/covid-19-essential-workers-in-the-states.aspx.
42. National Bureau of Economic Research, "Measuring the Virus Risk of Essential Workers and Dependents," (March 2021). Accessed from www.nber.org/digest-202103/measuring-virus-risk-essential-workers-and-dependents; and Mary Van Beusekom, "Study Details Covid's Toll on Essential Workers, Health Workers," *CIDRAP News* (April 11, 2022). Accessed from www.cidrap.umn.edu/news-perspective/2022/04/study-details-covids-toll-essential-workers-health-workers.
43. Ibid.
44. Ibid.
45. Monica Anderson, "Racial and Ethnic Differences in How People Use Mobile Technology," *Pew Research Center* (April 30, 2015). Accessed from www.pewresearch.org/fact-tank/2015/04/30/racial-and-ethnic-differences-in-how-people-use-mobile-technology.
46. Jamilah King, "How Big Telecom Used Smartphones to Create a New Digital Divide," *Colorlines* (December 6, 2011). Accessed from http://colorlines.com/archives/2011/12/the_new_digital_divide_two_separate_but_unequal_internets.html.
47. "Videotaping without Guilt," *The New York Times* (January 22, 1984), section 4:1.
48. Sam Costello, "Court Orders Napster to Stay Shut" (March 25, 2002). Accessed from www.pcworld.com/article/91144/court_orders_napster_to_stay_shut.html.
49. Apple Press Info, "Apple's App Store Downloads Top Ten Billion," (January 22, 2011). Accessed from www.apple.com/pr/library/2011/01/22Apples-App-Store-Downloads-Top-10-Billion.html.
50. The American Assembly, "The Copy Culture Survey: Infringement and Enforcement in the US" (November 15, 2011). Accessed from http://piracy.ssrc.org/the-copy-culture-survey-infringement-and-enforcement-in-the-us.
51. Felix Oberholzer-Gee and Koleman Strumpf, "The Effect of File Sharing on Record Sales: An Empirical Analysis." Accessed from www.utdallas.edu/~liebowit/intprop/OS%202006-12-12.pdf.

52. RIAA, "U.S. Music Revenue Database," Accessed from www.riaa.com/u-s-sales-database; and RIAA "Year-End 2022 RIAA Revenue Statistics." Accessed from www.riaa.com/wp-content/uploads/2023/03/2022-Year-End-Music-Industry-Revenue-Report.pdf.
53. The most influential book on this issue is Robert Putnam, *Bowling Alone: The Collapse and Revival of American Community* (New York: Simon and Schuster, 2000).
54. Raymond Williams, *The Long Revolution* (New York: Penguin Books, 1965), pp. 26–27.

Chapter 17

Issues in Information and Communications Technologies (ICTs)

As we said in the beginning of this section on ICTs, communications technologies are central to human societies. Like most technologies, they have a reciprocal relationship with social change and affect more than just their direct users. New technologies also reflect the issues of technologies that have come before. There are also new issues, as ICTs penetrate more and more domains of human interaction. We will review some current research on games and gaming; cryptocurrencies; mental health, self-image, and bullying; and disinformation in both political and health sectors. We will close this unit by looking at the rapidly changing areas of artificial intelligence (AI) and algorithms as forms of surveillance and control, and AI's intersections with learning. While smartphones may be the ubiquitous example of ICTs and changes to daily life, we also will examine other, perhaps less obvious places where ICTs are rapidly transforming sociocultural arenas, while at the same time being transformed by them.

Video Games and Gaming

Social media sites are a major component of the internet. Also significant are online games, which account for a large portion of internet use. To these can be added the games downloaded on smartphones, tablets, and computers, along with the consoles and handheld gaming devices. About three quarters of all U.S. households have someone who plays video games.[1] Originally limited to arcades or early adopters of personal computers, on-screen games now constitute a significant portion of leisure activity for many individuals. The technical development of video games has proceeded rapidly in recent years, providing amazingly realistic depictions of everything from bowling alleys to combat zones to rock concerts. Commentary about games generally focuses on issues of violence, and to a much lesser extent the opportunity costs of investing time and money in games and gaming.

Once comprising little more than a cluster of cottage industries, the video game industry has become a very big business, generating about $90 billion in economic activity and over 400,000 jobs annually.[2] Globally, revenues generated by electronic games are now double those of the entire music industry and are one-quarter greater than those of the magazine business.[3] In 2022, there were about 3.2 billion players globally, providing revenue of about $184.4 billion dollars.[4] Although the typical video gamer is often envisaged as an adolescent boy, the audience for video games has gone well beyond this stereotype. The average age of gamers is 33, with 49 percent between the ages of 18 and 49. It is sometimes a multigenerational activity as the first cohorts of video game players are becoming parents. Women now comprise 48 percent of game players, and in the United States, the breakdown by ethnicity trends to Whiter (71 percent non-Hispanic) than the actual U.S. population (59.3 percent non-Hispanic White).[5]

Of games rated and sold in 2021, 12 percent received a "mature" rating, and 50 percent a rating for "everyone." Parents say they are very aware of their child's game use, and most, but not all, have specific rules for access, purchases, and communications.[6] There is no escaping the fact that games featuring violent action are an important source of revenue and profit. The most successful game of this genre, *Call of Duty: Modern Warfare 3*, brought in $775 million on the first five days of worldwide sales in 2011,[7] and new releases continue to be the top-selling video game franchise in terms of PC and console sales. *Candy Crush Saga* holds firm as the top mobile purchase.[8]

Concerns about a possible connection between violent video games and real-life violence bring us back to some of the issues voiced by critics of televised violence. As is the case with televised violence, the influence of violent video games on behavior varies according to age, gender, and other individual and social variables. Gaming preferences also differ. Not only are males somewhat more likely to be video gamers, the types of games they choose to play differs from those of female players. Males are drawn to sports and action games (not all of which are violent), while their female counterparts generally prefer strategic and puzzle games. Recent studies show that women definitely engage in adventure gaming (but not tactical shooter games) as well as have a strong presence as mobile "match-three" users, such as *Candy Crush* and the many variations of related puzzle games.[9]

Other studies have found that girls and women tend to play for social interaction and to form relationships. In contrast, boys and men prefer games where winning is the most important outcome.[10] As has been the case of television, the American Psychological Association has noted a clear correlation of playing violent video games and aggressive behavior.[11] Once again, inferring causality from correlation can be problematic. For the majority of players, participation in violent on-screen activities may be "just a game," but these games may catalyze aggressive and even violent acts for those who are already disposed toward this kind of behavior.

Today's video games have reached impressive levels of realism, so much so that some are now being used to prepare soldiers and the police to operate in dangerous and frightening situations. It is certainly possible that graphic on-screen violence can have a stronger and more lasting effect than watching TV on the players of these games.[12] Along with the realism that they bring, video games also require a level of attention and involvement that television viewing does not. This greater degree of participation was highlighted in a report issued in 2001 by the U.S. Surgeon General, which speculated that video games might stimulate more aggressive and violent behavior than television viewing because the player is a participant in on-screen violence and not simply a vicarious consumer of it.[13]

"Speculated" is the proper term here. A considerable amount of methodologically sophisticated research will be required before the consequences of violent gaming are fully comprehended. These reservations aside, an interesting *negative* correlation between the spread of video games and society-wide violence merits our consideration. The hypothesized connection between violent behavior and violent video games is belied by a substantial *decrease* in the rate of violent crime in the United States and most other countries in recent decades. The reasons for the drop in violent crime have been the subject of much debate, but this trend makes it hard to argue that violent video games have triggered a wave of violence across the globe.

Although the American Psychiatric Association decided that existing research did not justify including video game addiction in the most recent edition of its *Diagnostic and Statistical Manual of Mental Disorders (DSM-5)*, it is a problem for some individuals.[14] Psychologists who have treated video game addiction note that the majority of their clients are young men under 30 with poor self-esteem and limited real-life social networks. They are especially drawn to role-playing games that allow them to assume a new identity as they inhabit a virtual world that is more appealing than the one in which they live. As with other addictions, video game addiction is characterized by the need to devote large amounts of time to gaming, withdrawal from other activities, and feelings of anger and depression when one is unable to play or prevented from doing so. Addiction to technology in general may be an inescapable part of modern life, but with excessive video game play, it may become self-destructive. Like many other activities, casual users generally underestimate the costs and overestimate their ability to control their engagement with games and online gambling. Outside of speculations about violence, we should also make note of the opportunity costs: time spent playing video games is time *not* spent on homework, exercise, employment, or household engagement. The average U.S. player spent approximately 13 hours gaming per week, with 53% playing more than 7 hours per week.[15] Plus there are expenditures for in-game purchases, such as add-ons and virtual "equipment," the expenditures for which might be better saved or allocated to household necessities.

On a more positive note, some organizations have begun to experiment online with applying gaming mechanics as motivational devices in behavior change—for example, providing real or virtual rewards as a user completes a sequence of stages on the way to achieving a goal, often within a framework that encourages competition and interaction. The term used for this is "gamification," which, according to its proponents, makes work and other activities more engaging and enjoyable.[16] The awarding of points, badges, or other rewards for doing certain things and completing particular tasks is

Figure 17.1 Esports have rapidly emerged as both competitive activities in and out of schools, through tournaments organized by game companies. Pictured here is an arena configured for an esports tournament. (Shutterstock/Roman Kosolapov)

not a novel technique, of course; psychologists have long known that frequent feedback generally improves performance, whether it be of workers, students, or other participants. What is new about gamification is that it depends on the existence of social media and the drive for constant connectedness. Gamification also taps into a universal attribute of human culture; in virtually every kind of society, people of all ages and genders play games of some sort. However, the extent to which human activities can be made to resemble games is open to question. The history of business strategies is replete with many examples of fads that achieve considerable popularity and influence, only to eventually fade away as their impracticality, triviality, and ineffectiveness become evident. Gamification may fall into this category, or it may flourish because its intended group of users has been involved with electronic games for much of their lives.

The growth of esports in schools, and at large, has been rapid—to the point where there are regional and state rankings and numerous state associations, and the National Federation of High School Sports (NFHS) recognizes the activity.[17] The benefits of esports are touted to be an engagement with STEM (Science, Technology, Engineering, and Mathematics) and CTE (Career and Technical Education, formerly "vocational" schooling) in U.S. high schools. Gaming is said to encourage participation in coding, and hardware-software interfaces and systems design. But beyond STEM, esports engagement is argued to enhance interest in English and Language Arts, and Arts more generally for the development of narrative, storytelling, character development, and graphic design in games.[18] Esports may keep otherwise disengaged or marginalized student interested in schooling.

University esports participation and curriculum has also transformed rapidly. At the university level, esports teams grew from 70 school-recognized programs in 2018, to 175 in 2022.[19] Many student recreation centers support esports as part of their programming. Universities have also added esports curricula, sometimes as management degrees, or certificates and minors from programs in Schools of Information, or Computer Science, and Business.[20] There are also several global dimensions: Global gaming curricula is also rising.[21] Gaming has the potential to enhance global connections and awareness of issues, as well as collaboration, problem-solving and enhanced communication skills.[22] Gamification or game-based learning are projected to continue growing rapidly, coupled with the intensified availability of online learning.[23]

Cryptocurrencies

As we were preparing this chapter, a major player in the cryptocurrency universe, Sam Bankman-Fried, was arrested for fraudulent business practices and later convicted in November 2023.[24] Translated into U.S. dollars, the global crypto market as of August 2023 was about $1.09 trillion. In 2022, there were several shocks and rapid devaluations of many crypto platforms, with a total value loss of 61% in 2022.[25] But cryptocurrencies, based on blockchain technology, promise independent financial opportunities not tied to governments and financial institutions.[26] On one hand, cryptocurrencies challenge Western institutions such as regulated banking. But that is seen by some as a good thing, promoting innovation, independence, and lack of government control.[27] As alternative and deregulated ways of paying for goods and services in global transactions, cryptocurrencies have been a highly speculative enterprise.[28] Like most technologies, the benefits of cryptocurrencies are unevenly distributed, as are the downsides. Blockchains are not completely secure, and are trackable and hackable, subject to private and public surveillance.[29] Because of the deregulation and proliferation, cryptocurrencies are unverified and not interchangeable or universally accepted.[30] For the skeptical, cryptocurrencies are very similar to pyramid schemes, as are MLM (multilevel marketing) ventures, only holding value based on the optimism that others will sign up and perpetuate worth. Both of these activities require the enrollment of new participants to expand the client base, but later users rarely benefit from the sales or commission structures.[31] Wealthier people and institutional investors are more likely to absorb shocks of market volatility, and small investors face the greatest risks.

Our Chapter 8 discussion outlined the environmental consequences of cell phones. Crypto similarly has a relatively hidden energy consumption costs: "As of August 2022, published estimates of the total global electricity usage for crypto-assets are between 120 and 240 billion kilowatt-hours per year, a range that exceeds the total annual electricity usage of many individual countries, such as Argentina or Australia."[32] Whether phone or server-based, cryptocurrency applications rely on the energy infrastructure that is burdened by household energy "vampires," all of the electronics that draw electricity, even when not in use. While small for any given household, in the aggregate even the little blue power indicators and the clock on a home microwave draw energy which has measurable global impacts. In the United States, up to one quarter of energy use is from passive draws, which is a significant cost to consumers and a strain on energy production.[33] Globally, data centers that support "the cloud" and internet services use about an estimated 190.8 Terawatts of energy per year, although, remarkably, this value has been stable because of attention to green design and energy use.[34]

Social Media and Harms

ICTs damage the health of the environment and social media use impacts individual health, both physical and mental. Whether it is strange, self-destructive fads on platforms such as TikTok; excessive attention to body image; or cyberbullying, individual engagement with the content of social media can negatively impact health. For example, self-injury and pro-eating-disorder videos are technically banned by the TikTok app, which has over one billion users globally.[35] But even anti-self-harm videos can paradoxically influence vulnerable users by increasing awareness of harmful activities.[36] The circulation of dangerous fads, perhaps meant to be ironic (Tide Pods or Nyquil chicken[37]), nonetheless has resulted in real hospitalizations and many calls to poison control centers.[38]

While social media use, per se, does not clearly predict mental health difficulties, the importance of social media to an individual, and whether or not they are using social media to signal difficulties, such as "vaguebooking" (attention-seeking posts with vague reports of distress) can be an indicator of struggles with well-being.[39] More ominously, some users of Facebook and other social media sites have been subjected to cyberbullying, where cruel comments and ugly photographs have been posted for all to see. Cyberbullying is a common phenomenon, especially prevalent among teens.[40] Estimates range from 20 to 37 percent of teens experiencing cyberbullying, and the variations in reporting are associated with age, income, the platforms being used, and increased encounters because of increased screen time due to COVID-19 lockdowns.[41]

The interpersonal aggression amongst peers is different in scope and scale, but not completely disconnected from the online harassment of public figures, such as doxxing (targeted malicious release of private information) and threats. Many public figures receive intense negative commentary, whether body-shaming celebrities, or derogatory language and threats of violence to political figures. People can suddenly find themselves the subject of coordinated campaigns of harassment and denigration, such as "Gamergate," which was a 2013–2014 campaign of threats and misogynistic commentary directed against women in and around the video game industry.[42] While Gamergate may seem too outmoded, it connected dispersed figures with "alt-right" attitudes and demonstrated cohesiveness and the efficacy of targeted online threats.[43] The direct connection between video game violence and destructive events in real life remains speculative,[44] the connectivity and potential for radicalization is measurable.[45] Anonymity also removes accountability and takes the brakes off aggressive behaviors. As algorithms (more on this below) continually reinforce perceived preferences, people's associations with hate groups and manipulation by demagogues and negative influencers can become all-consuming.

Facebook aggression can also have lethal consequences for large groups of people. Beginning in 2012, Facebook posts (and traditional media) were used to spread scurrilous stories about the Rohingya, a Muslim ethnic group living in Myanmar (previously known as Burma). Infuriated by widely circulated but bogus accusations, Burmese citizens and militias attacked Rohingya villages, burning 700,000 homes to the ground while killing and raping their residents and driving them into exile in neighboring Bangladesh. Although they have been free from further attacks, Rohingya exiles face grave shortages of food and medical care, while work opportunities are nonexistent.[46]

The usefulness of social media mobilizing people in times of crisis is well recognized, whether using "marked-safe" designations on Facebook or public safety announcements on X (formerly known as Twitter).[47]

But these uses stand in tension with the nefarious use of social media, such as by a state perpetrating ethnic violence.[48] This is nothing new for media of any sort. One could make the argument that "it's just how you use it" and that these platforms are somehow neutral. But the limits to communication, say established by character counts and text boxes, produce very specific kinds of capacity for messaging, such as preventing nuanced argument and the presentation of data. The speed of mobilization is also new. Information, or misinformation, travels at the click of a button.

While social media was a central part of movements such as the Arab Spring of 2011, "Umbrella Movement" pro-democracy street demonstrations in Hong Kong in 2014, or the 2022 "Women, Life, Freedom" movement in Iran, social media mobilization is only the first step in organizing, motivating, and defining the scope of a social movement, and engaging the mechanisms of political and cultural power.[49] In each of those cases, the visibility of social media helped hold those in power accountable to a larger global community, as well as providing communication and organization for the protestors. But the means of *controlling* social media, through censorship and through the spread of (mis)information is also present, but different, for new media.

Information and Disinformation: From Politics to Health

The case of the Rohinga, where government-sponsored information and media led to violence against an ethnic minority and produced a refugee crisis, is political disinformation. But disinformation, beyond just ordinary political framing and candidate-bashing, is again slightly transformed with new media. Although violence was not involved, Facebook posts were a major source of malicious mischief during the American presidential election of 2016. The perpetrators were foreign "internet trolls," many in the employ of the Russian government, who used Facebook accounts to flood targeted groups (like African Americans or unmarried women without college degrees) with Facebook postings, many of them presenting fake news, doctored photographs and videos, manufactured rumors, and recitations of wild conspiracy theories.

"On any given day," Brad Parscale, the digital director of the Trump campaign, explained, "the campaign was running 40,000 to 50,000 variants of its ads, testing how they performed in different formats, with subtitles and without, and static versus video, among other small differences."[50] The reach of Facebook news stories was most impressive. During the final three months of the U.S. presidential campaign, the top-performing fake-election news stories on Facebook were more closely followed than the top stories from major news outlets such as *The New York Times*, *The Washington Post*, *The Huffington Post*, and NBC News.[51]

This was not the only way in which Facebook affected electoral politics. Facebook's influence was not confined to the dissemination of dubious news reports. The site accumulates a large amount of data from its users that can be weaponized for political campaigns. Much of this information is acquired during regular Facebook operations, but in at least one instance, information was deliberately extracted from Facebook users. In 2015, an English company known as Cambridge Analytica acquired personal data of up to 87 million[52] Facebook users via the 270,000 users who used a Facebook app called "This Is Your Digital Life." By giving this third-party app permission to acquire users' data, this app also opened up access to information on the user's friend network, thereby adding information from about around 87 million users, the majority of whom had not explicitly given Cambridge Analytica permission to access the data being collected. The information was then employed to tailor campaign pitches to voters according to the information gathered: where they lived, how much formal education they had, with whom they worked, their psychological profiles, and other attributes that might affect their electoral preferences.

The 2016 presidential election was probably the first one in which massive amounts of data were used to sway voters' decisions. But can the strategy be credited (or blamed) for Trump's unexpected victory? The answer to this question will likely never be known. It probably would be an oversimplification to take as definitive the statement of a Facebook executive that his company was "responsible"for Donald Trump being elected as US president.[53] He further noted that Mr. Trump was not elected because of "misinformation," but "because he ran the single best digital ad campaign I've ever seen from any advertiser. Period."[54] Even so, the influence of data collection and its application cannot be ignored. Most importantly, the election was won by a very small margin. As Democrats are fond of pointing out, 107,000 votes added to the Democratic vote totals in Michigan, Pennsylvania, and Wisconsin would have put these states in the Democrats' column in the all-important Electoral College.[55] Second, although the election offered a clear-cut choice of candidates and platforms, as noted in the previous chapter, a large number of voters made their decisions very late in the campaign. Given the margin of victory very thin in a few states coupled with the winner-take-all structure of the Electoral College, even if they swayed only a few votes, the digital tricks described above may have been an important contributor to Trump's unexpected 2016 victory.

This playbook—of voter identification, circulation of disinformation, and media framing—was well established and has been reproduced for subsequent presidential and midterm elections. Most recently, the addition of a layer of health disinformation and the political reframing of scientific issues has affected not only the political outcomes, but health outcomes in the United States as well.

The 1918 Influenza Pandemic had similar health disinformation dimensions to what we see in the present, with similar consequences. The misnaming of the disease as a "Spanish Flu" engaged xenophobic fears of outsiders and shaped how seriously people took the illness and required health interventions. Similar distrust of public health officials led to a flu resurgence in 1920 and led to higher loss of life as people did not take pandemic restrictions seriously.[56] The strategy of not counting cases, hiding disease clusters, and misidentification of causes of death is not a new form of disinformation.[57] Conspiracy theories abounded.[58] What has intensified with social media is the availability of multiple sources of both reliable and unreliable health information, and specific fear-mongering and trust-erosion tactics. Most of us are not particularly good at statistical thinking, nor distinguishing between anecdotal and scientific data.[59]

COVID-19 disinformation has had deadly consequences. As of November 2023, the United States had 109,257,531 Coronavirus cases, with 1,181,723 deaths reported, which is about 16.9 percent of the global total, despite being only 4 percent of the world's population.[60] While our mortality rate hovers at about 1 percent, we have the highest number of reported deaths,[61] and approximately 23 million people experiencing "long covid" for periods of 3 or more months post-infection.[62] The circulation of misinformation through social media is one important dimension of this disproportionate outcome. People in the United States were easily convinced not to trust public health experts and initiatives—both masking and vaccinations. Without a background advantage of a public-health-conscious culture, such as the Japanese, where mask-wearing when concerned about illness has been long practiced, this has led to much unnecessary suffering and loss of life and livelihood.

As a technology, vaccines have led to the extinction of smallpox, which since the early 1980s only lives in a handful of laboratories.[63] And while there are adverse effects to vaccines, they do not cause autism.[64] Polio is extinct as an endogenous circulating virus in the Western hemisphere since the early 1990s,[65] appearing only when global travelers bring it from abroad. Without vaccines, we might have expected 5 million cases of paralytic polio from 1960–1987, and 24 million cases between 1988 and 2021.[66] Global eradication has been hampered by complex distrust in the vaccines in Afghanistan, Pakistan, and Nigeria, where the vaccines were framed as dangerous and as population control measures by Western imperialists.[67]

We could go on at length about quackery and fads, and engage more cases of disinformation in medicine, but we are going to focus in this next section on the role of algorithms, which help to produce the echo chambers that propagate and stabilize disinformation. It is understandable that people read sources that confirm their preconceptions, but the nearly invisible automation of new selection processes is an important difference occurring with ICTs as opposed to older media.

Algorithms: Echo Chambers, Privacy, Marketing, and Algorithmic Inequalities

Algorithms are mathematical tools which make decisions or take actions based on rules derived from analysis of data. If a person watches an advertisement for a specific product, they are probably going to be interested in a related product. An algorithm will review the people in a social network and predict their likely beliefs, and infer that those people will be interested in this product or that candidate. We live in dense webs of information and the ways in which we engage those webs is itself a source of information for marketers, designers, planners, and the government. Algorithms also extend our echo chambers. Because algorithms identify what we like, their filter bubbles propagate things like fake news and disinformation rather than diverse messages which might contradict pre-existing beliefs. Because these algorithms are profitable, it is unlikely code will be modified without regulation.[68]

There's an interesting gray area, as scholar Ruha Benjamin notes, between providing people with information about products they might like and stereotyping and profiling people.[69] Algorithms shape which ads one sees, what job opportunities might be shared, and what news someone might see. As stereotypes and expectations become written into code, they become invisible and seem to be "objective" assessments of people's capacity and character. For example, one's credit score is used to check on suitability for housing rentals or mortgages, and other financial risks. Credit scores correlate with health outcomes[70] and reproduce pre-existing biases.[71]

Consider the example of shopping at a convenience store. This act could signal risky behavior if Person A buys lottery tickets, bad food, and cigarettes. Or it could signal that Person B is living in a "food desert"—is one of the 13 million people in the United States without access to grocery stores. Person B could be entirely creditworthy, but the fact of shopping in a convenience store could be held against them because of its association with other risky customers.[72] The credit score was only invented in 1989, yet has become an essential indicator, even impacting eligibility for enlistment in the U.S. Armed Forces.[73] It seems to serve as a proxy for "character" and has race, class, and gender biases built into it.

As has been noted several times in this book, few technologies are unalloyed blessings; problems usually accompany advances. For example, we saw in Part 3 that modern genetics, in addition to serving as the basis of new medical treatments, also can provide intimate information about individuals, everything from hitherto unknown family relationships to propensities to contract certain diseases. Such is also the case with the wired world. All of the communications technologies covered in this chapter have the potential to undermine personal privacy, sometimes seriously. As users of email messages know (or should know) an email message is more like a postcard than a letter. Its contents can be easily accessed and used against senders or receivers to embarrass or even blackmail them. The use of social media and internet search engines leaves a record that contains quite a lot of information about the user. These can create new marketing possibilities, but sometimes at the expense of individual privacy. Social media is no different. Facebook users were outraged in 2007 when they discovered that the social networking site was tracking their activities on participating websites and reporting them to their Facebook friends.[74] This eventually resulted in a settlement with the Federal Trade Commission whereby Facebook agreed to a number of stipulations aimed at protecting the privacy of its users and to monitoring by the FTC of its practices for 20 years.[75]

One need not even be part of a social media network to have one's privacy threatened. Recent years have seen the rapid expansion of data mining firms which use easily accessible records to gather detailed information on individuals. This information is then sold to data brokers, who use sophisticated algorithms to determine the likelihood that an individual will buy a certain product or support a candidate for political office. Once this has been determined, these individuals can be targeted by advertisements that play to their particular interests and beliefs.[76] It can be debated whether or not this practice is effective in getting people to buy these products or vote for these candidates, but the danger that all this information can be misused by hackers and "legitimate" businesses is all too evident. Keeping private information out of the hands of marketers, potential blackmailers, and others will pose continual challenges for everyone who lives in the digital era. For example, simply using a mobile phone may expose the user to a variety of privacy breaches.[77]

The emerging technology of Radio Frequency Identification (RFID) gives governments, businesses, and individuals the ability to track the movements of people who have RFID tags incorporated into their passports, drivers' licenses, credit cards, employee access badges, student ID cards, and even library cards.[78] Word processing documents may contain hidden "metadata" such as the name of the writer and the type of computer that was used.[79] The installation of a GPS device on a person's car also offers new opportunities to know the whereabouts of an individual, although the U.S. Supreme Court, in the first decision regarding digital searches, ruled that such an action requires a court order.[80] Finally drones, also known as unmanned aerial vehicles (UAVs), offer many opportunities for spying on people and places. One computer entrepreneur has famously claimed, "You have zero privacy anyway. Get over it."[81] It can be safely countered that this statement is not entirely true today, for there are many defenses against privacy-invading technologies. But those of a pessimistic disposition may argue that this obituary for personal privacy, while not completely applicable today, is not wrong, only premature.

Algorithms are also part of surveillance, whether through facial recognition and its deeply flawed algorithms,[82] or associations generated from names and interactions with the police becoming fixed in databases.[83] Algorithmic governance is the term for policing and surveillance of people based on algorithmic models, which can degenerate into stereotypes. People who are poor face more and different surveillance than people who are wealthy: in public housing the concern is with crime, in a gated community it is about conformity.[84] Filming the police is a form of counter-surveillance, facilitated by cell phones, social media, and the availability of cloud storage.

One might accept surveillance because one has nothing to hide. But that presumes that the people, or organizations, doing the surveillance have a right to do so. As the old saying goes: *quis custodiet ipsos custodes*—who guards the guards themselves? Author David Brin has made a cogent argument for absolute transparency, imagining a society with no secrets.[85] As internet activists say: information wants to be free. But unless one has the technical skills and copious free time to watch everyone who might be watching oneself, surveillance and algorithmic governance remains asymmetric.[86]

We want to mention one more current event from the world of algorithms and AI that is impactful for us as educators. In late 2022 an AI chatbot "ChatGPT" became available as a tool for writing essays.[87] So far, it will not be caught by anti-plagiarism algorithms. It does not provide adequate bibliographic information or integrate sources, nor does it provide specific, factual information.[88] Because of the aphorism "garbage-in/garbage-out," it is no better than the data set of materials from which it draws to generate text, and can produce wildly contradictory papers full of disinformation. But it does ask us to think about the adequacy of the standard college essay to demonstrate student learning and critical thinking. In other domains, such as art and music, there are serious concerns about copyright violation and theft of intellectual property because of the way that the systems draw from databases of images.[89]

Genies and Bottles

It is a challenge, in fairy tales, to put genies back in their bottles. While technologies are often abandoned, ones that have taken hold and proved themselves to be useful, even if flawed, often linger. ICTs, whether social media or algorithms, are rapidly changing. The case studies and examples in this chapter have been prefaced by "as of this writing," but based on comparisons from other domains of technological change, there are things that we can expect, as heuristics, which would suggest interventions in design, and possibly in regulation or the cultural scripts with which we engage technologies. For example, it is pretty clear that the non-users and others—whether people or ecosystems—not directly engaged in a technology can face consequences. Cognizance of those effects can be influential for shaping people's behavior or acceptance of technologies, and to reprioritize design. We can also expect that the effects of the technology, risks or benefits, will fall in ways that generally reinforce existing inequalities and social patterns. Social media has the potential to organize and make visible a revolution, or the need for one, but implementation will mean finding access to the halls of power. Taking potential consequences into account might shape design practices or spur new regulations.[90] For all of our cases, however, sunlight is the best disinfectant, and we expect that transparency in knowledge production, about consequences throughout the commodity chain, and directed toward the design and governance of technological systems is necessary step in the right direction.

QUESTIONS FOR DISCUSSION

1. Are esports real sports or just competitive activities? What other activities are competitive but possibly not sports? How would you define the distinction, or would you?
2. What do you think about the association between video games, violence, and radicalization? Do you think there is a connection between social media and internet use and forms of bullying online and offline?
3. In what ways are you concerned or unconcerned about your privacy online and in real life? Do you think that you are being identified and stereotyped due to your online activities? Who do you think has access to your data and online behavior? Would you like things to be different?
4. While freedom of speech is a central value to liberal democracies, lies, libel or slander, or speech meant to harm others is not protected. The First Amendment protects speech from government interference but does not apply to the activities taking place on privately owned platforms. How does the distributed nature of ICTs create new technical and legal challenges for regulating speech that might be harmful or illegal?
5. What is the relationship between privacy and freedom? Safety and freedom? What level or kind of surveillance is acceptable?

Notes

1. Entertainment Software Association, "2020 Essential Facts About the Video Game Industry." Accessed from www.theesa.com/resource/2020-essential-facts.
2. Entertainment Software Association, "2020 Economic Impact Report." Accessed from www.theesa.com/industries/economic-impact.
3. Tim Cross, "All the World's a Game," *The Economist* (December 10, 2011). Accessed from www.economist.com/node/21541164.
4. See: Stephen Totilo, "Video Game Revenue Shrinks in 2022, Snapping Growth Streak," *Axios* (November 15, 2022). Accessed from www.axios.com/2022/11/15/global-video-game-market-decline-2022; and Simon Read, "Gaming Is Booming and Is Expected to Keep Growing. This Chart Tells You All You Need to Know," *World Economic Forum* (July 28, 2022) www.weforum.org/agenda/2022/07/gaming-pandemic-lockdowns-pwc-growth.
5. Entertainment Software Association, "2022 Essential Facts About the Video Game Industry." Accessed from www.theesa.com/resource/2022-

essential-facts-about-the-video-game-industry/ and https://usafacts.org/data/topics/people-society/population-and-demographics/our-changing-population for comparison demographics in the U.S.

6. Ibid.
7. Keith Stuart, "Modern Warfare 3 Smashes Records: $775 Million in Five Days," *The Guardian* (November 18, 2011). Accessed from www.guardian.co.uk/technology/2011/nov/18/modern-warfare-2-records-775m39.
8. Entertainment Software Association, "2022 Essential Facts About the Video Game Industry."
9. Melissa Telecki, et al., "Sex Differences and Similarities in Video Game Experience: Implications for the Gaming Industry," *Current Psychology*, 30, 1 (December 2010): 22–33. More recently: Colin Campbell, "Which Games Are Women and Girls Playing?" Polygon (January 20, 2017). Accessed from www.polygon.com/2017/1/20/14337282/games-for-women-and-girls; and Karolina Cieślak, "Report: Women and Games – How Do Modern Female Gamers Play?" *Try Evidence* (February 7, 2022). Accessed from tryevidence.com/blog/report-women-and-games-how-do-modern-female-gamers-play.
10. C. Heeter, et al., "Alien Games: Do Girls Prefer Games Designed by Girls?" *Games and Culture*, 4, 1 (2009): 74–100.
11. American Psychological Association, "APA Review Confirms Link Between Playing Violent Video Games and Aggression," (August 13, 2015). Accessed from www.apa.org/news/press/releases/2015/08/violent-video-games.aspx.
12. George Comstock and Erica Scharrer, *Media and the American Child* (Amsterdam: Elsevier, 2007), pp. 234–238.
13. Ibid., p. 236.
14. Science Daily, "American Psychiatric Association Considers 'Video Game Addiction'" (June 25, 2007). Accessed from www.sciencedaily.com/releases/2007/06/070625133354.htm; Also see: American Psychiatric Association, "Internet Gaming Disorder," at www.psychiatry.org/File%20Library/Psychiatrists/Practice/DSM/APA_DSM-5-Internet-Gaming-Disorder.pdf.
15. Entertainment Software Association, "2022 Essential Facts About the Video Game Industry."
16. J. P. Mangalindan, "Play to Win: The Game-based Economy" *Fortune* (September 3, 2010). Accessed from https://fortune.com/2010/09/03/play-to-win-the-game-based-economy.
17. USA Today HSS Staff, "High School Esports 2022 Fall Regional/State Rankings: 'Rocket League.'" (October 24, 2022). Accessed from usatodayhss.com/lists/high-school-esports-competitive-gaming-2022-fall-regional-state-rankings-rocket-league-week-1.
18. Gregory Rothwell and Michael Shaffer, "eSports in K-12 and Post-Secondary Schools," *Education Sciences* 9, 2 (2019): 105. You can track, even in two short years, by comparison to Jason A. Engerman and Robert J. Hein, "ESports Gaming and You." *Educational Technology* 57, 2 (2017): 62–64.
19. Inven Global, "The Rise of Esports in Academic Institutions In the US," (June 26, 2022). Accessed from www.invenglobal.com/articles/17550/the-rise-of-esports-in-academic-institutions-in-the-us.

20. Michelle Burton, "10 Colleges with Esports Degree Programs," *Animation Career Review* (August 28, 2019). Accessed from www.animationcareerreview.com/articles/10-colleges-esports-degree-programs, but this is already out of date.
21. Seth Jenny, Joey Gawrysiak, and Nicolas Besombes, "Esports.edu: An Inventory and Analysis of Global Higher Education Esports Academic Programming and Curricula," *International Journal of Esports*, 1, 1 (2021). Retrieved from www.ijesports.org/article/59/html.
22. Honor Moorman, "Gaming: Leveling Up Global Competence," *Asia Society*. Accessed from asiasociety.org/education/gaming-leveling-global-competence.
23. See: "Global Game-Based Learning Market Size, Share & Industry Trends Analysis Report By Component, By End User, By Deployment Type, By Game Type, By Regional Outlook and Forecast, 2021–2027," at www.globenewswire.com/news-release/2022/04/04/2415587/0/en/Global-Game-Based-Learning-Market-Size-Share-Industry-Trends-Analysis-Report-By-Component-By-End-User-By-Deployment-Type-By-Game-Type-By-Regional-Outlook-and-Forecast-2021-2027.html.
24. David Gura, "Sam Bankman-Fried Is Found Guilty of All Seven Charges" *NPR* (November 2, 2023). Accessed from www.npr.org/2023/11/02/1210364969/sam-bankman-fried-is-found-guilty-of-all-seven-charges.
25. Cheyenne DeVon, "Next Gen Investing Bitcoin Lost over 60% Of Its Value in 2022—Here's How Much 6 Other Popular Cryptocurrencies Lost," *CNBC* (December 23, 2022). Accessed form www.cnbc.com/2022/12/23/bitcoin-lost-over-60-percent-of-its-value-in-2022.html.
26. Walid Al-Saqaf and Nicolas Seidler, "Blockchain Technology for Social Impact: Opportunities and Challenges Ahead," *Journal of Cyber Policy*, 2, 3 (2017): 338–354.
27. Andrea O'Sullivan, "Ungoverned or Anti-Governance? How Bitcoin Threatens the Future of Western Institutions." *Journal of International Affairs*, 71, 2 (2018): 90–102.
28. Q. K. Nguyen, "Blockchain—A Financial Technology for Future Sustainable Development," 2016 3rd International Conference on Green Technology and Sustainable Development (GTSD), 2016: pp. 51–54.
29. Eswar Prasad, "Cryptocurrencies Could Help Governments and Businesses Spy on Us," *The Washington Post* (April 1, 2022). Accessed from www.washingtonpost.com/outlook/2022/04/01/cryptocurrency-privacy-mainstream.
30. W. Scott Frame, et al., "Technological Change and Financial Innovation in Banking: Some Implications for Fintech," *Federal Reserve Bank of Atlanta: Working Paper Series*, Oct. 2018, https://doi.org/10.29338/wp2018-11.
31. See the documentary on the MLM scandal surrounding LuLaRoe sportswear: Jenner Furst and Julia Wllougby Nason, Dir. Cinemart/Amazon Studies. (2021). *LuLaRich* (4 parts).
32. The White House, "Climate and Energy Implications of Crypto-Assets in the United States," (September 8, 2022). Accessed from www.whitehouse.gov/ostp/news-updates/2022/09/08/fact-sheet-climate-and-energy-implications-of-crypto-assets-in-the-united-states/?utm_source=link.

33. Brian Palmer, "Energy Vampires: Keep Your Devices from Wasting Energy and Money," *NDRC* (May 2, 2022). Accessed from www.nrdc.org/stories/keep-your-devices-wasting-energy-and-money; see also: Crystal Ponti, "'Vampire Energy' Is Sucking the Life out of Our," Planetwww.wired.com/story/vampire-energy-climate-environment-earth-day/ and https://smartenergy.com/save-energy-by-defeating-your-energy-vampires.
34. Paul Kirvan, "How Much Energy Do Data Centers Consume?" *TechTarget* (April 25, 2022). Accessed from www.techtarget.com/searchdatacenter/tip/How-much-energy-do-data-centers-consume.
35. Mansoor Iqbal, "TikTok Revenue and Usage Statistics (2023)," Business of Apps (October 31, 2023). Accessed from (www.businessofapps.com/data/tik-tok-statistics and www.statista.com/statistics/1327116/number-of-global-tiktok-users/ for estimates. The number of times an app is downloaded is not a completely accurate measure of how many people actually use the platform with any regularity.
36. Giuseppe Logrieco, Maria Rosaria Marchili, Marco Roversi, and Alberto Villani, "The Paradox of Tik Tok Anti-Pro-Anorexia Videos: How Social Media Can Promote Non-Suicidal Self-Injury and Anorexia," *International Journal of Environmental Research and Public Health* 18, 3 (2021): 1041.
37. Patty Huang, "Tide Pod Challenge: The Latest Dangerous Fad," *Children's Hospital of Philadelphia* (January 15, 2018). Accessed from https://injury.research.chop.edu/blog/posts/tide-pod-challenge-latest-dangerous-fad; For NyQuil chicken see: www.healthline.com/health/dont-eat-tiktoks-nyquil-chicken.
38. Lisa Nico, "Utah Teens Take the Deadly 'Tide Pod Challenge,'" *KUTV* (January 16, 2018). Accessed from https://kutv.com/news/local/utah-teens-taking-the-deadly-tide-pod-challenge. The deaths are generally from babies or adults with cognitive disabilities, mistaking the pods for candy. Teens get sick and seek help before becoming critically ill.
39. C. Berryman, C. J. Ferguson, and C. Negy, "Social Media Use and Mental Health among Young Adults," *Psychiatric Quarterly*, 89 (2018): 307–314.
40. Meng-Jie Wang, et al., "How Common Is Cyberbullying Among Adults? Exploring Gender, Ethnic, and Age Differences in the Prevalence of Cyberbullying," *Cyberpsychology, Behavior, and Social Networking* 22, 11 (2019).
41. Security.org Team, "Cyberbullying: Twenty Crucial Statistics for 2023," (May 10, 2023). www.security.org/resources/cyberbullying-facts-statistics. See also:www.dosomething.org/us/facts/11-facts-about-cyber-bullying.
42. Caitlin Dewey, "The Only Guide to Gamergate You Will Ever Need to Read," *The Washington Post* (October 14, 2014). Accessed from www.washingtonpost.com/news/the-intersect/wp/2014/10/14/the-only-guide-to-gamergate-you-will-ever-need-to-read; and Aja Romano, "What We Still Haven't Learned from Gamergate," *Vox* (January 7, 2021). Accessed from www.vox.com/culture/2020/1/20/20808875/gamergate-lessons-cultural-impact-changes-harassment-laws.

43. Emanuel Maiberg, "Under Trump, Gamergate Can Stop Pretending It Was About Games," *Vice News* (February 9, 2017). Accessed from www.vice.com/en/article/bm5wd4/under-trump-gamergate-can-stop-pretending-it-was-about-games; and David Neiwert, *Alt-America: The Rise of the Radical Right in the Age of Trump*. (London: Verso, 2017), p. 215.
44. Dalton Cooper, "Ox News Suggests Mass Shootings Have 'Gotten So Much Worse' Because of Violent Video Games," Game Rant (May 16, 2022). Accessed from https://gamerant.com/fox-news-violent-video-games-buffalo-shooting.
45. Galen Lamphere-Englund and Jessica White, "The Buffalo Attack and the Gamification of Violence," *RUSI* (May 16, 2022). Accessed from www.rusi.org/explore-our-research/publications/commentary/buffalo-attack-and-gamification-violence.
46. Council on Foreign Relations, "The Rohingya Crisis," (January 23, 2020). Accessed from www.cfr.org/backgrounder/rohingya-crisis.
47. Fatma Elzahraa Elsayed, "Social Media Role in Relieving the Rohingya Humanitarian Crisis" *New Media and Mass Communication*, 87 (2020).
48. Ronan Lee, "Extreme Speech in Myanmar: The Role of State Media in the Rohingya Forced Migration Crisis," *International Journal of Communication*, 13 (2019).
49. Raluca Moldovan, "The Social Media Revolution That Failed: Lessons from the Arab Spring," *Redefining Community in Intercultural Context*, 9 (2020): 257–269; See: Fei Shen, Chuanli Xia, Marko Skoric, "Examining the Roles of Social Media and Alternative Media in Social Movement Participation: A Study of Hong Kong's Umbrella Movement," *Telematics and Informatics*, 47 (2020): 101303; For Iran see: https://theconversation.com/how-gen-z-is-using-social-media-in-irans-women-life-freedom-movement-195783.
50. Issie Lapowsky, "Here's How Facebook *Actually* Won Trump the Presidency," *Wired* (November 15, 2016). Accessed from www.wired.com/2016/11/facebook-won-trump-election-not-just-fake-news.
51. Alexis C. Madrigal, "What Facebook Did to American Democracy," *The Atlantic* (October 12, 2017). Accessed form www.theatlantic.com/technology/archive/2017/10/what-facebook-did/542502.
52. Scott Detrow, "What Did Cambridge Analytica Do during the 2016 Election," *National Public Radio* (March 20, 2018). Accessed from www.npr.org/2018/03/20/595338116/what-did-cambridge-analytica-do-during-the-2016-election.
53. Craig Timberg and Nitasha Tiku, "Facebook Executive Says Company Was Responsible for Trump's Victory But Warns Against Policy Changes," *The Washington Post* (January 8, 2020). Accessed from www.washingtonpost.com/technology/2020/01/07/facebook-exec-says-company-was-responsible-trumps-victory-warns-against-policy-changes.
54. *BBC*, "Facebook Ad Campaign Helped Donald Trump Win Election, Claims Executive," (January 8, 2020). Accessed from www.bbc.com/news/technology-51034641.

55. Tim Mako, Denise Lu, and Lazaro Gamlo, "How Trump Won the Presidency with Razor-Thin Margins in Swing States," *The Washington Post* (November 11, 2016). Accessed from www.washingtonpost.com/graphics/politics/2016-election/swing-state-margins.
56. Wendy E. Parmet and Mark A. Rothstein, "The 1918 Influenza Pandemic: Lessons Learned and Not-Introduction to the Special Section," *American Journal of Public Health*, 108 (2018): 1435–1436.
57. Robert Kessler, "Outbreak: Lies and Misinformation," *EcoHealth Alliance*. Accessed from www.ecohealthalliance.org/2018/05/outbreak-lies-and-misinformation; and Becky Little, "As the 1918 Flu Emerged, Cover-Up and Denial Helped It Spread," *History* (October 4, 2023). Accessed from www.history.com/news/1918-pandemic-spanish-flu-censorship.
58. Medical News Today, "The Flu Pandemic Of 1918 And Early Conspiracy Theories." Accessed from www.medicalnewstoday.com/articles/the-flu-pandemic-of-1918-and-early-conspiracy-theories.
59. See Björn Mederand and Gerd Gigerenzer. "Statistical Thinking: No One Left Behind." In: E. J. Chernoff and B. Sriraman (eds.), *Probabilistic Thinking* (Dordrecht, Netherlands: Springer, 2014).
60. See Worldometer's Coronavirus count at www.worldometers.info/coronavirus/country/us.
61. See Johns Hopkins' Mortality Analyses page at https://coronavirus.jhu.edu/data/mortality. They stopped collecting data on March 10, 2023.
62. United States Government Accountability Office, "Science & Tech Spotlight: Long COVID." Accessed from www.gao.gov/products/gao-22-105666; and Cecelia Smith-Schoenwalder, "How Many People Have Long COVID? The Statistics Are 'Pretty Scary,'" *U.S. News & World Report* (July 8, 2022). Accessed from www.usnews.com/news/health-news/articles/2022-07-08/how-many-people-have-long-covid-the-statistics-are-pretty-scary.
63. Donald A. Henderson, "The Eradication of Smallpox—An Overview of the Past, Present, and Future," *Vaccine*, 29 (2011): D7–D9; and Hermann Meyer, Rosina Ehmann, and Geoffrey L. Smith, "Smallpox in the Post-Eradication Era," *Viruses*, 12 (2020): 138.
64. Frank DeStefano and Tom T. Shimabukuro, "The MMR Vaccine and Autism." *Annual Review of Virology*, 6 (2019): 585–600.
65. Ciro A. de Quadros, Jon K. Andrus, Jean-Marc Olive, Carlyle Guerra de Macedo, and Donald A. Henderson, "Polio Eradication from the Western Hemisphere." *Annual Review of Public Health*, 13 (1992): 239–252.
66. Kamran Badizadegan, Dominika A Kalkowska, Kimberly M Thompson, "Polio by the Numbers—A Global Perspective," *The Journal of Infectious Diseases* 226 (2022): 1309–1318.
67. James O. Olufowote and Dalaki J. Livingston, "The Excluded Voices from Africa's Sahel: Alternative Meanings of Health in Narratives of Resistance to the Global Polio Eradication Initiative in Northern Nigeria," *Health Communication* 37 (2022): 1389–1400.

68. Samuel C. Rhodes, "Filter Bubbles, Echo Chambers, and Fake News: How Social Media Conditions Individuals to Be Less Critical of Political Misinformation," *Political Communication* 39 (2022): 1–22; and see Daron Acemoglu, Asuman Ozdaglar, and James Siderius, "A Model of Online Misinformation," NBER Working Paper Series, (National Bureau of Economic Research, 2022). Accessed from www.nber.org/paers/w28884.
69. Ruha Benjamin, *Race After Technology: Abolitionist Tools for the New Jim Code* (Cambridge, UK: Polity Press, 2019).
70. Lorraine T. Dean and Lauren Hersch Nicholas, "Using Credit Scores to Understand Predictors and Consequences of Disease," *American Journal of Public Health* 108 (2018): 1503–1505.
71. Kimberly Adams, "Credit Scores and the Bias Behind Them," *Marketplace* (July 1, 2022). Accessed from www.marketplace.org/shows/marketplace-tech/credit-scores-and-the-bias-behind-them, citing: Safiya Umoja Noble, *Algorithms of Oppression: How Search Engines Reinforce Racism* (New York: NYU Press, 2018).
72. Ibid.
73. Rod Powers, "U.S. Military Enlistment and Credit Checks," *LiveAbout*, www.liveabout.com/us-military-enlistment-and-credit-checks-3354005; and Beth Braverman, "What You Need to Know About Military Credit Checks," *CreditCard.com* (March 9, 2022). Accessed from www.creditcards.com/credit-management/military-recruiting-credit-checks-1270.
74. Daniel Solove, "The End of Privacy?" *Scientific American* 299 (2008): 104.
75. Robert Hof, "What Facebook's FTC Privacy Settlement Means to Marketers," *Forbes* (November 29, 2011). Accessed from www.forbes.com/sites/roberthof/2011/11/29/what-facebooks-ftc-privacy-settlement-means-to-marketers.
76. Evan Halper, "Privacy at Risk in Hunt for Data on Voters," *Los Angeles Times* (January 28, 2016, A1, A9.
77. Michael Cooney, "10 Common Mobile Security Problems to Attack," *PC World*, (September 21, 2012). Accessed from www.pcworld.com/article/2010278/10-common-mobile-security-problems-to-attack.html.
78. Katherine Albrecht, "RFID Tag—You're It." *Scientific American* 299 (2008): 84–85.
79. Simson L. Garfunkel, "Information of the World Unite!" *Scientific American* 299 (2008): 84–85.
80. Joan Biskupic, "Supreme Court Rules GPS Tracking Requires Warrant," *USA Today* (January 24, 2012), p. 2A.
81. Polly Sprenger, "Sun on Privacy: 'Get Over It,'" *Wired* (January 26, 1999). Accessed from www.wired.com/1999/01/sun-on-privacy-get-over-it.
82. See: "Wrongfully Accused by an Algorithm," from www.nytimes.com/2020/06/24/technology/facial-recognition-arrest.html; and for London: "Do I Know You? How Flawed is Facial Recognition?" from www.electropages.com/blog/2022/08/do-i-know-you-how-flawed-facial-recognition.

83. Leila Miller, "The LAPD Branded Them as Gang Associates. But They Fought Back and Got Removed from Calgang Database," *Los Angeles Times* (February 13, 2020). Accessed from www.latimes.com/california/story/2020-02-13/california-gang-database-lapd-calgang; and Anita Chabria and Leila Miller, "Reformers Want California Police to Stop Using a Gang Database Seen as Racially Biased," *Los Angeles Times* (June 24, 2020). Accessed from www.latimes.com/california/story/2020-06-24/california-police-urged-to-stop-using-gang-database-deemed-biased.
84. Torin Monahan, "Electronic Fortification in Phoenix: Surveillance Technologies and Social Regulation in Residential Communities," *Urban Affairs Review* 42 (2006): 169–192.
85. David Brin, *The Transparent Society* (New York: Perseus Press, 1998).
86. See Rob Reich, Mehram Sahami, and Jeremy M. Weinstein, *System Error: Where Big Tech Went Wrong and How We Can Reboot* (New York: Harper Collins, 2021).
87. Jonathan Vanian, "Why Tech Insiders are So Excited about ChatGPT, A Chatbot that Answers Questions and Writes Essays," *CNBC* (December 13, 2022). Accessed from www.cnbc.com/2022/12/13/chatgpt-is-a-new-ai-chatbot-that-can-answer-questions-and-write-essays.html; and Chris Stokel-Walker, "AI Bot ChatGPT Writes Smart Essays—Should Professors Worry?" *Nature* (December 9, 2022). Accessed from www.nature.com/articles/d41586-022-04397-7.
88. Matt O'Brien, "Explainer: What is ChatGPT and Why are Schools Blocking It?" *Associated Press* (January 6, 2023). Accessed from https://apnews.com/article/what-is-chat-gpt-ac4967a4fb41fda31c4d27f015e32660.
89. Charlie Sorrel, "Why It Matters That AI Photo Apps Are Infringing on Artists' Copyrights," *Lifewire* (December 26, 2022). Accessed from www.lifewire.com/why-it-matters-that-ai-photo-apps-are-infringing-on-artists-copyrights-6892284.
90. Ruja Benjamin, author of *Viral Justice: How to Grow the World We Want* (Princeton, NJ: Princeton University Press, 2022), argued at a recent lecture that global social media like Twitter and Facebook might be better regulated if they were treated like essential public utilities. See https://libcal.library.arizona.edu/event/10066479. See also: Reich, et al., *System Error: Where Big Tech Went Wrong and How We Can Reboot* (NY: HarperCollins, 2021).

PART 6

The Tools of Destruction

War presents the dark side of technological advance. It is a sad fact of human history that a large portion of human ingenuity has been directed toward improving the ability to kill and destroy. Chapter 18 traces the evolution of weapons from ancient times to the present, paying particular attention to the reciprocal relationships between weaponry on the one hand, and military organization, cultural orientations, and political structures on the other.

Chapter 19 picks up where the previous chapter ended with a description of today's arsenal of cruise missiles, smart bombs, spy satellites, and drones. The development and deployment of these devices has come at considerable expense, yet for all of their improvements in accuracy and destructive power, these devices have not conferred unchallenged military supremacy. Relatively simple weapons continue to be used to deadly effect by forces with limited financial resources. Moreover, a high level of technological sophistication increases vulnerability to cyber attacks that can cripple a modern economy.

Chapter 20 reprises a now-familiar theme: the influence of culture, social arrangements, economics, and politics on the development and application of technology. It describes how the advance or the delay of particular weapon systems has been associated with particular social arrangements and organizational interests. The chapter ends with a consideration of one of the greatest imperatives of our time: the reduction and eventual elimination of nuclear weapons. The history of earlier effort to eliminate or control new weapons shows that efforts of this sort have had mixed results at best.

Finally, while technologies for warfare are intentionally designed to produce damage, technologies can be inadvertent sources of destruction when they fail and become sources of accidents and disasters. As noted in the first chapter, the *Challenger* and *Columbia* space shuttles, the massive oil spill in the Gulf of Mexico, the catastrophic failure of the Fukushima nuclear plant in Japan, and the disastrous breaching of the levees in New Orleans in the wake of Hurricane Katrina are highly prominent technologically intensive disasters. Thus, in Chapter 21 we will examine a framework that articulates the scale and scope of risks and consequences. While we recognize that technologies can be used to increase safety and respond to disasters, there is ambiguity as to the effectiveness of relying on technological fixes.

Chapter 18

Weapons and Their Consequences

For most of human existence, the baser instincts of humanity were checked by the limitations of the weapons that could be used. Clubs and stone knives could be used at close range, and rocks and other missiles could be hurled at a target a few dozen yards away. There are still places in the world where warfare is conducted as it was thousands of years ago. Unfortunately, however, human ingenuity has not always been confined to activities that make our lives better. On many occasions our most inventive minds have used their talents to find new ways of wreaking havoc on other human beings. Warfare mocks the notion that technological progress is synonymous with human betterment. Although technological advances have improved many aspects of human life, they also have led to a terrifying expansion in the ability to massacre people and harm everything that sustains them.

Military Technology in the Ancient World

The technological changes that marked the beginning of civilization's development were from the start used for martial purposes. The production of metals, first bronze and then iron, resulted in more lethal weapons, as well as armor to protect against them. Around 4,000 years ago, armies in parts of the Middle East began to deploy archers in horse-drawn chariots, greatly increasing the mobility and striking power of their attacking forces. So equipped, the armies of Assyria and Persia successfully invaded their neighbors' territories and established extensive empires.

Effective as they were, military forces based on the chariot and bow did not conquer everything in their path. Horses and chariots operated with great effectiveness in the open expanses of Central Asia and the Middle East, but they were far less useful in mountainous areas. Equally important, horses had to be fed, and in the absence of adequate forage an army using large numbers of horses quickly outran its supply lines. In contrast, the armies of ancient Greece relied on the hoplite, a foot soldier armed with a spear and a short sword and protected by a shield, helmet, breastplate, and shin guards. The effectiveness of a Greek fighting force derived from its being organized into a phalanx, a mass formation of infantrymen at least eight rows deep. Deployed in close ranks, the warriors' shields offered a modicum of protection, while the spears projecting out from the first three rows presented a fearsome offensive threat. A collision between two phalanxes was the essence of Greek warfare, with victory usually going to the side that demonstrated the greater degree of discipline and cohesion.[1]

The Romans brought a greater degree of mobility to infantry combat by deploying their soldiers in smaller units of 60 to 80 men each, who in turn were aggregated into larger units that together formed a legion of 3,600 to 4,600 infantry and 300 cavalry.[2] Although armaments were little changed from the time of the Greeks, the Romans carved out a vast empire by employing superior tactics and, above all, because they could rely on the discipline and coordination of their legions. Mobility, after all, is no advantage if it leads to the dispersal of troops and the lack of unified action. The

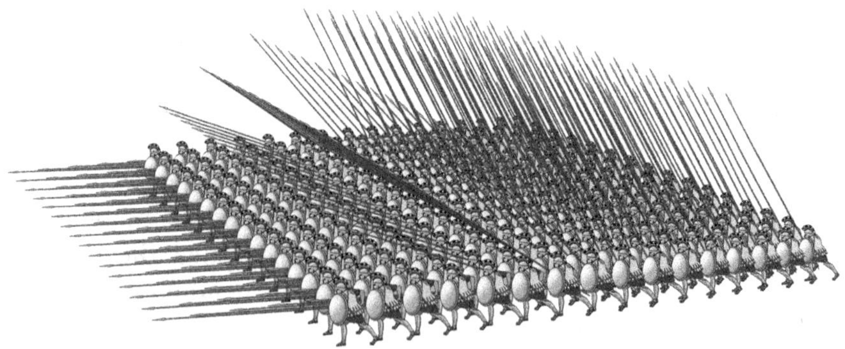

Figure 18.1 The wall of spears of the Greek phalanx was a fearsome sight for enemy combatants. (Source: Shutterstock/Massimo Todaro)

cohesion of the Greek phalanx depended on its being composed of groups of warriors belonging to social networks of friends and relatives. The cohesion of the Roman legions derived from two elements unknown to the Greeks: extensive drilling and the leadership of a permanent officer class, the centurions.[3] The effectiveness of Rome was at least as much organizational as it was technological, and it gave the Romans a military superiority that lasted for centuries.

While the weapons of the infantry soldier changed little in the ancient world, significant strides were made in the development of siege machinery. The most fearful weapons of antiquity were catapults, first used in Sicily during the fourth century BC. They were then employed with considerable success by Philip II of Macedon, the father of Alexander the Great, and played an important role in many of the battles waged by the Greeks and Romans, as well as by their enemies. It is also significant that siege machinery was the creation of some of the first specialists in technology; indeed, the word "engineer" is derived from the Latin *ingenium*, an ingenious device used for siege warfare.[4]

Catapults hurled rocks or large arrows by releasing the stored-up energy of a torsion spring, made of twisted cords of animal sinew and human hair. Hair was much prized as a military supply; there is an ancient report of thousands of pounds of hair being sent as a present from one king to another.[5] The most powerful catapults were capable of hurling a bolt or a small stone a distance of 800 yards (although 400 yards was a more normal range), while a 60-pound rock thrown a distance of 150 yards produced the best effect when fortifications were the target. A smaller catapult (known as an *onager*) used by the Romans for battlefield service was capable of flinging an eight-pound projectile up to 500 yards.[6] The main limitation of all catapults was that hair and sinew stretch when wet, so in damp climates the springs lost their resiliency and therefore much of their power. Despite this limitation, these devices played an important role in siege warfare. Each Roman legion was supported by 55 catapults, and 300 catapults were deployed by Rome during the siege of Jerusalem in AD 70. Even after the invention of gunpowder and the cannon, it was many decades before the destructive power of these weapons was surpassed.

The use of such siege machinery conferred great strength on offensively minded empire builders like the Romans. The crude fortifications of the "barbarians" usually provided scant defenses against invading armies armed with these early missile launchers. Conversely, Roman military superiority was less evident when Rome itself was put on the defensive during the (barbarian) invasions. Roman defensive technology was inferior to the offensive weaponry of the time; this incapacity, when coupled with the many internal problems besetting Rome, resulted in the eventual disintegration of one of the world's greatest empires.

Military Technology and the Feudal Order

Military technologies made few advances during the centuries following the fall of Rome. But in the sixth century a new technology emerged that added a new dimension to warfare. This innovation was not itself a weapon, but it made existing weapons far more effective. The device was the stirrup, an invention of Chinese origin that had diffused to Europe by the ninth century. Horses had long been used in battle; as we have seen, many battles were won through the use of horse-drawn chariots from which arrows could be shot and spears thrown. But a warrior mounted directly on a horse was precariously perched. Although cavalry charges were sometimes employed in battle, in most cases when it came time to fight, the warriors dismounted and fought on foot.

The stirrup greatly amplified the destructive potential of a warrior by fixing him firmly to his horse. The horse was no longer merely a means of conveyance; its power could now be used to augment the striking force of a spear or lance. As we shall see, the horse-mounted knight was by no means invincible, but his determined charge could be devastating. Although mounted combat was not the dominant mode of medieval warfare, in many places the local populace suffered considerable insecurity if they lacked the protection of a mounted warrior.

The central importance of the mounted knight produced some significant economic and political changes. A horse was an expensive item, and its maintenance required considerable outlay. At the same time, the mounted knight was the product of years of specialized training, during which time he could not be engaged in productive activities. The maintenance of a horse, knight, his assistants, and their training required 300 to 400 acres of land.[7] Even kings lacked the financial means and administrative capacity to raise and support an army of knights on their own. Instead, they gained the military support of mounted warriors by granting them tracts of land in return for their military support. This was the basis of the feudal order, the dominant form of political organization throughout the Middle Ages.

Feudalism was an inherently decentralized system, and armies often were temporary and disorderly assemblies. Knights relied on their individual skill and courage, and were disinclined to submit to centralized discipline and strategic planning. Battles were often chaotic affairs involving a great deal of individual combat and little in the way of careful tactical planning. On occasion, opposing forces might spend several days just trying to find each other so that a battle could take place. The military technologies of the Middle Ages thus led to a retreat from the principles of discipline and precise organization that had made Roman armies so formidable.

Horse-mounted combat also generated the culture of chivalry. The word itself derives from the French *cheval*, or horse. The horse-mounted knight was different from ordinary soldiers by virtue of his noble birth and the long period of training that was required to develop his proficiency in combat. The chivalrous knight lived by a code that regulated certain aspects of warfare. Noncombatants—provided that they were Christians—were to be left in peace, and combatants were supposed to adhere to specific rules of battle. Above all, war was seen primarily as a contest between members of a distinctive warrior class, in which individual glory could be as important as the attainment of military objectives. War still remained a grisly business, but as long as it was considered to be a contest between participants in a common culture, some limits were set on the manner in which it was waged, and the land was usually spared the total devastation that has been a hallmark of more modern conflicts.

New Weapons and the Decline of Feudalism

Knightly combat was only one aspect of medieval warfare, and usually not the most important one. Sieges were the most common military engagements, while in pitched battles foot soldiers outnumbered mounted knights by five or six to one.[8] Cavalry charges against numerically superior infantrymen were rarely successful and beginning in the fourteenth century the utility of the mounted warrior was further undermined by the widespread deployment of the pike. Used to especially good effect by the Swiss, the 19-foot-long pike was wielded by a solid formation of soldiers with four rows of pikemen extending their weapons to the front line—a formation that resembled the ancient Greek phalanx in many respects. United by an iron discipline and a high morale that came from being free men in the service of their canton (a mini-state that often had its own army), the Swiss pikemen became the most feared soldiers in late medieval Europe. Hard-charging knights were dispersed by pikemen and then assaulted by other soldiers wielding halberds (long battle axes).[9]

The vulnerability of mounted knights was further increased by the use of another infantry weapon, the longbow. Archers had been part of combat since Neolithic times.[10] But they were limited by the design of the weapons they used. The bow was transformed into a decisive weapon simply by making it longer. At the same time, however, the longbow was a difficult weapon to use properly, for it required considerable strength and skill, and that came only with extensive practice.[11]

First used by the Welsh during their resistance to English conquest, the longbow was adopted by King Edward I during the early fourteenth century. In skilled hands, this bow, which was six feet or more in length, could be a devastating weapon, capable of rapidly firing arrows that could do damage up to 650 feet away.[12] Drawn back to the ear instead of the chest, as was the case with smaller bows, the longbow was capable of driving an arrow through several inches of oak, and although it could not penetrate the best plate armor, it could pierce chain mail or the joints in plate armor.[13] Horses, having less armor protection, were especially vulnerable, and when a horse went down its rider was in serious trouble. The prowess of English archers was clearly demonstrated at the Battle of Crécy in 1346. Despite a two-to-one numerical inferiority, the English forces won a decisive victory. A rain of arrows penetrated the chain-mail armor of the French knights and wrought even greater destruction on their horses.

The mounted knight was also threatened by an eleventh-century invention, the crossbow. A mechanically sophisticated device, the crossbow used a lever or a crank-and-ratchet assembly to draw the string. The crossbow had a lot of stopping power, as it fired a bolt that weighed a half pound or more. Its range and accuracy were superior to that of the longbow, but the cumbersome process of winding it restricted its rate of fire to only two bolts per minute. It was thus most effective when the archer could find cover where he could safely reload.[14] Despite this deficiency, the crossbow was a formidable weapon that made the life of the knight all the more precarious.

One answer to these new offensive weapons was the replacement of chain mail by plate armor. This armor became increasingly thick, until by the late sixteenth century a mounted knight was weighted down with as much as a hundred pounds of armor. His horse, too, required protection, for a felled horse often meant death for its dismounted rider. The result was a crippling loss of mobility for the knight and his steed. Maneuverability and speed, hitherto the great virtues of the mounted warrior, were lost, and the knight ceased to be the mainstay of the European battlefield. Much later, mounted soldiers played a significant role in combat, not as lance-wielding knights but as cavalry armed with pistols and sabers. If carefully coordinated with artillery salvos, a cavalry unit could force the enemy to maintain solid ranks, thus presenting a convenient target for artillery.[15] But the traditional knight was no longer a major player in the game. At best he was an auxiliary, at worst a vulnerable target.

The castle, the other great symbol of the Middle Ages, was also challenged by new military technologies. Early medieval fortifications were little more than mounds of earth surrounded by wooden walls, a far cry from the elaborate defenses constructed by the Romans. During the Crusades, Europeans learned a great deal by observing the castles constructed by their Muslim opponents, and this knowledge diffused through Europe after they returned home.[16] Throughout the early Middle Ages, castles stood as virtually impregnable redoubts that conferred as much security as could be hoped for during those turbulent times; about the only way to conquer a fortified position was to starve it into submission.

Attackers attempted to breach the walls of castles and other fortifications with weapons little changed from Roman times, until in the twelfth century besieging armies began to employ a device known as a trebuchet. This weapon consisted of a long arm that was unequally balanced on a fulcrum. The long end, which held the projectile, was held in place by a catch, while the short end was loaded with heavy weights. When the catch was released, the long arm flew up and hurled the projectile in the direction of the enemy. These could be formidable weapons, capable of delivering a 300-pound projectile a distance of up to 300 yards and weights of more than a ton for shorter distances.[17] Although some trebuchets had less range than the catapults of antiquity, the greater weight of the projectile made it a more effective siege weapon.[18] Also, since it did not depend on torsion springs made from hair as the older missile launchers did, it was effective in all kinds of weather. While improved siege engines were not always up to the task of battering down sections of walls, they could be used to demoralize a castle's inhabitants by lobbing incendiaries, live snakes, and dead horses over the walls. And on occasion they were used to speedily return prisoners to their place of origin over the walls.[19]

Figure 18.2 The medieval trebuchet used a counterweight and some human assistance to hurl a projectile placed at the end of the long lever arm. (Source: EMLACH/CC BY 4.0)

The Gunpowder Revolution

The crossbow, longbow, pike, and trebuchet posed a severe challenge to the knight and the feudal system that maintained him. New weapons based on gunpowder sealed his fate. Incendiary weapons had long been used in combat. Flaming arrows; vases filled with a mixture of pitch, sulfur, and boiling oil; and primitive grenades filled with naphtha were all used in battle. One of the most terrifying weapons used in antiquity was Greek fire. Although its ingredients are a matter of debate even today, there can be little doubt of its effectiveness. Like modern-day napalm, it clung to whatever it came into contact with and burned fiercely, even on water. First used in seventh-century Byzantium, it served as a decisive weapon in naval engagements and as an antisiege weapon. Despite the tactical advantage that it offered, Greek fire was used only sparingly, in part because it was more effective as a defensive weapon than as an offensive one. No less important, the rulers of Byzantium were reluctant to make widespread use of it for fear that it would fall into enemy hands, a concern that has limited the diffusion of many other "secret weapons" throughout history.[20]

The First Gunpowder Weapons

Early incendiary devices were generally employed as auxiliaries to armies whose main weapons continued to be swords, spears, bows, and crossbows. With the invention of gunpowder, the nature of warfare underwent a profound transformation. As with many of the inventions that transformed medieval European society, gunpowder originated in the East. First used for medicinal purposes by its Chinese inventors, gunpowder became the propellant for simple rockets during the Tang Dynasty (AD 618–907) in the eighth century.[21] In succeeding dynasties these were complemented by an impressive array of gunpowder weapons, starting with incendiary arrows and bombs, and followed by increasingly sophisticated cannon and small arms.[22]

When first used in Europe during the early fourteenth century, gunpowder was used to fire large arrows and stone cannonballs. At first, these artillery pieces were little more than a nuisance. They had less destructive power than a good trebuchet, and their chief advantage was that they could be built more cheaply and be more easily transported than mechanical missile launchers. Their barrels were often made of wood; when iron was used, the barrel was made from parallel stripes that were welded together. In neither case could the cannon withstand repeated firing, and their projectiles were stone balls that often shattered when fired.

Artillery began to be used against fortifications by about 1420, and by 1453 Turkish attackers successfully used cannon to batter down the walls surrounding Constantinople prior to taking the city. In the succeeding years, a number of improvements resulted in larger, more accurate and reliable weapons. Iron and bronze gun barrels were cast through the use of techniques that had been used for the production of church bells—an ironic application of a peaceful technology to a martial one. Iron cannonballs replaced stone ones. "Corned" gunpowder, which was made into coarse grains, produced a more consistent and rapidly burning charge. As a result of these innovations, by the middle of the sixteenth century the fortified castles that had long been immune to direct attack were now vulnerable.

Still, the age of the castle was by no means at an end. Military engineers followed earlier Chinese practices by backing castle and town walls with large amounts of dirt, which absorbed much of a cannonball's force, and by constructing bastions protruding from castle walls, from which their own cannon could be fired in multiple directions. Such fortifications provided a good deal of security until mobile artillery developed during the late eighteenth century concentrated the firepower of several cannon, once again tipping the balance in favor of the besiegers. Until then, castles conferred a great deal of security against attack and did much to preserve the independence of smaller states, thereby preventing the political consolidation of Europe.[23]

Smaller, more mobile artillery pieces were also used to good effect on the battlefield. Of particular importance were the innovations of Sweden's Gustavus Adolphus during the Thirty Years' War (1618–1648). Instead of deploying the large field pieces commonly used, Gustavus relied on cannons that fired nine- or four-pound balls. Light in weight, these could be handled by a small crew and rapidly deployed during the course of battle. Artillery took on its classic role in battle: softening up the enemy's lines in preparation of a charge by the cavalry and infantry.

As we have seen, technological advances often occur through the expansion of a small-scale device or process. In the case of firearms, however, the process was reversed. Handguns were at first nothing more than small cannon, lacking even a stock that would allow them to be tightly held and accurately aimed, and it was not until a century after the introduction of cannon that useful handheld firearms began to appear in battle. Still, they had many shortcomings. Large artillery pieces were aimed at a fixed target and then had their charge lighted through a touchhole. In contrast, small guns had to be used against rapidly moving targets. They could not be aimed with any degree of precision if a soldier had to shift his gaze to the weapon's touchhole. The answer to this problem was the matchlock, which used a smoldering wick held by a trigger-operated clamp to ignite the powder. Even so, firing this gun was a cumbersome process, requiring no fewer than 28 separate procedures to load, fire, and clean the gun.[24]

Despite these shortcomings, by the first quarter of the sixteenth century small arms had graduated from being battlefield auxiliaries to indispensable articles of combat.[25] Subsequent improvements made firearms all the more significant. The cumbersome matchlock eventually gave way to the wheellock and then the flintlock musket, which was capable of firing three rounds per minute.

The firepower and accuracy of these weapons greatly augmented the offensive power of armies, provided that they were used in conjunction with proper battlefield procedures. European, Chinese, Korean, and Japanese military commanders learned to deploy their soldiers in rows, so that a soldier could fire his weapon and then retreat to the back row where he could reload in relative safety. Still, troops could not always be depended on to use their weapons effectively. Although a musket could fire a ball several hundred yards, it was accurate only to a distance of about 80 yards. The familiar "don't fire until you see the whites of their eyes" was a wise admonition, but one difficult to adhere to under combat conditions. In the heat of battle, soldiers were inclined to fire prematurely, and many were so distracted that they completely botched up the necessary routines used for loading these single-shot weapons, as one report from the

American Civil War indicates. The official report of the examination of the arms collected upon the battlefield of Gettysburg, states that:

> "Of the whole number received, 27,574, we found at least 24,000 of these loaded; about one-half of these contained two loads each, one-fourth from three to ten loads each, and the balance one load each. In many of these guns from two to six balls have been found, with only one charge of powder. In some, the balls have been found at the bottom of the bore with the charge of powder on top of the ball. In some cases as many as six paper regulation caliber '58 cartridges have been found, the cartridges having been put in the gun without being torn or broken (preventing them from being exploded by the percussion cap). Twenty-three loads were found in one Springfield rifle-musket, each loaded in regular order. Twenty-two balls and 62 buckshot with a corresponding quantity of powder, all mixed up together, were found in one percussion smooth-bore musket."[26]

This sort of bungling could be expected, given the panic and paralysis often experienced by men in combat. In order to counter it, successful military leaders had to institute rigid battlefield procedures that had been deeply imbued through countless parade-ground exercises. The numerous steps required for the effective use of firearms were incessantly driven into the troops until they became virtually automatic, even in the face of withering attacks from the enemy. In complete contrast to the unruly egocentricity of knightly combat, warfare had become a routinized procedure, and soldiers "became replaceable parts of a great military machine just as much as their weaponry."[27]

Earlier chapters have stressed the complementarity of organizational and technological changes, and nowhere is this better illustrated than in the military realm. A well-drilled army always has an advantage over one that is lax in its ways, and the technological development of warfare made this even more the case. The handling of firearms required a precise series of operations; incessant drill was required if soldiers were not to be totally unhinged during the course of battle. Adhering to an iron discipline created by constant drilling, and blindly obedient to a rigid chain of command, European armies of the eighteenth century had become some of the most grimly efficient organizations the world had ever seen.

The creation of these armies had a significance that transcended their importance in the military sphere, for they served as the prototype of new forms of civilian organization: routinized, regimented, and hierarchical. Obedience to authority is nothing new in human societies; it is, after all, one of the most important lessons learned by young children. But this sort of authority is embodied in a particular person—first a parent, and then another authority figure. Soldiers of this era obeyed the authority of a king and his delegates, but, equally important, they had to submit to standard procedures drummed in through constant repetition. In time, churches, schools, and factories made abundant use of these principles, and even in our own times this mode of organization is predominant in many areas of life.

War and the Centralized State

Artillery and firearms expanded the scope of warfare. Battles were more complex, organized affairs, while at the same time sieges became more elaborate. Both forms of warfare intensified logistical problems; according to one calculation, maintaining an army of 50,000 required the daily provision of 475 tons of food for soldiers and horses.[28] Bullets and powder also had to be readily available. This meant that an army required long supply lines, giving rise to continual strategic problems of maintaining these lines, as well as considerably enlarging the zone of military operations.

New forms of warfare greatly increased the cost of military campaigns. The feudal nobility had the means to pay for their own arms and armor, and a longbow did not entail large expenditures. In contrast, large contingents of musketeers and artillerymen required heavy financial outlays if they were to be adequately trained and equipped. In most cases only the monarch of a country could meet these demands.[29] At the same time, firearms gave centralized states a decisive advantage in expanding their territory and increasing control throughout their domain.[30] New weapons technologies thus reinforced the rising tide of nationalism, centralization, and royal absolutism that characterized European history from the sixteenth century onward. The converse was also true; the growth of the nation-state generated a large demand for improved weaponry. As Charles Tilly has summed up this positive feedback loop, "War made the state and the state made war."[31] The active involvement of centralized governments was also evident in the standardization of weaponry. Prior to the seventeenth century, most soldiers did not wear uniforms and the weapons they used were of great variety. Successful military leaders such as Gustavus Adolphus and Oliver Cromwell strove to standardize weaponry and the other accoutrements of war. Standardization, however, often acts as a brake on technological progress. Once a gun or cannon becomes an army's standard weapon, its replacement or substantial modification requires a multitude of parallel changes. Not only will the weapon's auxiliaries, such as bullets, have to be changed, but so will a host of procedures, as well as the culture that has grown up around it. Soldiers, like the rest of us, can become quite attached to doing things in a particular way, and they may be especially prone to adhere to established routines when engaged in such an inherently risky business as warfare.

Accordingly, the revolutionary changes brought on by the use of gunpowder lost momentum. During the eighteenth and early nineteenth centuries, improvements in weapons were few and far between. One important exception was the invention of rifling, cutting a spiral groove inside the barrel to impart a spin on a bullet, which greatly improved the accuracy and range of firearms, but they did not supplant the traditional musket. The "Brown Bess" flintlock musket remained the standard English infantry weapon for 160 years before it was replaced by breechloading rifles (in which the bullet is loaded from behind the barrel, rather than through the muzzle) during the mid-1800s. Artillery fared no better; according to A. Rupert Hall, "The guns of Queen Victoria's wooden ships were capable of little more accurate practice than those of Drake's fleet which defeated the Armada."[32]

Yet after decades of stagnation, military technologies changed dramatically during the long reign of Queen Victoria (r. 1837–1901). One important invention was the Minié bullet that expanded as it was fired, making it easier and quicker to put a bullet in muzzle-loaded firearms because they did not have to be laboriously rammed into the bore. The later development of breechloading weapons further accelerated the speed of loading while having the added benefit of allowing soldiers to remain prone while reloading instead of standing upright, where they presented a conspicuous target. Improved powders produced higher muzzle velocities, allowing the use of smaller bullets and lighter firearms. These new powders also had the advantage of being smokeless, so a soldier did not give away his position when he fired. But none of these changes can be attributed to developments exclusively within the realm of weapons production; they were products of an emerging industrial order, and as such, they will be considered in greater depth in Chapter 19.

Technological Change and Naval Culture in the Era of the Battleship

While land warfare in the nineteenth century was being transformed by the use of new weapons, a similar revolution was taking place on the sea. One key element of this revolution was the use of steam engines to propel ocean-going vessels so they were no longer affected by the vagaries of the winds. Moreover, steam propulsion was the perfect complement to steel construction, since the new energy source allowed the construction of much larger and heavier ships. The ships used by England's Admiral Nelson at the beginning of the nineteenth century displaced 2,000 tons at most; by the 1860s, such ships were up to 9,000 tons displacement, and by the end of the century they displaced 20,000 tons.[33]

Ships of this size were able to carry a formidable array of guns, made more accurate and rapid firing by rifling, breechloading, and improved powders and projectiles. All of these developments reinforced a trend that had begun when significant numbers of cannons first began to be mounted on ships during the sixteenth century. Before this time, the traditional pattern of naval combat had been based on closing with the enemy vessel (ramming it if possible), boarding it, and then subduing its crew through the naval equivalent of battlefield combat. Ship-mounted cannon made these tactics obsolete; opposing ships now attempted to sink or disable their foe through long-distance cannon fire. Countries such as Spain and the city-states of Italy that clung to the old methods of combat put themselves at a great disadvantage, and over time the naval balance of power began to shift to England and the other nations that embraced the use of shipboard artillery.[34] By the beginning of the twentieth century, the steam-powered battleship with its collection of heavy ordnance had become the very embodiment of modern weaponry, the most fearsome military artifact the world had ever seen. Embodied by Britain's *Dreadnaught*, battleships were more than simply weapons systems; they came to symbolize a nation's armed might for both traditional European naval powers and aspiring ones like the United States and Japan.

As with any profound technological change, the development of the battleship presented numerous problems of adjustment. Modern navies became dependent on the availability of coaling stations located along their areas of operation, and this gave a strong impetus to imperial expansion, an expansion that was facilitated by growing naval power. Navy officers basked in the reflected glory of modern warships, but at the same time, the modern naval vessel forced changes in military routines that were accepted with great reluctance. In the days of sail, a seaman's life revolved around a unique set of routines that had imbued navies with a distinct culture. Steam changed many of these; many sailors were now little more than floating factory workers—stoking coal, monitoring gauges, and keeping the machinery in good repair.[35] The result was a serious erosion of traditional maritime virtues, and the eventual replacement of old-guard officers with more technically minded ones. Just as the longbow and the gun destroyed the medieval knight and the culture of chivalry, advances in naval technology put an end to a way of life that had been common to generations of sailors.

Weapons and the Making of the Modern World

All these military innovations changed the conduct of war markedly. Although the full implications of new military technologies were not always quickly grasped by military planners in Europe, they were put to devastatingly good use abroad by the European armies and navies that brought large portions of Asia and Africa under imperialist domination. Despite China's centuries of military parity with the West, by the late eighteenth century it had fallen far behind. The technological superiority of Western military forces was clearly demonstrated during the first Opium War (1839–1842), when the cannon of British gunboats battered down Chinese fortifications and the superior firepower of British soldiers routed a numerically superior Chinese army. A single British ship armed with two 32-pound cannons destroyed nine war junks, five forts, a shore battery, and two military stations—all in one day.[36] China, the originator of gunpowder and many other advanced martial technologies, had fallen woefully behind the Western world —and paid the price in its subjugation by the West.

By the second half of the nineteenth century, the technological gap between the Western powers and the rest of the world had grown even larger. Improved weaponry and military auxiliaries like the steamboat, railroad, telegraph, and medicines for the treatment of tropical diseases sharply divided the world into winners and losers. Nations with industrial economies were able to expand their domains to the point that, by 1914, European nations directly or indirectly controlled most of the world's landmass. The disparity in military capability was most marked in Africa, where breechloading repeating rifles allowed small military contingents to prevail over native troops that at best were armed with flintlock muskets, and often with only spears and shields. Numerous instances could be cited of a few hundred European soldiers utterly routing native armies that numbered in the thousands.[37] To take the most prominent example, in 1898 at the Battle of Omdurman in the Sudan, a British force, assisted by a gunboat on a nearby river, confronted 40,000 Dervishes. After five hours of combat, 48 British soldiers and their Egyptian allies had been killed. For the Dervishes, the battle was more costly; more than 10,000 of them lay dead.[38]

The most effective weapon of European imperial advance was the machine gun. Used sparingly during the American Civil War, the machine gun came into its own during the colonial wars of the late nineteenth and early twentieth centuries, when a single gun crew could cut to shreds any native force that attacked them. But little did the colonial powers realize that this product of Western ingenuity would soon be used against their own kind. Smug with the belief that the machine gun took its lethal toll only when directed against "lesser breeds of men," Europeans were slow to realize that the machine gun had transformed the conditions of warfare.[39] That realization came with frightening force during the first years of World War I, as generals on both sides continued to hurl their troops against enemy lines fortified by machine gun nests. The result was an appalling slaughter, as attested by 60,000 British casualties during *the first day* of the Battle of the Somme.

While the machine gun was rewriting the rules of land warfare, the submarine was doing the same thing to war on the sea. Slipping beneath the ocean's surface and firing its torpedoes at enemy ships blind to its presence, the submarine defied all of the conventions of naval combat, as well as the international laws that covered the attack of commercial vessels. Instead of adhering to established prize rules that stipulated that passenger and crew be allowed to board lifeboats before the sinking of their ship, submarines remained beneath the surface and summarily sunk their unsuspecting prey. The submarine was the centerpiece of Germany's naval strategy, and during World War I German U-boats wreaked havoc on Allied shipping. Germany's success was fleeting; unrestricted submarine warfare ultimately brought the United States over to the side of Britain and France in 1917, decisively tipping the military balance in their favor.

At the same time that improved artillery and the machine gun were bringing new horrors to the battlefield, the airplane was beginning to demonstrate what modern technologies could do to civilian populations. First confined to observation and reconnaissance, by the third year of World War I aircraft were being employed as strategic weapons. Bombs dropped on England from German airplanes and Zeppelins killed about 1,400 people and injured another 3,400, as well as doing a fair amount of material damage.[40]

Later wars were to eclipse these figures by huge margins. Even before the first atomic bomb was dropped, American bombing attacks

Figure 18.3 A World War I machine gun crew wearing gas masks deployed during the catastrophic the Battle of the Somme.

Figure 18.4 The atomic bomb killed an estimated 140,000 people in Hiroshima, Japan on August 6, 1945. Approximately 80,000 died instantly, with the rest succumbing to radiation exposure over ensuing months. (Source: National Archives)

on Japanese cities during World War II killed at least 260,000 and injured 412,000, and destroyed 40 percent of 66 cities' built-up areas.[41] A single incendiary attack on Tokyo destroyed more than a quarter-million buildings, killed 84,000, injured more than 40,000, and left more than a million homeless.[42] On August 6, 1945, the first atomic bomb was dropped on Hiroshima, killing an estimated 80,000 people, injuring as many more, and destroying half of the city. A second A-bomb had similar results when it was detonated over Nagasaki three days later.[43]

Aerial bombardment also devastated much of Germany. Tens of thousands of bombs were dropped on urban industrial centers, leaving thousands dead and many more homeless. A single raid on Berlin that took place toward the end of the war may have killed as many as 25,000 civilians.[44] Yet it is evident that for all the destruction it wrought, the bombing of industrial centers with large civilian populations during World War II was not the decisive factor in winning the war in Europe. In part, this was due to the inaccuracy of strategic bombing; only about 30 percent of the bombs dropped from American bombers landed within 1,000 feet of their intended targets.[45] Although large portions of cities were leveled and many industrial plants were damaged or destroyed, German industry was never brought to its knees. Many factories were put quickly back into operation after apparently having been destroyed, and substitutes were found for components that could no longer be produced. German industrial production increased each year until 1944, reaching its highest level in the year when Allied bombardment was at its greatest. Despite massive bombardment in 1944, German industry produced three times as many military aircraft, five times as many armored vehicles, and eight times as many artillery weapons as it had in 1941.[46]

Equally important, civilian morale never broke down in the face of continual air raids. It may even have been the case that bombardment resulted in a stiffening of the German and Japanese resolve to continue the war. Although there is ample room for debate concerning its contributions to winning World War II, there is little doubt that strategic bombing failed to live up to the extravagant claims made by its proponents prior to the war. The official U.S. Strategic Bombing Survey that was conducted after the war noted the accomplishments of strategic bombing, but it also called attention to its costs and shortcomings. In Walter Millis' words, "It would probably be fair to summarize the net conclusion as showing that strategic bombing, when armed only with TNT and incendiary weapons, had involved much greater (sic) casualties and had produced much smaller military results than had been expected."[47]

Part of the explanation for this ineffectiveness lies in the nature of long-term bombing attacks. According to B. H. Liddell Hart's analysis, strategic bombing with conventional weapons puts pressure on an enemy but does not produce the decisive result that comes with a sudden shock. Instead of producing quick, widespread results, strategic bombing, even if it results in considerable devastation, only increases pressure slowly. But, as Liddell Hart noted, "Human beings have an almost infinite power of accommodation to degradation of living conditions, so long as the process is gradual."[48]

Yet once again technology has altered the rules of war. Missiles with nuclear warheads are shock weapons of the greatest magnitude. A nuclear war would be like no other war of the past and would result in death and destruction on a scale that can scarcely be imagined. A single one-megaton bomb would dig a thousand-foot-wide crater to a depth of 200 feet. No structures would be standing within a radius of nearly two miles, and damage would be heavy for a considerable distance beyond this. If the area had a normal urban population density, at least 200,000 people would be killed immediately, and half a million would be injured.[49] Fires would spread through the city, killing thousands more. After this initial shock, radiation would spread through the area, killing many more in a few weeks or months. And, of course, this would not be an isolated attack. Through the late 1980s, the arms race between the United States and the Soviet Union culminated in the two countries collectively amassing an arsenal of over 4,000 nuclear-tipped ballistic missiles, along with hundreds of bombers and cruise missiles with nuclear capabilities. If even a small number of them had been used in anger, devastation would have been immense.[50] The basic structures of society would undoubtedly have snapped under the strain, leaving civilization perched on the threshold of chaos and barbarism. Had this happened, it might have been better if the long history of technological advance had never occurred.

Wars of the past, whatever their horrors, at least left some room for individual valor; a nuclear war would be depersonalized, push-button destruction on a massive scale. In this way, nuclear warfare would be a culmination of an important trend of military technology. Increasingly sophisticated weapons have opened the psychological distance between the warrior and his victims. To some extent at least, modern warfare can be seen as an extreme example of the rational mindset discussed in Chapter 1, the psychological distancing of oneself and one's actions from the people and things that suffer the consequences of these actions. In the past, the inhabitants of a village might be slaughtered in face-to-face conflict by troops wielding swords and axes; today the same thing can be accomplished by dropping an incendiary bomb from an altitude of 40,000 feet. The result is the same (or worse), but soldiers, sailors, and aviators, as well as the engineers who have designed their weapons and the political leaders who sent them into battle, are removed from the fray. As Charles Lindberg reflected after flying a World War II bombing sortie:

> You press a button and death flies down. . . . How can there be writhing, mangled bodies? How can this air around you be filled with unseen projectiles? It is like listening to a radio account of a battle on the other side of the earth. It is too far away, too separated to hold reality. . . . In modern war one kills at a distance, and in so doing he does not realize that he is killing.[51]

QUESTIONS FOR DISCUSSION

1. Can any equivalents of medieval knights be found today? In what ways do they resemble medieval warriors? How do the technologies they use shape their activities and attitudes?
2. To make effective use of firearms, soldiers had to be well-drilled and subjected to rigid discipline. How have military models of organization influenced nonmilitary social institutions? What are the advantages and disadvantages of military styles of organization when they are applied elsewhere?
3. For all of its horrors, war historically has provided opportunities for people to demonstrate some real virtues, such as resourcefulness, initiative, and courage. Have modern military technologies made these virtues irrelevant? If so, does the loss of these virtues make war more absurd than ever before?
4. Military analysts sometimes refer to a military force's ratio of "teeth to tail," that is, the ratio of combat personnel to support staff. Throughout the past and present centuries, the ratio of the latter to the former has steadily increased. Why do you think this has happened? What does it tell you about the nature of modern warfare?
5. To what extent, if any, should an engineer feel a sense of individual responsibility for designing a weapon that is used to kill large numbers of people? What about the combatant who is ordered to use the weapon? Do the citizens whose taxes were used to design and build the weapon also bear some responsibility?

Notes

1. Victor Davis Hanson, *The Western Way of War: Infantry Battle in Ancient Greece* (New York: Knopf, 1989).
2. Stephen Morillo, Jeremy Black, and Paul Lococo, *War in World History: Society, Technology, and War from Ancient Times to the Present* (New York: McGraw-Hill, 2009) p. 59.
3. James K. Finch, *Engineering and Western Civilization* (New York: McGraw-Hill, 1951), p. 22.
4. Doyne Dawson, *The Origins of Western Warfare: Militarism and Morality in the Ancient World* (Boulder, CO: Westview, 1996), p. 112.
5. William Reid, *The Lore of Arms: A Concise History of Weaponry* (New York: Facts on File, 1984), p. 15.
6. E. W. Marsden, *Greek and Roman Artillery: Historical Development* (Oxford: Clarendon Press, 1969), pp. 86–91.

7. Robert Laffont, *The Ancient Art of Warfare*, vol. 1 (Greenwich, CT: New York Graphic Society, 1968), pp. 121, 128–129.
8. Geoffrey Parker (Ed.), *The Cambridge Illustrated History of Warfare: The Triumph of the West* (Cambridge: Cambridge University Press, 1995), p. 84.
9. Charles William Chadwick Oman, *A History of the Art of War: The Middle Ages from the Fourth to the Fourteenth Century* (London: Methuen, 1898), pp. 73–115.
10. Hugh D. H. Soar, *The Crooked Stick: A History of the Longbow* (Yardley, PA: Westholme, 2004), pp. 1–17.
11. Archer Jones, *The Art of War in the Western World* (New York and Oxford: Oxford University Press, 1987), pp. 156–157.
12. Gervase Phillips, "Longbow and Hackbutt: Weapons Technology and Technology Transfer in Early Modern England," *Technology and Culture* 40, 3 (July 1999): 579.
13. Richard A. Preston, Sydney F. Wise, and Herman O. Werner, *A History of Warfare and Its Interrelationships with Modern Society* (New York: Frederick A. Praeger, 1956), p. 85.
14. Vernard Foley, George Palmer, and Werner Soedel, "The Crossbow," *Scientific American* 252, 1 (January 1985): 104–110.
15. Laffont, *The Ancient Art of Warfare*, p. 444.
16. Darke, Siana (2010). *Syria*, see especially p. 204. Bradt Travel Guides. See also: Howard, Jonathan (2011). *The Crusades: The History of One of the Most Epic Military Campaigns of All Time*. Golgotha Press.
17. Lynn White, Jr., *Medieval Technology and Social Change* (New York: Oxford University Press, 1966), p. 102.
18. Paul E. Chevedden, Les Eigenbrod, Vernard Foley, and Werner Soedel, "The Trebuchet," *Scientific American* 273, 1 (July 1995).
19. Martin van Creveld, *Technology and War: From 2000 B.C. to the Present* (New York: The Free Press, 1989).
20. Alex Roland, "Secrecy, Technology, and War: Greek Fire and the Defense of Byzantium, 678–1204," *Technology and Culture* 33, 4 (October 1992).
21. China Science and Technology Museum, *China's Ancient Technology* (Beijing: China Reconstructs Magazine, 1983), pp. 18–20.
22. Tonio Andrade, *The Gunpowder Age: China, Military Innovation, and the Rise of the West* (Princeton: Princeton University Press, 2016), pp. 29–87, 135–187.
23. William H. McNeill, *The Pursuit of Power: Technology, Armed Force, and Society since A.D. 1000* (Chicago: University of Chicago Press, 1982), p. 91.
24. For a listing of all of these procedures, see Kenneth Chase, *Firearms: A Global History to 1700* (Cambridge: Cambridge University Press, 2003) p. 25.
25. Michael Howard, *War in European History* (London: Oxford University Press, 1976), p. 33.
26. Quoted in Walter Buehr, *Firearms* (New York: Thomas Y. Crowell, 1967), pp. 134–135.
27. McNeill, *The Pursuit of Power*, p. 141.
28. van Creveld, *Technology and War*, p. 107.

29. H. W. Koch, *The Rise of Modern Warfare: From the Age of Mercenaries through Napoleon* (New York: Crescent, 1982), p. 18.
30. Richard Bean, "War and the Birth of the National-State," *Journal of Economic History* 33, 1 (March 1973): 203–221.
31. Charles Tilly, "Reflections on the History of European State-Making," in Charles Tilly (Ed.), *The Formation of National States in Western Europe* (Princeton, NJ: Princeton Univ. Press, 1975), p. 42.
32. Quoted in Bernard Brodie and Fawn Brodie, *From Crossbow to H-Bomb* (Bloomington: Indiana University Press, 1973), p. 61.
33. Howard, *War in European History*, p. 123.
34. Carlo M. Cipolla, *European Culture and Overseas Expansion* (Harmondsworth, England: Penguin, 1970), pp. 70–72.
35. Elting E. Morison, *From Know-How to Nowhere: The Development of American Technology* (New York: New American Library, 1977), pp. 139–152.
36. Geoffrey Parker, *The Military Revolution: Military Innovation and the Rise of the West, 1500–1800* (Cambridge: Cambridge University Press, 1988), p. 154.
37. See Daniel R. Headrick, *The Tools of Empire: Technology and European Imperialism in the Nineteenth Century* (New York: Oxford University Press, 1981), pp. 117–119.
38. Morillo, Black, and Lococo, *War in World History*, pp. 468–469.
39. John Ellis, *The Social History of the Machine Gun* (Baltimore: The Johns Hopkins University Press, 1975), pp. 111–147.
40. James L. Stokesbury, *A Short History of Air Power* (New York: Morrow, 1986), p. 78.
41. Brodie and Brodie, *From Crossbow to H-Bomb*, p. 224.
42. Theodore Ropp, *War in the Modern World* (New York: Collier Books, 1962), p. 379.
43. "The Atomic Bombings of Hiroshima and Nagasaki," *Atomic Archive.com*. Accessed from www.atomicarchive.com/Docs/MED/med_chp10.shtml.
44. Noble Frankland, *Bomber Offensive: The Devastation of Europe* (New York: Ballantine, 1970), p. 149.
45. Stephen L. McFarland, *America's Pursuit of Precision Bombing, 1910–1945* (Washington and London: Smithsonian Institution Press, 1995), p. 203.
46. Ian Hogg, *The Weapons That Changed the World* (New York: Arbor House, 1986), p. 144.
47. Walter Millis, *Arms and Men: A Study in American Military History* (New York: New American Library, 1956), p. 277.
48. B. H. Liddell Hart, *The Revolution in Warfare* (London: Faber and Faber, 1946), p. 25.
49. See: https://science.howstuffworks.com/tsar-bomba.htm on the most powerful nuclear weapon, and www.atomicarchive.com/science/effects/energy.html as a primer of the science of nuclear weapons.
50. Randall Forsberg, "A Bilateral Nuclear-Weapon Freeze," *Scientific American* 247, 5 (November 1982): 61.
51. Quoted in Michael S. Sherry, *The Rise of American Air Power: The Creation of Armageddon* (New Haven, CT: Yale University Press, 1987), pp. 209–210.

Chapter 19

The Era of Smart Weapons

The last chapter noted that advances in military technologies have created a spatial and often a psychological gulf that has separated combatants from the consequences of their actions. This trend has intensified in the opening years of the twenty-first century. To be sure, much of the burden of fighting still falls upon infantry soldiers close to the action, but a great amount of killing and destruction is now done by remote control. Moreover, although many combatants were detached from the consequences of their actions in past wars, their own lives were at risk. In World War II, aerial combat was a highly dangerous enterprise in which 19,876 American aviators lost their lives in the European Theater of Operation alone, a toll greater than the number of U.S. Marines killed in the Pacific Theater.[1] Today's weapons have removed much of that risk, but their use has been the source of many strategic, political, legal, and ethical issues.

Cruise Missiles

Although some of its functions have been taken over by unmanned aerial vehicles (see below) cruise missiles remain crucial components of a modern nation's arsenal. Cruise missiles are relatively small pilotless craft that can be launched from land, submarines, surface vessels, and aircraft. Through the use of sophisticated navigation devices, cruise missiles can home in on their target while eluding radar detection by flying close to the ground. Some cruise missiles are capable of supersonic speeds, and many can deliver nuclear weapons.

As with many other technologies, the concept underlying cruise missiles has been around for a long time, but its practical realization required the development of many complementary technologies. Back in 1917, Charles Kettering—who later headed General Motors research laboratories—demonstrated an "Aerial Torpedo" that came to be known as the "Kettering Bug." Built largely out of papier-mâché with cardboard wings, the Bug was powered by a 40-horsepower engine and carried 300 pounds of explosives. Although several dozen were manufactured, none saw action during World War I. The first operational cruise missile appeared during World II in the form of Germany's V-1. Powered by a simple pulsejet engine and stabilized by a gyroscope, the V-1 had a top speed of about 400 mph and a range of 200 miles. From 1944 to the end of the war, about 10,000 were launched at England, of which 2,410 reached London, where they killed more than 6,000 and injured nearly 8,000.[2] Unlike the rocket-powered V-2, the V-1 flew at subsonic speeds, which made it vulnerable to attacks by fighter pilots, who either shot it down or sent it out of control by putting their plane's wingtip under its wing and flipping it.

Although a single cruise missile can cost more than a million dollars, their use has been ordered on several occasions by American presidents who wanted to take action against a country or an insurgent group but were reluctant to put American soldiers, sailors, and airmen in harm's way. On several occasions, cruise missile strikes were directed against Serbian forces during the conflicts that followed the breakup of Yugoslavia in the early 1990s. Since then, the major theater of operations for cruise missiles has been the Middle East. During the 1991–92 Gulf War, which was waged to repel Iraq's occupation of Kuwait, 288 Tomahawk cruise missiles were launched. Cruise missiles were also fired from U.S. Navy vessels in the Arabian Sea in 1998 to attack suspected Al-Qaeda sites in Sudan and Afghanistan in retaliation for the bombing of U.S. embassies in Kenya and Tanzania. Cruise missiles were used again against Iraq when 725 were launched by the United States when it invaded that country in 2003.[3] More recently, cruise missiles have been used by the United States, Russia, and France to attack Islamic State (ISIS) positions in Syria and Iraq.

Figure 19.1 The destroyer *Dewey* conducts a Tomahawk flight test while underway in the western Pacific. (US Navy/MCS 2nd Class Devin M. Langer)

Smart Bombs

As we saw in the previous chapter, the expectation that the strategic bomber would be the decisive weapon of war was thwarted by the inaccuracy of bombing attacks. What was needed was an unpowered explosive device that could be accurately guided to its target. Germany developed such a weapon during World War II, the FX-1400 "Fritz," a four-winged glide bomb that targeted armored ships. With a warhead containing 660 pounds of explosives, it was dropped from an airplane and steered by radio. In 1943 it succeeded in sinking an Italian battleship that was attempting to defect to Allied forces after Italy's surrender. It went on to sink or damage several more ships, but the vulnerability of the airplane carrying the flying bomb limited its deployment.[4]

During the 1950s advances in solid-state electronics created new possibilities for remote-controlled ordnance. The first "smart bombs," or "Precision Guided Munitions" (PGMs), to use the military's preferred term, employed a television camera to provide a bomb's-eye view of the terrain below. The bomb could then be guided by steerable fins as it zeroed in on its target. In the 1960s the emerging technologies of lasers and integrated circuits were employed in a new generation of PGMs. These achieved a fair measure of success during the Vietnam War, when strategic targets such as bridges that had withstood dozens of conventional bombing attacks were brought down by a few laser-guided bombs. The capabilities of PGMs were further demonstrated during the first Gulf War when television viewers got a real-time view of targets being destroyed by a smart bomb. These attacks made for good television, but in fact most bombing attacks used conventional "iron bombs."

Whether guided by television or a laser, PGMs were ineffective when adverse weather conditions obscured an intended target. This deficiency began to be overcome through the use of another emerging technology, Global Positioning Systems (GPS), for the guidance of a bomb's trajectory. Yet, as ever, defensive countermeasures were adopted; the intended victims of a PGM were often able to jam the radio signals guiding the bomb, forcing it to depend on its less accurate internal navigation system to guide it to its target.

Cruise missiles and smart bombs seem to have realized the prophecies made back in the 1920s and 1930s by the early apostles of air power. With the emergence of solid-state electronics, onboard computers, and GPS navigation, the dream of precision bombing has become a reality. But pinpoint accuracy is of no value if military intelligence fails to accurately identify a target's location. During the invasion of Iraq led by the United States in 2003, at least 50 "decapitation strikes" using precision-guided weaponry were launched with the intent of killing Saddam Hussein and members of his leadership cadre. None of the targeted individuals was hit, but dozens of civilians died or were injured as a result of these attacks.[5] Although his country's military was decisively routed after a few days of fighting, Saddam Hussein was able to evade capture for many weeks, until he was finally caught by ground troops acting on information derived from the interrogation of his former bodyguards and members of his family.

A tragic example of smart weaponry and defective intelligence came in 1988 when crewmen aboard the USS *Vincennes* in the Persian Gulf identified what they thought was an Iranian Air Force F-14A. Attempts at radio contact were unsuccessful, and under the impression that they were about to be attacked, two SM-2MR surface-to-air missiles were fired at the presumed attacker. The missiles accurately homed in on their target and destroyed what turned out to be a civilian Airbus A300 bound for Dubai. All 290 passengers and crew, including sixty-six children, were killed. Although there had been some ambiguity about the identity and intentions of the aircraft, it is evident that the crew of the *Vincennes* made critical mistakes when assessing the situation, errors that were compounded by the attitude of the ship's captain, who had a known penchant for aggressive and even reckless actions. The United States, while not apologizing for what had happened, eventually paid reparations amounting to $131.8 million. Two years after the incident, the Navy awarded the retiring captain with the Legion of Merit for his service in the Persian Gulf.

High-Tech Surveillance

The tragic fate of the crew and passengers on the airliner brought down by a "smart" weapon is one of untold numbers of breakdowns in military intelligence. Throughout history, military operations have been hampered by an inability to effectively identify and track an enemy. One technological response to this age-old problem has been the development of satellites capable of providing highly detailed views of the earth's surface. Although much of the information about the performance of these satellites is cloaked in secrecy, a general understanding of their capabilities can be derived from nonclassified sources. The first "reconnaissance satellites," to use their official name, used film cameras to record features on the earth's surface. The exposed film was ejected and captured in mid-air by specially equipped aircraft. Subsequent generations of satellites have used digital imagery that can be directly relayed to stations on earth. Satellite-based spying by the United States is the responsibility of the National Reconnaissance Office, an agency so secretive that its existence was not publicly acknowledged until 1992, and much of its budget is "black," known to only a small number of elected representatives and government officials.

At the outset, America's satellite program was aimed at the surveillance of the Soviet Union, although it was originally presented as a means of collecting scientific data. Officially known as the Discovery Project, this supposedly scientific endeavor was in reality a facade for the Corona spy satellite program, which over the years sent many highly detailed pictures of militarily significant sites.[6] In recent years, satellites and the ability to launch them have developed to the point where commercial firms are able to deploy satellites that can send back images nearly as good (and often at lower cost) as those produced by the National Reconnaissance Office. This has led to considerable disagreement within the government regarding how much satellite surveillance should be handed over to private firms.[7]

Drones

Although it is possible that the most current spy satellites are capable of retrieving images even when clouds obscure the earth's surface, their orbits limit the amount of time they can be within range of an area of interest. Consequently, they cannot be expected to provide up-to-the-minute images of particular locations. For these, the military has made increasing use of low-flying unmanned aircraft that are collectively known as drones, or Unmanned Aerial Vehicles (UAVs) in military parlance. Some of these craft are the size of a conventional airplane, while others are no bigger than a radio-controlled model airplane. Classified research is also being done on drones that are no bigger than a large insect but are equipped with sensors and miniaturized cameras that offer the prospect of getting inside a building and obtaining information about its interior.[8] In addition to being used for aerial surveillance, some drones are able to fire missiles, and others are intended for "suicide missions" in which they crash into a target.

The absence of a pilot and a support system means that UAVs can be lighter than a crewed airplane, allowing them to fly higher and at greater distances while staying aloft for lengthy periods of time. The use of reconnaissance drones goes back to the Vietnam War, where the Fire Fly drone flew 3,435 missions, but insufficient research was done to accurately adjudge their effectiveness.[9] Widespread awareness of drones and their capabilities came with the American invasion and occupation of Iraq. There, Predator drones logged hundreds of thousands of air miles annually. So extensive has been the use of UAVs in recent years that in 2009 Secretary of Defense Robert Gates stated that the U.S. Air Force was training more drone pilots than pilots of conventional aircraft.[10]

As noted above, in addition to their ability to gather accurate and up-to-date information, drones have the added attraction of serving as offensive weapons. Hellfire missiles carried by Reaper UAVs have been effectively used as antitank weapons. More significantly, drone strikes also have been used in Somalia, Yemen, Afghanistan, and the tribal areas of Pakistan to kill suspected Al-Qaeda and Taliban insurgents. These attacks have also resulted in the deaths of civilians, as admitted by John Brennan, the Obama administration's counter-terrorism advisor.[11] The extent of these deaths has been a matter of contention. One rough estimate, made in 2015 based on averaging casualty counts of three independent organizations, indicated that 522 drone strikes resulted in 3,852 deaths, of which 476 were civilians.[12] Thousands more civilians have been killed or injured since then, although how many attacks have been the work of the Taliban and other militant groups is in dispute.[13] One recent drone strike in fall 2022 was an assassination of a 9/11 perpetrator, Ayman al-Zawahiri, and U.S. DOD News sources assert that no other casualties were reported.[14] Although drone strikes are widely supported by the American public, many critics believe that they have engendered considerable anti-American sentiment within the Middle East[15] as

Figure 19.2 A widely used drone, the Reaper is 36 ft. long with a wingspan of 66 ft. It can carry 3,800 lbs. of reconnaissance gear or 14 Hellfire air-to-ground missiles. (U.S. Air Force photo/Staff Sgt. Brian Ferguson)

well as other places where drones have been deployed.16 Drones of various sizes and capacities are being deployed widely in the Russia-Ukraine War, perhaps making this one of the first "drone wars."[17]

One of the striking things about the use of drones as offensive weapons is that they are flown by pilots who are situated thousands of miles from their targets. Drone attacks on Afghanistan and Pakistan are controlled from Nellis Air Force Base in Nevada, where pilots guide their craft and fire their onboard missiles while sitting in front of television monitors. At a superficial level at least, this ability to engage in remote-control warfare has made combat seem like just another high-tech job, but this is hardly the case for drone pilots, who complain about fatigue, elevated levels of stress, and poor prospects for promotion.[18]

The Cost of Technological Sophistication

The ethical and legal issues engendered by the use of UAVs and other advanced military technologies will be taken up later in this chapter, but first we will consider a more mundane yet still significant aspect of high-tech weaponry, the financial burden that they place of governmental budgets. In the opening decades of the twenty-first century, the United States has at its command the most formidable collection of weaponry in human history. But do cruise missiles, smart bombs, imaging satellites, and UAVs guarantee security in an unstable world? Military strength is undeniably important, but a healthy economy is also essential for national well-being. The decision to orient government policies toward either military strength or a thriving civilian economy has been (erroneously) attributed to the nineteenth-century German Chancellor, Otto von Bismarck, as the choice of "guns or butter." From the 1940s onward, the United States has not stinted when it has come to the acquisition of guns and other weapons of war. In absolute terms, the defense budget of the United States is enormous. In 2024, for example, total Defense spending was set at $826.15 billion.[19] This amounts to about 4.35 percent of the country's Gross Domestic Product, less in percentage terms than defense expenditures at the height of the Cold War, but it is still a lot of money, more than a third of the world's collective military expenditures.[20]

There are many reasons for such a large budget, but there is no question that technology has been a major driver. As we saw in Chapter 3, a single Air Force F-22 fighter costs U.S. taxpayers $412 million when R&D and testing costs are taken into account. This is very different from the World War II era, when a P-51 Mustang cost about $650,000 in today's dollars. Much of the difference between the two airplanes can be attributed to vast improvements in engine power, materials, armaments, avionics, ability to evade radar, and all the other technological advances that separate the World War II era from the present day. But can such cost increases be sustained indefinitely? This question was answered by Norman Augustine, an aerospace executive and former Under Secretary of the Army, who plotted the increasing cost of fighter planes over time. Extrapolating into the future, he quipped: "In the year 2054, the entire defense budget will purchase just one aircraft. This aircraft will have to be shared by the Air Force and Navy three and a half days each per week, except for leap years, when it will be made available to the Marines for the extra day."[21] Other examples of high-priced military hardware are not hard to find. For fiscal year 2019 the Defense Department put in a request for approximately $5.6 billion for more than 70,000 precision-guided missiles.[22] A single M1A2 Abrams battle tank carries a price tag of $8.58 million.[23] A fleet of twelve ballistic missile submarines is expected to cost around $100 billion.[24]

A less obvious cost of a technologically advanced military force is the large number of scientists and engineers who are engaged in defense research and development (R&D). In a world in which scientific discoveries and technological innovations are crucial for economic advance, a lot of talent is expended on projects and activities with little or no relevance to the civilian economy. There have always been spinoffs from the defense to the civilian sector—DARPA's early sponsorship of what became the internet is an outstanding example—but most defense R&D remains confined to that sector.

It is an irony of modern technology that the military sector, which is tasked with death and destruction, bears a number of similarities with medical care, which is supposed to be about health and healing. In both cases there is a strong tendency to be satisfied with only the most sophisticated technologies, which usually means the most expensive technologies. A fighter pilot does not want to engage an enemy who is flying the world's best combat aircraft when he is in the seat of the world's second-best airplane, just as a patient suffering from a life-threatening disease will not be happy with the second-best medical treatment. The two sectors also receive much of their funding from third parties—insurance companies and local, state, and federal governments in the case of medical care, and the federal government for military expenditures. It also has to be noted that both the medical and the defense sectors are prolific spenders on lobbying efforts when they interact with Congress and other agencies. More politics are involved when some projects owe their existence to their location in states and districts represented by influential members of Congress.

Asymmetrical Warfare

High defense expenditures may be justified as the cost of protecting a country and its inhabitants, but does modern weaponry guarantee security? Military engagements in recent decades, most notably the Vietnam War, showed that wars are not won by advanced technologies alone. Since the Korean War, the United States has been engaged in a number of conflicts that can be characterized as "asymmetrical warfare," engagements where the United States was vastly superior in the quantity and quality of its materiel. In regard to quantity, during the Vietnam War the United States dropped 7.8 million tons of bombs in Indochina, far more than the 2.7 million tons dropped by the Allies against Axis targets during World War II.[25] The United States also was vastly superior to North Vietnam in military technology. Infrared goggles gave American soldiers the ability to see in the dark, smart bombs destroyed their targets with pinpoint accuracy, and surveillance satellites and first-generation drones provided detailed views of the terrain below. More recently, during the invasion of Iraq in 2003, Saddam Hussein and his military forces were quickly routed by a "shock and awe" campaign waged with cruise missiles, smart bombs, depleted uranium ordnance, and many other examples of advanced military technology. But these weapons did little to quell the subsequent insurgency that took place during the coalition occupation, resulting in the deaths of 4,431 Americans and 320 British and others. There were 31,944 U.S. personnel reported wounded in action (WIA).[26] For Iraqis the toll was much higher. An estimated 10,000–12,000 died as a result of the air campaign (2,300 of them civilians according to the Iraq government), while the ground war produced up to 10,000 casualties.[27]

In Iraq and Afghanistan, the insurgents' principal weapon has been the improvised explosive device, or IED, planted in a roadway. Early versions consisted of nothing more than explosives connected by wire to a detonator. Later examples were more sophisticated; some used timers or wireless triggers such as modified car door lock openers and cordless telephones so a wire could not be followed to where the insurgents were hiding. Over time, Iraqi insurgents devised more than ninety ways of setting off an IED. Coalition forces were eventually able to deploy robots to disarm IEDs, but even these could be thwarted by placing roadside bombs so they were out of the robots' reach.[28]

Figure 19.3 Improvised explosive devices (IEDs) may be triggered by a pressure plate, mobile phone, radio, or various timers. (Source: NCO Journal, U.S. Army)

Technology and Terrorism

In some quarters, justification for the invasion of Iraq rested on the assertion that Saddam Hussein had conspired with Al-Qaeda on the September 11 attacks. This supposed conspiracy was unlikely from the start, but the belief persisted long after it was proven to be a fabrication. It did, however, focus the world's attention on a new set of security issues. While the spread of nuclear weapons among established nation-states remains a grave potential threat, the world is now focused on a different source of danger—global terrorism. Terrorism is closely related to asymmetrical warfare in that a militarily weaker side is able to challenge a stronger one through the use of simple but effective weapons such as roadside bombs and car bombs. Terrorism is not a new phenomenon; a depressing litany of examples can be found throughout human history. What has changed in recent years is the increasing technological sophistication of terrorists and their use, or potential use, of weapons of vastly greater destructive power.

Nuclear weapons are unparalleled in their destructive power, and the acquisition of a multimegaton atomic bomb would provide terrorists with the ultimate weapon. At the same time, however, the likelihood of this happening is relatively low. The construction of a nuclear bomb requires a high degree of expertise and access to specialized equipment, along with the production of sufficient quantities of fissionable materials like plutonium or uranium-235. The production of a workable bomb is likely to be beyond the capabilities of terrorist groups lacking the assistance of a nation that already has nuclear weapons, but fissionable material for a crude bomb could be bought or stolen. A terrorist group does not need a military-grade nuclear bomb to induce widespread terror and panic; the detonation of a "dirty bomb," a conventional explosive device that would spread radioactive materials like cesium-137, iridium-192, or cobalt-60, could induce radiation sickness on a significant scale.[29] More importantly, it would engender panic over a wide area even if actual loss of life were not great. Since the primary goal of terrorist organizations is to spread fear and disruption, a bomb of this sort would be a highly effective weapon.

Much easier than the development of nuclear weapons are the manufacture and use of chemical weapons. For reasons that will be discussed in the next chapter, poison gas was not deployed as a combat weapon during World War II, although it was used in the 1920s and 30s by Spain and Italy to subdue indigenous populations while securing and expanding their colonial possessions.[30] Poison gas re-emerged as an offensive weapon when Saddam Hussein used it during the Iran–Iraq war (1980–1988) and then against the Kurdish population of his own country in 1988. Saddam's weapon of choice was a combination of mustard gas and one or more nerve agents: sarin, tabun, and VX.[31]

In general, chemical weapons fall into four main categories: nerve agents; blood agents, notably hydrogen cyanide and cyanogen chloride; blister agents, like the mustard gas that was first used in World War I; and choking and incapacitating agents, such as chlorine, another gas used in that war.[32] Up to now, the most destructive use of a chemical weapon by a terrorist organization occurred in March 1995 when a Japanese religious cult known as Aum Shinrikyo (Aum Supreme Truth) released sarin gas into the Tokyo subway system, killing 12 and injuring more than 5,000. In the years leading up to the attack, the sect had accumulated considerable expertise in the manufacture of sarin and other chemical weapons, an endeavor that was facilitated by having a number of scientists and engineers in its ranks. At the same time, however, the mode of delivery was distinctly low-tech, entailing nothing more than placing plastic bags containing sarin in five subway cars and then puncturing them, allowing the volatile liquid to disperse as a lethal vapor.

The third category in the unholy trinity of terror weapons consists of biological weapons. The list of these weapons is extensive, including bacteria that are the sources of anthrax, tularemia, and plague, and viruses that cause smallpox, yellow fever, Dengue fever, and Ebola. All of these can produce death and grave injury, but the most effective weapon employed so far has been the bacteria that cause salmonella, a much less lethal disease. This was used by a group of religious extremists in Oregon in 1984 to contaminate salad bars in local restaurants in the hope of incapacitating a large number of people prior to a local election.

As with chemical weapons, biological weapons are easy to manufacture, and they certainly do not require the assistance of a "rogue state." Biological weapons can be made in small facilities using little

Figure 19.4 After the 1995 sarin gas attack perpetrated by Aum Shinrikyo, Japan Defense Force (JDF) Chemical troops arrived to help decontaminate the Tokyo subway.[45]

in the way of sophisticated equipment, and the basic materials can be obtained from commercial sources or cultivated from naturally occurring sources.[33] Effective distribution is more problematic, however. Even if there were effective means of spreading the pathogens, contaminating a large area would be difficult because most of the released organisms would not survive for long (anthrax is an important exception because its spores can survive for an indefinite period). Most biological agents are not highly infectious and generally require close contact or exposure to bodily fluids for transmission. But biological weapons do not have to operate at a high level of effectiveness and efficiency; as with terror weapons in general, their significance lies in their ability to spread fear on a massive scale. The novelty of biological weapons and the fear engendered by an invisible attacker, coupled with the difficulty of administering antidotes or immunization on a large scale, makes them particularly attractive terror weapons.

Cyberterrorism and Cyberattacks

While modern technologies have enhanced the capabilities of terrorists, they also have created new vulnerabilities for these terrorists to exploit. Technologically sophisticated societies are highly dependent on a variety of infrastructures, such as electrical grids run by computers and other digital devices. In the defense sector, a nation's ability to protect itself rests heavily on computer networks and equipment to detect an imminent enemy attack and take measures against it. The large-scale crippling of a country's computers would undercut the use of radar and other detection technologies, cut off military communications, and prevent the use of most weapons. In the civilian realm, the disabling of computers and their associated networks would bring chaos to transportation systems, communications of all sorts, banking and finance, manufacture, retail trade, healthcare, and virtually every other aspect of modern society.

Dependency on computers and their networks can be exploited by terrorists using tactics that fall into two broad categories: information attacks and infrastructure attacks.[34] Information attacks are attempts to access, destroy, or alter the information retained, processed, and distributed by computers and their networks. Some of these attacks are fairly trivial—such as the defacement of a webpage—and can be easily remedied. Others could pose a serious threat to national security if, for example, a terrorist gained access to vital information for espionage, or simply to destroy or alter it.

Infrastructure attacks are potentially more dangerous than information attacks because they can damage or disable critical services. Obvious targets are the computer networks essential to the distribution of electricity, gas, and water; air traffic control; and critical government services like policing and firefighting. Although much of this infrastructure is protected by the use of redundant networks, it would be overly optimistic to believe that these systems are invulnerable. There have been numerous occasions when individuals have done millions of dollars of damage by hacking into computer networks. As with the dangerous and expensive mischief caused by these hackers, cyberterrorists also can infect computer programs and files with disabling and self-replicating viruses, worms, and Trojan horses.

A cyberattack planned and implemented by a sovereign country began in 2008 when the United States working in collaboration with Israel was able to introduce a computer worm called Stuxnet into the computers running centrifuges used by Iran for the enrichment of uranium. As a result, at least a thousand centrifuges were put out of commission. The extent to which these attacks put a serious crimp in Iran's nuclear program is still a matter for debate, but they certainly did not put an end to it. Critics of the attack have pointed out that the unleashing of Stuxnet will lead to a new kind of arms race based on expanded abilities to launch and repel cyberattacks.[35] By launching a cyberattack, the United States and Israel gained a temporary advantage over a potential nuclear adversary, but it may have come at the cost of long-term security.

Another striking example of a major attack on a government network, but not one directly tied to military matters, occurred in 2015, when the computer network used by the U.S. government's Office of Personnel Management was breached, resulting in the exposure of personal information of 19.7 million federal employees who had undergone background checks, as well as 1.8 million others, including neighbors, friends, and spouses. The information collected included Social Security numbers, home addresses, past foreign travel, financial transactions, and even information about acquaintances.[36] The source of the attack, if known, has not been revealed, although China is a prime suspect. If this indeed had been the case, it would not have been a terrorist attack per se, since by definition terrorism is the work of actors not directly connected to a legitimate state.

A modern nation-state's ability to launch a cyberattack can be seen as a continuation of a high-tech arms race that has been going on since the invention of gunpowder. In contrast, the development of this capability by terrorist groups poses a certain irony. A cyberattack requires technological sophistication for its planning and execution. In this sense, terrorists are effective participants in the modern world. At the same time, however, in a number of countries and regions, the rapid development of a globalized and networked world has been viewed as a threat to established cultures and religious beliefs.[37] The response of some disaffected groups and individuals has been to attack those nations that have been the source of the ideas and images diffused throughout the world via the internet and other electronic media. When it was at the peak of its influence, the so-called Islamic State was particularly adept in employing digital media; every day on average it released 38 items of digital propaganda into cyberspace—documentaries, photo essays, pamphlets, and audio clips in a variety of languages.[38]

In considering the threats posed by cyberterrorism, along with nuclear, chemical, and biological weapons, it is important to keep in mind that effective terrorist attacks do not require the use of high-tech weapons. The destruction of the Alfred P. Murrah Federal Building in Oklahoma City in 1995 by a white supremacist was accomplished by a bomb made from nothing more sophisticated than a mixture of ammonium nitrate fertilizer and fuel oil. The attacks on three railway stations in Madrid that killed nearly 200 people and wounded 1,450 on March 11, 2004, were done with conventional explosives. The horror of 9/11 entailed the destructive use of the embodiment of modern technology, a jet airliner, but the terrorists were able to hijack three planes by incapacitating their crews with nothing more sophisticated than a few small-edged weapons.

Although chemical, biological, and even nuclear weapons do pose potential threats, it can be argued that modern technology's greatest contribution to terrorism has been the electronic media. Just as the internet, social media, and mobile phones have helped to peacefully unseat repressive dictators, they also can be employed to promote the goals of violent terrorists. Mobile phones have facilitated coordination of terrorist operations and have been used to detonate bombs, and the internet has been an invaluable means of recruiting terrorists to serve the Islamic State, presenting detailed information on how to make weapons, and publicizing grievances and demands while extoling life in the Caliphate. Above all, the global reach of television has brought terrorist acts into the consciousness of the whole world. The hundreds of millions of people who watched broadcasts of the collapse of New York's Twin Towers after the 9/11 attacks were suddenly aware of the intent of Osama Bin Laden and Al-Qaeda to remake the world by waging holy war.

Military Technologies in a Changing World

Although the use of today's weaponry has extended the ability to kill and destroy to a level scarcely conceivable to earlier generations of warriors, many of the issues presented by advanced military technologies have been around for centuries, if not millennia. New kinds of weapons promise impressive tactical and strategic gains, but they also have called into question traditional martial virtues as they redefined what it means to be a warrior. As one Spartan king lamented when he witnessed the demonstration of an early catapult, "O Hercules, the valor of man is at an end!"[39] A similar attitude was expressed during the Middle Ages, when military leaders were infuriated by the insidious way in which crossbowmen and musketeers could pick off their soldiers from afar. Their anger was reflected in their brutal treatment of captured soldiers using these weapons, which often extended to cutting off their hands and having their eyes put out.

To be sure, punitive measures and an occasional disdain for new weapons did not stem the advance of military technology. We continue to struggle today to find ways to limit the consequences of advances in military technology for soldiers and civilians alike. The development and spread of smart bombs, drones, cruise missiles, and the other components of high-tech warfare has raised many questions about the conduct of wars and individual warriors. One of the most striking changes centers on the ability to wreak havoc from halfway around the globe. The use of remotely controlled drones has put their pilots out of harm's way, but it has also engendered a psychological as well as spatial remoteness from the realities of warfare; a drone pilot may launch an attack on a suspected terrorist base, witness the ensuing destruction, and then go to a PTA meeting a few hours later.[40] The use of drones and other high-tech weapons also presents some thorny legal issues. Are drone pilots bonafide combatants, and if so, would that justify an enemy killing them in cold blood as they walked out the front door of their homes? Who is at fault when a drone or smart bomb is guided to the wrong target; is it the officer in the field who ordered the attack or the pilot in Nevada who flew an incorrect course? And who is to blame if innocent civilians are killed because the program used to control a weapon had a glitch of some sort? Should the programmer who made the error be blamed for the loss of lives?[41] The use of drones and other robotic weapons may be counterproductive if it convinces an opponent that their adversaries are cowards who are unwilling to put their lives on the line in combat situations, thereby increasing the enemy's will to fight on.[42] Also, the effects of these kinds of weapons reprise a familiar theme of winners and losers amid technological change; younger soldiers who grew up with video games may take over the jobs and responsibilities of older soldiers whose skills have become less relevant.[43] Finally, and most disturbingly, by promising lower casualty rates, remote-control warfare may lower the inhibitions for going to war.

While high-tech weapons have presented a number of new issues and dilemmas, changing configurations of power have called into question their relevancy. For more than four decades after the conclusion of World War II, an immense amount of money and expertise was expended on defending the United States and its allies from a nuclear attack carried out by missiles and bombers. The dissolution of the Soviet Union in 1991 lowered global tensions and reduced the threat of a nuclear war, but in some respects the world is a more uncertain and dangerous place than it was during the Cold War. Instead of facing one powerful foe, the United States and its allies have to contend with a multitude of potential adversaries. Some of them are nation-states while others are terrorist organizations. None of them has anything like the military might once possessed by the Soviet Union, but many have the capacity to produce significant damage through the use of biological, chemical, and perhaps nuclear weapons.

Despite ongoing efforts to settle disputes without recourse to war, the world remains an unsettled place, torn by a multitude of grievances and disputes between and within nations. Many of these could erupt into wars made more destructive than ever through the use of increasingly lethal weaponry. While the possession of modern weapons may have a deterrent effect in cases of state-to-state conflict, the threats posed by terrorists are more problematic. Drones and reconnaissance satellites can identify areas that harbor terrorists, and missiles launched from drones can kill and injure suspected terrorists. At the same time, however, the dispersal of terrorist cells can make the whole enterprise seem like playing Whack-a-Mole, the arcade game in which a mechanical mole is knocked back into its hole, only to be replaced by another one popping out of another hole.

In addition to the tactical difficulties of dealing with terrorists, there is the larger problem of living on a planet that includes people whose beliefs, anger, and grievances are the underlying source of terrorist attacks. There is no technological fix, military or otherwise, for this situation. As one former Army officer mused, "The great paradox of this high-tech age is that its security problems arise from the human heart and soul, domains which remain opaque to technology (and to those who worship it)."[44]

For all the present dangers engendered by terrorism, it is still the case that nuclear proliferation poses the greatest long-term threat to humanity. Despite, or perhaps because of, mutual hostility, the United States and the Soviet Union were able to forge several arms control agreements from the 1960s onward. At the same time, the prospect of Mutually Assured Destruction (MAD) further restrained whatever aggressive tendencies the two nations may have had. But while the Soviet Union is now a receding memory, other nations have advanced their ability to manufacture and deliver nuclear weapons. Preventing or at least limiting the further spread of these weapons will be one of the major tasks in the twenty-first century. Some of these efforts are explored in the next chapter, but before taking up this subject, we should know more about the forces propelling technological advance in weaponry. This will be the theme of the first part of the next chapter.

Questions for Discussion

1. "Collateral damage," in which innocent people are killed or injured and their homes destroyed, has been an inevitable feature of warfare. World War II was particularly lethal, as millions of civilians died as a result of enemy actions. Do modern weapons have the potential to reduce civilian casualties? If so, what prevents them from doing so?
2. Drones are now used extensively for reconnaissance in combat areas, and they can also be used by the police and private parties to gain information about activities below them. Do they pose a serious threat to privacy? Should anything be done to prevent or limit the use of drones?
3. A major goal of antiterrorist operations is to deny safe havens in countries or parts of countries that support or tolerate terrorists. Does harboring terrorists justify military actions that amount to waging an undeclared war on a sovereign power?
4. Could the United States survive an extensive cyberattack? What can be done when no "return address" can be associated with the source of the attack? What sort of things could be done to blunt the force and mitigate the consequences of a cyberattack?
5. In recent wars, the number of uniformed men and women engaged in support activities far outnumber those who are involved in combat. Are they really soldiers? What about drone aircraft operators who are located far from the field of combat?

Notes

1. Martin W. Bowman, *The USAAF Handbook 1939–1945* (Gloustershire, England: Sutton, 1997) p. 232.
2. Luftwaffe Resource Center, "Fi-103/V-1 'Buzz Bomb,'" accessed on May 1, 2012, at www.warbirdsresourcegroup.org/LRG/v1.html.
3. Sharon Weinberger, "Cruise Missiles: The Million Dollar Weapon," *The Huffington Post* (May 25, 2011), accessed on May 1, 2012, at http://www.huffingtonpost.com/2011/03/25/cruise-missiles-missile_n_840365.html.
4. "Ruhrstahl X-1" *Wehrmacht History 1939 to 1945*, accessed on May 2, 2012, at www.wehrmacht-history.com/luftwaffe/missiles/ruhrstahl-x-1-anti-ship-missile.htm.
5. Fred Kaplan, "Smart Bombs, Dumb Targets," *Slate* (December 16, 2003), accessed on May 3, 2012, at www.slate.com/articles/news_and_politics/war_stories/2003/12/smart_bombs_dumb_targets.2.html.
6. Robert A. Gurriero, "Space-Based Reconnaissance," accessed on May 2, 2012, at www.armyspace.army.mil/spacejournal/Article.asp?AID=13.
7. James Risen, "A Military and Intelligence Clash over Spy Satellites," *The New York Times* (April 19, 2012), accessed on May 2, 2012, at www.nytimes.com/2012/04/20/us/politics/spy-satellite-clash-for-military-and-intelligence-officials.html?pagewanted=all.
8. P.S. Singer, *Wired for War: The Robotics Revolution and Conflict in the Twenty-First Century* (New York: Penguin Press, 2009), pp. 117–118.
9. Ibid., pp. 54–55.
10. Robert M. Gates, "Remarks by Secretary Gates at the United States Air Force Academy," U.S. Department of Defense, *News Transcript*. Accessed from www.defense.gov/transcripts/transcript.aspx?transcriptid=4779.
11. Brian Bennett and David S. Cloud, "Obama's Counter-Terror Advisor Defends Drone Strikes," *Los Angeles Times* (April 30, 2012). Accessed from http://articles.latimes.com/2012/apr/30/world/la-fg-brennan-drones-20120501.
12. Scott Shane, "Drone Strikes Reveal Uncomfortable Truth: U.S. Is Often Unsure About Who Will Die," *The New York Times* (April 23, 2015). Accessed from www.nytimes.com/2015/04/24/world/asia/drone-strikes-reveal-uncomfortable-truth-us-is-often-unsure-about-who-will-die.html?_r=0.
13. Farooq Jan Mangal and Fahim Abed, "U.S. Drone Killed Afghan Civilians, Officials Say," *The New York Times* (December 1, 2019). Accessed from www.nytimes.com/2019/12/01/world/asia/drone-civilians-afghanistan.html.
14. Jim Garamone, "U.S. Drone Strike Kills Al-Qaida Leader in Kabul," U.S. Department of Defense (August 2, 2022). Accessed from www.defense.gov/News/News-Stories/Article/Article/3114362/us-drone-strike-kills-al-qaida-leader-in-kabul.
15. See Shane, footnote 12, Ibid.
16. For global tracking of known drone strikes, see: www.aljazeera.com/tag/drone-strikes.

17. See: Isabelle Khurshudyan, Mary Ilyushina, and Kostiantyn Khudov, "Russia and Ukraine Are Fighting the First Full-Scale Drone War," *Washington Post* (December 2, 2022). Accessed from www.washingtonpost.com/world/2022/12/02/drones-russia-ukraine-air-war; and Amos Chapple, "The Drones Of The Ukraine War," *Radio Free Europe/Radio Liberty* (November 17, 2022). Accessed from www.rferl.org/a/ukraine-russia-invasion-drones-war-types-list/32132833.html as well as https://www.bbc.com/news/world-62225830.
18. W. J. Hennigan, "Push to Elevate Drone Pilots," *Los Angeles Times* (November 9, 2015), pp. A1, A13.
19. See: House Approves H.R.4365, The Department of Defense Appropriations Act (September 28, 2023) at https://appropriations.house.gov/news/press-releases/house-approves-hr-4365-department-defense-appropriations-act
20. Sarah Tully, "U.S. Defense Spending vs. Global Defense Spending," *The Center for Arms Control and Non-Proliferation* (May 14, 2015). Accessed from http://armscontrolcenter.org/u-s-defense-spending-vs-globaldefense-spending.
21. "Defense Spending in a Time of Austerity," *The Economist* (August 26, 2010). Accessed from www.economist.com/node/16886851.
22. U.S. Department of Defense, "DoD Releases Fiscal Year 2019 Budget Proposal," Accessed from https://comptroller.defense.gov/Portals/45/Documents/defbudget/fy2019/fy2019_Press_Release.pdf.
23. "How Much Does a Common Battle Tank Cost?" *Quora*. Accessed from www.quora.com/How-much-does-a-common-battle-tank-cost.
24. Franz-Stephan Gady, "US Navy Secretary: Cost of New Ballistic Missile Sub Is 'Eye Watering,'" *The Diplomat* (March 13, 2018). Accessed from https://thediplomat.com/2018/03/us-navy-secretary-cost-of-new-ballistic-missile-sub-is-eye-watering.
25. James P. Harrison, "History's Heaviest Bombing," in Jayne S. Werner and Luu Doan Huynh (Eds.), *The Vietnam War: Vietnamese and American Perspectives* (Armonk, NY: M. E. Sharpe, 1993), p. 133.
26. Iraq Coalition Casualty Count, "Coalition Military Fatalities by Year." Accessed from http://icasualties.org/IRAQ/index.aspx; See also: www.defense.gov/casualty.pdf.
27. "Iraqi Death Toll," *Frontline*. Accessed from www.pbs.org/wgbh/pages/frontline/gulf/appendix/death.html.
28. Singer, *Wired for War*, pp. 218–19.
29. Gilmore Commission First Annual Report, "Reasons and Rationales behind Potential CBRN Terrorism," in Yonah Alexander and Milton Hoenig (Eds.), *Super Terrorism: Biological, Chemical, and Nuclear* (Ardsley, NY: Transnational Publishers, 2001), p. 16.
30. Daniel R. Headrick, *Power over Peoples: Technology, Environments, and Western Imperialism, 1450 to the Present* (Princeton and Oxford: Princeton University Press, 2010), pp. 322–27.
31. Christine M. Gosden, "The 1988 Chemical Weapons Attack on Halabja, Iraq," in Alexander and Hoenig, *Super Terrorism*, p. 8.

32. Jessica Stern, *The Ultimate Terrorists* (Cambridge, MA: Harvard University Press), p. 24.
33. Joseph F. Pilat, "The Bioterrorism Threat: Technological and Political Considerations," in Alexander and Hoenig, *Super Terrorism*, p. 64.
34. Robert W. Taylor, Tory J. Caeti, D. Kall Loper, Eric J. Fritsch, and John Liederbach, *Digital Crime and Digital Terrorism* (Upper Saddle River, NJ: Pearson Prentice-Hall, 2006), pp. 23–28.
35. Misha Glenny, "A Weapon We Can't Control," *The New York Times* (June 24, 2012). Accessed from www.nytimes.com/2012/06/25/opinion/stuxnet-will-come-back-to-hauntus.html?_r=1&ref=stuxnet.
36. Julie Hirschfeld Davis, "Hacking of Government Computers Exposed 21.5 Million People," *The New York Times* (July 9, 2015). Accessed from www.nytimes.com/2015/07/10/us/office-of-personnel-managementhackers-got-data-of-millions.html?_r=0.
37. Manuel Castells, *The Rise of the Network Society* (Malden, MA: Blackwell, 2000).
38. Brendan I. Koerner, "#Jihad," *Wired*, 24, 4 (April 2016): 78.
39. B. H. Liddell Hart, *The Revolution in Warfare* (London: Faber and Faber, 1946), p. 30
40. Singer, *Wired for War*, p. 347.
41. Ibid, pp. 382–87
42. Ibid., p. 312
43. Ibid., p. 364.
44. Quoted in ibid., p. 296.
45. This is discussed in Sugiyama et al. (2020) "The Tokyo subway sarin attack has long-term effects on survivors: A 10-year study started 5 years after the terrorist incident." *PLoS One* 15(6): e0234967.

Chapter 20

How New Weapons Emerge—and How They May Be Contained

Chapter 19 outlined some of the consequences of technological change for the conduct of wars and the effects of advances in military technology on the world as a whole. But what is the source of technological change in the military realm? Why has the development of weapons accelerated at such a frightening rate within our own lifetimes? Why have some weapons been adopted while others have languished? How have social and cultural conditions stimulated the development and use of some weapons and not others? In the first part of this chapter we will try to address some of these questions, and at the same time provide some background for the last sections, which deal with past and present efforts to limit the use of new weapons.

Action and Reaction

The most obvious reason for the development of new weapons is that combatants or potential combatants continually strive to gain the upper hand over their enemies through superior weaponry. When one succeeds in doing so, its enemies are strongly motivated to develop new weapons that can neutralize this new threat. New offensive weapons are countered by new defensive weapons, which in turn stimulate the development of better offensive weapons, and so on. The process is one of constant action and reaction, as rival nations thrust and parry, each seeking to neutralize their opponent's capabilities, while at the same time augmenting their own. The development of military technologies thus becomes an endless cycle in which each new weapon stimulates the invention of another, and the capacity to kill and destroy proceeds without limit. It might be considered a "co-evolutionary" model—where it is in the best interests of a gazelle to be slightly faster than its neighbors, there is evolutionary pressure on cheetahs to catch the faster gazelles.

History supplies us with many examples of this process. We have seen how the offensive power of the mounted knight was checked by the pike and longbow. Cannon threatened the security of medieval castles, until their destructive force was temporarily diminished by improved fortifications. The machine gun produced a stalemate during much of World War I, but the armored tank made the machine gun a much less potent weapon and tipped the balance back in favor of the offense. Since then, however, a number of military actions have powerfully demonstrated the vulnerability of tanks to precision-guided weapons.[1]

Useful as this action–reaction schema is, it does not tell the whole story. In the first place, it isn't always a simple matter to categorize military tactics as either offensive or defensive. When a war is being fought, the contestants do not think neatly in terms of offense and defense. Wars are won by defeating an enemy, which means offensive action, but at the same time defensive measures are necessary to protect the capacity to wage an offensive campaign. And even if a country seeks only to protect itself from an aggressor, it may still find it necessary to go on the attack, for as the old maxim has it, "The best defense is a good offense."

In similar fashion, many weapons can be used for either offensive or defensive purposes. For example, the possession of a more accurate, faster-firing gun may allow an army either to rapidly advance against the enemy or to more easily defend their positions in the face of enemy attack. Also, a strong defensive capability enhances the capacity to engage in successful offensive actions. Nineteenth-century imperialist powers were able to conquer less technologically advanced lands—an obviously offensive purpose—because their superior firepower allowed them to easily hold off the attacks of indigenous populations.

The close relationship between offensive and defensive capabilities is a particularly important matter today as billions of dollars are being spent to develop an antiballistic missile defense system. Should a workable system be implemented—which is unlikely, given the immense technical obstacles to be overcome—this defense would allow the United States to launch its own missiles with impunity, for an enemy would be incapable of retaliation. A defensive system thus becomes the basis of an overwhelming offensive superiority.

Social Structure and the Development of Military Technologies

Be they offensive or defensive, weapons are not developed and put into service simply because of their intrinsic superiority. As has been noted throughout this book, the creation and use of technologies are social constructions to a significant degree, shaped by larger processes that involve existing social patterns, cultural orientations, and the motivations of individual people. The development of military technologies is no different. A weapon is part of a larger system, and its use reflects the basic features as well as the strengths and deficiencies of that system. This is well illustrated by the adoption of firearms and the eclipse of the longbow as the basic infantry weapon.[2] When viewed solely in terms of effectiveness, the longbow was by no means inferior to early muskets. Fired by a competent bowman, an arrow could penetrate the armor of that era just as easily as a bullet propelled by a firearm. Its range was greater, and it could be shot far more rapidly. In addition, the longbow was less expensive, it was not subject to the continual misfires that plagued early firearms, nor was its effectiveness diminished by wet weather, as was the case with early matchlocks. Why then was the bow replaced by firearms despite all their imperfections?

In this description of the virtues of the bow, it was necessary to preface the list with "fired by a competent bowman." A bowman was, of course, an essential element in the technological system of medieval archery, and the level of his skills determined the bow's effectiveness. This skill could not be taken for granted; proficiency with the bow required some natural ability and a great deal of practice. In medieval England there was a strong expectation that men and boys would devote a considerable amount of time to archery practice; indeed, it amounted to the national sport of England.

Even so, not everyone was willing to submit to the rigors of archery practice. Many preferred to spend their spare time engaged in card-playing, bowling, shooting dice, and the medieval version of football. On several occasions royal edicts banned these activities, in the hope, as one Elizabethan act put it, that "archery may be revived and practiced and that kind of ancient weapon whereby our nation in times past has gotten so great honor may be kept in use."[3] As might be expected, these laws were almost impossible to enforce, and the quality of English archery underwent a steady decline.

In other nations matters took a different course. During the late fourteenth century, the king of France also promoted archery by banning all other diversions. The quality of French archery surpassed that of England, to such a degree that the ruling elite began to fear that their bowmen could pose a challenge to their rule. Consequently, mass archery was superseded by requirements that a limited number of bowmen be cultivated in each town and district, and the masses went back to their traditional pastimes.[4] In general, rulers were not inclined to support the military prowess of their citizens, which could just as easily be used in the service of rebellion. Rather, autocratic states were predisposed to build up permanent professional armies composed of soldiers lacking high levels of individual skill.

Firearms fit perfectly into this system, for they allowed the use of poorly paid mercenary soldiers drawn from the poorest and least educated stratus of the society. Unlike archery, the use of firearms did not require high levels of skill; as Bert Hall has noted, "It was far easier to learn to use guns . . . than it had been to use most of the weapons that had preceded them, especially longbows, and this reduction in the fundamental skill component meant that masses of untrained men could be made into competent soldiers in six months or less."[5] To be sure, the use of early firearms was by no means a simple matter, requiring numerous operations in precise sequence. But none of these, not even aiming and firing the weapon, required a great deal of skill. Nor was there much opportunity to instill higher levels of proficiency even if it were desired; gunpowder and bullets were expensive, and as a result practice time was sharply limited, amounting to the firing of only a few dozen rounds per year. In any event, individual skill in handling weapons was far less important than maintaining a concentrated rate of fire. Battles were won by the discharge of withering volleys, not by individual sharpshooting.

What counted, therefore, was iron discipline. The main requirement for effective use of arquebuses and muskets was that soldiers have the proper procedures drummed into them through rigorous drill, and that they adhere to these procedures in the heat of battle. Ranks of soldiers were expected to advance as a single line toward the enemy and to maintain formation despite the prospect that as many as a third of them would fall victim to enemy gunfire.

Mercenary soldiers of this sort were effective in battle because they had been transformed into virtual robots, who, as Frederick the Great of Prussia wanted, feared their own officers more than they did the enemy.[6] Firearms were thus particularly well suited to armies raised by centralized, bureaucratically organized states where discipline was much more highly valued than freedom and individual ability. Troops of archers made up of independent yeomen were appropriate to the political and social setting of late medieval England, while gun-wielding mercenary armies were well suited to the social structures found in most of Europe from the sixteenth century onward.

It should also be noted that the advantage of firearms was not confined to their appropriateness to a particular mode of political organization. Technologies are sometimes embraced because of the psychological needs they meet. Although early firearms had a number of deficiencies when compared with the bow, they did have the advantage of producing a great deal of noise. Despite being inaccurate and cumbersome to use, early firearms certainly conveyed the impression that they were dangerous and terrifying weapons. And one could also speculate on the sexual connotations that have long been attached to guns; one needn't be a close follower of Sigmund Freud to realize that handguns and cannons can easily serve as symbols of male virility and potency.[7]

As the adoption of firearms shows, the motives for using one military technology instead of another do not necessarily stem from the objective capabilities of a weapon. The fit of a military technology with established interests and ways of doing things may be of paramount importance for its adoption. At the same time, stubbornness and short-sightedness can delay the adoption of a weapon and retard its effective use when this finally occurs. Although political and military leaders might be expected to employ new military technologies that confer specific advantages, some weapons have languished due to misperceptions of their potential.

A failure of this sort was evident in the slow adoption of the machine gun by late nineteenth-century armies. This was a particularly ironic situation in the United States, for the first practical machine guns, the Gatling gun and the Maxim gun, were American inventions. But France and England were equally slow in recognizing the machine gun's potential. The problem with these weapons was that they were heavy, cumbersome devices that had to be supported by large gun carriages. Since they looked like conventional artillery pieces, they were treated as such. The leaders of the armies of late nineteenth-century England, France, and the United States did not grasp the unique capabilities of the machine gun, and they made no effort to develop new tactics that could take advantage of the machine gun's rapid rate of fire.[8]

In contrast, the German General Staff had a strong interest in the machine gun because a central element of their military plans was the massive use of reservists whose marksmanship was necessarily poor. In this way, the German army at this time was quite different from the small, professional armies of England, France, and the United States. Given this basic difference, the machine gun was particularly appealing to the Germans since its firepower more than made up for the deficiencies of its operators.[9]

As we saw in Chapter 18, the capabilities of the machine gun were soon demonstrated in the early battles of World War I. Combat bogged down into trench warfare, with neither side capable of advancing in the face of machine gun fire. Decisive breakthroughs became possible only after the deployment of the armored tank, which was largely impervious to machine gun fire. But again, military planners were slow in realizing how a new technology drastically altered the nature of warfare. Just as the machine gun caught World War I military leaders unprepared, the military value of the weapon developed in response to it, the tank, was not always appreciated during the years leading up to World War II. This situation was particularly evident in England, France, and other countries where military officers tended to be recruited from the traditional elite—the figurative, and in some cases literal, descendants of the knights of the feudal era. A good bit of their lives still centered on the horse, and prowess on horseback was an essential requirement of aristocratic manliness. In this culture, noisy, smelly machines were no match for a noble steed. Thus, a successful 1910 demonstration of an armored car in Austria came to naught because the vehicle had frightened the horses upon which the observing generals were mounted; worst of all, the emperor's own mount came close to bolting.[10]

Although the use of tanks during the final years of World War I gave a strong indication that land warfare was entering a new phase, the military establishment was reluctant to alter battlefield tactics in order to take advantage of these new weapons. In the years that followed the war, armored vehicles usually remained under the control of long-established combat branches within the army—the infantry and the cavalry—which hardly provided a supportive environment for their development. Convinced that the old rules of combat still applied, infantry and cavalry officers were not eager to see the foot soldier and cavalryman displaced by armored vehicles and other advanced weapons. At best, tanks were combat auxiliaries; as the U.S. Army's chief of the infantry put it in 1928, "The tank is a weapon and as such it is an auxiliary to the infantryman, as is every other arm or weapon."[11] Even as late as 1938, the Army's cavalry chief was arguing that mechanization "has not yet reached a position in which it can be relied upon to displace the horse cavalry . . . . For a considerable period of time [mechanization is] bound to play an important but minor role while the horse cavalry plays the major role so far as our country is concerned."[12] In Britain, Field Marshal Douglas Haig, whose tactics resulted in hundreds of thousands of casualties suffered in hopeless assaults against enemy lines protected by machine guns, was quite clear on this point: "I am all for using aeroplanes and tanks, but they are only auxiliaries to the man and the horse."[13] As a consequence of this sort of thinking, twice as much money was spent in Britain in 1935 on cavalry as on tanks.[14] In France the situation was no better. According to the official *Directive on the Use of Tanks*, "In battle, tanks, whether singly or in groups, act within the framework of the infantry. Tanks are merely support weapons, which may be assigned to the infantry for a limited time; they considerably augment its striking force, but they do not replace it."[15] Accordingly, tanks were to be commanded by infantry officers and were not intended to serve as independent strike weapons. Throughout the 1920s and 1930s, tanks were designed to move at the speed of the foot soldier. They were expected to rip through barbed wire and neutralize the weapons that threatened the infantry's advance, but they were not to play an independent role in combat.

Figure 20.1 British soldiers leave their trench to go "over the top" at the 1916 Battle of the Somme. Assaults such as these gained little if any enemy territory at the cost of enormous numbers of casualties.

Figure 20.2 Although inferior to the German Tiger and the Russian T-34, the Sherman tank was produced in huge numbers and was the mainstay of U.S. armored forces during World War II.

In contrast, the army of the Soviet Union, which had been formed in the wake of the Russian Revolution and was led by an entirely different breed of officer, was far more alert to the new opportunities presented by these weapons. In similar fashion, the military forces of Nazi Germany were commanded by a new generation of officers, for Germany's defeat in 1918 had discredited the old military leadership, and the post-World War I disarmament of Germany left no entrenched military structure to resist new weapons and tactical concepts. The military staff that guided Germany's rearmament in the 1930s was highly receptive to the deployment of new weapons, as can be seen in the circulation of 30,000 copies of a translation of General J. F. C. Fuller's prescient book on tank operations, *Provisional Instructions for Tank and Armoured Car Training*. In Fuller's native Britain, only 500 copies were printed.[16] Fuller's lessons were taken to heart by the German army, which made good use of the tank in the blitzkrieg attacks that conquered much of Europe during the first two years of World War II.

Organizational Interests and the Air Weapon

The resistance to new technologies by entrenched military establishments was also evident in the development of a weapon that had begun to demonstrate its capabilities during World War I, the airplane. Here, however, the story diverges from that of the tank. During the interwar years members of fledgling air corps defied conventional military thinking and aggressively pushed the idea that the airplane had initiated a new era in warfare. According to these proponents of "air power," most notably General Billy Mitchell of the U.S. Army Air Corps, the airplane had rendered conventional warfare obsolete. According to this new doctrine, bombers now could wipe out ground armies and sink navies at little risk to themselves. Equally important, if horrendous in its implications, bombers could attack cities and industries with impunity, paralyzing economies and breaking the enemy's will to resist.

As was noted in Chapter 19, the expectations of these prophets of air power were at best only partially realized. Although aerial bombardment produced appalling devastation, bombers equipped with conventional bombs did not singlehandedly win World War II. The expectations that aerial bombardment could play a decisive wartime role rested on a number of largely untested assumptions. More ominously, the doctrine of strategic bombing also served as a rationale for advancing particular personal and organizational interests.

During the years between the two world wars, members of the U.S. Army Air Corps were eager to see their organization transformed into an independent military arm equal, if not superior, to the Army and Navy. This would require a new theory of war based on "air power"—and a weapon capable of turning theory into practice. The Air Corps found that weapon in the B-17 Flying Fortress. First flown in 1935, the four-engine B-17 was not defined as a strategic bomber, since the isolationist climate of the time was not receptive to the idea of getting involved in conflicts with far-off nations; rather, it was portrayed as a weapon to be used against a seaborne invasion. It was designed to fly at very high altitudes and have great range so it could seek out an enemy fleet far out at sea. The bomber's ability to survive the attacks of intercepting aircraft was questionable, for no escort fighters of the time could begin to match its range, but it was believed that the massed machine gun fire of a tight formation of B-17s would stave off all attackers.

Figure 20.3 The B-29 Superfortress was the larger and faster successor to the B-17 Flying Fortress. A B-29 dropped the first atomic bomb on Hiroshima in 1945.

None of these expectations were met. As combat experiences during World War II proved, the B-17 was just about worthless as an antinaval weapon and had severe operational shortcomings, most notably a vulnerability to antiaircraft fire and an inability to defend itself from a determined fighter attack. It flew many missions and inflicted a great deal of damage on the enemy, but it wasn't the weapon that vindicated the extravagant claims made for air power, nor could any airplane carrying conventional bombs have done so.

In one sense, the mentality of Billy Mitchell was diametrically opposed to the mindsets of the hidebound officers who failed to perceive the value of armored vehicles; an air force, unlike the infantry or cavalry, represented the future and not the past. But there was still a basic similarity; both were incapable of viewing weapons with much objectivity. Instead, personal interest and commitment to a particular service arm strongly influenced the way they looked at the utility of a particular weapons technology. The traditional military resisted new weapons and clung to the established ways of doing things, often with disastrous results. Adherents of air power, and especially of strategic bombardment, were correct in their assessment that their weapons could do an immense amount of damage, but they made the same mistake as the traditionalists in believing that their personal and organizational interests were identical with the military needs of their countries.

Social Revolution and the Enlargement of War

We have just seen how personal and organizational interests can affect the course of technological change. It is to be hoped that this discussion has illuminated a recurrent theme in this book: that technologies are shaped by the societies in which they develop, and as such they reflect culture, the distribution of power, and economic, organizational, and social relationships. At this point it may be appropriate to expand on this theme by considering how social and political changes on a massive scale have altered not only weapons but the very nature of war.

During the late eighteenth and early nineteenth centuries, a series of social and political revolutions profoundly changed the conduct of warfare. The effects of these revolutions can clearly be seen by briefly examining warfare in the period immediately prior to this era. Throughout the eighteenth century most European wars were rather limited affairs. War was the ultimate "sport of kings," fought to advance the interests of a ruler rather than the country as a whole. Most of the populace remained aloof from these conflicts; the dirty work was done by small contingents of professional soldiers, social outcasts who were drilled to perfection by noncommissioned officers equally removed from the larger society. Commissioned officers came from the upper class of society and had little emotional connection with their soldiers. But these officers and their ruler valued the troops all the same, for they knew that well-drilled, effective fighting men could not be quickly replaced. As a result, soldiers were used with great care. When a major battle did occur, it produced frightful casualty rates that could decimate an army. This in turn could fatally weaken a country. Military and political leaders therefore preferred to avoid these confrontations and were eager to arrange compromise peace settlements instead of engaging in go-for-broke military actions.[17]

At the same time, war was constrained by the temper of the times. The wars between Catholics and Protestants that had convulsed Europe during the previous centuries were frightful examples of what happened when wars were fought over great principles. In place of zealotry, eighteenth-century political thinkers preached a code based on reason, or at least reasonableness. Wars might be an unfortunate consequence of human nature or political policy, but their worst effects could be constrained by the mutual agreement of the combatants to observe proper rules of behavior. Civilians were to be spared, hatred of the enemy kept in check, and terms of peace kept moderate.[18]

The great forces of democracy and nationalism changed all that. In an earlier time, the Swiss had shown the power of a disciplined, armed citizenry motivated by a fierce desire to fight for their homeland. The American Revolutionary War also showed the power that could be unleashed when large numbers of combatants took up arms in support of national self-determination and a more democratic form of governance.

These principles, and their military consequences, were driven to new heights by the French Revolution (1789–1799). This movement was energized by a desire to overthrow France's traditional ruling class and substitute democratic rule. But along with the rights engendered by democracy there came responsibilities. The people of France were no longer passive subjects of the king; they were now citizens of their nation, a nation that now faced the hostility of countries governed in the traditional manner. At the same time, the revolution had produced a zeal to extend the new order across the boundaries of France. Territorial conquest was therefore undertaken in order to extend "liberty, equality, and fraternity" throughout Europe. Warfare now involved a newly emergent citizenry, who fought for a cause far greater than the personal ambitions of traditional monarchs. The mercenary army had been replaced by "the nation in arms," where, as a report to the French Assembly in 1789 put it, "every citizen must be a soldier and every soldier a citizen."[19]

Not everyone subscribed to these lofty principles. But for those who were unwilling to voluntarily take up arms in the service of their nation, there was always the draft. Partial conscription had been employed in many places at many times, but these limited efforts were completely overshadowed by the universal conscription that was proclaimed by France's revolutionary government in 1793, which expanded the French army to nearly 750,000 men. In turn, the increased number of soldiers changed the ways that wars were conducted. Moderation and practical restraints on warfare were swept away, and military actions became far more massive in scope. Whereas the key battles of the eighteenth century had been fought by armies rarely exceeding 80,000, in 1812 Napoleon was able to invade Russia with a force of 600,000.[20]

These huge armies were expensive to maintain. A necessary consequence of the vast expansion in the size of armies was a substantial increase in taxation. Governments have always used their power to tax, and tax revenues have been disproportionately used for military purposes. But after the French Revolution, the close alignment between the individual and the nation expanded the willingness and ability of governments to tax their citizens, while large-scale military operations increased their need to do so.

Large armies required huge quantities of guns and other materiel. In revolutionary France these needs were met by mobilizing the population to produce weapons and equipment. The government drafted craftsmen of every description, as well as legions of workers with no prior experience in armament making. A good deal of waste resulted, and many of the muskets were of poor quality, but armament production was taken to new heights. While France's government arsenals never produced more than 25,000 muskets per year during the last decades of the monarchy, the Saint-Etienne armory alone produced 85,000 muskets in a 12-month period extending from 1793 to 1794.[21]

The days of restrained, limited warfare were over. The spread of democracy and nationalism brought with it mass-based warfare. And at the same time, this expansion of warfare created a receptive environment for the development of new weapons that made warfare far more destructive and terrifying than ever before. Equally important, the greatly enlarged scope of warfare stimulated the development of new production technologies that allowed the large-scale production of these weapons, as we shall see next.

Industrial Technology in the Service of War

Until the nineteenth century, military technology developed slowly. Science made few inputs into invention, and most work was done on a trial-and-error basis. As a result, guns were inaccurate and not very powerful because metallurgy was still in its infancy and chemistry had only recently separated from alchemy. This situation began to change dramatically during the nineteenth century as military technologies advanced rapidly. Smokeless powders resulted in more accurate gunnery and gave gunners a clear field of vision. The new powders also allowed the production of guns with smaller bores, making weapons lighter and allowing soldiers to carry larger loads of ammunition. The discovery of fulminates made the percussion cap possible, rendering the flintlock obsolete. Improved machining techniques stimulated the production of better weapons, such as breechloading cannon and small arms. Accurate machine tools and the use of new alloy steels also made possible the mass production of guns with rifled barrels, which greatly increased their accuracy. By the middle of the century, these new technologies also facilitated the production of practical repeating weapons, culminating with the machine gun.

For much of the nineteenth century, most of these improvements occurred with little direct support from central governments. Invention was the work of individuals and firms that attempted to sell their devices to the armies and navies of their own nation, or failing that, to anyone who was willing and able to pay for them. By the end of the century, some of these firms had become quite large and powerful, with steadily expanding political influence. Enterprises such as Armstrong-Vickers in Britain and Krupp in Germany employed scientists and engineers who had as their sole responsibility the development of new weaponry. Derided as "merchants of death," these firms gave a powerful impetus to the invention and diffusion of new military technologies that further increased the destructiveness of warfare.

By the twentieth century, the nation-state had ceased being a passive consumer of military technologies developed by private firms. Government sponsorship of technological innovation in the military sector resulted in the rapid advance of martial technologies. This reached new heights during World War II, when the United States government employed 30,000 scientists and engineers for the war effort.[22] During the Cold War that followed, the military continued to be a voracious consumer of scientific and engineering output. In the United States, by the mid-1980s, 70 percent of federal research and development funds was going to military projects,[23] while nearly one-third of the nation's total R&D) expenditures were absorbed by the defense sector.[24] Defense-related R&D had been less prominent. For example, in 2014, nearly half of the federal government's research and development budget was absorbed by the Department of Defense.[25] The 2023 Consolidated Appropriations Act, known as an omnibus bill, includes a $195 billion dollars in R&D spending, with just over 50% assigned to defense R&D. The United States spends a far greater proportion of its gross domestic product (GDP), and total R&D budget, on defense R&D and defense in general than any other developed nation.[26]

The technological advances that occurred during the last 150 years transformed armaments. But of equal or greater importance was the development of industrial technologies that allowed great increases in their production. Gunsmiths organized along traditional craft lines could never produce weapons in the quantities that became commonplace during the second half of the nineteenth century. Nor could conventional means of producing uniforms, boots, and foodstuffs meet the immense requirements of modern armies. Expanding needs of the military were met through the development of mass production. The essence of mass production is the assembling of a product from standard parts so that no individual fitting is required. This in turn requires precision tools, assembly jigs and fixtures, accurate gauges and other measurement devices, and, in general, high degrees of accuracy.[27] Mass production also requires a large market for its product; to put it another way, mass production requires mass consumption.

The military was an ideal consumer of the increased output made possible by mass production, and its needs generated a strong stimulus for the development of new production technologies. As we have seen in Chapter 13, the manufacture of pulley blocks for the ships of the Royal Navy was one of the first instances of mass production. The technique was soon adapted to the fulfillment of other military needs, most notably the production of firearms. According to popular belief, Eli Whitney was the first to employ mass-production techniques when, at the beginning of the nineteenth century, he undertook the large-scale manufacture of muskets at his factory in Mill Rock, Connecticut. In fact, Whitney never fully accomplished what he intended to do, and other enterprises have a better claim to being the initiators of mass production.[28] But whatever the source, by the middle of the nineteenth century weapons were being manufactured according to the principles of mass production, allowing huge armies to be sent into battle. At the same time, industrial technologies supplied them with mass-produced provisions, such as canned rations. Gone were the days when armies lived off the lands they marched through and fought on, often destroying more in the process than they did on the battlefield.

Large armies created immense logistical problems; the best army is worthless if it cannot be rapidly moved into a theatre of war and then be continuously supplied with arms and provisions. By the middle of

Figure 20.4 The application of modern industrial technology greatly increased the production of war materiel, as exemplified by the Rolls-Royce aircraft engine—the Eagle. A 1914 pre-World-War I factory photo.

the nineteenth century the railroad and the steamship made the movement of troops and the provision of their materiel a more regular and rapid process. Troops were no longer worn out by lengthy marches, and during the course of a campaign they were much better supplied. The wounded and exhausted could be quickly evacuated and replaced by fresh troops. This in turn expanded the scope of war, as railroads allowed military leaders to relentlessly use troops and supplies in numbers that hitherto had been impossible. Improved transportation technologies thus moved the world closer to an era of total war, in which the performance of industries operating on "the home front" took on at least as great an importance as the performance of soldiers on the battlefield.[29]

Improvements in medical technologies reinforced the trend to larger armies, for far fewer soldiers were removed from combat by disease. Throughout the history of warfare many more soldiers had died of disease than from combat wounds. This situation began to change noticeably toward the end of the nineteenth century. Before 1870, for every soldier that died of wounds, five died of disease; by 1918 this ratio had been inverted.[30]

The technological achievements of the nineteenth and early twentieth centuries thus allowed a great expansion in the scope of warfare. In this way, technological development was a perfect complement to the social and political changes that put large armies at the disposal of rival states. All of this became painfully evident as unparalleled death and destruction followed in the wake of World War I. Enormous armies faced one another across fronts that stretched for hundreds of miles, and there were more casualties in a day's battle than were suffered in the course of many eighteenth-century wars. The optimistic belief of the nineteenth century that technological advance would necessarily lead to a better life had been cruelly mocked by the second decade of the twentieth century.

Controlling Military Technologies

During the twentieth century, military technologies developed at an accelerating pace. When the century began, the battleship was the most technologically sophisticated weapon, infantrymen had only recently been armed with repeating rifles, and many thought heavier-than-air flight to be an impossibility. Today, manned bombers fly well over twice the speed of sound, the battlefield has been transformed by drones and precision-guided weapons, and a single intercontinental ballistic missile (ICBM) has the capability of destroying a large segment of a major city. On the not-too-distant horizon lies the possibility of laser-based weaponry, biological weapons, enhanced radiation bombs, and many other ghastly instruments of mass murder. Is there any hope that technological progress can be arrested in the military sphere so that we will not forever live under the shadow of increasingly lethal products of human ingenuity? The following sections will provide some perspectives that help to answer this question.

Past Attempts to Limit New Weapons

The threat posed by new military technologies is nothing new. Horrible as our latest weapons are, they are only the latest in a long series of "advances." On numerous occasions, people have surveyed the emergence of new weapons and fervently attempted to stop their spread. Pope Innocent II was aghast at the destructive power of the crossbow, and in 1139 the Second Lateran Council banned its use, although the ban did not extend to its use against non-Christians, so "infidels" such as Muslims continued to be fair game.

Figure 20.5 The Virginia-class ballistic missile submarine is expected to be deployed in 2031 and will be a stealth vehicle housing 16 Trident II D5 missiles.

Leonardo da Vinci kept secret the details of a submarine he designed, "on account of the evil nature of men, who would practice assassinations at the bottom of the seas by breaking the ships to their lowest parts and sinking them together with the crews who are in them."[31] In similar fashion, Niccolo Tartaglia, the first person to make a systematic study of ballistics, suppressed his own calculations because "it was a thing blameworthy, shameful and barbarous, worthy of severe punishment before God and man, to wish to bring to perfection an art damageable to one's neighbor and destructive to the human race."[32] But when the Turks threatened Tartaglia's homeland, he put aside his moral scruples, and his treatise on gunnery was quickly published.[33]

Each succeeding advance in weaponry has raised the stakes of warfare. At the same time, many of these advances have elicited a desire to limit the spread of new weapons. The percussion cap, an important but not epochal change in armament, was viewed with great alarm by one correspondent to *The Gentleman's Magazine* in 1837:

> If . . . this new system were applied to the military, war would shortly become so frightful as to exceed all bounds of imagination, and future wars would threaten, within a few years, to destroy not only armies, but civilization itself. It is to be hoped, therefore, that many men of conscience, and with a reflective turn, will militate most vehemently for the suppression of this new invention.[34]

A Successful Example of Arms Control

As things turned out, the nineteenth century had many nasty surprises that far exceeded the percussion cap. And, to their credit, political leaders made some effort to limit the spread of new weaponry. In 1899, delegates from 26 nations participated in a conference held at The Hague in the Netherlands that banned the use of asphyxiating gas and dum-dum bullets (which caused large internal wounds by expanding as they entered a body) and put a five-year moratorium on aerial bombardment (which at that time meant dropping projectiles from balloons).

World War I demonstrated the terrible consequences of new military technologies, and in the early postwar period several international conferences attempted to control one of the worst of them:

poison gas. All but one of the signers of the 1922 Washington Conference on the Limitation of Armaments agreed to refrain from the use of poison gas in warfare. A similar resolution was contained in the Geneva protocol of 1925 and was ultimately ratified by 42 nations, although 19 of them reserved the right to use gas if they were attacked by an enemy that made first use of this weapon.[35]

As with all such resolutions, there was no way of ensuring compliance, since there was no international organization with the power to do so. Still, these resolutions struck a responsive chord, for the terrible potential of poison gas had been amply demonstrated during World War I. Chlorine and then mustard gas killed an estimated 90,000 soldiers on both sides (more than half of whom were Russian), and permanent lung damage led to the premature deaths of many veterans after the war.[36] During the 1920s and 1930s, the threat of poison gas was quite pronounced, as military planners, novelists, and journalists presented scenarios of huge civilian populations annihilated by gas shells dropped from high-flying aircraft. During the early months of World War II the fear of gas attacks was widespread, and in Britain every man, woman, and child was supplied with a gas mask that they were supposed to carry with them at all times.

Figure 20.6 Even little children had to undergo gas mask drills during the early months of World War II.

But the attacks never came. Over the course of World War II, made increasingly horrifying by the use of new military technologies, which culminated with the dropping of the first atomic bombs, gas was never used on European battlefields. The declarations of the Washington Arms Conference and other such pronouncements were not the main reason; its provisions were violated in a number of other areas. Part of the reluctance to use gas was due to inherent logistical problems. Troops had to be protected from the gas that they used, and shifts in the wind could upset a military operation. A conquered area would require detoxification before it could be occupied. But by themselves, these factors do not explain why gas was not used; any new weapon requires adjustments if it is to be successfully used, and ways could have been found to make gas a practical weapon. Yet civil and military leaders were not inclined to do so because they had every expectation that if they used gas, their opponent would follow suit, thus nullifying whatever advantage the use of gas conferred in the first place. The nonuse of poison gas indicates that in some instances deterrence really works; a combatant may refrain from using a weapon if he is certain that his opponent will use it. But the deployment of poison gas would have subjected its user to a great deal of suffering in return for a transitory advantage.

Fear of retaliation was not the only reason that gas was eschewed. It is possible that military planners would have taken the risks inherent in the use of gas if they felt a greater affinity for it. In fact, they never felt comfortable with the use of gas. Poison gas was unlike all other weapons—silent, insidious, and even dishonorable. As one World War I German commanding general put it, "I must confess that the commission for poisoning the enemy just as one poisons rats struck me as it must any straightforward soldier; it was repulsive to me."[37] At the same time, poison gas represented the sort of scientific advance that rendered obsolete many of the classic martial virtues. A technologically superior enemy with sufficient industrial capacity could triumph over superior training and leadership.[38] Some advances in military technology could be fit into existing molds—an armored tank could be seen as a modern embodiment of the cavalry—but poison gas was something altogether different. Irrespective of their nationality, military leaders were disinclined to use poison gas, and their reticence helped to prevent modern warfare from being even more horrible than it actually was.

Gun Control in Old Japan

The nonuse of poisonous gas is not the only example of a potentially significant military technology falling by the wayside. During the middle of the sixteenth century European traders introduced firearms into Japan, and guns rapidly diffused through the country during the succeeding decades and became dominant weapons on Japanese battlefields. But during the early seventeenth century they began to disappear, and for 250 years Japanese soldiers reverted to swords, spears, and bows. The eclipse of firearms was not caused by a lack of technological prowess. Although pistols and muskets were foreign inventions, Japanese craftsmen quickly learned to duplicate them, making some useful improvements in the process. Not long after the introduction of firearms into their country, Japanese gunsmiths were producing weapons that were the equal of any in the world. And they produced them in large numbers; in one battle 10,000 soldiers on one side were armed with matchlock guns.[39] The Japanese had already demonstrated their technological virtuosity in building construction, metallurgy, agriculture, and a host of other areas; they certainly were not stymied by the demands of gun manufacture.

The reasons that firearms disappeared from Japan were cultural and political.[40] As had been the case in feudal Europe, the political elite of Japan were warriors, and Japanese culture was deeply suffused with a martial ethic. Even after the Tokugawa shogunate initiated a long period of peace, the nation was dominated by the samurai, a warrior class that comprised almost 8 percent of the total population. The ethos of this class centered on the military virtues of strength and courage. Their favored weapon was the sword, and only they had the right to wear and use it. Even if it was never used, the sword was the symbol of their special status, and it was much valued for its aesthetic appeal.

In contrast, firearms were foreign imports, with no connection to traditional Japanese ways. Of greater importance, the use of firearms in combat destroyed the time-honored way that the Japanese had gone about the business of warfare. War was a highly stylized affair that began with the ritual introduction of the main combatants. Its essence was hand-to-hand combat waged by heroic opponents who had ample opportunities to display their bravery and prowess. Firearms were highly disruptive to this form of warfare. As had been the case in early modern Europe, a line of soldiers armed with muskets could quickly decimate the bravest and most skillful warriors. Adding insult to injury, soldiers with firearms were social inferiors to the soldiers that they picked off with such ease and impunity.

If Japan had been threatened by foreign armies, it undoubtedly would have been forced to adopt the same weaponry as its adversary. But as an island nation, Japan was able to close its doors to the outside world and avoid foreign invasion. Only in the middle of the nineteenth century did foreign powers begin to threaten Japanese independence. When that happened, the Japanese wasted no time in arming themselves with modern weaponry.

But before this happened, Japan had enjoyed two and a half peaceful centuries in which the most dangerous weapon was the samurai's sword. The government achieved this by centralizing the production of guns in one city and then strictly controlling their distribution. Purchasers of guns were required to have permission from the Commissioner of Guns, but in fact the commissioner granted permits only to the government. Since the government bought very few, gun production virtually ended in Japan during the early seventeenth century, and the use of firearms faded from the scene.

Japan's experience from the seventeenth to the early nineteenth centuries shows that a superior military technology is not always destined to be universally accepted; here, as elsewhere, technological determinism shows its inadequacies. In the case of firearms in Japan at least, it was possible to "turn back the clock." At the same time, it must be remembered that the disappearance of firearms was the result of a conscious political decision, a decision that the government did not make through altruistic motives. These weapons vanished after the Tokugawa shoguns succeeded in bringing Japan under centralized control. Local power holders had been neutralized, and the shoguns wanted to maintain this state of affairs. The spread of firearms would have made it much easier to mount local rebellions. At the same time, these weapons would have allowed the common folk to pose an effective challenge to the martial class that formed the core of the shogun's military power. It was therefore obviously to the advantage of the ruling elite to reverse the course of technological development by eliminating firearms from their domain.

The case of Japan during this era leaves us with a certain sense of optimism, for it shows that military technologies do not necessarily have an unstoppable momentum; the development and use of increasingly deadly weapons can be arrested. At the same time, however, it is necessary to temper this optimism with a dash of cynicism; this case also shows that disarmament is most likely to take place when it works to the advantage of a dominant group. Today's nuclear weapons make arms control advantageous to all. But can we effectively formulate and enact arms control programs that allow us to realize our own best interests? It is to this crucial issue that we will next turn.

The Control of Nuclear Weapons

There is widespread agreement among both civilian and military leaders that an all-out nuclear war would be unwinnable. Since no combatant could expect to profit from the use of nuclear weapons, it is to the advantage of all to prevent these weapons from being used. Effective arms control agreements are not generated by one nation's willingness to make one-sided sacrifices; they are the product of the realization that all parties would gain from them. Nor is it necessary that the parties to an agreement be friends; after all, the very need to come to an agreement implies that there are basic conflicts between the two parties. Nations enter into arms control agreements not because they are on the best of terms with the other parties, but because they have good reason to fear the consequences of unshackled military power. In sum, arms control treaties are signed when nations see them as being in their best interest, and as long as their interests are preserved, they have no reason to violate the agreements.[41]

The United States and Russia (including the Soviet Union) have a long history of efforts to limit nuclear weapons.[42] The first of these was a 1963 agreement to cease nuclear tests in the atmosphere, outer space, or under water, and to restrict the detonation of nuclear devices underground. This was followed by several treaties to reduce the number of ballistic missiles in each country's arsenal. One important example occurred in 1987 when Mikhail Gorbachev and Ronald Reagan signed the Intermediate-Range Nuclear Forces Treaty (INF), an agreement to eliminate ground-launched, mid-range nuclear missiles (ranging from about 300 to 3,400 miles). This was followed by the New Strategic Arms Reduction Treaty (START), which was approved by the U.S. Congress in late 2010 and by the Russian Parliament in early 2011. This set a 30 percent reduction on deployed warheads and lowered caps on deployed and nondeployed intercontinental ballistic missile launchers, submarine-launched ballistic missile launchers, and heavy bombers equipped for nuclear weapons.

Efforts to build on these and other treaties have been hindered by disagreements over the development of missile defense systems, and by accusations by each country that the other party was not adhering to the terms of the treaties. In 2019, the Trump administration suspended its obligations under the INF Treaty after Russia continued to deny that it possessed a prohibited cruise missile, leaving the New START agreement as the last remaining treaty aimed at the limitation of the two countries' nuclear weapons.[43] While Russia remains in conflict in the Ukraine and facing global sanctions, as of this writing (November 2023), the New Start Treaty remains in effect through February 2026.[44]

Deterrence, but No More

Despite a high level of tension between the United States and Soviet Union during the Cold War era, neither side was inclined to take on the other in a full-scale war. Much of the reason for this reluctance lay in the fact that both nations had the capacity to utterly ruin each other by launching a nuclear attack in retaliation for the other's attack. This policy is known as mutually assured destruction (MAD), and although not pleasant to contemplate, it has had some positive results. Despite all of the insecurity it engendered, the development of nuclear military technologies helped to keep the peace because neither nation dared attack the other for fear of retaliation.

Although the Soviet Union no longer exists, our defense strategy continues to rest on this principle. However, deterrence through mutually assured destruction has its obvious perils. To have a credible capacity to retaliate is similar to keeping a gun in your house for protection. A gun is of little use unless it is loaded and readily accessible. But at the same time, this availability increases the level of danger, for the gun could easily be used by a small child, a disturbed acquaintance, or an enraged spouse. In similar fashion, nuclear weapons must be kept ready for immediate use, but this readiness increases the chances that they will be launched as a result of an accident, miscalculation, or mechanical failure.[45] Military planners are thus confronted with the inherently conflicting demands of peace and war. In peacetime, an accidental or unauthorized use of nuclear weapons is a constant peril; at the same time, an effective deterrent requires the ability to launch retaliatory attacks without excessive delay or deliberation.[46]

The maintenance of a credible yet reasonably safe deterrent is difficult in its own right, and recent technological developments have made it even harder. If one country has the capacity to destroy an enemy's retaliatory forces in a first-strike attack, the ability to deter is lost. Although it is impossible to make completely accurate predictions about the survivability of ICBMs, bombers, and missile-carrying submarines (the "strategic triad") in the event of a preemptive attack, there can be little doubt that advances in delivery systems, such as the use of multiple warheads known as MIRVs (multiple independently targeted reentry vehicles) on missiles, have increased the vulnerability of at least the land-based component of retaliatory forces.[47] Even if sufficient numbers of retaliatory weapons survived such an attack, the disruption of civil and military communications facilities might make the launching of a retaliatory strike difficult or even impossible.[48]

This vulnerability to a first-strike attack creates a dangerous incentive for a nation to mount an attack on an enemy before damage is suffered, or indeed, before it is entirely certain that an attack is under way. Once a potential enemy possesses a first-strike capability it instills a "use 'em or lose 'em" mentality in other nations. If these nations' retaliatory weapons are vulnerable to a preemptive strike there will be a strong incentive to strike first in a crisis situation; otherwise, most of these weapons could be destroyed before they were launched.

No present nuclear power has deliberately sought a first-strike capability nor used it to achieve a position of military superiority. Rather, official policy has stressed maintaining parity with potential adversaries. Even so, devising mutually acceptable arms control agreements that preserve parity is very difficult. "Parity" is an elusive concept, more so now than at any time before. In the past, it might have been possible to maintain parity by ensuring that one nation did not get ahead of others by acquiring a numerical superiority in the quantity of weapons, soldiers, and war materiel. Today, the situation is different. In addition to considering sheer numbers, successful arms control agreements have to take into account the technological level of the respective military forces. These qualitative assessments are inherently more difficult than those that center on the mere counting of weapons. To cite one example, the first Strategic Arms Limitation Treaty (SALT I, which was signed into law in 1972) between the United States and the Soviet Union resulted in restrictions on the number of missiles, but it failed to halt technological advances that produced missiles with greater accuracy. This produced a new kind of arms race, one that was qualitative instead of quantitative.[49] Further complicating matters, modern technologies are inherently dynamic, so changes in military technology can render an existing arms control agreement irrelevant. It also makes it even more difficult to negotiate a new one. As one group of scholars has pointed out, "The negotiation of arms control agreements takes time, and the development of technology can run ahead of the ability of arms control negotiators to come up with effective means of identifying, counting and limiting weapons."[50]

This problem is illustrated by the case of cruise missiles. An arms control agreement requires the ability to verify how many weapons have been deployed; cruise missiles and mobile launching facilities are extremely difficult to identify through nonintrusive inspection measures. Additionally, it is difficult to determine through direct observation if a cruise missile is a tactical weapon with a conventional warhead or a strategic weapon with a nuclear one.

The Perils of Proliferation

During the Cold War, military planners were largely concerned with the prospect of a nuclear war between the United States and the Soviet Union. Much less attention was devoted to an issue that was taking on growing importance, the development of nuclear capabilities in other parts of the world. There are certainly no scientific obstacles to many nations "going nuclear," as there are no real secrets about the construction of atomic weapons any more. The present situation, where eight nations (the United States, Russia, Great Britain, France, China, Pakistan, India, and North Korea) have already tested nuclear weapons, is dangerous enough; the prospects of a world full of nuclear powers is frightening beyond description.

Still, some encouragement can be taken in other nations' willingness to cooperate in limiting the spread of nuclear weapons. The most visible manifestation of this spirit has been the Nuclear Nonproliferation Treaty, which by 2018 had 191 signatories, including five nations with known nuclear capabilities. Its key provisions enjoin established nuclear powers from transferring nuclear weapons to other countries or aiding in their production. Nonnuclear signatories agree to refrain from acquiring or producing nuclear weapons.[51] In another effort at nonproliferation, all of the nations of Latin America except Cuba are signatories to the Treaty of Tlatelolco (in force since 1968), which mandates that their region be kept free of nuclear weapons.

Up to now, the spread of nuclear weapons has been kept in check as most of the nations of the world have observed the Nonproliferation Treaty. At the same time, however, several nations have not signed the treaty, including two existing nuclear powers, India and Pakistan, as well as Israel, which almost certainly has a nuclear capability, even if it has not yet tested a nuclear device. North Korea, which has tested nuclear weapons on six occasions, withdrew from the treaty in 2003.[52] The treaty's signatories have the right to withdraw on three months' notice; thus, they could easily produce nuclear weapons secretly and then begin to test them after withdrawing from the treaty. One thing that has prevented the nations that have signed the treaty from actively flouting it has been the willingness of the nuclear nations to assist the nonnuclear ones in the development of nonmilitary applications of nuclear energy. As things have turned out, this may not have produced great benefits, for nuclear power is not well-suited to the needs of many developing countries. And even "peaceful" applications of nuclear technology can be applied to warlike purposes. It is also likely that the nonnuclear countries may be less inclined to continue practicing self-restraint if the major powers make only halting progress in reducing their own nuclear arsenals.

Is warfare, defined as armed conflict between nation states, inevitable? Or can we learn to get along? There are several heuristics on both sides. For example, the "McDonald's Hypothesis" suggests that nation-states do not go to war if they both have McDonalds as a corporate actor. In some sense this is true, but not in a preventative way. For example, with the 2022 Russian invasion of Ukraine, McDonalds withdrew corporate operations in Russia shortly after the invasion began.[53] This model is closely related to ideas about economic exchange—that it is less likely for nations with closely intertwined economies to escalate to direct conflict. This model is often implied when discussing the unwillingness of the United States to interfere with China, despite human rights violations and global economic interference. Because China produces about 2/3 of the products sold in the U.S. consumer electronics market, it's unlikely that warfare will be a first-choice option for international conflict.[54]

But arguments abound that predict the continuation of war—that it's big business, for example.[55] President Eisenhower warned of the military–industrial complex and its potential to direct national interests for its own profits.[56] Or, as we noted, cyber-systems and drones depersonalize warfare and reduce the psychological brakes on harming others. And because military careers are most rapidly advanced in times of conflict, there are internal mechanisms which promote interest in armed engagements. Finally, arms races, whether nuclear or conventional, embrace a logic that supports the continuation of armed conflict.

On November 9, 1989, the Berlin Wall was breached and the process of German reunification began. By late 1991, the USSR was disaggregated into independent states, and the "Cold War" which drove four decades of much military planning and expenditures was over.[57] People discussed the possibility of a "peace dividend" whereby the funds expended on the military could be redistributed toward infrastructure, research, and the general welfare.[58] Despite fiscal concerns, the United States rarely reduces military spending,[59] even when disengaging from direct international conflict, such as from Afghanistan in 2021. And the question remains as to how to address terrorism, civil wars, and intervention in autocratic regions which threaten human rights and global stability, as these conflicts involve nonstate actors.

QUESTIONS FOR DISCUSSION

1. In what ways have warfare and military preparation stimulated technological development? Do civilian "spinoffs" from the military sector partially justify today's tremendous expenditures on military technology?
2. During World War I, high-ranking officers sent hundreds of thousands of men to their deaths in futile attempts to gain

territory defended by machine guns, artillery, and barbed wire. Why did they do so? Could their social origins and past educational experiences have blinded them to the realities of modern warfare?

3. Firearms were suppressed in Tokugawa Japan because it was in the interests of the ruling elite to do so. Are the political elites of modern nations similarly motivated to prevent the development and use of new weapons? Do individual nations differ from one another in this regard?
4. Is mutually assured destruction (MAD) an inherently immoral policy? Can it continue to be the cornerstone of our military policy? Are there any alternatives other than the forging of effective arms control treaties?
5. The history of arms control agreements presents, at best, a mixed record; a few successful agreements are more than offset by many failures. But nuclear weapons have been held in check up to now. Is there anything about these weapons that makes them better subjects for successful arms control agreements? Is it reasonable to hope that we will succeed in controlling these weapons, or are we living on borrowed time?
6. The threat of retaliation has often prevented attacks from hostile countries. Today, however, an attack may come from an amorphous terrorist group with no significant territorial base. How can retaliation be a credible threat in such circumstances? Are there other ways to prevent terrorist attacks?
7. What are the factors that support the idea that war between nation states is inevitable? What are the factors that suggest that war between nation states might become a thing of the past? What about other kinds of armed conflicts?

Notes

1. See, for example, Richard P. Hallion, *Storm over Iraq: Air Power and the Gulf War* (Washington and London: Smithsonian Institution Press, 1992), pp. 59–61.
2. The narrative is based on Thomas Esper, "The Replacement of the Longbow by Firearms in the English Army," *Technology and Culture* 6, 3 (Summer 1965): 382–393.

3. Ibid., pp. 392–393. See also Hugh D. H. Soar, *The Crooked Stick: A History of the Longbow* (Yardley, PA: Westholme, 2004), pp. 82, 99.
4. Philipe Contamine, *War in the Middle Ages* (Oxford: Basil Blackwell, 1984), p. 217.
5. Bert S. Hall, *Weapons and Warfare in Renaissance Europe: Gunpowder, Technology, and Taxes* (Baltimore and London: Johns Hopkins University Press, 1997), p. 148.
6. Geoffrey Parker (Ed.), *The Cambridge Illustrated History of Warfare: The Triumph of the West* (Cambridge: Cambridge University Press, 1995), p. 178.
7. William H. McNeill, *The Pursuit of Power: Technology, Armed Force, and Society since A.D. 1000* (Chicago: University of Chicago Press, 1982), p. 83.
8. Peter Browning, *The Changing Nature of Warfare: The Development of Land Warfare from 1792 to 1945* (Cambridge: Cambridge University Press, 2002) p. 100.
9. David A. Armstrong, *Bullets and Bureaucrats: The Machine Gun and the United States Army, 1861–1916* (Westport, CT: Greenwood Press, 1982), p. 173.
10. Armin Hall, *Tanks: An Illustrated History of Fighting Vehicles* (NY: Crescent Books, 1971), p. 28.
11. David E. Johnson, *Fast Tanks and Heavy Bombers: Innovation in the U.S. Army, 1917–1945* (Ithaca and London: Cornell University Press, 1998), p. 98.
12. Ibid., pp. 136–137.
13. Witold Rybczynski, *Taming the Tiger: The Struggle to Control Technology* (New York: Viking/Penguin, 1985), p. 171.
14. Richard A. Preston, Sydney F. Wise, and Herman O. Werner, *A History of Warfare and Its Interrelationships with Western Society* (New York: Frederick A. Praeger, 1970), p. 281.
15. Quoted in Hall, *Tanks*, p. 79.
16. Preston, Wise, and Werner, *A History of Warfare.*
17. Walter Millis, *Arms and Men: A Study of American Military History* (New York: New American Library, 1956), p. 16.
18. B. H. Liddell Hart, *The Revolution in Warfare* (London: Faber and Faber, 1946), pp. 40–45.
19. Millis, *Arms and Men*, p. 48.
20. Michael Howard, *War in European History* (London: Oxford University Press, 1976), p. 99.
21. Ken Alder, *Engineering the Revolution: Arms and Enlightenment in France, 1763–1815* (Princeton, NJ: Princeton University Press, 1997), pp. 173, 333.
22. Preston, Wise, and Werner, *A History of Warfare*, p. 320.
23. Seymour Melman, "Swords into Plowshares: Converting from Military to Civilian Production," *Technology Review* 89, 1 (January 1986): 64.
24. Stockholm International Peace Research Institute, SIPRI Yearbook, 1984 (Stockholm: SIPRI, 1984), p. 170.
25. Mark Hourihan, "A Primer on Recet Trends in R&D Budgets,"The American Association for the Advancement of Science, (January 15, 2015). Accessed from www.aaas.org/sites/default/files/R%26D%20Budget%20Overview.pdf.

26. Congressional Research Service (2020). *Government Expenditures on Defense Research and Development by the United States and Other OECD Countries: Fact Sheet.* https://sgp.fas.org/crs/natsec/R45441.pdf; and 2023 appropriations: https://appropriations.house.gov/news/press-releases/house-passes-2023-government-funding-legislation and estimations of R&D activities: https://www.aaas.org/news/fy-2023-omnibus-highlights-and-rd-estimates.
27. Simon Winchester, *The Perfectionists : How Precision Engineers Created the Modern World* (New York: Harper Collins, 2018).
28. Robert S. Woodbury, "The 'American System' of Manufacturing," in Edwin T. Layton, Jr. (Ed.), *Technology and Social Change in America* (New York: Harper & Row, 1973), pp. 47–63.
29. Maurice Pearton, *Diplomacy, War, and Technology since 1830* (Lawrence: University Press of Kansas, 1984), pp. 64–76.
30. Howard, *War in European History*, p. 116.
31. Bernard Brodie and Fawn Brodie, *From Crossbow to H-Bomb* (Bloomington: Indiana University Press, 1973), p. 10.
32. Ibid.
33. W. Reid, *The Lore of Arms: A Concise History of Weaponry* (New York: Facts on File, 1984), p. 88.
34. Quoted in Daniel R. Headrick, *The Tools of Empire: Technology and European Imperialism in the Nineteenth Century* (New York: Oxford University Press, 1981), p. 86.
35. James E. Dougherty, *How to Think about Arms Control and Disarmament* (New York: Crane, Russak, 1973), pp. 43–44.
36. Marek Pruszewicz, "How Deadly Was the Poison Gas of WW I?" *BBC World Service* (January 30, 2016). Accessed from www.bbc.com/news/magazine-31042472.
37. Quoted in Frederic J. Brown, *Chemical Warfare: A Study in Restraints* (Princeton, NJ: Princeton University Press, 1968), p. 41.
38. Ibid., p. 40.
39. Noel Perrin, *Giving Up the Gun: Japan's Reversion to the Sword, 1543–1879* (Boston: David R. Godine, 1979), p. 19.
40. The narrative is based on Ibid., pp. 33–45.
41. John H. Barton, *The Politics of Peace: An Evaluation of Arms Control* (Stanford, CA: Stanford University Press, 1981), p. 105.
42. For a chronological summary of nuclear limitation treaties between the United States, the Soviet Union and Russia, see Council on Foreign Relations, "U.S.-Russia Nuclear Arms Control 1949-2019." Accessed from www.cfr.org/timeline/us-russia-nuclear-arms-control?gclid=EAIaIQobChMIo8uV4s2k5gIVkhh9Ch0F4AnUEAAYASAAEgLG1fD_BwE.
43. See: Council on Foreign Relations, "U.S.-Russia Nuclear Arms Control," at www.cfr.org/timeline/us-russia-nuclear-arms-control?gclid=EAIaIQobChMIo8uV4s2k5gIVkhh9Ch0F4AnUEAAYASAAEgLG1fD_BwE.
44. See: www.state.gov/new-start.

45. Harvard Nuclear Study Group, *Living with Nuclear Weapons* (Cambridge, MA: Harvard University Press, 1983), p. 34.
46. John Steinbruner, "Launch under Attack," *Scientific American* 251, 1 (January 1984): 38.
47. Dietrich Schroeer, *Science, Technology, and the Nuclear Arms Race* (New York: John Wiley & Sons, 1984), pp. 158–159.
48. Steinbruner, "Launch under Attack."
49. Herbert Scoville, Jr., "The SALT Negotiations," *Scientific American* 237, 2 (August 1977): 24.
50. Harvard Nuclear Study Group, *Living with Nuclear Weapons*, p. 153.
51. United Nations Office for Disarmament Affairs, "Treaty on the Non-Proliferation of Nuclear Weapons (NPT)." Accessed from www.un.org/disarmament/WMD/Nuclear/NPT.shtml.
52. "North Korea: What Missiles Does It Have?" Accessed from www.bbc.com/news/world-asia-41174689.
53. See McDonalds' statement: "McDonald's To Temporarily Close Restaurants & Pause Operations in Russia," at https://corporate.mcdonalds.com/corpmcd/en-us/our-stories/article/ourstories.Russia-update.html.
54. See: Consumer Electronics—China at www.statista.com/outlook/cmo/consumer-electronics/china; and Hugh R. Morley, "China Dominates Growing Us Electronics Imports," Journal of Commerce (2018): www.joc.com/breakbulk/china-dominates-us-electronics-imports-%E2%80%93-big-and-small-screens_20180525.html.
55. Peter G. Peterson Foundation, "The United States Spends More on Defense than the Next 10 Countries Combined," (April 24, 2023). Accessed from www.pgpf.org/blog/2022/06/the-united-states-spends-more-on-defense-than-the-next-9-countries-combined.
56. See President Dwight D. Eisenhower's farewell address at www.archives.gov/milestone-documents/president-dwight-d-eisenhowers-farewell-address.
57. See more info at the Office of the Historian: https://history.state.gov/departmenthistory/short-history/berlinwall.
58. Steve Chan, "Grasping the Peace Dividend: Some Propositions on the Conversion of Swords into Plowshares," *Mershon International Studies Review* 39, 1 (1995): 53–95; See also: Hamid Davoodi, Benedict Clemens, Jerald Schiff, and Peter Debaere, "Military Spending, the Peace Dividend, and Fiscal Adjustment," International Monetary Fund Working Paper (1999) at www.imf.org/external/pubs/ft/wp/1999/wp9987.pdf
59. Jonathan Guyer, "Why the US is paying more for the military after the Afghanistan war is over," *Vox* (December 22, 2021). Accessed from www.vox.com/22840615/us-defense-spending-increase-afghanistan-withdrawal; and Jeffrey Mervis, "Rise in Defense Spending Means Smaller Increase for Science in New U.S. Budget," *Science* (March 16, 2022). Accessed from www.science.org/content/article/rise-defense-spending-means-smaller-increase-science-new-u-s-budget.

CHAPTER 21

Accidents, Disasters, and the Negative Consequences of Working Technologies

While the chapters on military technologies trace the trajectory of innovations meant to threaten or do harm to intervene in political disputes, most technologies, when working properly, are not expected to produce direct harm to their users. When a technology fails, rather than works as expected, harms are produced through injury, environmental degradation, and economic losses. In this chapter we look at cases of technological failures, such as the Three Mile Island nuclear power plant accident, the massive oil spill in the Gulf of Mexico, the catastrophic failure of the Fukushima nuclear plant in Japan, and the disastrous breaching of the levees in New Orleans in the wake of Hurricane Katrina. We introduced several of these cases in Chapter 1. The Challenger launch decision will also make an appearance in Chapter 23, where we discuss ethics, professions, and in the case of the 1986 space shuttle launch failure, whistleblowing and the responsibility and challenges of professionals to disclose the inner working of technological decision making. We will also discuss the negative consequences of technologies that are working for some users but producing harms for others. We start with a framework for understanding technological failures, particularly large-scale systemic failures in complex systems with distributed consequences.

"Simple" Failures

So you dropped your cell phone into a puddle, both cracking the case and shorting out the electronics. While that seems to be a pretty straightforward user error that produced a nonworking technology, the consequences could be significant beyond just the cost and inconvenience of a broken phone. What if you missed an invitation to a job interview? Or lost important data? Or missed an important life event or emergency? Or couldn't reach out for help? All of these are complex and personally meaningful and context-specific outcomes that emerge from a relatively small-scale accident. Even a simple failure can have a cascade of consequences that can range from the merely inconvenient to the personally devastating.

Is a dropped phone any different from a cell phone that catches on fire due to a battery malfunction?[1] Most of the consequences might be similar, but there is a different question about responsibility for the accident and for reparations. We can look to the commodity chain as a framework for analyzing a basic consumer product and the potential sources of error or poor manufacture. For our fiery cell phone, the lithium battery failure could be the outcome of a flawed manufacturing process or poor design, or user error. Does information get back to the various actors in the multiple phases of the commodity chain to improve safety and reliability? Where does responsibility for repairs or reparations lie?

> "A little neglect may breed great mischief. For want of a nail, the shoe was lost; for want of a shoe, the horse was lost; for want of a horse, the battle was lost; for want of the battle, the war was lost."

This version of the old adage was included by Benjamin Franklin in the 1758 *Poor Richard's Almanac*, while other versions sometime include mentions of lost messages, and ultimately, lost kingdoms. Whichever the version, we can see in this adage the idea of the cascading effects of a technological failure. Small errors or malfunctions, even those anticipated with safety devices and procedures, can escalate into large-scale events with catastrophic outcomes.

Normal Accidents

Sociologist Charles Perrow was tasked with an "organizational analysis" report following the 1979 nuclear accident at Unit 2 of the Three Mile Island nuclear power plant, near Harrisburg, Pennsylvania. The "President's Commission on the Accident at Three Mile Island,"[2] was assembled as a 12-person team with expertise but without strong pro- or anti-nuclear views. Called into action by then-President Carter, the commission was tasked with finding an explanation of the causes and a survey of the effects of the disaster. Three Mile Island's cascade of failures produced a partial meltdown and measurable radioactive leaks, although it has not been possible to demonstrate with any great certainty that these leaks produced illnesses in humans. The accident did intensify the decline of the already-slowing nuclear industry. Unit 2 is expected to be decommissioned by 2037.[3]

From the study of Three Mile Island, Perrow (1984) coined the term "normal accidents" to describe systems of high complexity and coupling that *will* fail and *may* fail in catastrophic ways. Complexity refers to the ways that subsystems are interconnected, with feedback loops and many control systems that interact in multiple ways. Complex systems stand in contrast to linear systems that have easily isolated components, few feedback loops, and direct single-parameter control systems with few interactions. A paper-making facility would be an example of a linear processing system, while a nuclear power plant is a complex system.

Coupling is an independent factor, and on the surface sounds like complexity. A school or university is an example of a loosely coupled system: success or failure in one classroom or department has little impact on the overall functioning of the school, at least in the short term. Problems are often solved by departments or subunits without centralized control. The paper-making facility mentioned above is linear, but tightly coupled. An error or failure might shut down the entire paper-making-process because each step is dependent on the one before. For a tightly coupled organization and system, outputs are tightly dependent on other steps in the process. Decisions and priority-setting are centrally controlled and need to be conducted in a very timely manner for ordinary operations, and in times of crisis.

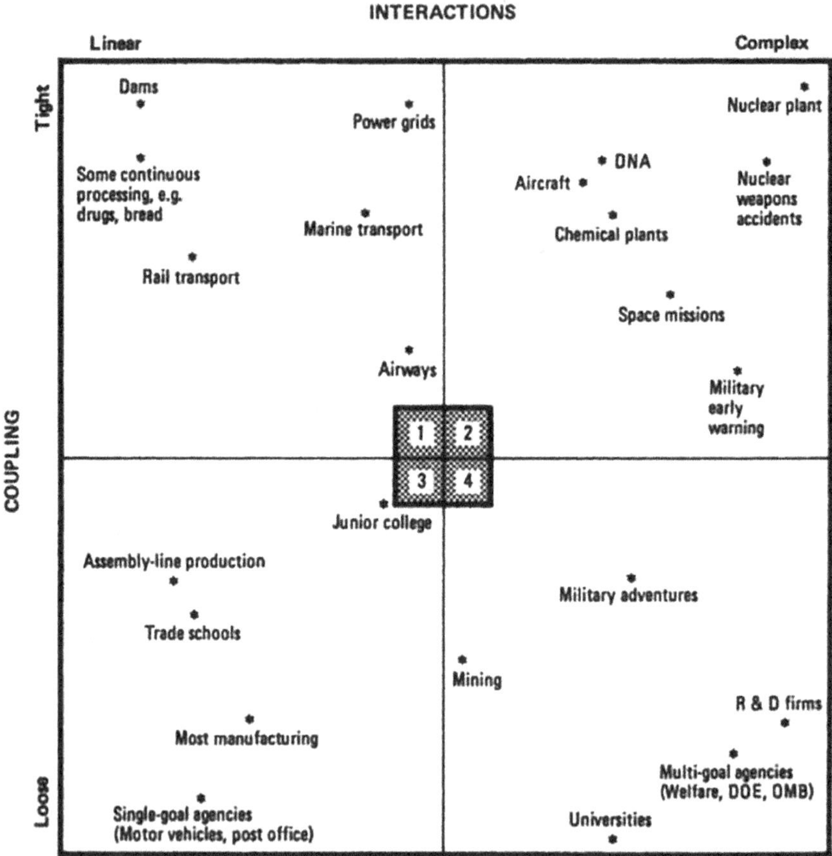

Figure 21.1 Charles Perrow's (1984, p. 97) chart outlining examples of enterprises with varying degrees of coupling and interactions. Each produces different kinds of system failures and organizational responses.

What is important about Perrow's complexity–linearity and tight–loose coupling framework is what it means for cascading failures, and organizational responses to failures. Are there back-up systems? Can a subsystem be separated from the larger process to maintain operations and still solve problems without total shutdown? Does one subsystem provide critical information or processes for another subsystem? Do operators have reliable information, sufficient training, and timely authority to interpret and act on information? When a component fails, how much of operations is shut down? Do the cascading effects of a failure remain at the same scale, or do they continue to grow in scale and scope producing a nonlinear outcome?

Let us review the barest outline of the Three Mile Island cascade of failures, which started with a very minor cooling circuit failure in Unit 2 at about 4 a.m. on March 28, 1979. That failure caused a temperature rise, which then caused the reactor to shut down automatically, within seconds. At this point there are two problems that emerged: A relief valve failed to close, and instrumentation did not indicate that fact. Operators based their procedurally correct responses on this flawed information. The relief valve's failure to close allowed coolant to drain, which then allowed the reactor core, which continually produces heat, to overheat and become damaged. Working on the assumption that a pressurizer level was a good indicator of water level, operators further reduced the water inflow to the reactor, which turned out to be an error. That error allowed steam to form, which stressed pumps with severe vibration. Operators shut down pumps to avoid damaging them, which allowed the core to get hotter because it became uncovered. In an attempt to stop more coolant loss, at 6:22 a.m. operators closed a valve, but steam and gas blocked water flow. Eventually, operators injecting high-pressure water collapsed the blocking steam bubbles and a pump was able to

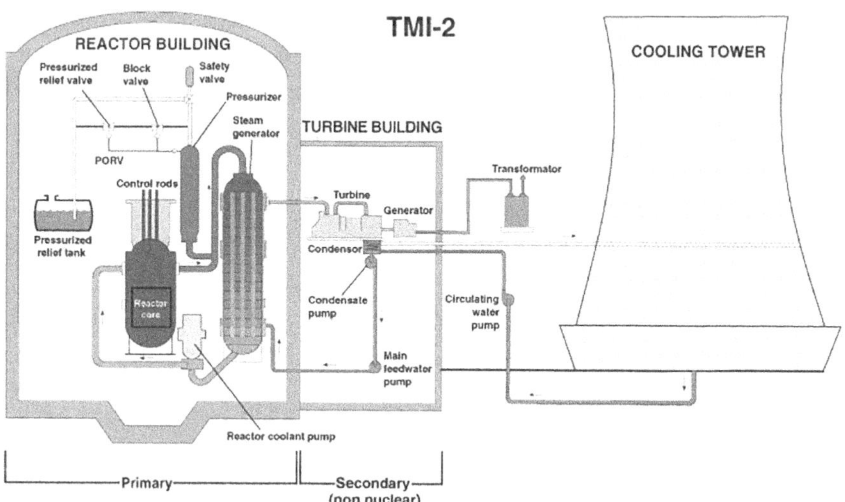

Figure 21.2 A schematic of the Three-Mile Island Reactor #2. (Source: U.S. Nuclear Regulatory Commission)

run without severe vibrations (7:55 a.m.). While the reactor's core was uncovered, it started forming a hydrogen bubble, which at first officials and operators were worried could explode. Here the time scale changes from seconds, then minutes, to days; the hydrogen bubble was released from March 30 through April 1. In late March and through April, radioactive gases were released, although because of a very short radioactive half-life, this was perceived to not be a health hazard. It took until April 27, 1979, for operators to stabilize cooling through natural convection.[4]

Instrumentation and component failures, operator missteps, and organizational errors contributed to the accident. And each of those elements—components, design, instrumentation, operator knowledge, and organizational procedures—kept it from being worse. The Fukushima Daiichi nuclear incident shares much with the Three Mile Island cascade, and yet differs in important ways. On March 11, 2011, Japan was shaken by an offshore earthquake. This earthquake produced a tsunami, which crested at over 10 meters (33 feet or more), and in some locations, over 15 meters. The earthquake had a magnitude of about 9.0, and through a cascade of problems, produced a major accident at the Fukushima Daiichi Nuclear Power Station.[5]

When the Daiichi plant was designed and sited in the 1960s, the assumption was that a tsunami would be somewhat less than 10 meters. The original design basis was for a surge of 3.1 meters,[6] possibly with the typical "engineering factor of safety" multiplier of 3. But even by the early 1990s, the possibility of a higher tsunami, such as 15.7 meters, had been calculated. Like the "500-year flood," which has a 0.2 percent chance of occurring in any given year,[7] tsunami estimates are developing probabilities which can be adjusted as new events occur and models are refined. But the outdated low estimate of a catastrophic post-earthquake surge, and the time and expense of refitting the powerplant, meant that the Daiichi pumps for cooling were not upgraded and were overwhelmed with seawater and failed, leading to a meltdown. Hydrogen accumulation produced a series of explosions over the days after March 11, with a peak radioactive release on March 15, 2011.[8]

The areas within 20 km (just over 12 miles) of the Daiichi facility were subject to extensive evacuation, with decontamination work taking place over the following decades. Over 40,000 people remained displaced from their homes in 2020.[9] Three workers lost their lives in the immediate aftermath of the meltdown, while across Fukushima and nearby prefectures, almost 20,000 people lost their lives due to the earthquake and tsunami. There were fewer than 10 direct injuries or deaths reported from radiation. There are, however, lingering indirect effects, due to the stress of dislocation, or lack of access to medical care due to the destruction, producing more deaths than expected during ordinary times—the "excess deaths" estimation.[10]

Few technological or organizational errors occurred in the immediate unfolding of the Fukushima events. It was a poorly anticipated seismic event that exceeded design parameters, although changing knowledge and expectations were available to organizational leadership, but not acted upon. Timely action on revised surge estimates might have prevented the failure of the cooling system and subsequent meltdown. Inaction proved to be costly, as the entire facility now must be decommissioned, and evacuees have been awarded close to 10 *trillion* yen (approximately $70 billion) as of January 2021, not including the decommissioning costs; costs of using other energy sources; increased costs of insurance; and compensation for lost wages, agricultural production, and transport.[11]

Perrow helps us to articulate some heuristics for the assessment of complex and tightly complex systems with the potential for catastrophic failures. First of all, there is certainly a need for training and oversight of operators to reduce operator error. But in conditions with high coupling and complexity, overly rigid procedures and managerial control might reduce flexibility and responsiveness by sticking to known scripts for attending to problems that may require out-of-the-box thinking and comprehensive, rather than rote, understanding of the system. Too few procedures lead to operator error—this might be obvious. But too many rote procedures may be just as bad. Secondly, a system cannot sensor itself into safety. Adding layers of technology adds complexity. Sensors, for example, can become a source of error themselves. Thirdly, redundancy may increase safety, but it is not necessarily cost-efficient in the short run. Further, redundant systems may become sources of problems: interactions among redundant components with the primary systems may themselves become a source of error. And finally, complexity means that an incident may accelerate and quickly grow to a catastrophic disaster. But technical and economic simplification might not be possible.

Considering Victims: Knowledge, Proximity, and Responsibility

Unlike dropping your own cell phone in a puddle, which probably has a rather narrow number of potential victims, when complex systems fail they often produce a complex array of victims. Perrow also provided a framework for helping us think about the consequences of disasters on human, and perhaps nonhuman, victims.[12] There are three primary dimensions: knowledge, responsibility, and proximity. A first-party victim to a technological disaster is a user who is an owner/operator, who has both the knowledge of and responsibility for the technology. Such first-party victims are often close to the scene. For a consumer product, such as an automobile or airplane, that would be the driver of the vehicle. Someone in a supervisory position at a nuclear power plant, for example, who has both the knowledge and responsibility over system operations might be the first party victim in the case of a nuclear accident. For first party operators, we can identify the causal chain of events with high certainty. The number of people affected is often small, and while it may still be substantial, the cost of repair or reparations is relatively fixed.

Second-party victims are willing participants in a technological system, although they may not have the authority to intervene in system operations. The passenger in an automobile, or a lower-level employee in a nuclear power plant, say janitorial services, would have some idea about the risks involved in taking a ride or going to work, but not have much responsibility for operations. For this class of users, we can often identify with high certainty what happened to inflict harm or damage. The number of people may be higher than first-party victims, and the cost of repair or reparations also increasing in comparison to first party operators.

Third-party victims are innocent bystanders, who may have no idea and no choice about engaging in a risky venture, and they have no control over the process. Kids waiting by a bus stop who are run down by an impaired driver, to continue our automobile example, or long-term residents near a hazardous industry might be representative of this category of disaster victim. The number of people might be very large, and the cost of repairs or reparations may escalate very quickly for third party victims. There's a bit of ambiguity here as to the amount of choice that people have: if it's the best or only job in town, is taking a job at a hazardous job site really a free choice? Or is living near a nuisance or hazard a chosen risk, especially if residency pre-dated the emergence of the toxic facility or is the only source of affordable housing? Innocent bystanders may find it difficult to establish causal linkages: the question of why *now*.

Finally, there are the fourth-party victims. They are often either geographically remote, say babies harmed by nitrogen fertilizer wastes flowing downstream from potato farming.[13] Harms may also be distributed in time far away from a disaster. It is difficult to ascertain some harms with certainty: is someone's thyroid cancer near the Chernobyl region of Ukraine merely bad luck or the result of long-term but small exposure to radioactivity?[14] Was there an actual childhood leukemia cluster in Sierra Vista, Arizona, due to metals, volatile organic compounds, or other toxins from old mining sites and/or an active military base? Or were the cases of leukemia really just a statistical anomaly, perhaps an effect of a small community with dense reporting that made illnesses more visible.[15] For fourth-party victims, uncertainty is high; costs to fix or provide compensation may become astronomical; and the number of people potentially affected is very large, difficult to identify, and perhaps will grow for quite some time.

Table 21.1 Table of the relative dimensions of variation across categories of victims of technological disasters.

Category of Victim	Number	Proximity	Knowledge	Responsibility	Certainty	Cost to Fix
First Party						
Owner/ Operator	Low	Nearby	High	High	High	Specific, low
Second Party						
Willing Participant	Low but increased	Nearby	High	Low	High	Specific, Increased
Third-Party						
Innocent Bystander	Higher	Increased Distance	Low	Low	Decreasing	Increasing Rapidly
Fourth-Party						
Remote Bystander	Very High	Distant in Time and/ or Space	Low	Low	Very low	Unbounded, can be very high

If the 1979 Three Mile Island nuclear disaster accelerated the decline of the nuclear power industry, the April 26, 1986, Chernobyl nuclear disaster in the Ukrainian Soviet Socialist Republic (at that time), cemented negative assessments of nuclear power in the global public imagination. Occurring during a safety test, operators accidentally reduced power output, and in the attempts to stabilize the reactor, control rods were removed which produced localized increases in reactor activity when attempting to shut down the reactor as a whole. This resulted in ruptured fuel channels and produced a positive feedback loop that produced steam explosions and a core meltdown. While the specifics of the operator and system failures are unique to the situation, we can see the general process of cascading effects of a failure in a complex and tightly coupled system that produces a nonlinear, catastrophic response.[16]

In the case of Chernobyl, we can see the four categories of victims in fairly clear outlines: first-party victims were operators and workers at the plant immediately on site and conducting the safety test which went awry. Second-party victims were the heroic pilots and firefighters who gave their lives in the process of damping fires with about 5000 tonnes (11 million pounds) of sand, lead, clay, and boron dropped by helicopter, attempting to slow down the core meltdown. Twenty-eight people died of acute radiation syndrome in the weeks following the accident. Certainty of cause of death is very clear in the first- and second-party victims. But the potential impacts on health for more distributed groups of third- and fourth-party victims, such as from thyroid cancers across the Baltic region or excess suicides of clean-up workers,[17] are harder to determine with certainty as to causality.

We need to consider, as well, the nonhuman victims of technological disasters. Soviet soldiers exterminated many abandoned pets immediately following the evacuations of the Chernobyl region out of fear that they would transport radiation. The few that survived have become important in studying longer-term genetic effects of the disaster.[18] Studies of local wildlife suggest a general reduction of biodiversity and animal health. The effects on animal populations are very visible in a disaster such as the 2010 Deepwater Horizon oil spill. While 11 human workers perished when the oil drilling rig at the British Petroleum Deepwater site exploded in the Gulf of Mexico, the damage to many species, interconnected in a complex ecosystem, is nearly incalculable. In that disaster, the well spilled 4 million barrels of oil over 87 days until it was finally capped. The economic costs of penalties to the petroleum companies, financial holding companies, and operators are approaching a 50 billion dollar value ($50,000,000,000) in both criminal and civil rulings.[19] These financial penalties only partially offset the economic impacts on the fisheries, tourism, and the health of communities.[20] While we can consider the costs in economic terms in terms of human livelihoods, it is more difficult to quantify the health of human and animal communities, and as we have discussed elsewhere, things that are *priceless* have a tendency to disappear from cost-benefit analyses which rely on consensus on their calculated current and future value.[21]

The Chronic Disaster: Not an Oxymoron

Sociologist Kai T. Erikson studied the social consequences of catastrophic events. His 1978 book *Everything in Its Path*, was a groundbreaking study of the February 26, 1972, Buffalo Creek, West Virginia, dam collapse. A series of three mining impound dams collapsed under heavy rains, in sequence, basically scouring a valley bare of vegetation and human habitation and destroying 16 small downstream towns with 132,000,000 gallons (or 500,000 cubic meters) of black coal-tailings water. Four thousand of the 5000-plus residents were left homeless. One hundred twenty-five were killed, and an additional thousand were injured. Immediately following the dam failures, the Federal Emergency Management Administration (FEMA) was on site to provide medical support, housing through portable trailers, and food and emergency management assistance.

Despite significant federal and state aid, the communities along Buffalo Creek were slow to recover, and collective chronic trauma compounded the individual trauma of the loss of life and livelihood. A lack of community and individual resilience prolonged recovery, heightened by the unforeseen consequences of the random assignment of people into trailer parks irrespective of town of origin. This disrupted already vulnerable social networks that had provided stability, accountability, and, above all, shared resources and knowledge. People were stressed out, having lost everything, and then parked in close proximity to relative strangers whom they had no reason to trust. Even nearly two decades later, former residents of washed-out

Figure 21.3 An aerial view of homes washed away in the Buffalo Creek Dam failure. (Courtesy of *The Herald-Dispatch*)

towns expressed more "trait anxiety," meaning undefined fear of potential future events, than people from nearby undamaged towns.[22] It took nearly 50 years for the water quality to be capable of supporting trout fishing again.[23]

The term *disaster* suggests a sudden, catastrophic event—an acute disruption with transformative destruction. Erikson coined the term "chronic disaster"[24] to underscore the pressures already eroding town health and cohesion which inhibited the recovery of the towns of Logan County and other West Virginia communities. Mining in Appalachia has had numerous boom and bust cycles. At the time of the dam failures, mining communities already faced population loss due to outmigration, particularly of the younger generation. As the mining companies moved elsewhere, they stopped investing in local communities or supporting the local tax base. Dwindling economic opportunities had produced multiple generations of stress and loss of optimism, creating a lack of resilience.

We can identify the economic marginalization of rural West Virginians as a form of economic precarity: people may be getting by, but just barely. As the aphorism goes, being poor is being one paycheck away from eviction.[25] Precarity is a term that emerged to describe the vulnerability of economically marginalized groups. More than half of all Americans can't afford a $1000 emergency expense.[26] But it is more than money, it is also the vulnerability where problems, such as an illness or a car breakdown, have cascading effects, such as employment loss, then housing loss, with subsequent disruptions becoming more and more difficult to recover from. Precarity is not only a phenomenon in the United States, but it takes different forms in different global contexts. Refugees live precariously; people forced into urban slums when modern agricultural practices reduce the need for their labor live precariously. Admonitions to just work harder, save better, or not waste money on frivolous expenses do not address unfair compensation, work insecurity, or the lack of worker rights or protections.[27]

People facing high socioeconomic precarity are thus poorly positioned to be able to recover from disasters that might have acute and short-term consequences for more secure populations. The combination of precarity and disasters can produce intergenerational harms that persist over decades, if not longer. Consider the Bhopal disaster of December 1984. The Union Carbide India Limited plant in Bhopal, Madhya Pradesh, India, had a series of leaks and system failures that led to the release of 40 tons of deadly MIC (methyl isocyanate). Immediate deaths were estimated to be between 2,200 and 3,700, with perhaps 8,000 people dying in the first 48 hours. Approximately 574,400 people were injured, many suffering permanent damage to their lungs. Estimates vary, and we can track the first-, second-, and third-party victims of the disaster. But it is now becoming clear that there are intergenerational, fourth-party victims,[28] from those exposed while in utero carrying forward direct injuries and mutagenic harms. Thirty years onward, there are increased disabilities and higher cancer rates among those exposed in utero. In addition, research suggests that the scope of harm was larger than originally thought; there may be exposure consequences nearly 100 km away (62 miles or so), rather than just the immediate vicinity. Weak worker protections and environmental regulations intersected with the high population density, pre-existing urban poverty, and a weak governmental response, to produce a chronic disaster beyond the acute devastation of the arguably worst industrial accident to date.

Beyond Natural Disasters: Into the Anthropocene

Global climate change is both a chronic disaster that is unfolding over time, and produces localized, acute disasters. The climate system is tightly coupled and complex, and it remains difficult to predict local weather conditions over longer periods of time (say, months in advance). This complexity and the stochastic, randomly determined processes generate multiple levels of uncertainty. Meteorological forecasting produces statistical ranges rather than determinant values—a kind of uncertainty. Uncertainty, however, is not the same as nonknowledge, and we can claim, with high certainty, that global climate patterns are highly influenced by human activities.

While it might be considered anthropocentric, identifying the deep and comprehensive human-driven changes to the environment has led to identifying contemporary society as the Anthropocene era. This means that human activities have measurable impacts on large scale global climate and ecological systems, distinct from the already complex natural processes which produce variation and trends in the weather.[29] All organisms have an impact on their ecosystems, but only a few directly and intentionally change their environment like, say, beavers or elephants. The intentionality of human interventions into ecosystems blurs the boundaries between nature and culture. When humans transform waterways and estuaries and shift water-flow patterns, the potential catastrophic flooding that might occur after a severe rainfall because of construction means that the flooding cannot be dismissed as a natural disaster. Similarly, a heatwave cannot be considered a natural disaster, because it is the human organization (or disorganization) in response to the heatwave that is directly causal, producing higher than usual mortality due to excessive heat.[30] Human interventions into global climate systems that produce extreme weather means that there really is no such thing as a "natural" disaster, if there ever was such a thing, at least since humans developed the capacity to choose how to respond to actual and potential environmental threats.

The idea of the Anthropocene, and the notion there is no such thing as a natural disaster, means that effects of major climatological events, such as Hurricane Katrina (an acute disaster in late August 2005) or rising global sea levels (a chronic disaster), cannot be dismissed as somehow outside human political and ethical responsibility, as just "natural" occurrences that people have the bad luck to encounter. Responses to earth systems events, such as a hurricane, are comparable to other technological disasters, such as the failure of a city's water supply or the toxic fumes of a chemical plant explosion, because they are all predicated on the consequences of the choices of infrastructure and organization.

Weather-based disasters are an example of the differential effects of technologies: already marginalized and vulnerable groups face higher rates of both the incidence of sociotechnological failures associated with intense weather events, and more difficult circumstances for recovering from such disasters. For example, in late August of 2005, Hurricane Katrina traversed the Caribbean and slammed into the Gulf Coast of the United States.[31] It arrived in New Orleans as a Category 3 hurricane with winds reaching 174 miles per hour. It had a direct death toll of 1,392 in the United States, with about a third due to drowning. The severe storm was exacerbated by the technological failure of the levee system. More than 50 levees failed, inundating 80 percent of New Orleans.[32]

The Lower Ninth Ward was particularly hard hit. It is an area known for its high proportion of African American residents,[33] and long neglected by city officials.[34] In relation to Hurricane Katrina, the intersection of race and class affected evacuation rates, and the persistence of stress related to the disaster.[35] Orleans, Jefferson, and St. Bernard parishes also had mortality rates which showed the vulnerability of the elderly, with 49 percent of the fatalities over age 75, and higher death rates for Black, and male, residents. In Orleans Parish, mortality was at least 1.7 times higher for Blacks than Whites.[36] Black elders faced higher rates of cardiovascular disease than Whites, before, during, and after Katrina, but the rates of hospitalizations and mortality increased even more during and after the hurricane.[37]

Excess death assessments over subsequent years post-Katrina seem paradoxical. New Orleans had a pre-existing high mortality rate due to the population's poor health and weak health infrastructures, a sign of precarity, so when residents left the area due to the hurricane, they generally went to areas with lower baseline mortality rates.[38] Nevertheless, long-term stress-related mental and physical effects are becoming apparent, with pre-disaster psychological distress compounded by the specific stressors of Hurricane Katrina.[39]

Both the technological features of the levee system and its long-term neglect, and the inept government responses to the disaster contributed to the intensity of the disaster for communities of color in New Orleans, who already faced great precarity and vulnerability.[40] As sociotechnical assemblages, a city's infrastructure has both material and organizational dimensions which embody the legacy of past political action and inaction. Few African Americans assess the government's response to Katrina as effective, drawing very different lessons on the persistence of racial inequality in America from their White counterparts.[41] The antecedents and government responses to Katrina are often considered a contemporary case study for environmental justice.[42]

Infrastructure

Infrastructure systems reflect large commitments and are generally the product of complex historical and political processes. While an essential feature of complex societies, these systems are often invisible and taken-for-granted, especially when working effectively. This often leaves infrastructure systems vulnerable to neglect. Consider the case of Flint, Michigan. In 2014, the Flint municipality shifted its water supply from Lake Huron water treated by the Detroit Water and Sewerage Department to the Flint River and the Flint Water Treatment Plant. The different water supply and lack of capacity led to problems such as bacterial growth, lead contamination, and trihalomethane exposures as the new water supply corroded pipes because of poor treatment management. This is an unfortunately "typical" confluence of technology, politics, and disasters. As Pauli notes:

> A variety of causal explanations for the crisis have been put forward, some focusing on water treatment, regulation, and infrastructure, others on the political context created by state takeover of the city, and others on historical factors like structural racism, deindustrialization, and depopulation.[43]

Some have emphasized this as a regulatory failure.[44] But even a detailed technical analysis makes note that the Flint Water Treatment Plant had not been fully operational for *50 years*, and was chronically understaffed and undertrained.[45] Harms from the degraded water supply included elevated lead in the bloodstream of children of Flint; an outbreak of Legionnaires' disease with at least 13 fatalities; and the possibility of higher rates of miscarriages and fetal deaths, and low birth-weight babies.[46] A crisis lasting eight years might be transitioning from an acute to a chronic disaster.[47] The backdrop of deindustrialization and population and economic losses from the declining auto industry makes the crisis of Flint look similar to the crisis that Erikson described in the Buffalo Creek disaster: high levels of precarity make communities and individuals more vulnerable when crises do happen, and less resilient in terms of recovery from sociotechnical failures. While different in causality, the recent collapse of the Jackson, Mississippi, water system due to unexpected weather events suggests a pattern of infrastructure neglect that is intensified in marginalized communities.

One question that lingers across many of the cases we are examining is the role of government regulations. Clearly, weak regulations or enforcement failures become contributing factors to the effects of technical and ecological disasters. For example, on February 6, 2023, there was a major earthquake in south Turkey and northwestern Syria. The 80 seconds of 7.8 magnitude seismic activity collapsed buildings, destroyed electrical and water distribution, and made about 1.5 million people homeless. The confirmed *direct* death toll was 59,259, with 50,783 in Turkey. The excess deaths caused by the stress and lingering public health issues caused by a disrupted infrastructure have yet to be evaluated. The severity of the disaster is caused by an unpredictable geological process, compounded by poor quality construction, a national response hampered by weather, road damage, and poor communications, and a conflicted global humanitarian response.[48] Poor enforcement of building codes and weak construction regulations are considered a significant factor in the building collapses that produced many of the fatalities.[49]

Similarly, deregulation and privatization are considered the background social factors that produced the February 3, 2023, East Palestine, Ohio, train derailment[50] and complicated responses to it.[51] While there were no apparent first-party operator fatalities, the derailment led to the spilling of hazardous materials including acrylate compounds, isobutylene, and vinyl chloride. These materials caught fire, producing hazardous fumes, and were also spilled into the local water supply. It is, like the disaster in Turkey, still too early to determine the geographically extended and time-dependent consequences of this failure.

But clearly the U.S. railway system is in disarray. On June 8, 2023, a freight train derailment in Norther Arizona damaged a delivery of new cars.[52] A June 24, 2023, derailment in Montana put molten sulfur and asphalt into the Yellowstone River. Railway officials asserted that there were no negative effects.[53] There have been so many train accidents in 2023 that a safety company has provided a map of accidents through May of 2023.[54]

The railways were nationalized for the duration of World War I (1917–1920), but then returned to private sector control shortly after. While many argue that privatization and deregulation of railroads since the 1980 Staggers Rail Act has "saved" the American railway system,[55] a new conversation about railway safety, regulation, and national infrastructure may be in order.[56] Because safety technologies and regulations, plus staffing for inspections and testing, take away from profits, private sector interests have a very narrow scope on the way that risks and harms might be evaluated. There is a lack of incentive for considering long-term consequences, many of which, such as the costs to fourth-party victims of accidents, can be externalized. Our point is that *someone* is going to have to pay, whether through extensive taxes for regulatory purposes; taxes to pay for cleanups; or through decreased profits and potentially increased consumer costs to pay for fines, cleanups, safety systems, and compliance costs via private sector activities. Most distressing to consider is the way that companies and investors can take advantage of disasters to acquire property, disrupt and privatize services such as schools or health systems, or otherwise profit from disasters and failures. There is some concern that the August 2023 Maui fires will produce land loss for individual families as investors sweep in. It is clear that an ongoing legal battle over water rights is part of the backdrop of infrastructure challenges which inhibited access to water to fight the fires.[57] Those living precariously may face the worst consequences of system failures but have the least political and economic power to shape the recovery process. While there are numerous regulatory failures that cause or exacerbate current disasters, there are clear limits to expecting the private sector to act morally above and beyond basic financial self-interests.

From the Commodity Chain to the Spiral of Consequences

As we saw in Chapter 8, the commodity chain provides a tool for examining the distribution of costs and benefits of a working technology. We have encountered many examples of the four categories of victims already, among participants and bystanders facing the distributed effects of hazards along the commodity chain for working technologies. Globally complex commodity chains can produce hybrid victims, who may be users, say of a cell phone, but are also facing the consequences of lithium extraction in Chilean wetlands—both first-party users and fourth-party victims of a technological system.[58]

The people experiencing harms produced by the production and use of a working technology are victims of the technology's success. It is easy to dismiss such harms as "side-effects," as they are not the intended outcome of technological innovation or commodity chains. But such a framing means that the harms, and the people suffering the harms, are considered inconsequential relative to the aggregate benefits of the technology. The harms might certainly be significant to the people experiencing them. The notion of a side-effect also inhibits thinking about how these consequences may grow and spread over time.

In the 1980s, and even into the present, rBST (recombinant bovine somatotropin; also known as rBGH, or just "bovine growth hormone") has been a controversial treatment for cows to improve productivity. It increases milk production by 10–15 percent, which has been argued to mean fewer cows, improved environmental outcomes with less silage needed, less cow excrement and methane gas production,[59] and improved profits. But milk is not a scarce product, it is sometimes dumped by farmers to decrease supply and increase prices.[60] It may be that rBST is a technology in search of a problem, from a consumer perspective, although the ways that it leads to farm consolidation and higher profits for larger farms suggests that the end-users of milk (ordinary consumers) are not the intended audience of the technology.

BGH is not *directly* harmful to human health, despite episodic panics about its presence in the milk supply. Nevertheless, we need to take issue with Neil DeGrasse Tyson's endorsement of GMOs and biotechnologies for food production.[61] We agree with Professor Tyson that GMOs are not hazardous for the human body, at least in some directly measurable way. But they are harmful for human communities, and in the case of rBST, terrible for cows,[62] and indirectly bad for human health because rBST use increases the incidence of mastitis in cows, increasing the prevalence of antibiotics in the milk supply. There is some evidence that rBST use decreases a cow's useful life, leading to earlier culling for meat.[63]

Like rBST, GMOs (genetically modified or genetically engineered organisms) can be the site of analysis of the negative consequences of a *working* technology. As we have seen elsewhere, GMOs are a technological success, with widespread acceptance in the Western Hemisphere and much of Asia, but banned elsewhere. The market size is estimated to be about $22 billion U.S. dollars in 2022,[64] but the apparent success of GMOs holds only if the boundaries of analysis are tightly drawn. GMOs are often *more* energy- and water-intensive crops, and often disrupt traditional farming and community practices. They produce a set of cascading externalities where the negative consequences are born by producers, which remain rather invisible to end users—consumers. Amongst the negative side-effects, GMO use is premised on monocropping, which reduces variability and can leave an entire crop susceptible to a new threat. In the spirit of the *Jurassic Park* franchise, "life finds a way," and pests or weeds start to develop resistance to the treatments which give GMOs their advantages.[65] Similarly, GMO use decreases competition amongst firms, producing oligopoly or even monopoly conditions. Industry consolidation can produce higher prices and less choice for the farmers which bear the economic risks of GMO use, and ultimately less choice for consumers.

Ehrenfeld suggests that we consider a "spiral" of consequences for evaluating products such as rBST.[66] While large corporations might want us to stay focused only on assessing technologies such as GMOs or growth hormones direct effects on human health, this focus keeps us from considering other, more widely distributed effects. The spiral is a twist on the commodity chain analysis (Chapter 8) and adds a sense of scale and longer time spans for considering the negative externalities of technologies on a broad range of human and ecological concerns.

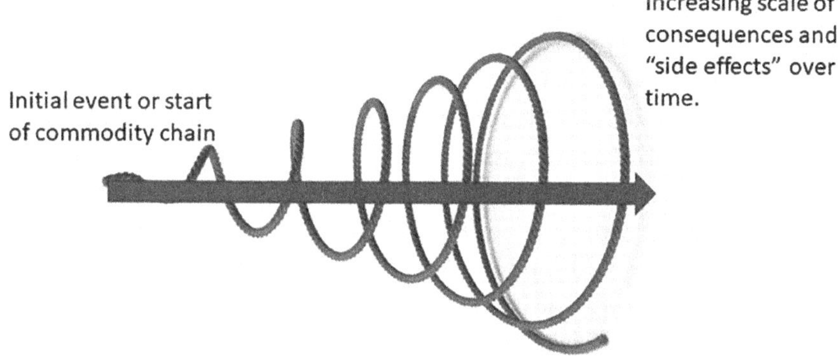

Figure 21.4 Spiral of consequences. (After Ehrenfeld, 2002)

Perrow's ultimate argument is there are some complex and necessarily tightly coupled systems that we should probably avoid, ones that can fail with cataclysmic, nonlinear outcomes that affect a great many people over time and in geographically distributed areas. Strategies internal to a system could help improve system safety and decrease the scope of many system failures. Decreasing complexity, improving organizational capacities for managing anomalies, balancing training and flexibility for operators, systems design for fail-safe modes, and system and organizational transparency might improve safety and cost-effectiveness for production and reduce the possibility of catastrophic failures. Regulations, oversight, and expectations for responsibility for failures can be placed on the owners of complex systems such as corporations or public utilities. Holding owners responsible instead of shuffling the cost of disasters onto the public as economic externalities would also greatly incentivize safer systems or technological alternatives, which are necessarily expensive.

There can be no completely fail-safe or fool-proof system: components and indicators fail, and people make mistakes. Who bears the costs of those failures? And who bears the costs of the "side effects" of accepted and "working" technologies? As discussed in Chapter 6, on social construction, the *working* of a technology is determined by complex social processes that assess not only the mechanical functioning of a technology, but its semiotic effectiveness in providing meaning for its users. Whose perspectives on working and meaning are valued? Various kinds of social and economic power, whether informal or formal, shape the development and impacts of a technology. It should be clear by now that technological accidents and the harmful consequences of working technologies are interrelated, and that systems of inequality, such as colonialism, economic inequalities, or racialized social systems, undergird the distribution of the costs and benefits of working and failing technological systems. In the final chapters we will look to mechanisms for shaping the development of technology and for wrestling with the decisions about the distributions of the risks and profits from technological systems.

QUESTIONS FOR DISCUSSION

1. Are there technological systems so complex and difficult to make safe and cost-effective that they are probably not worth continued development? Besides nuclear power, are there other systems or technologies which warrant discussion?
2. Describe some examples where economic precarity makes people both more vulnerable to technological failure and their recovery from harm more difficult. What might be done to level the playing fields surrounding complex technological system failures?
3. Does the concept of a chronic disaster make sense to you? Why or why not?
4. Electric cars are a very individualistic solution to the problem of automobile emissions and global climate change. Philosopher and social critic Ivan Illich famously wrote that mandating a vehicular speed limit of 15 miles an hour would transform civilization, from urban planning to environmental and human health. How does this suggestion exemplify connections between society and technology? Could it work?
5. In countries across the world, railway systems are reliable, cost-effective, fast, and popular. What changes would be needed to be made socially and technologically for intracity and intercity rail to be a viable sociotechnical system for transportation in the United States?
6. Can you think of other technologies besides GMOs that may have indirect effects on human health and general well-being? What could be done to improve the safety of such technologies?

Notes

1. Arthur Shi, "One Major Way Smartphone Batteries Can Catch Fire—and How to Prevent It," *iFixit* (November 17, 2022). Accessed from www.ifixit.com/News/69041/how-batteries-can-catch-fire-and-how-to-prevent-it.
2. Charles Perrow, *Normal Accidents: Living with High Risk Technologies* (New York: Basic Books, 1984), pp. vii. See also: https://en.wikipedia.org/wiki/Three_Mile_Island_accident for an account of the accident, and the NRC backgrounder: www.nrc.gov/reading-rm/doc-collections/fact-sheets/3mile-isle.html.
3. Constellation, "Three Mile Island Decommissioning." Accessed from www.constellationenergy.com/our-company/locations/decommissioning.html; and see the U.S. NRC page at www.nrc.gov/info-finder/decommissioning/power-reactor/three-mile-island-unit-2.html.
4. See the World Nuclear Association at https://world-nuclear.org/information-library/safety-and-security/safety-of-plants/three-mile-island-accident.aspx; and Perrow op cit. 16–31.
5. See the World Nuclear Association page for the Fukushima Daiichi Accident at https://world-nuclear.org/information-library/safety-and-security/safety-of-plants/fukushima-daiichi-accident.aspx.
6. Ibid.
7. See the FEMA website for more info on flood zones: www.fema.gov/glossary/flood-zones.
8. See the World Nuclear Association page for the Fukushima Daiichi Accident at https://world-nuclear.org/information-library/safety-and-security/safety-of-plants/fukushima-daiichi-accident.aspx.
9. Ibid.
10. For a general review of calculating expected mortality rates, see Statistics Explained, "Excess Mortality—Statistics," from https://ec.europa.eu/eurostat/statistics-explained/index.php?title=Excess_mortality_-_statistics; For COVID-19 in particular see National Center for Health Statistics, "Excess Deaths Associated with COVID-19," from www.cdc.gov/nchs/nvss/vsrr/covid19/excess_deaths.htm, and Our World in Data, "Excess Mortality: Cumulative Deaths from All Causes," https://ourworldindata.org/grapher/cumulative-excess-deaths-covid for global estimates.
11. See the World Nuclear Association page for the Fukushima Daiichi Accident at https://world-nuclear.org/information-library/safety-and-security/safety-of-plants/fukushima-daiichi-accident.aspx.
12. Perrow (1984), pp. 66–70.
13. See National Academies of Sciences, Engineering, and Medicine, *Reducing the Health Impacts of the Nitrogen Problem: Proceedings of a Workshop—in Brief* (Washington, DC: The National Academies Press, 2021); and Frankie Barnhill, "Watering Idaho: Nitrate Health Risks For Children In Rural Idaho," *NPR* (September 21, 2016). Accessed from www.boisestatepublicradio.org/health/2016-09-21/watering-idaho-nitrate-health-risks-for-children-in-rural-idaho.

14. Burton Bennett, Michael Repacholi, and Zhanat Carr, *Health Effects of the Chernobyl Accident and Special Health Care Programmes* (Geneva, Switzerland: World Health Organization, 2006); and F. Cetta, G. Montalto, M. Petracci, and A. Fusco, "Thyroid Cancer and the Chernobyl Accident. Are Long-Term and Long Distance Side Effects of Fall-Out Radiation Greater Than Estimated? *The Journal of Clinical Endocrinology and Metabolism* 82, 6 (1997): 2015–2016.
15. See: CDC, "Cancer Clusters," from www.cdc.gov/nceh/clusters/sierravista.
16. See: "Chernobyl Accident 1986," from https://world-nuclear.org/information-library/safety-and-security/safety-of-plants/chernobyl-accident.aspx#:~:text=From%20the%20second%20to%20tenth,the%20release%20of%20radioactive%20particles.
17. K. Rahu, M. Rahu, H. Zeeb, et al, "Suicide and Other Causes of Death among Chernobyl Cleanup Workers from Estonia, 1986–2020: An Update," *European Journal of Epidemiology* 38 (2023): 225–232.
18. Julia Jacobo, "Chernobyl Nuclear Disaster Altered the Genetics of the Dogs Left Behind, Scientists Say," *ABC News* (March 3, 2023). Accessed from https://abcnews.go.com/Technology/chernobyl-nuclear-disaster-altered-genetics-dogs-left-scientists/story?id=97493621; Nick Vinocur, "Chernobyl Animals Worse Affected than Thought: Study," *Reuters* (March 17, 2009). Accessed from www.reuters.com/article/us-chernobyl-radiation/chernobyl-animals-worse-affected-than-thought-study-idUKTRE52H09020090318.
19. See: EPA, "Deepwater Horizon – BP Gulf of Mexico Oil Spill," from www.epa.gov/enforcement/deepwater-horizon-bp-gulf-mexico-oil-spill.
20. U. R. Sumaila, A. M. Cisneros-Montemayor, A. Dyck, L. Huang, W. Cheung, J. Jacquet, et al., "Impact of the Deepwater Horizon Well Blowout on the Economics of US Gulf Fisheries. *Canadian Journal of Fisheries and Aquatic Sciences* 69 (2012): 499–510; and Christa Court, Alan Wade Hodges, Kara Coffey, Cameron H. Ainsworth, and David Yoskowitz, "Effects of the Deepwater Horizon Oil Spill on Human Communities: Catch and Economic Impacts," In: *Deep Oil Spills* (Springer, Cham: Switzerland, 2020).
21. See: NOAA Fisheries, "Sea Turtles, Dolphins, and Whales - 10 years after the Deepwater Horizon Oil Spill," and The Center for Biological Diversity, "A Deadly Toll: The Devastating Wildlife Effects of Deepwater Horizon—and the Next Catastrophic Oil Spill."
22. Simpson-Housley, Paul; De Man, Anton, "Flood Experience and Post-traumatic Trait Anxiety in Appalachia," *Psychological Reports* 64, 3 (1989): 896–898.
23. John Raby, "Coal Disaster 50 Years Later: W. Va. Creek Teeming with Fish," *Associated Press* (February 27, 2022).
24. Kai Erikson, *A New Species of Trouble: Explorations in Disaster, Trauma, and Community* (Norton: New York, 1994).
25. According to the Schwab Corporation "Modern Wealth Survey," in 2019, about 59% of people in the U.S. were living "paycheck-to-paycheck." Accessed from www.aboutschwab.com/modernwealth2019.

26. Ivana Pino, "57% Of Americans Can't Afford a $1,000 Emergency Expense, Says New Report. A Look at Why Americans Are Saving Less and How You Can Boost Your Emergency Fund," *Fortune* (January 25, 2023). Accessed from https://fortune.com/recommends/banking/57-percent-of-americans-cant-afford-a-1000-emergency-expense.
27. See: Clara Han, "Precarity, Precariousness, and Vulnerability," *Annual Review of Anthropology* 47 (2018): 331–343; and Louise Waite, "A Place and Space for a Critical Geography of Precarity?" *Geography Compass* 3, 1 (2009): 412–433; and C. K. Lee & Y. Kofman, "The Politics of Precarity: Views beyond the United States," *Work and Occupations* 39, 4 (2012): 388–408.
28. Gordon C. McCord et al., "Long- Term Health and Human Capital Effects of in Utero Exposure to an Industrial Disaster: A Spatial Difference- In-Differences Analysis of the Bhopal Gas Tragedy," *BMJ Open*. Accessed from https://bmjopen.bmj.com/content/bmjopen/13/6/e066733.full.pdf; and Rose Goldman and Gabriel C. Gaviola, "Chapter 1.7 - Methyl isocyanate—Bhopal, India, 1984," *History of Modern Clinical Toxicology* (2022): 85–96; and Rhitu Chatterjee, "The World's Worst Industrial Disaster Harmed People Even before They Were Born," *NPR* (June 17, 2023). Accessed from www.npr.org/sections/goatsandsoda/2023/06/17/1181244389/the-worlds-worst-industrial-disaster-harmed-people-even-before-they-were-born.
29. Simon L. Lewis and Mark A. Maslin, "Defining the Anthropocene," *Nature* 519 (2015): 171–180; and Steffen Will, Grinevald Jacques, Crutzen Paul, and McNeill John, "The Anthropocene: Conceptual and Historical Perspectives," *Philosophical Transactions of the Royal Society* 369, 1938 (2011).
30. See: "Dying Alone: An Interview with Eric Klinenberg." Accessed from https://press.uchicago.edu/Misc/Chicago/443213in.html; and Eric Klinenberg, *Heat Wave: A Social Autopsy of Disaster in Chicago*, 2nd ed. (Chicago: University of Chicago Press, 2015).
31. See: National Weather Service, "Extremely Powerful Hurricane Katrina Leaves a Historic Mark on the Northern Gulf Coast." Accessed from www.weather.gov/mob/katrina; and Sarah Gibbens, "Hurricane Katrina, Explained," *National Geographic* (January 16, 2019). Accessed from www.nationalgeographic.com/environment/article/hurricane-katrina; and for background information see Kathryn Reid, "2005 Hurricane Katrina: Facts and FAQs," *World Vision* (November 25, 2019). Accessed from www.worldvision.org/disaster-relief-news-stories/2005-hurricane-katrina-facts.
32. Sarah Gibbens, "Hurricane Katrina, Explained," *National Geographic* (January 16, 2019). Accessed from www.nationalgeographic.com/environment/article/hurricane-katrina.
33. See: The Data Center, "Lower Ninth Ward Statistical Area," from www.datacenterresearch.org/data-resources/neighborhood-data/district-8/lower-ninth-ward.
34. Juliette Landphair, "The Forgotten People of New Orleans": Community, Vulnerability, and the Lower Ninth Ward," *Journal of American History* 94 (2007): 837–45.

35. James R. Elliott and Jeremy Pais, "Race, Class, and Hurricane Katrina: Social Differences in Human Responses to Disaster," *Social Science Research*, 35 (2006): 295–321.
36. J. Brunkard, G. Namulanda, and R. Ratard, "Hurricane Katrina Deaths, Louisiana, 2005," *Disaster Medicine and Public Health Preparedness* 2, 4 (2008): 215–23.
37. N. A. Becquart, E. N. Naumova, G. Singh, K .K. H. Chui, "Cardiovascular Disease Hospitalizations in Louisiana Parishes' Elderly before, during and after Hurricane Katrina," *International Journal of Environmental Research and Public Health* 16 (2019): 74.
38. T. Deryugina, D. Molitor, "Does When You Die Depend on Where You Live? Evidence from Hurricane Katrina," *American Economic Review* 110, 11(2020): 3602–3033.
39. Ethan J. Raker, et al. (2019). "Twelve Years Later: The Long-Term Mental Health Consequences of Hurricane Katrina," *Social Science & Medicine*, 242 (2019): 112610.
40. R. M. Zoraster, "Vulnerable Populations: Hurricane Katrina as a Case Study," *Prehospital and Disaster Medicine* 25, 1 (2010): 74–78.
41. Carroll Doherty, "Remembering Katrina: Wide Racial Divide over Government's Response," *Pew Research Center* (August 27, 2015). Accessed from www.pewresearch.org/short-reads/2015/08/27/remembering-katrina-wide-racial-divide-over-governments-response.
42. Cassidy Krulewitch, "Hurricane Katrina and Environmental Justice," (May 15, 2021). Accessed from https://storymaps.arcgis.com/stories/83645ea7281e44a3aa94e6c7e5ea724f; and Robert D. Bullard and Beverly Wright (Eds.) *Race, Place, and Environmental Justice After Hurricane Katrina* (Boulder, CO: Westview Press, 2009). See also: Julie Sze, "Toxic Soup Redux: Why Environmental Racism and Environmental Justice Matter after Katrina," (June 11, 2006). Accessed from https://items.ssrc.org/understanding-katrina/toxic-soup-redux-why-environmental-racism-and-environmental-justice-matter-after-katrina.
43. Benjamin J. Pauli, "The Flint Water Crisis," *WIREs Water* 7, 3 (2020): e1420.
44. Lindsey J. Butler, Madeleine K. Scammell, and Eugene B. Benson, "The Flint, Michigan, Water Crisis: A Case Study in Regulatory Failure and Environmental Injustice," *Environmental Justice* 9, 4 (2016): 93–97.
45. Susan J. Masten, Simon H. Davies, and Shawn P. Mcelmurry, "Flint Water Crisis: What Happened and Why?" *American Water Works Association* 108 (2016): 22–34.
46. Op. cit Masten and Mcelmurry (2016).
47. Ron Fonger, "Bottled Water Dries Up in Flint as Water Crisis Fallout Continues in New Year," *MLive* (January 6, 2023). Accessed from www.mlive.com/news/flint/2023/01/bottled-water-dries-up-in-flint-as-water-crisis-fallout-continues-in-new-year.html.
48. Guardian Staff," Before and after Satellite Images Show Scale of Earthquake Destruction in Turkey," *The Guardian* (February 7, 2023). Accessed from www.theguardian.com/world/2023/feb/08/before-and-

after-satellite-images-pictures-turkey-syria-earthquake-destruction; and Ruth Michaelson, "Earthquake in Turkey and Syria Kills Thousands and Devastates Cities," *The Guardian* (February 6, 2023). Accessed from www.theguardian.com/world/2023/feb/06/turkey-earthquake-2023-gaziantep.

49. Martin Chulov and Ruth Michaelson, "Erdoğan Faces Backlash over Building Standards in City Wrecked by Quake," *The Guardian* (February 19, 2023). Accessed from www.theguardian.com/world/2023/feb/19/erdogan-faces-backlash-over-building-standards-in-city-wrecked-by-quake; and Zeynep Bilginsoy and Suzan Fraser, "Turkey's LAX Policing of Building Codes Known before Quake," *The Associated Press* (February 10, 2023). Accessed from https://apnews.com/article/politics-2023-turkey-syria-earthquake-government-istanbul-fbd6af578a6056569879b5ef6c55d322.
50. See: U.S. Environmental Protection Agency, "East Palestine, Ohio Train Derailment," at www.epa.gov/east-palestine-oh-train-derailment.
51. Josh Funk, "Response to Fiery Ohio Derailment Frustrated by Poor Communication and Incomplete Information," *The Associated Press* (June 22, 2023). Accessed from https://apnews.com/article/norfolk-southern-derailment-east-palestine-ohio-ntsb-bc2531012afcac4ee54e6b3999e31ed9.
52. The Associated Press, "23 Freight Cars, New Vehicles Heavily Damaged in Train Derailment in Northern Arizona," (June 8, 2023). Accessed from https://apnews.com/article/freight-train-derailment-northern-arizona-d75ac1da5d9d6f67f4d5d2ca298ff822.
53. Eduardo Medina, "Montana Derailment Sends 10 Freight Cars Into the Yellowstone River," *The New York Times* (June 24, 2023). Accessed from www.nytimes.com/2023/06/24/us/montana-train-derailment-yellowstone.html.
54. See: Aden Tate, "Special Report: 2023 U.S. Train Derailments and Spills Map," from www.mirasafety.com/blogs/news/train-derailments-and-spills-map
55. See: Association of American Railroads, "Chronology of America's Freight Railroads," from www.aar.org/chronology-of-americas-freight-railroads.
56. Kari Lydersen, "The Case for Nationalizing the Railroads," *In These Times* (February 16, 2023). Accessed from https://inthesetimes.com/article/nationalize-the-railroads-workers-on-strike-biden-wages.
57. Naomi Klein, "How Power Profits from Disaster," *The Guardian* (July 6, 2017). Accessed from www.theguardian.com/us-news/2017/jul/06/naomi-klein-how-power-profits-from-disaster; and Naomi Klein, *The Shock Doctrine: The Rise of Disaster Capitalism*. Toronto, ON: Knopf Canada, 2007); and Adeel Hassan and Tim Arango, "Maui Wildfires Latest: Lahaina Reopens to Residents," *The New York Times* (Sept. 13, 2023). Accessed from www.nytimes.com/article/maui-wildfires-hawaii.html; and Naomi Klein and Kapua'ala Sproat, "Why Was There No Water to Fight the Fire in Maui?" *The Guardian* (August 27, 2023). Accessed from www.theguardian.com/commentisfree/2023/aug/17/hawaii-fires-maui-water-rights-disaster-capitalism.
58. Bnamericas, "Spotlight: Chile's Lithium Landscape," (June 15, 2022). Accessed from www.bnamericas.com/en/features/spotlight-chiles-lithium-landscape;

and Nicole Greenfield, "Lithium Mining Is Leaving Chile's Indigenous Communities High and Dry (Literally)," *NRDC* (April 26, 2022). Accessed from www.nrdc.org/stories/lithium-mining-leaving-chiles-indigenous-communities-high-and-dry-literally.

59. Judith L. Capper, Euridice Castañeda-Gutiérrez, Roger A. Cady, and Dale E. Bauman, "The Environmental Impact of Recombinant Bovine Somatotropin (rBST) Use in Dairy Production," *Proceedings of the National Academy of Sciences* 105 (July 2008): 9668–73.
60. John Lauritsen, "Dairy Farmers Resort to Dumping Thousands of Gallons of Milk, with Processing Plants Unable to Keep Up," *CBS News* (June 12, 2023). Accessed from www.cbsnews.com/minnesota/news/dairy-farmers-resort-to-dumping-thousands-of-gallons-of-milk-with-processing-plants-unable-to-keep-up.
61. See: "Neil Degrasse Tyson Gets to the Bottom of GMOs," from www.youtube.com/watch?v=aMDhUsxom0U; and Joseph Stromberg, "Neil Degrasse Tyson Rips Apart One of the Silliest Anti-GMO Arguments," *Vox* (July 31, 2014). Accessed from www.vox.com/2014/7/31/5954941/neil-degrasse-tyson-gmo-food.
62. I. R. Dohoo, et al., "A Meta-Analysis Review of the Effects of Recombinant Bovine Somatotropin," *Canadian Journal of Veterinary Research* 67, 4 (October 2003): 252–264.
63. Judith L. Capper & Roger A Cady, "The Effects of Improved Performance in the U.S. Dairy Cattle Industry on Environmental Impacts between 2007 and 2017," *Journal of Animal Science*, 98, 1 (January 2020): skz291.
64. The Business Research Company, "Genetically Modified Crops Global Market Report." Accessed from www.thebusinessresearchcompany.com/report/genetically-modified-crops-global-market-report.
65. Dan Charles, "As Biotech Crops Lose Their Power, Scientists Push for New Restrictions," *NPR* (October 29, 2020). Accessed from www.npr.org/2020/10/29/927111009/as-biotech-crops-lose-their-power-scientists-push-for-new-restrictions; and Matt Shipman, "We Modified Crops to Kill Pests, and Pests Evolved to Adapt. Now What?" *NC State University* (November 10, 2020). Accessed from https://news.ncsu.edu/2020/11/preserving-utility-of-bt-crops.
66. David Ehrenfeld, "The Cow Tipping Point," *Harper's Magazine* (October 2002), pp. 15–19; and Tony Hiss, "How Now Drugged Cow: Biotechnology Comes to Rural Vermont," *Harper's Magazine* (October 1994), pp. 80–87, 90.

PART 7

The Shaping and Control of Technology

Technology is a human creation; as noted in Chapter 1, one of the distinguishing characteristics of the human race is that its members have consciously developed and used technologies to extend and surpass their natural capabilities. Yet, we as individuals seem to have little control over the technologies that affect our lives. Although crucial decisions about technology appear to be made anonymously, some individuals and groups are in fact making these decisions. Who are they, and how do they operate?

The final set of chapters extends and grapples with the question we asked in Chapter 1: Do we control technology or does technology control us? In these chapters, we look at the social forces that help to shape technology, and how they do so. We look at the role experts, the individuals who play a particularly important role in the creation and diffusion of new technologies. We then consider how the aggregation of these individuals into organizations affects the process of technological change, and how organizations and professions shape ethics and efforts to promote safe technologies, avoid disasters, and promote equitable inputs and outputs to technological change. We'll conclude with a look at the governmental structures and processes that strongly influence the overall environment for the individuals and private organizations that create and sustain our technologies.

CHAPTER 22

Organizations and Technological Change

Organizations are a dominant part of modern life. Most of us are born in organizations, are educated in them, spend our working lives in them, and, when we finally depart from this world, one organization takes care of our mortal remains while another attempts to ease our passage into the hereafter. The pervasive influence of organizations has stimulated a considerable amount of research on their relationship to technological change. We will first consider how organizations are affected by technology. After that, we will turn things around and look at how organizational structures and processes can influence the course of technological change.

Technology as a Cause of Organizational Structure

Much of the research on the interaction between organizations and technology has been concerned with the influence of technology on organizational structures and processes. Many of these studies have come to the conclusion that an organization's structure—the extent to which authority is centralized, the extent of worker specialization, and the number and importance of formal rules—is strongly influenced by the principal technology that the organization uses.

"Organizational structure" is an abstract term that may seem remote from actual working lives. In fact, organizational structures can be of supreme importance; some promote workplaces that are challenging, involving, and sources of satisfaction, while work performed under different structural conditions is "just a job" that provides a paycheck and little else. The same can be said about the kinds of technologies found in different work environments. Some workplace technologies have the potential to empower workers and make their work a meaningful activity, while other technologies have the opposite effect.

Beginning in the 1960s, quite a lot of research on the relationship between technology and organizational structures has been conducted. One early study was carried out by Robert Blauner.[1] His research focused on four different industries that exemplified particular types of technologies and how these affected organizations and their employees. His research provided some useful ways of considering the relationship between production technologies and organizational patterns. By examining printing, textile production, assembly-line automobile manufacture, and chemical processing, he found widespread and systematic differences in necessary skills, worker autonomy, work pace, and degree of worker involvement. All of these strongly reflected how different workplace technologies required distinctive roles and activities for the workers.

Figure 22.1 Workers in the 1950s inspecting paper at Newton Falls Paper Mill. (Source: Mark Fridén, Clifton Town Historian; from the Ernie Bognar Collection, donated by Randy Clark)

Whereas Blauner's study focused on the relationship between the technology used by particular industries and its effects on workers, Joan Woodward's inquiries centered on how different technologies affected organizational structures.[2] Of particular concern to her were such structural variables as the number of levels in the managerial hierarchy, the ratio of managers to other personnel, the number of people supervised by first-line managers, and the flexibility of the managerial system. Woodward was not concerned with individual technologies, but with general manufacturing processes. Her mode of analysis made use of three broad categories: unit and small-batch production (such as shipbuilding or the manufacture of large transformers), large-batch and mass production (as exemplified by automobile manufacture), and process production (such as making chemical or petroleum products).

According to Woodward, each of these categories was distinguished by differences in technological complexity and the extent to which uncertainty could be reduced through the application of routine procedures. These differences, in turn, resulted in different types of organizational structures, each with their distinctive processes. For example, mass-production technologies required routinized processes, narrowly defined job duties, and a clear hierarchy of authority that kept a firm grip on things. Continuous process production was much less rigid in its operations and had a regular flow of worker–management communication that was not filtered according to hierarchical position.

Other kinds of organizational differences also reflected their particular production technologies. Managerial specialization occurred more frequently in mass-production firms, and process production technologies required a higher ratio of managers to other personnel than was the case in organizations using different productive technologies. More relationships could be cited, but Woodward's essential point is that organizational forms must be matched to production technologies. In practical terms, an effective organization is one that uses structures and procedures that are appropriate for the productive technology employed.[3]

Woodward's study was an early contribution to a mode of organizational analysis known as "contingency theory," which postulated that there was no optimal mode of organization that could be applied to every type of activity. This insight is still relevant after the passage of five decades; but it needs updating. Organizations today deal with products and processes that barely existed when Woodward was conducting her research. Her characterization of the relationships between technology and organizational structure is useful but it may need to be modified when dealing with post-industrial technologies such as software, biomedicine, and material technology.[4]

A third example of an organizational theory guided by a consideration of technology comes from Charles Perrow. Perrow's theory is pitched at a more general level than that of Blauner or Woodward. Instead of considering specific production technologies, Perrow considers two basic aspects of the work undertaken within an organization: the degree of variability of the raw materials processed by the organization, and the extent to which problem-solving procedures can be routinized. By raw materials, Perrow does not mean just inanimate objects; iron ore is a basic raw material for a steel mill, but for an employment agency, people are its major raw material. In "processing" its raw materials, an organization will confront different degrees of difficulty. Perrow's analysis of the problem-solving process centers on the number of exceptional cases that must be confronted and the extent to which the problem-solving procedures are "analyzable"—that is, the extent to which they can be reduced to a routine.[5]

Perrow's implicit definition of technology is quite general and abstract. Technology is simply defined as a way of doing things, and no attention is paid to the material artifacts that are part of these technologies. What is important is the nature of the raw material and methods needed to convert it into the organization's products. Above all, Perrow is concerned with organizational effectiveness. A major conclusion of his analysis is that the organizational structure should reflect the particular qualities of the raw materials being used and the technology used for their transformation. To be effective, an organization must use technologies appropriate to its tasks. His conclusions are thus similar to Woodward's: there is no such thing as an optimal organizational form that is universally valid. Everything depends on how things are produced and the materials from which they are made.

These studies, and others like them, have sensitized organizational theorists to the influence of technology on organizational structures and processes, albeit at the risk of drifting toward technological determinism. Yet at the same time, few hard-and-fast conclusions can be asserted. Some empirical studies have found no relationship between an organization's technology and its structure, but taken as a whole, research into organizations has shown that their structures and processes are influenced by the technologies that they use, although the relationship is not particularly strong.[6]

The lack of a definitive relationship between technology and organizational structure is understandable, given the numerous variables that shape organizations. The way that an organization gets its work done will likely affect its structure, but technology is not the only factor involved. As many students of organizations have pointed out, an organization's size also will affect how it is structured.[7] The same can be said about complexity; rather than making use of a single technology, organizations employ a variety of technologies, especially when they have numerous functional divisions. An R&D laboratory and an accounting office may be components of the same organization, but their technologies will differ substantially. Accordingly, the linkage between technology and organizational structure is likely to be more evident when the basic element of analysis is the work unit rather than the organization as a whole.

What does seem clear, however, is that a crucial variable connecting technology and organizational structure is the degree to which an organization's operations consist of routine activities. What is meant by "routine" in this context is the kind of activities that can be done by following standardized procedures.[8] The job of a cook in a fast-food restaurant is mostly a matter of cooking a few items according to a preset timer and doling out portions of ketchup from a dispenser that always supplies the same quantity. Contrast this with the work of a chef in a high-end restaurant, who uses a variety of cooking techniques, develops new dishes, and makes use of seasonally available fruits and vegetables. Technologies that reduce work to preset routines are a good fit with formalized organizational structures and their well-articulated rules, clearly defined responsibilities, and fixed lines of authority. In contrast, forcing routinization on the employees of an organization that requires creativity is likely to be counterproductive.

Although size, the extent of routinization, and their technologies influence the way that organizations are structured, this is not the end of the story. Organizational structures are created by individual men and women, and as such they reflect the cultural traits that these individuals bring to the organization.[9] These, in turn, will affect the relationship between technology and organizational structure. One study of manufacturing plants in Britain, France, and Germany found that the relationship between technology and organizational structure found by Woodward did not hold when a cross-national analysis was undertaken. No matter what the level of technological complexity, German firms had higher levels of worker expertise, autonomy, and flexibility than French firms, while British firms were in an intermediate position. In similar fashion, a study of American and Japanese factories found that there was less specialization and taller hierarchies in the Japanese firms even when the technologies used were substantially identical to the ones employed in the American factories.[10]

The influence of national cultures on organizational structures and processes illustrates an enduring problem in ascertaining the relationship between technology and organizational structure. Although many questions remain unanswered regarding this relationship, the studies that have been conducted have led to a significant rethinking of the nature of organizations and their management. As has been noted above, what works with one production technology will not necessarily work with another. In order to be successful, an organization needs to develop and use organizational structures and processes that are appropriate for the technologies it employs. As a result, many different kinds of organizational structures have been developed and used—matrix organizations, "ad-hocracies," task forces, project teams, loose aggregations of professionals, and "skunk works."[11] And all the while, plain old bureaucratic organization is alive and well in places where it is still appropriate (and sometimes where it is not).

Technology as a Consequence of Organizational Structure

Up to now we have considered how the technologies employed influences organizations. Yet the reverse can also be true. Organizations can shape technological change through their ability to affect the supply and demand for a particular technology. We are a long way from Adam Smith's perfect market, where no individual firm is able to significantly influence the supply of a product or the demand for it. In regard to supply, an organization that accounts for a significant portion of an industry can strongly influence the technological development of the industry as a whole when it creates (or refrains from creating) new products. The choices of a few firms thus determine the kinds of cars we drive, the way steel is made, the operating systems used by our computers, and the links we get when we do a Google search. In similar fashion, when a few organizations are the major purchasers of particular products (as when a few aerospace firms buy highly specialized machine tools), their requirements will determine the nature of the product and the technology it embodies.

On what basis do organizations make technological choices? This is an important question, given the importance of organizational decisions for the course of technological change. Unfortunately, researchers have not paid much attention to this issue. Most of the studies centering on the connection between technology and organizational structures and processes simply take technology as a given. In the studies just summarized, the technologies used by organizations appeared as independent entities. These studies tacitly imply that organizations and their personnel react passively to the presumed dictates of technology, and in so doing they fall prey to technological determinism.

Robert Thomas reminds us in *What Machines Can't Do* that technological choices in organizations frequently happen in the middle of an organization. Rank and file workers may use technologies and have insights into potential innovations, but they are not often tapped to contribute to technological change either in the organization or coming in from outside. People at the very top of an organization often have few insights into direct working conditions or technological processes and similarly rarely contribute to technological innovations and generally take the advice of specialists in terms of choosing new systems in which to invest. But in the middle are managers and people for whom changes in technology can be means by which they can advance their own careers by promoting innovations. In a university, for example, while faculty may be the "line workers" (and this makes students the product), they are rarely the source of innovations in teaching technologies. Specialists in instructional offices, and in the IT hierarchy, promote new technologies to improve teaching but in the process also advance their organizational visibility and power. No one asked the authors which learner management system they preferred, for example. But we learned to live with it, as do students. These systems then set up path dependencies and sunk costs. For example, one of your authors has developed a 1,100-item question library which is built into the current learner management system at their institution. It would be a ruinous amount of work to have to convert to a new system or format if it were not something that might be automated.

Thomas fundamentally argues that the configurations of power within the organization shape the selection and success of new technologies. This is in line with the perspective presented earlier in this book that technologies can be, and often are, employed by individuals or groups in order to advance their own interests. Charles Perrow, whose ideas were described a few pages ago, has recognized this, and in so doing has expressed some disenchantment with his earlier work in which he took technology to be an independent influence on an organization's structure. Perrow forcefully makes the point that, after all is said and done, organizations are powerful tools for achieving goals. Much therefore depends on who controls the tools, and for what purposes.[12] A technology might be selected not because of its innate superiority but because it meets the needs of the power holders within that organization.

One example of this process can be found in the early history of electrical power in the United States.[13] Until 1915, the most common sources of electrical power were small-scale generating plants that had been set up to serve individual homes, apartment buildings, or business enterprises. In the years that followed, these facilities went into decline as large generating stations serving a wide area came to be the dominant mode of generating electrical power. Some of the success of these stations can be attributed to more efficient operation through economies of scale, but more was involved than the pursuit of efficiency. Of considerable importance to the shaping of the American electrical power industry were two industry-wide organizations, the National Electric Light Association (NELA) and the Association of Edison Illuminating Companies (AEIC). These organizations were, in turn, strongly influenced by a few men who had close personal and organizational ties to Samuel Insull, the entrepreneur who played the leading role in the construction and operation of large-scale urban generating plants. Insull's associates used their strategic position within NELA and AEIC to determine the kinds of topics discussed at meetings of the two associations and the papers they published, and to take the lead in prescribing the equipment and procedures deemed best for the industry. Most importantly, the group was able to strongly influence the drafting of government regulations that worked to the advantage of privately owned, centralized generating plants, and to the detriment of local, small-scale facilities. In sum, the basic technological configuration of electricity generation was not solely the result of technical and economic requirements, but it also reflected the use of organizational power to advance the interests of particular individuals and firms.

Another noteworthy example of the choice of technology being affected by the exercise of power in an organizational setting is provided by David Noble's study of the development of numerically controlled machine tools.[14] During the early 1950s, the Servomechanism Laboratory at the Massachusetts Institute of Technology, with the financial support of the United States Air Force, produced the first numerically controlled (NC) machine tools. Instead of being manipulated by skilled machinists, these devices were guided by programs stored on magnetic tape. NC technology increased productivity and allowed the precise machining of the complex shapes used in modern aircraft. At the same time, NC machine tools significantly reduced the amount of skilled labor required and removed a great deal of decision-making power from the machinists. These consequences could be taken as an unavoidable by-product of technological advance; that is, the technology determined the skills, work patterns, and hierarchical relationships found in a machine shop using NC machine tools.

The actual story is quite a bit more complicated. Most importantly, the use of numerical control was not the only way to make machine tools run automatically. During the late 1940s, a "record-playback" method for machine-tool guidance was developed. This technology recorded the motions of a machinist making a part. These motions were recorded on magnetic tape, which was then used to run the machine tool. The machinist, and not a programmer removed from the workplace, was thus the source of the machine's instructions. As it turned out, this was the greatest drawback of this technology. Numerical control was far more appealing to managers because it allowed them to gain more control over the work process. NC tools promised a reduction of the decision-making powers of workers and vested it with managers, engineers, and programmers. In the words of one manager:

> Look, with record-playback the control of the machine rests with the machinist—control of feeds, speeds, number of cuts, output; with NC there is a shift of control to management. Management is no longer dependent upon the operator and can thus optimize the use of their machines. With NC, control over the process is placed firmly in the hands of management—and why shouldn't we have it?[15]

As things turned out, NC did not completely fulfill management's hopes for greater control. As we have seen earlier, computer-controlled production systems have their drawbacks, and the fully automated factory is still a long way off, even with today's generation of computer numerical control (CNC) machines. Although the growth in the number of jobs for machinists is only about average when compared with all other occupations,[16] their services are still vital; a worker lacking the necessary skills can do tens of thousands of dollars of damage in a few seconds.

Figure 22.2 CNC machine tool operator is a high-demand trade but does not require direct hands-on-skill with operating manufacturing tools, such as the lathes and cutting tools a machinist might use. (Shutterstock/Dmitry Kalinovsky)

The relationship between technology, worker skills, and organizational authority has become especially salient as a result of the incorporation of computers and automated systems into production processes. Modern information systems can be used to enhance the power and responsibilities of shop floor workers. At the same time, however, they can produce the opposite effect by giving management a greater capacity to centralize operations and restrict the discretion of the workers. As one manager asserted, "The more we can control with the computer, the less dependence we will have on operators with in-depth knowledge. When there is a problem, we will have computer programs, logic, electronic sensors. We will be able to understand the problems."[17]

A certain degree of computer-based centralization may be justified in terms of its supposed contribution to coordination and efficiency. But this is hardly its only rationale. Many managers worry that computerized information systems can just as easily dilute managerial authority by making information more readily available to ordinary workers. A top executive from the same firm as the manager quoted above described the potential erosion of managerial authority:

> The classic managerial role has been that of the handler, manipulator, dealer, and withholder of information. An issue that the technology is forcing us to face involves the loss of managerial control. . . . Suddenly you are vesting in folks at the control station, folks who are interfacing with a tremendous technology power—power to see all the functions of the operation. That is kind of scary to some managers.[18]

Managers realize that the control of information has been an essential element of their authority. Consequently, in the words of one line manager, "Managers perceive workers who have information as a threat. They are afraid of not being the "expert." They are used to having everyone come to them and ask what to do."[19] This use of information was duly noted by a worker in the same firm: "[Managers] can't share information with us, because only by holding on to knowledge that we don't have can they maintain their superiority and their ability to order us around."[20]

In this setting, computer-based information systems are not neutral instruments; rather, they are likely to be used in ways that are congenial to those occupying positions of authority within organizations. As with NC machine tools, a technology may be employed because managers see it as a way of extending their own power. Conversely, computerized information systems may be used to empower workers, enhance their responsibilities, and give them a claim to higher wages and salaries. There is no simple technological logic governing the use of these systems. The use of an existing technology may reflect the existing distribution of power in an organization, while the installation of a new technology may become an occasion for conflict over how it will be wielded in the future.

Organizational Size and Technological Innovation

Technological innovation requires the commitment of personnel, money, and equipment if it is to be successful. It therefore seems logical to expect that large organizations, which are well endowed with these resources, should be the prime sources of technological changes. This expectation has been clearly articulated by John Kenneth Galbraith:

> A benign Providence . . . has made the modern industry of a few large firms an excellent instrument for inducing technical change. It is admirably equipped for financing technical development. Its organization provides strong incentives for undertaking development and for putting it to use.[21]

For Galbraith and others who hold to this view, the day has long passed when independent inventors and small firms were the prime source of important new technologies. Their place has been taken by the well-financed laboratories and workshops of large corporations and government agencies.

On the face of it, this seems reasonable. It is hard to imagine how jumbo jets and mainframe computers could be produced in basement workshops. But the large-scale production of an innovation is not the same thing as its original conception and creation. In point of fact, many important inventions were first conceived and produced by small firms. Only when their technical feasibility had been demonstrated did they become products of large enterprises. Key patents on the turbojet were held by Frank Whittle, an Englishman who produced some of the world's first jet engines in a small workshop that was chronically short of funds and personnel. The first digital computer, the ENIAC, was put together at the University of Pennsylvania's Moore School of Electrical Engineering by John Mauchly and J. Presper Eckert with the assistance of ten engineers. Hewlett-Packard's startup in a garage and Apple's modest financial underpinnings during its early years are the stuff of Silicon Valley legend.

Examples like these are interesting but hardly definitive, and in fact efforts to determine with some precision the connection between organizational size and inventive activity have had mixed results. The assertion that large organizations play a preeminent role in the process of technological advance has not been supported, but neither has it been completely repudiated.[22] Attempts to find a simple correlation (either positive or negative) between firm size and inventive activity are doomed to failure because factors other than size have to be taken into account. Of at least equal importance is the perceived developmental potential of the technologies relevant to an industry's products and operations. R&D efforts will be extensive in industries where key technologies are judged to have high potential for further development, and less so when technological prospects appear to be limited. To take two examples, in 2017 the U.S healthcare industry spent an average of 17 percent of its revenues on R&D, while firms making consumer staples spent 3 percent of their revenues for this purpose.[23]

Even if a correlation between size and inventive activity existed, it would tell us nothing about how effective the research efforts of large firms have been. Much of the research conducted by large organizations with secure markets is not oriented toward the production of radically new things but is directed at the creation of modest, incremental improvements. Large firms that control a substantial share of their market are more likely to produce inventions that maintain the status quo and to avoid engaging in research that might radically alter the industry in which they already have a comfortable niche. Conversely, small firms or independent units of larger firms have been more successful in creating, developing, and marketing inventions that disrupt the technological status quo.[24]

To be sure, invention is not the same thing as innovation. To actually bring an invention successfully into the marketplace may require the resources of a large organization. As we have already seen, the development of new technologies can be a lengthy and expensive process. Large organizations are often better equipped to do development work, where scaling up and making relatively small but nonetheless crucial improvements are the key tasks. This may be their most important contribution to technological innovation.

These generalizations of course do not apply to all industries. In some industries such as nuclear power, the bulk of both the research and development work has been done by large firms and government agencies. In other industries, such as smartphone and tablet apps, a great amount of innovative effort has come from individuals and small firms. About the safest statement that can be made is that technological advance requires the efforts of a wide variety of organizational types and sizes. The initiation and development of some technologies require the massive efforts that only large organizations can provide. At the same time, many technologies have been—and will continue to be—the result of the efforts of people working in small organizations or just by themselves.

Organizations and New Information Technologies

Some of the issues surrounding the relationship between technological change and the evolution of organizational structures appear in sharper focus when we take a look at digital information technologies. Numerous scholars have pointed out that the essential work of organizations centers on the manipulation of information, broadly construed. Whether it is within an employment office, a bank, or a school, members of an organization spend a great deal of their time gathering information, analyzing it, deciding what to do in the light of what they have learned, communicating with others, and evaluating and reporting what has been done.

During the second half of the nineteenth century, organizations were able to expand, improve their operations, and operate at higher levels of efficiency and effectiveness through the use of what were then novel technologies for gathering, recording, analyzing, distributing, and storing information—typewriters, telegraphs, dictating machines, telephones, pneumatic tubes, standardized forms, carbon paper, and filing systems.[25] Recent decades have seen more revolutionary changes in information technology, and few of today's organizations can get along without computers, database management programs, word processing, spreadsheets, fax machines, the internet, email, photocopy machines, and voicemail, while growing numbers of organizations are making extensive use of more recently introduced technologies like smartphones, intranets for internal communications, and videoconferencing for linking participants from all over the globe in real time.

The nearly universal use of electronic information technologies has motivated a substantial number of researchers to conduct inquiries into how these technologies have affected the structure and functioning of contemporary organizations. In general, this research has indicated that advanced information technologies have speeded up operations, reduced costs, increased the amount and accessibility of information, and more extensively linked members of organizations.[26] Less certain, however, are the effects of new information technologies on organizational structures and procedures. As with all significant technological advances, there is rarely a simple, straightforward connection between technological and organizational changes. A new information technology, no matter how radical, is only one of many influences on a particular organization, and not necessarily the most important one; other organizational features, such as size, the composition of the workforce, and the nature of the clientele may be more significant than the information technologies in use.

Figure 22.3 A stenographer and her cutting edge tools, from the 1890s.

One much-discussed issue has to do with the role of new information technologies in promoting organizational decentralization. As some students of organizations have argued, a primary reason for the concentration of organizational authority in times past was the difficulty of obtaining and analyzing information. Information was a scarce resource, something that could be provided for only a few members of an organization, those at the top of the managerial pyramid. In similar fashion, only those at the summit of the organization were deemed capable of intelligently acting on the information made available to them. In contrast, today's digital technologies make information and the ability to analyze it readily available at low cost, allowing more decentralized and participatory modes of decision making. This does seem to be happening; one study of 400 large firms found that greater levels of information technology were in fact associated with more delegation of authority to individuals and teams.[27]

The easy availability of information makes it possible to decentralize organizational authority and to empower workers, but as has been noted before, the way that a technology is used is not entirely determined by the technology itself. Whether a powerful information technology promotes centralization or decentralization may have more to do with existing distributions of power and authority than with presumed technological imperatives. Instead of democratizing the workplace, today's information technologies may be used to enhance managerial control. As we have seen in an earlier chapter, some new workplace technologies, far from empowering workers, may enhance top-down control through the use of electronic monitoring.[28] To take another example, it has sometimes been stated that the use of email has eroded organizational hierarchies by allowing direct communications between the upper and lower ranks. This may come to pass when the top organizational echelons value and encourage bottom-up communications, but many managers would prefer to avoid such "distractions." Even when managers are receptive to communications from employees lower down in the organization's hierarchy, there is always the likelihood that these messages will be lost in the blizzard of email communications that plague many organizations. In any event a lot depends on what the members of the organization, especially the managerial cadre, want to do with new communications technologies.

As a final point, it should be kept in mind that new communications technologies do not necessarily supplant existing ones. For example, it might be expected that the growing use of the telephone in the early twentieth century reduced the need for traditional forms of communication, but in fact the opposite occurred, as telephone use seems to have stimulated the desire for more face-to-face communication, not less.[29] The same thing may be happening today. New electronic media, such as email and videoconferencing, certainly have their place, but they fail to provide some crucial elements of communication, such as body language and indications of whose turn it is to speak. Old-fashioned, face-to-face communication may not be of much importance for many organizational tasks, but it is usually essential for activities that have an important social-psychological component, such as negotiating, maintaining relationships, and building trust.[30]

Interorganizational Relations and Technological Development

No business firm, no matter how large, develops all of the technologies that are embodied in its products. Every firm depends on a network of suppliers for the new materials, components, and even the ideas that are incorporated into its own products. This is especially true of mature industries, where supplier firms are likely to be the principal source of substantial technological breakthroughs.[31] Without the efforts of outside suppliers to develop and sell innovative products, it is likely that the technological level of many established industries' products would stagnate. Technological changes are therefore shaped not only by the structure and dynamics of individual organizations but also by the pattern of relationships between organizations.

The automobile industry exemplifies the transfer of technology from supplier organizations to the automobile manufacturers themselves. It also demonstrates that interorganizational relationships may act as obstacles to technological advance within the industry. From its earliest years, the automobile industry has depended on thousands of outside suppliers for many of its components—everything from nuts and bolts to complete engines. These suppliers have provided some of the most important technological innovations for the automobile: disc brakes, radial tires, fuel injection, electronic ignition, and digital engine management systems, to name but a few. These innovations did not sell themselves, however. Before they appeared in automobiles, they had to be sold to the automobile manufacturers (also known as "OEMs," or original equipment manufacturers).

Throughout most of the history of the industry, the relationship between suppliers and OEMs was that of independent buyers and sellers; cooperative approaches were generally absent. Suppliers were often left in the dark about the actual application of their products; a supplier might not even know which cars used its products or that they were about to be replaced by something totally different. The OEMs did not inform their suppliers about long-range intentions, which meant that their suppliers could not effectively plan for the future. Research done by suppliers usually took place without direct consultation with the OEMs, and the job of application came only after the component had been developed. There were few attempts to foster joint efforts between OEM and supplier in order to devise and develop new technologies.

The lack of supplier involvement in the design process was a major reason why in the 1970s and 1980s American automobiles fared poorly in competition with those made in Japan, where there was a greater inclination to incorporate suppliers in the design process at an early stage of a car's development.[32] However, faced with stringent competition from automakers in Japan and other countries, in recent years American car manufacturers have begun to initiate joint activities with their suppliers. This is a logical and necessary step for an industry that needs to constantly upgrade the technological level of its products.

These cooperative activities stand in sharp contrast to traditional buyer–seller relationships, where both parties are only concerned with their own narrow interests. These relationships work well enough when standard commodities are bought and sold, and price is the paramount concern of the parties to the transaction. In contrast, the development and transfer of new technologies involve unique products and are fraught with many uncertainties. These uncertainties make it difficult to evaluate transactions on a precise cost–benefit basis. Moreover, marketplace exchanges by their nature do not foster the long-term vision necessary for the development and application of modern, sophisticated technologies. As the

Japanese experience has indicated, the creation of effective working relationships between separate organizations is an essential element of technological and economic vitality.

Entrepreneurs and Organizations

Organizations are essential for the operation of established technological systems like airlines or oil refineries. Their fundamental purpose is to coordinate the efforts of individuals in order to accomplish essential tasks. When routine tasks are involved, this coordination is usually accomplished through the application of basic bureaucratic principles: specialization, formalization (the use of rules, regulations, and standard operating procedures), and hierarchical authority. Although bureaucracies are sometimes thought to be inherently inefficient, there is no reason that they should be. In fact, when it works as it should, a bureaucracy is like a well-oiled machine that faithfully does the job for which it created.

But bureaucratic organization has its limitations. As has been noted above, bureaucracies thrive on routine, working best in stable environments where nothing new comes along to disturb their machinelike operations. At the same time, however, these bureaucratic routines are difficult or impossible to reconcile with the inherent uncertainty of creating something new. Furthermore, many unresolved issues surround not just the technology itself but also all of the ancillary changes that may be required to use it effectively. All sorts of uncertainties may be involved in such matters as the acceptance of the technology by customers, the retraining of personnel, legal liabilities, performance of component suppliers, and changing government regulations.[33]

Bureaucracies rely on rules, regulations, and formal managerial practices to get their employees to do what needs to be done. In most cases, activities and projects are closely scheduled and subject to careful financial scrutiny. These tendencies are reinforced by the "not invented here" (NIH) syndrome (mentioned previously) which is the resistance to innovations coming from outside of an organization. But innovators usually need a lot of elbow room and have an indefinite time horizon, which puts them at odds with bureaucratic procedures. A large corporation that is organized according to bureaucratic principles can be a difficult environment for inventors. As Jack Kilby, the co-inventor of the integrated circuit, has noted:

> There is a basic incompatibility of the inventor and the large corporation. Large companies have well-developed planning mechanisms which need to know at the beginning of a new project how much it will cost, how long it will take, and above all what it's going to do. None of these answers may be apparent to the inventor.[34]

As we have seen in Chapter 3, technological innovation is often the work of entrepreneurs. By definition, entrepreneurs are people who disrupt normal activities and procedures by vigorously pushing new ways of doing things. Promoting innovation is risky business, and routine-seeking bureaucratic organizations are generally disinclined to take risks. Not only is a bureaucracy's structure attuned to routine activities, but its personnel also usually seek to protect their positions and maintain their customary modes of operation. Under these circumstances, an entrepreneur's efforts to promote innovation within an organization may encounter a fair amount of resistance and even hostility.

Although entrepreneurial activities are usually associated with the private sector, entrepreneurship may emerge as a force for change in government agencies, charitable organizations, schools, and religious institutions. The transformative power of an entrepreneur in the face of stiff opposition is well illustrated by an episode in the history of one of the most hidebound institutions in nineteenth-century America, the United States Navy.[35] At that time, gunnery was highly inaccurate; of the 9,500 shots fired from American naval vessels during the Spanish–American War, only 121 found their target. No attempt was made to follow a target; gunners used their cannons' elevating gear to give the shell the proper trajectory, fired after the roll of the ship brought the target into their sights, and hoped for the best. Then, during the last years of the nineteenth century, a British naval officer modified the elevating gear of his ship's guns so that they could be rapidly moved, allowing them to continuously track their target. He also modified the gun's telescopic sight so that it wouldn't poke the gunner in the eye when the gun recoiled.

The improvements in accuracy were spectacular, and before long an American officer named William S. Sims embarked on a personal mission to bring these improvements to American ships. Sims was the quintessential entrepreneur; although he hadn't invented the new system, he was utterly convinced of its value and passionately devoted to seeing it installed. And he was willing to rattle a lot of cages in the course of promoting it. For its part, the Navy acted as a bureaucracy might be expected to act. At first, it ignored Sims' reports on the inadequacies of American gunnery technology and what could be done to improve it. The reports were simply filed away, where they provided a few good meals for the cockroaches that inhabited the file cabinets. When Sims became too insistent to be ignored, the Navy arranged a series of inappropriate tests that "proved" the inadequacy of the new system. Sims finally prevailed only after he succeeded in getting the attention of President Theodore Roosevelt, who installed him in a position that allowed him ultimately to become "the man who taught the Navy to shoot."

The case of Sims versus the United States Navy is an admittedly extreme example of conflict between an entrepreneur and an entrenched bureaucracy. There are, after all, many examples of successful working relationships between innovative individuals and the organizations that employ them. Not every organization is structured along classic bureaucratic lines, and as we have already seen, an organization committed to innovation will likely have different structural characteristics than those engaged in routine activities. Still, there is no getting around the fact that organizations may find it hard to accommodate employees who take on the role of entrepreneur. These individuals can be arrogant, stubborn, and single-minded to the point of fanaticism. But at the same time, their vision and energy may be the most important element in their organization's success. To take a recent example, Steve Jobs embodied the positive and negative qualities of the classic entrepreneur. He was an exceedingly difficult person in many ways, but without his vision and guidance, Apple Computer Company would lie buried along with the dozens of defunct personal computer firms that are barely remembered today.[36] For many organizations, technological dynamism has required continual efforts to find a balance between the needs of would-be entrepreneurs and established organizational structures and processes. We cannot discard bureaucratic organizations, but we need to find ways that allow them to continue doing what they do best, while at the same time preventing them from obstructing the individuals whose efforts are a major source of technological advance.

Complex societies could not exist without organizations, bureaucratic or otherwise. Organizations are no less important for the initiation, development, and application of most technologies. At the same time, however, inappropriate organizational structures can stifle technological advance. Equally important, the controllers of powerful organizations can select and cultivate technologies that serve only their own interests. This problem is not confined to business organizations; it can also manifest itself in the public sector. In the next chapter, we will look at the relationships between individuals and organizations in relation to promoting public safety and welfare, broadly defined, as matters of individual ethics. We will then turn our attention to a very large and powerful organization, the federal government of the United States and related systems of state governance of science and technology.

QUESTIONS FOR DISCUSSION

1. How would you organize work if you were the director of a research laboratory? How would you create a climate that would stimulate the creativity of your personnel and at the same time prevent them from going off on unproductive tangents?
2. How might you design a research project that investigates the extent of managerial control over the selection of a firm's technologies? What sort of industries would be particularly suitable for such a project?
3. Many key industries are dominated by a handful of large firms. This tendency has been increasingly evident in recent years, as many separate firms have merged into larger ones. How might this trend affect technological innovation? Could it pose a threat to the overall development of the economy?
4. Why have many organizational analysts fastened on information as an organization's key resource? To what extent do your experiences with organizations reflect this idea? How does the ability to control the flow of information reflect and reinforce power relations within an organization?

5. Who comes to mind when you think of present-day entrepreneurs? To what extent are they similar to the classic entrepreneurs of the past who propelled industrial innovation? Does today's economic and political environment promote or inhibit entrepreneurial efforts?

Notes

1. Ron Westrum, *Technologies and Society: The Shaping of People and Things* (Belmont, CA: Wadsworth, 1991), p. 57.
2. Alex Inkeles, *One World Emerging? Convergence and Divergence in Industrial Societies* (Boulder, CO: Westview Press, 1998).
3. Recent assessments of Woodward's work along with some personal reminiscences appear in Nelson Phillips, Dorothy S. Griffiths, and Graham Sewell (Eds.), *Technology and Organization: Essays in Honour of Joan Woodward, Research in the Sociology of Organizations*, vol. 24 (Bingley, UK: Emerald Group Publishing, 2010).
4. Markus Perkman and Andre Spicer, "What Are Business Models? Developing a Theory of Performative Representations" in ibid, pp. 265–75.
5. Adapted from Charles Perrow, "A Framework for the Comparative Analysis of Organizations," *American Sociological Review*, 32 (1967): 194–208.
6. W. Richard Scott, *Organizations: Rational, Natural, and Open Systems* (Upper Saddle River, NJ: Prentice-Hall, 1981), p. 37.
7. Stephen P. Robbins, *Organization Theory: Structure, Design, and Applications*, 3rd ed.(Englewood Cliffs, NJ: Prentice Hall, 1998), p. 239.
8. Stephen P. Robbins and Timothy A. Judge, *Organizational Behavior*, 13th ed. (Upper Saddle River, NJ: Pearson Prentice Hall, 2009), pp. 536–37.
9. Richard Daft, *Organization Theory and Design*, 10th ed. (Mason, OH: South-Western Cengage Learning, 2010), p. 398.
10. Karlene Roberts and Martha Grabowski, "Organizations, Technology, and Structuring," in Stewart R. Clegg, Cynthia Hardy, and Walter R. Nord, *Handbook of Organization Studies* (London and Thousand Oaks, CA: Sage, 1996), p. 415.
11. The original "Skunk Works" was a small, semiautonomous part of the Lockheed Aircraft Corporation that developed, among other things, the U-2 spy plane and the record-setting SR-71 bomber. See the memoirs of its first director, Clarence "Kelly" Johnson, Kelly: *More than My Share of It All* (Washington, DC: Smithsonian Institution Press, 1985).
12. Charles Perrow, *Complex Organizations: A Critical Essay*, 2nd ed. (Glenview, IL: Scott, Foresman, 1979), pp. 13–16.

13. Mark Granovetter and Patrick McGuire, "The Making of an Industry: Electricity in the United States," in Michel Callon (Ed.), *The Laws of the Markets* (Oxford: Blackwell, 1998), pp. 147–73.
14. David F. Noble, *Forces of Production: A Social History of Industrial Automation* (New York: Oxford University Press, 1986).
15. David F. Noble, "Social Choice in Machine Design: The Case of Automatically Controlled Machine Tools," in Andrew Zimbalist (Ed.), *Case Studies in the Labor Process* (New York: Monthly Review Press, 1979), p. 34.
16. United States Bureau of Labor Statistics, "Machinists and Tool and Die Makers," *Occupational Outlook Handbook* (December 17, 2015). Accessed from http://www.bls.gov/ooh/production/machinists-and-tooland-die-makers.htm.
17. Shoshana Zuboff, *In the Age of the Smart Machine: The Future of Work and Power* (New York: Basic Books, 1988), p. 268.
18. Ibid., p. 250.
19. Ibid., p. 252.
20. Ibid., p. 264.
21. John Kenneth Galbraith, *American Capitalism: The Concept of Countervailing Power* (Boston: Houghton Mifflin, 1952), p. 86.
22. For a review of research on the relationship between organizational size and technological innovation, as well as an empirical study of this relationship in one industry, see Gregory N. Stock, Noel P. Greis, and William A. Fischer, "Firm Size and Dynamic Technological Innovation," *Technovation* 22, 9 (September 2002).
23. National Science Foundation, "Industrial Research and Development Information System, Historical Data 1953–2007," *Table 50*. Accessed from www.nsf.gov/statistics/iris/search_hist.cfm?indx=7.
24. Clayton M. Christensen, *The Innovator's Dilemma: When New Technologies Cause Great Firms to Fail* (Boston, MA: Harvard Business School Press, 1997).
25. JoAnne Yates, *Control through Communication: The Rise of System in American Management* (Baltimore and London: Johns Hopkins University Press, 1989).
26. Janet Fulk and Gerardine DeSanctis, "Articulation of Communication Technology and Organizational Form," in Jay M. Shafritz and J. Steven Ott (Eds.), *Classics of Organization Theory*, 5th ed. (Belmont, CA: Wadsworth, 2001), pp. 500–01.
27. Erik Brynjoliffson and Lorin Hitt, "Beyond Computation: Information Technology, Organizational Transformation and Business," *The Journal of Economic Perspectives* 14, 4 (Autumn 2000): 35.
28. Darin Barney, *Prometheus Wired: The Hope for Democracy in the Age of Network Technology* (Chicago: University of Chicago Press, 2000), pp. 155–63.
29. Claude S. Fischer, *America Calling: A Social History of the Telephone to 1940* (Berkeley, CA: University of California Press, 1992).
30. Frank Levy and Richard J. Murnane, *Windows on the Workplace: Technology, Jobs, and the Organization of Office Work*, 2nd ed. (Princeton, NJ: Princeton University Press, 2004), pp. 83–92.

31. James M. Utterback, "The Process of Technological Innovation within the Firm," *Academy of Management Journal* 14, 1 (March 1971); Donald A. Schon, *Technology and Change: The New Heraclitus* (New York: Dell, 1967), pp. 161ff.
32. Robert E. Cole, "Quality Control Practices in the Auto Industry: United States and Japan Compared," in Robert E. Cole (Ed.), *The Japanese Automobile Industry: Model and Challenge for the Future* (Ann Arbor: University of Michigan, Center for Japanese Studies, 1981), p. 92.
33. Christopher Freeman, *The Economics of Industrial Innovation*, 2nd ed. (Cambridge, MA: MIT Press, 1982), p. 176.
34. T. R. Reid, *The Chip: How Two Americans Invented the Microchip and Launched a Revolution* (New York: Random House, 2001), p. 250.
35. This account is based on Elting E. Morison, *Men, Machines, and Modern Times* (Cambridge, MA: MIT Press, 1966), pp. 17–44.
36. Walter Isaacson, *Steve Jobs* (New York: Simon and Schuster, 2011).

CHAPTER 23

Professions, Expertise, and Ethics

In Chapter 1 we discussed the ways in which technical means can be used for monstrous ends. Steven B. Katz describes a 1942 memo between two Nazi soldiers as an exemplary piece of technical writing. The memo is from a person named W. Just to a Section Commander named W. Rauff, of the German Reich, requesting technical changes in existing vans and new vans made to order. As a memo, it outlines a problem, suggests potential solutions, and recommends scope of implementation. There are many euphemisms and indirect references: "the load," "pieces," even "merchandise" that may shift during "processing." Language about "screaming" and description of how "the load" "rushes to the light" gives us some indication that something more nefarious is going on. Improving efficiency and productivity is applied to a horrific end: the extermination of human beings. And yet this ethic of expediency, the value of convenience, is one common to Western ideologies.[1] What happens when the means themselves supersede the ends to which we might aspire?

We know little of the author of this memo, W. Just. Was he an engineer? Was he coerced into this position or a willing participant? What were his feelings about this project? This case, of applying technical expertise to unethical ends, while perhaps extreme, nonetheless illustrates quite a number of issues surrounding professions, expertise, ethics, and the organizational contexts which shape professional work.

Experts, Expertise, and the Shaping of Technology

At the end of the sixteenth century, the philosopher, essayist, and corrupt government official Francis Bacon formulated a famous maxim: "Knowledge is power." Since technology is ultimately based on knowledge, it stands to reason that the power to control technology will be wielded by those with the greatest technical knowledge. But is it so simple? Is the possession of the appropriate knowledge a sufficient basis for the exercise of control over the course of technological change?

During Bacon's time it might have been possible for a well-educated person to understand most of the basic principles underlying the science and technology of the era. The body of human knowledge was limited, and no intellectual barriers separated laypeople from scientists and technicians. Scientific inquiry was not the exclusive province of professional scientists, so amateurs could make significant contributions. Technological knowledge was more restricted, as many craft secrets were jealously guarded, but there were no intellectual reasons that these could not be comprehended by an outsider.

The situation today is vastly different. Knowledge has been growing at an accelerating rate of speed. Whereas a mere ten scientific journals existed in the middle of the eighteenth century, according to one estimate around 30,000 academic journals are now published each year, accounting for two million articles.[2] A university library in the United States can be expected to subscribe to several thousand different scholarly periodicals.

The great technological achievements of our time are to a significant degree a reflection of the enormous growth of knowledge that has taken place, but at the same time, this "knowledge explosion" has made all of us relative ignoramuses. Even well-trained scientists and technicians have to admit their own limitations when they leave their own field of specialization. In a world in which there is so much to know, nobody knows very much. This ignorance can limit the ability of citizens and their representatives to control the course of technological change. Most technologies are built on a base of specialized knowledge; if most people are incapable of comprehending that knowledge, there looms the danger that the direction of technology will be left to the small cadre who are in the know. When public choices depend on expert information, experts and not the electorate will supply and evaluate this information.[3] If this happens on a widespread scale, democracy becomes an illusion as vital issues such as climate change, nuclear power, genetic modification, and missile defense systems are not subject to public debate and citizen oversight. People can do nothing but accept the judgments of experts and hope that they are right.

But what does it mean to be "right"? Technologies cannot be judged purely according to their technical merit. Their social consequences have to be considered, and when this happens, value judgments must be made. These judgments necessarily take the expert well outside the realm of his or her expertise. To illustrate, biologists might agree that a new medical technology is effective in allowing a couple to choose the sex of the child they intend to have. More crudely, in some countries' sonograms can be used to identify the sex of a fetus, which may be aborted if it is of the "wrong" gender—that is, female. These may be effective technologies, but should they be used? If they are used, the likely result will be a lopsided sex ratio where boys and men substantially outnumber girls and women. This is already happening in some countries. In 2008, for example, China and India had births of males outnumber births of females by ratios of 119 and 115 to 100, respectively. Over time, this will result in a very large cohort of men with no hope of ever getting married, and a host of problems for these countries.[4]

Even when an issue seems to revolve solely around technical matters, hidden value judgments have a way of sneaking in. Many sophisticated tools used by experts, such as cost–benefit analysis, achieve apparent precision because they are based on simplifying assumptions that may reveal the analyst's own preferences. To take one straightforward example, some cost–benefit analyses require the determination of the value of a human life, which has some quantitative measures, such as expected future earning power, but also may be affected by a government's interest in promoting lifesaving regulations. The Environmental Protection Agency, under the presidency of George W. Bush, set $6.8 million as the value of a human life. When the Obama administration took over, the figure was raised to $9.1 million. At the same time, the Food and Drug Administration reckoned that a human life was worth $7.9 million, $2.9 million more than the number used at the end of the Bush administration. Meanwhile, the figure used by the Department of Transportation remained the same at $6 million for each life.[5]

Public participation in the shaping of technology may also be thwarted by the desire of some experts to keep their knowledge secret. An expert can bolster his or her importance by being the sole source of important knowledge. The arcane nature of scientific and technological knowledge makes this inevitable to a certain degree, but experts have been known to draw a veil of mystery over their work in order to enhance their positions. The organizations in which they work have done the same thing. As the great sociologist Max Weber pointed out many decades ago, organizations commonly seek to expand their power and insulate themselves from outside pressures by keeping their knowledge secret.[6]

In addition to undermining democratic participation, secrecy can also result in bad decisions because it promotes "groupthink." The only people deemed to be fully qualified experts are those who have access to secrets, and these secrets are accessible only to people within the same tight circle. As a result, everyone involved in a new technological project is an enthusiastic supporter, while potential opponents are excluded from the action. Major decisions can thus be made by a cadre of self-interested enthusiasts who are unchallenged by an informed opposition. The decisions to support the Concorde supersonic airliner and an ambitious atomic energy program in Britain occurred under precisely these circumstances.[7]

At this point, it is important to note that the power of expertise is not always based on the ability to offer superior knowledge; there are times when expert opinion is used to bolster an organization's self-interested claims.[8] As an old joke has it, individuals and organizations use expertise the way a drunk uses a lamppost—more for support than for illumination.

The commercial value of knowledge has at times created such a close relationship between profit-making enterprises on the one hand and scientific and technical experts on the other that free-standing experts are almost nonexistent in some areas of research. To take one notable example, until the practice was banned in 2004, researchers at the National Institutes of Health could receive payments for acting as consultants for pharmaceutical firms.[9] Close ties, financial and otherwise, between researchers and these firms have also motivated some medical journals to require that contributors divulge these connections when presenting the results of their research. Another example of independent technical expertise being undermined by the experts' entanglement with other organizations and programs is provided by a software systems analyst who was asked to serve on an advisory panel that was to make recommendations regarding the computer development program for the Strategic Defense Initiative (SDI; the "Star Wars" antiballistic weapon program):

> I have a project with the U.S. Navy that could profit from SDI funding and I suggested to the panel organizer that this might disqualify me. He assured me quite seriously that if I did not have such a conflict, they would not want me on the panel. He pointed out that the other panelists, employees of defense contractors and university professors dependent on Pentagon funds for their research, had similar conflicts. Citizens should think about such conflicts the next time they hear of a panel of "distinguished experts.[10]

Of course, even if some experts are not inclined to challenge technological policies, there may be other people who will. Ralph Nader, in calling attention to the shortcomings of American automobiles in the 1960s, especially in regard to safety (or the lack of it), probably has done more to affect their design than any single engineer. Nuclear power has been under assault from a variety of quarters, and it is not likely that the issue will be resolved on the basis of purely technical criteria. Decisions to develop new military technologies can be intensely political. Technical considerations are important but not preeminent, for the decision to develop and deploy a new weapons system usually has significant economic, political, and strategic ramifications.

When major policy decisions are at stake, freestanding experts will likely find that their technical knowledge cannot easily be converted into political influence. This has been particularly evident in the political battles that have been waged over climate change. There are two aspects to this issue. The first centers on the likelihood of rising future temperatures, which has been challenged by some members of the U.S. Congress, who claim that experts are divided on this issue. There in fact have been some dissenters, but one extensive study found that belief in the reality of climate change is closely correlated with the knowledge and research accomplishments of the respondents.[11]

The second issue, the extent of human contributions to global warming and climate change through the burning of carbon-based fuels, has been affirmed by 200 worldwide scientific organizations, ranging from the American Academy for the Advancement of Science to the U.S. National Academy of Sciences and from the Chinese Academy of Science to Britain's Royal Chemical Society.[12]

Yet for all of this scientific firepower, large numbers of the citizenry have been slow to accept the reality of climate change and its causes. According to a survey conducted in 2019, although 69 percent of those polled believed in the reality of global warming, only a slim majority of 55 percent of those polled believed that human activity is a major cause of this phenomenon.[13]

Public opinion is important for the shaping of policy, but for the short term at least, what really matters are the beliefs of those holding positions of power. It is here where expert knowledge has met with the most resistance; in 2016, large numbers of Congressional officeholders, including the chairman of the Senate Committee on the Environment and Public Works, vehemently denied the existence of climate change and the human causes of it.[14] More significantly, in 2017, President Donald Trump withdrew from the Paris Climate Agreement, which set future limits on emissions from energy sources to address global warming and climate change. Facts do not always speak for themselves, and ideological stances coupled with the influence of special interests can be very difficult to overcome.

In sum, experts are in a position to influence the course of technological change, but at the same time they face many difficulties in converting their knowledge into power. Being free from outside influences and having the correct technical answers are not enough; wielding power requires the mobilization of a constituency and the ability to create a broad base of support.[15] Even outside government, technological decisions have the same political character, with the same need for gathering support for projects and policies. This means that except for fairly trivial matters, issues often are not settled on the basis of purely technical considerations.

Engineers and the Control of Technology

To further explore the connection between expertise and the shaping of technological change, we will consider the working lives of the people who are most closely involved with technological innovation: engineers. As we shall see, more is involved in their work than the exercise of purely technical skills. In a modern economy, the design and development of new technologies is largely the work of engineers and other technical specialists. It seems natural, then, to expect that a great deal of control over the direction of technological development should be exercised by members of these occupations. Moreover, since so much of our world is shaped by technology, technical experts might be expected to have a powerful influence on society as a whole.

As we have seen in Chapter 2, some have argued that technocracy, the exercise of political authority by engineers and other technical experts, was the soundest and most rational way to govern. One of the most systematic and influential statements of the virtues of technocracy was the work of a sociologist/economist named Thorstein Veblen, who expounded his ideas during the first two decades of the twentieth century. According to Veblen, engineers and other technical experts were the "indispensable factor in the everyday work of carrying on the country's productive industry."[16] But at the same time, "they have nothing to say in the planning and direction of this industrial system, except as employees in the pay of financiers."[17] It was this contradiction between the technical acumen of the engineers and their subjugation to managerial control that Veblen saw as the greatest obstacle to continued economic progress. The only solution was a takeover of business enterprises by a "Soviet of Technicians" so that the economy would be continuously stimulated by technological advance instead of being retarded by the narrow financial interests of absentee owners and their delegated managers.[18]

Veblen's ideas provided inspiration for a technocratic political movement that flourished for a brief period during the 1930s and then expired.[19] But the idea that technical expertise should be the proper basis of control did not die with it. Several decades later, a number of social scientists advanced the idea that Veblen's prescriptions had been achieved; the enterprises of advanced industrial economies were now under the control of engineers and other technologically oriented personnel. For sociologist Daniel Bell, the key personnel in modern corporations were the "corporate organizers," a "special breed, often engineers, whose self-conscious task was to build a new economic form."[20] A similar theme was sounded by economist John Kenneth Galbraith in his influential book *The New Industrial State*. Galbraith argued that the owners of large business firms no longer ran them, as had been the case in the days of the traditional entrepreneur. Instead, the management of modern enterprises was now in the hands of the "technostructure": managers and engineers with the specialized knowledge and talent essential to planning and decision making.[21]

Many of the assertions presented in *The New Industrial State* have been largely refuted in the decades following the book's publication,[22] but one concept is at least worthy of further inquiry. Galbraith's invocation of the "technostructure" has directed our attention to the role of engineers and other technical experts in the management of modern firms. As subsequent research has shown, engineers as a group are deeply involved in management, but they do not form a technocratic elite, nor, at the other end of the spectrum, are they simply hired hands.

In understanding the role played by engineers, it is important to note that, along with other skilled occupational groups, engineers expect to be recognized as professionals. At the most basic level, the work of a professional centers on the possession and application of specialized knowledge that is found only within the ranks of the profession. This means that someone outside the profession does not have sufficient knowledge to pass judgment on working professionals; only other professionals can do so. Often, there is a strong link, legally defined, between schooling and occupations. Medicine has this feature: not just anyone, however knowledgeable, can "put up a shingle" and practice medicine. One must go to an accredited medical school and pass exams determined by the professional association and certifying boards.

Consequently, the essence of professionalism is autonomy.[23] This has been evident in the long-established professions, such as medicine and law, where doctors and attorneys have been accustomed to the insulation of their careers from outside forces. In part, this has been due to the fact that, until fairly recently, they have been in business for themselves as independent practitioners. With the rise of managed care in medicine, however, both professional autonomy and career satisfaction have been on the decline.[24] For engineers, except for a small number of private consultants, engineers have always been employees of an organization, usually some sort of business enterprise.

The National Society of Professional Engineers (NSPE) has a list of "fundamental canons." It states that:

Engineers, in the fulfillment of their professional duties, shall:

1. Hold paramount the safety, health, and welfare of the public.
2. Perform services only in areas of their competence.
3. Issue public statements only in an objective and truthful manner.
4. Act for each employer or client as faithful agents or trustees.
5. Avoid deceptive acts.
6. Conduct themselves honorably, responsibly, ethically, and lawfully so as to enhance the honor, reputation, and usefulness of the profession.

While it is the governing body for engineers, fewer than 20 percent of the people with the job title "engineer" are registered professionals.[25] In fields like civil engineering, where practitioners are often heads of firms which provide contracted construction and infrastructure services, the PE license is very important. Licensure is awarded after a four-year degree, four years of supervised employment by a PE, and two examinations, plus continuing education over the course of a career.[26] But in many other domains, one's skills, with coding (or analysis in the case of software engineers), hardware design, mathematical analysis and modelling, or applied physics, can open doors for work as an engineer no matter the university degree.

As we will discuss below, codes of ethics are protective of both clients and the boundaries of a profession by regulating who can claim expertise in a specific knowledge domain. Codes of ethics are part of the apparatus that allows a profession to self-regulate. Otherwise, the legal system might be called in to regulate practice. However, one can see a number of potential conflicts embedded even in these brief canons for engineers. For example, what if protecting the health and well-being of the public interferes in business decisions that benefit their client and employer? This is an example of what might be defined as a question of utilitarian ethics—the greatest good for the greatest number—conflicting with a duty-driven model of ethics, which emphasizes one's responsibilities in a particular role or contracted obligation.

While serving as employees, engineers do enjoy a considerable amount of freedom as they go about their work. They are not subject to rigid supervision, and they have considerable discretion regarding how they plan their activities and schedule their work. They have to respond to the requests of management and deal with the problems presented to them, but they have wide latitude in determining how the job is done.[27] At the same time, however, the work of engineers is closely constrained by the expectations of their employers. While they have the freedom to decide how a job is to be done, engineers have little influence over the choice of what is to be done. And, in truth, few engineers are concerned about their lack of influence in the setting of company policies or the purposes to which their talents are being put. Engineers' training and on-the-job socialization do not generally produce ambitions of this sort. The work of engineers exemplifies the rational search for the best way of attaining particular ends. But engineers rarely have the privilege of selecting these ends. Indeed, once they do, arguably they no longer act as engineers.[28]

While engineers are subject to managerial authority, they may seek to extend their influence by joining the ranks of management. At the same time, however, engineers in management positions paradoxically lead to a dilution of engineers' influence. For many engineers, career success means attaining a managerial position, not being an increasingly proficient engineer.[29] Engineering and management require different skills and orientations. As Samuel C. Florman summarizes, "The characteristics of a good manager—a feeling for people, politics and the bottom line—appear to conflict with those of a first-rate creative engineer—an aptitude for numbers, theorems, materials, and spatial relationships."[30] But while orientations and even fundamental values may conflict, with the prospect of a managerial career dangled in front of them, a practicing engineer is not likely to challenge managerial values, even when they may conflict with the engineer's professional values. Expert knowledge and a lifetime devotion to the acquisition and use of that knowledge are central to the professional's value system; in contrast, businesspersons and managers value loyalty and personal initiative. The prospect of moving from engineering into management can affect people's values and undercut their identity as engineers.[31]

Blowing the Whistle

The problems of engineers and other employees influencing company policy are illustrated by the dangers of whistleblowing. A whistleblower is someone who detects some sort of wrongdoing within his or her organization and brings it to the attention of the public. An engineer can be put in this position when management makes a decision that is so technically unsound or otherwise defective that it poses a threat to the user or to society as a whole. Since an engineer knows the technical facts of a case better than anyone else, he or she is in the best position to reveal whatever problems have appeared.

Engineers may be willing to call attention to serious problems because of their conviction that "the facts speak for themselves."[32] In this, they are sadly naive. Most engineers work in organizational settings with strong hierarchical structures. Decisions are made in accordance with vertical relationships; those at the upper levels can countermand those at the lower level, even if the latter have the "facts" in their favor. This is dramatically illustrated by the experiences of some engineers who worked for a major defense contractor. After a chief engineer designed an aircraft disc brake that was woefully deficient in stopping power, he enlisted the support of his superiors in falsifying crucial test data. When a lower-level engineer and a technical writer objected, they were told to keep their mouths shut and participate in the falsification. Subsequent testing by the Air Force confirmed the brake's dangerous shortcomings. Deeply disturbed by the whole process, the dissenting engineer and the writer resigned from their positions. All of the conspirators retained their jobs, except for two who were promoted to higher positions.[33]

There are occasions when prior complaints by whistleblowers come to light only after a tragedy validates their complaint. Two of them emerged following the crash of two Boeing 737 Max aircraft, one in Indonesia, the other in Ethiopia. The crashes, which took a total of 346 lives, occurred after the airplanes' new flight control system (known as the Maneuvering Characteristics Augmentation System) erroneously detected an impending stall and repeatedly pushed planes' noses down. Despite frantic efforts by the pilots to overcome the automated system, the planes flew into the ground at a high rate of speed, killing everyone aboard.

Curtis Ewbank, a Boeing engineer, repeatedly used Boeing's internal whistleblower communication channel to assert that the MCS needed a second airspeed indicator to ensure that the system was not being activated by a false indication of an impending stall. His complaint was ignored by Boeing's management, leaving him to conclude that management was "more concerned with cost and schedule than safety and quality."[34]

Boeing was the target of another whistleblower complaint when Ed Pierson, a senior manager, called attention to slipshod work due to overworked and weary production personnel. He raised concerns about the 737 Max's safety to the plane's program manager with an email message asserting that "Frankly right now all my internal warning bells are going off and for the first time in my life, I'm sorry to say that I'm hesitant about putting my family on a Boeing airplane." In addition to notifying Boeing of workplace issues, Pierson contacted the Federal Aviation Administration (FAA), the National Transportation Safety Board (NTSB), and the Department of Transportation (DOT). The FAA showed no interest in his concerns, while the NTSB claimed that addressing his issues was not among its responsibilities. Both aviation authorities met Pierson's complaints with "indifference," according to his lawyer. DOT never replied to him.[35]

The hierarchical structure of organizations also helps management to control communication; employees are expected to "go through channels." Facts that pose a challenge to a superior can thus be easily suppressed. The only alternative is circumventing the hierarchy; a concerned employee can attempt to get the attention of officials at much higher levels of the organization, or he or she can reveal malfeasance to the media or some other outside institution. These are risky courses of action. Top-level officials do not like disturbances of conventional channels of authority, and they are also likely to have doubts about an employee who has gone behind the boss's back. Revealing a problem to the media or an outside agency is an even worse breach of proper organizational behavior. Loyalty is highly prized in organizations, and an employee perceived as disloyal to their superior or to the organization as a whole is often someone to be scorned.

The perils of whistleblowing by an engineer were starkly revealed by the events following the breakup of the space shuttle *Challenger*, 73 seconds after it was launched on January 28, 1986.[36] A group of engineers from Morton-Thiokol, the manufacturer of the shuttle's solid-fuel booster rockets, had warned the night before the launch that cold weather could cause a failure of the O-rings that sealed the segments of the rockets. After the erosion of an O-ring resulted in the tragedy that many had feared, Roger Boisjoly, a specialist in seals who had worked on NASA projects for 20 years, provided the official government board of inquiry with an internal memo he had written six months earlier in which he had called attention to the danger of launching the shuttle in subfreezing weather. This soon led to his being ostracized and isolated at Morton-Thiokol. Suffering from depression, headaches, and double vision, he left his secure and remunerative job, sold his house, and moved out of the community in which he had lived for many years. Afterwards, he spoke to more than 300 universities and civic groups about corporate ethics and was awarded the American Academy for the Advancement of Science's Prize for Scientific Freedom and Responsibility, but he had paid a heavy price for calling attention to the reckless irresponsibility of some of his colleagues.[37]

The battle of an engineer against their organization is likely to be a lonely one. Especially unfortunate is the fact that the professional engineering associations usually are not very helpful to the whistleblower. These associations have often been dominated by businessmen who are more likely to be concerned with their industry than they are with the career concerns of individual engineers.[38] This is illustrated by the case of Ernest Fitzgerald, who was summarily fired after he had called attention to the huge cost overruns on the Lockheed C5A transport plane. His professional association, the American Institute of Industrial Engineers, spurned Fitzgerald's request to "investigate the professional and ethical questions involved." It refused on the grounds that it was a "technical organization" and not a "professional society." The inclusion of military contractors as members of the Institute made this a likely outcome.[39]

The Ford Pinto is a well-known case study. As journalist Mark Dowie tells the story, Ford Motor Company was aware for seven years that the Pinto, due to a flawed gas tank design lacking an $8 part, would potentially explode upon rear-end impact. In short—despite some protests by engineers—Ford decided that paying out for lawsuits because of the infrequent but often lethal fires was more cost-effective than a recall or redesign and retooling production. Engineers were largely unwilling to rock the boat after moving their concerns up into management. The Ford Corporation also spent significant effort in delaying and opposing emerging automobile safety regulations and blocked stricter penalties for knowingly advancing dangerous designs.[40] Here, again, the tension between an engineer's responsibilities to the public is in conflict with their responsibilities to their employer.

Because the educational origins of people with computing jobs is more diverse than engineering, in terms of fields and professional associations, software ethics is an even less-settled field than engineering ethics. People with degrees in MIS (Management Information Systems), Computer Systems Engineering, Computer Science, and Mathematics, and Information Sciences, as well as people with all sorts of other degrees in the social and natural sciences, can develop computing skills; all contribute to coding for complex systems. There are two especially important case studies to consider. The Therac-25 was a linear accelerator designed to provide radiation treatment for cancer patients. It was a complex system that allowed for computer-controlled positioning, and variable time and strength applications of radiation therapy. However, by re-using code from one model to another, lack of code verification allowed for interlocks which prevented problems with dosing and patient positioning to be overridden. Six patients suffered serious radiation injuries and died. What had been a matter of hardware checks on safety was transferred to software, without sufficient attention to conflicting information within the system. Like scientific replication, software verification is nowhere near as "sexy" as producing new code or new knowledge. It is difficult to measure "productivity" for code verification. But inattention to buggy code means more than dissatisfied customers: it can produce lethal results.[41]

The Therac-25 problem was a matter of coding errors and poor system integration. What about when algorithms work as planned but produce inequitable outcomes? As discussed in Chapter 17, on advanced ICT issues, algorithmic bias—such as algorithmic redlining—is especially important as an example of how the *perceived* neutrality of technologies, enhanced by the relative invisibility of their inner workings and the social construction of their design, produces the illusion of objectivity.[42] Redlining is the now-illegal grading of neighborhoods and their perceived "riskiness" for extending bank loans for individual borrowers. "Undesirable" neighborhoods would be coded "red" by banks on actual maps, with ethnic diversity being a negative mark in the years after World War II. The red lines around "bad neighborhoods" resulted in difficulties attracting new residents and investments, perpetuating cycles of urban decay, and depriving individuals of access to residential and business loans in what might have been desirable and low-cost neighborhoods.[43] Now, the inscrutability of algorithmic codes in financial decisions makes redlining a possibility again, and more difficult to examine and prove because the codes are proprietary and difficult to understand.[44]

Science: Institutional Norms and Ethics

Finally, we want to include here the role of scientists and mention one more set of ethical and professional issues in relation to the norms governing science, and the rights of human subjects in research. Science as an institution has often drawn strong boundaries between its work in basic research and the application of knowledge, deferring questions about the ethics of scientific knowledge to those (engineers) who might apply it. There are both internal and external ethical frameworks that help to govern science. One of the first to articulate contemporary scientific norms was R. K. Merton.[45] Merton believed that for science to work as a democracy-supporting institution, it needed to be disinterested, communal, universal, and engage organized skepticism. Disinterestedness, to Merton, recognized that profitability could produce fraud and misdirect knowledge. Communalism (originally communism) was the idea, that knowledge needed to be shared to be verified. Universalism, to Merton, represented the idea that the institutional standing of someone making a knowledge claim ought not matter. Knowledge from East Stroudsburg State ought to be as valuable as knowledge from Stanford, and from a graduate student as from a winner of a Nobel Prize. Finally, science needs error correction methods, such as through organized skepticism. Peer review, public presentations, and new things like preprint circulation represent that sort of error checking. Internet trolls need not apply, however: effective skepticism is organized through professional affiliations, not random commenting on social media platforms.

Writing in 1942, Merton was concerned with protecting science from the abuses of autocratic societies such as Nazi Germany and the Soviet Union. Those totalitarian systems produced distortions in science, such as Lysenkoism in the USSR,[46] or the rejection of Einstein's work in physics as "Jewish Science" by the Nazis.[47] And of course individual lives and careers were hurt by those systems, as well as producing a generation of lost knowledge. In the case of Lysenkoism, a belief in the direct heritability of traits, it produced a generation of poor farm production and famine and put Soviet scientists far behind their Western counterparts in moving the biological sciences forward.

While we can recognize the flaws and weaknesses of Merton's norms, we also can see how they are deeply embedded in contemporary institutions. For example, scientists need to declare conflicts of interest when applying for grants. Publication practices rely on peer review. These norms also help to generate other internal expectations against fraud, falsification, and plagiarism. But they have not always been so good in the outward facing ethics of science—its relationships to the public as users, and often subjects, of knowledge production. There are two particularly egregious case studies to be mentioned here: the Tuskegee syphilis study and the Henrietta Lacks (He-La) case.

The Tuskegee syphilis study was conducted by the United States Public Health Service and the Tuskegee Institute from 1932 to 1972. It studied the "natural history" of syphilis by allowing 399 incarcerated Black men to live with untreated syphilis so that the effects of the disease over time could be studied. Even when, in 1943, penicillin became available as a treatment, that treatment was withheld. Men were told they were being observed for "bad blood," and received free medical exams, meals, and burial insurance.[48] As a result of this clear abuse of informed consent, the 1979 Belmont Report emerged from the National Commission for the Protection of Human Subjects of Research. This report established the principles of beneficence: (1) seeking outcomes expected to be of positive use and active minimization of potential harms; (2) respect for persons, which is encapsulated in informed consent; and (3) justice, meaning that people should have equitable access to the burdens and benefits of research.[49] These are essential principles shaping the conduct of good scientific research on human subjects, although not generally part of engineering and innovation research.

The Henrietta Lacks case illustrates the complexity of refining research ethics over time. Henrietta Lacks was admitted to the Johns Hopkins Hospital in February 1952. While Mrs. Lacks had informed consent for her surgery, she did not give consent for the *use* of her tissues for future research. However, informed consent for tissue use was not typical. Cells from her cervical cancer turned out to be basically "immortal," and stable over time in culture, a rare property. HeLa cell lines are a multibillion-dollar enterprise, although Mrs. Lacks' descendants received no compensation.[50] Descendants of Henrietta Lacks recently sued Thermo Fischer Scientific and about 100 other pharmaceutical companies that have profited from the HeLa cells, which have been used for cancer, HIV, and many, many other scientific pursuits.[51] However, there is not much reason to be optimistic about more than a token settlement, like the undisclosed amount reached with Thermo Fischer Scientific in July 2023. Since the 1990 *Moore vs. the Regents of California* case (otherwise known as the "Hairy Cell Leukemia Case") determined that discarded tissue was no longer a patient's property and patients therefore had no claim to future intellectual property or profits,[52] it is unlikely that the courts will upend precedent and a multibillion-dollar industry. It is not yet clear what impact, if any, the recent confidential settlement between the Lacks Estate and Thermo Fisher Scientific Inc. will mean for the human tissue industry, but the specificity of the Lacks case, and the precedent of Moore, suggest that this will not extend the property rights of patients who have "waste" tissues removed. The Henrietta Lacks case has extended rules about patient privacy with regard to removed tissue, but time will tell if the problem of profiting from removed tissues will change directions.

Both the Tuskegee experiments and the use of Henrietta Lacks' cells concern the *use* of persons, and even animals. These cases both violate what would be Kant's categorical imperative to never treat a person as a means to an end. Persons are always ends in themselves. This is foundational to a rights orientation, which can be in conflict with a utilitarian orientation which seeks the greatest good for the greatest number.

Intents and Effects, Means and Ends

W. Bernard Carlson and Rosanne L. Welker of the University of Virginia developed a framework they called "The Whammy Line." The goal for innovators is to avoid an unexpected, disruptive jolt—the titular "whammy." It is the line where unexpected negative outcomes, especially, can be significantly harmful.[53] An actor can have a negative or positive intention. For the case of engineers, decisions or professional skills can be used with intent to do someone harm, or to do the right thing. But because we are humans with limited ability

Effects and Intents

Effect \ Intent	Negative	Positive
Intended	Ethically suspect.	Desired Outcomes.
Unintended	Difficult to imagine. (Complexity, enthusiasm, organizational constraints) * Attempt to minimize	Value unexpected benefits.

Figure 23.1 An adaptation of W. Bernard Carlson and Rebecca Welker's 2001) "whammy line." Engineers (and managers, designers, and really almost anyone) should strive to stay on the "positive side" of effects, avoiding negative outcomes.

to predict the future, and many aspects of science and technology are complex and difficult to predict, we face the problem of *unintended* consequences. Sometimes the consequences are technically difficult to predict. Sometimes we "fool ourselves" because we are enthusiastic about something and don't *want* to see the possible complications and negative consequences. This is a bit different from actively ignoring known harms (an intention) but given our difficulties with sunk costs and vested interests, we have emotional and organizational commitments to plans of action and simply fail to imagine consequences. This is an important reason diversity of perspective is extremely important in groups and organizations. It's not about being "negative," but about thinking creatively to avoid future harms that could be avoided with some foresight.[54]

Similarly, we often don't imagine that there are *positive* unintended consequences, and thus may not be in a position to capitalize on new opportunities. The first inventors of the telephone were not all that quick to recognize its positive potential. The case of Viagra, and other erectile dysfunction drugs, is an example of capitalizing on an unexpected benefit. The drugs were designed to combat chest pains. The little blue pills didn't work for that, but the patients *really* liked the side-effects, and so the drugs were "repurposed." That's gaining value from unexpected benefits.[55]

Our goal is to spend as much time as possible in the upper right quadrant of Carlson and Welker's framework, where our technoscience is working for the benefit of all, as planned. This assumes, however, that we *agree* what "good" and "bad" are. If you are committed to the idea that abortion is *always* wrong, there is no way to "perfect" that technology for treatment. With an anti-abortion commitment, the whole project of improving abortion care is flawed as there are only negative consequences. If you are committed to allowing medically necessary abortions, then you want technoscience innovations to protect the life of the pregnant person with minimal risk—harm to the adult patient being a negative consequence.

In effect, we are suggesting that the questions about the impacts of technologies on the users, and non-users, need to be foregrounded in engineering research, education, and the design activities of corporations in advance of the release of new technologies, with greater imagination and foresight. But we also recognize how difficult that will be for people embedded in complex organizations, which produce organizational blinders and what sociologist Diane Vaughan has called the "dark side" of organizations.[56]

From this brief survey, it should be apparent that engineers do not exercise commanding influence over the course of technological change. Their work, while essential to technological development, is constrained by the economic, political, and organizational environment in which engineering operates. Engineers are intensely focused on the *means* of developing new technologies. And technologies are *means* to achieving *ends*. Crucial decisions are made by management, and even if engineers are often recruited into its ranks, when this happens they necessarily change their orientations. Within institutional settings and public discourse, where do our discussions about the ends of science and technology take place? Managers must be primarily concerned with profit or some other organizational goal; and technical virtuosity should not be an end in itself. If we are looking for a major source of control over technology, and for the discussions of the ends to which we are applying technology, we must extend our focus beyond engineers to the organizations that employ them, and the governmental systems which provide the context for the development and regulation of technologies.

QUESTIONS FOR DISCUSSION

1. Some scientific and technological experts claim that nuclear power is safe and economical, and that catastrophes like Chernobyl and Fukushima were anomalies caused by highly unusual events. Other experts say just the opposite. How can a nonexpert choose between the two? What sort of procedures might be used in order to determine which group's claims are more valid?
2. In this chapter, the assertion is made that engineers cease to act as engineers when they are involved in the selection of the goals to be pursued by their organization. Do you agree? How might the professional abilities of engineers make them effective participants in the goal-selection process? How might these abilities hinder them?
3. What responsibilities do nonexperts have if we cannot completely rely on experts?
4. What are some morally or politically contentious issues which might have technological solutions? Are there ways to reach consensus on those issues in such a way as to allow for thoughtful advance of technological development?
5. Think about some case studies that had unintended consequences. What are potential mechanisms for addressing unanticipated negative consequences? Unanticipated positive consequences?

Notes

1. Steven B. Katz, "The Ethic of Expediency: Classical Rhetoric, Technology, and the Holocaust." *College English* 54, 3 (1992): 255–75. The full memos can be read at https://gombinsociety.org/jews-of-gombin-holocaust/gas-vans/technical-changes-to-improve-the-gas-vans.
2. Phillip G. Altbach and Hans de Wit, "Too Much Academic Research is Being Published," *University World News* (September 7, 2018). Accessed from www.universityworldnews.com/post.php?story=20180905095203579.
3. Duncan MacRae, Jr., "Science and the Formation of Policy in a Democracy," in Thomas J. Kuehn and Alan L. Porter (Eds.), *Science, Technology, and National Policy* (Ithaca, NY: Cornell University Press, 1981), p. 497.
4. Jeremy Hsu, "There Are More Boys than Girls in China and India: Preference for Sons Could Spell Trouble for China and India," *Scientific American* (August 2008). These numbers have come down, slightly over time, in China: www.statista.com/statistics/282119/china-sex-ratio-by-age-group, and in India: https://statisticstimes.com/demographics/country/india-sex-ratio.php.

5. Binyamin Appelbaum, "As U.S. Agencies Put More Value on a Life, Businesses Fret," *The New York Times* (February 16, 2011). Accessed from www.nytimes.com/2011/02/17/business/economy/17regulation.html.
6. H. H. Gerth and C. Wright Mills (Eds.), *From Max Weber: Essays in Sociology* (New York: Oxford University Press, 1958), p. 233.
7. David Collingridge, *The Social Control of Technology* (New York: St. Martin's Press, 1980), pp. 135–38.
8. Nico Stehr and Reiner Grundmann. *Experts: The Knowledge and Power of Expertise* (New York and London: Routledge, 2011), pp. 45–46.
9. David Willman, "NIH to Curb Scientists' Deals with Drug Firms," *Los Angeles Times* (June 23, 2004): A1.
10. David L. Parnas, "Why I Quit Star Wars," *Common Cause* 12, 3 (May–June 1986): 32, 34.
11. Debra Javeline, Jessica I. Hellman, Rodrigo Castro Cornejo, and Gregory Shufeldt, "Expert Opinion on Climate Change and Threats to Biodiversity," *BioScience* 63, 8 (2013). Accessed from https://bioscience.oxfordjournals.org/content/63/8/666.
12. NASA, "Scientific Consensus: Earth's Climate Is Warming" (February 11, 2016). Accessed from http://climate.nasa.gov/scientific-consensus.
13. Gayathri Vaidyanathan, "Big Gap Between What Scientists Say and Americans Think about Climate Change," *Scientific American* (January 30, 2015). Accessed from www.scientificamerican.com/article/big-gap-between-what-scientists-say-and-americans-think-about-climate-change; Anthony Leiserowitz, et al., "Climate Change in the American Mind: April 2019," *Yale Program in Climate Change Communication* (June 27, 2019). Accessed from https://climatecommunication.yale.edu/publications/climate-change-in-the-american-mind-april-2019/2.
14. David M. Herszenhorn, "Votes in Congress Move to Undercut, Climate Pledge," *The New York Times* (December 1, 2015). Accessed from www.nytimes.com/2015/12/02/us/politics/as-obama-pushes-climate-deal-republicans-move-to-block-emissions-rules.html.
15. Mark R. Berg, "The Politics of Technology Assessment," in Kuehn and Porter, *Science, Technology, and National Policy*, p. 494.
16. Max Lerner (Ed.), *The Portable Veblen* (New York: Viking, 1948), p. 440.
17. Ibid., p. 442.
18. Ibid., p. 463.
19. See William E. Akin, *Technocracy and the American Dream: The Technocrat Movement, 1900–1941* (Berkeley: University of California Press, 1977).
20. Daniel Bell, *The End of Ideology* (Glencoe, IL: Free Press, 1967), p. 43.
21. John Kenneth Galbraith, *The New Industrial State*, 2nd ed. (Boston: Houghton Mifflin, 1971), pp. 59–71.
22. Richard Parker, *John Kenneth Galbraith: His Life, His Politics, His Economics* (New York: Farrar, Straus, and Giroux, 2005), pp. 439–50.
23. Rudi Volti, *An Introduction to the Sociology of Work and Occupations* (Thousand Oaks, CA: Sage, 2012), pp. 153–72.

24. The emergence of managed care in the United States in the 1990s has produced both "top down" challenges to medical professional autonomy, and "bottom up" competition from non-MD medical practitioners, who while they may be eminently qualified are nonetheless presented as a low-cost alternative to MDs. See: J. J. Stoddard, J. L. Hargraves, M. Reed, and A. Vratil, "Managed Care, Professional Autonomy, and Income: Effects on Physician Career Satisfaction," *Journal of General Internal Medicine* 16, 10 (2001): 675–84; Perri A. Morgan, et al., "Impact of Physicians, Nurse Practitioners, and Physician Assistants on Utilization and Costs for Complex Patients," *Health Affairs* 38, 6 (2019).
25. PDH-Pro, "Becoming A Professional Engineer Provides Significant Advantages." Accessed from https://pdh-pro.com/pe-resources/becoming-a-professional-engineer-provides-significant-advantages.
26. National Society of Professional Engineers, "What is a PE?" Accessible from www.nspe.org/resources/licensure/what-pe.
27. Robert Zussman, *Mechanics of the Middle Class: Work and Politics among American Engineers* (Berkeley: University of California Press, 1985), p. 58.
28. Ibid., pp. 122–23.
29. Zussman, *Mechanics of the Middle Class*, p. 151.
30. Samuel C. Florman, "Up and Down the Dual Ladder," *Technology Review* 88, 6 (August–September 1985): 12.
31. Edwin T. Layton, Jr., *The Revolt of the Engineers: Social Responsibility and the American Engineering Profession* (Cleveland: Case Western Reserve University Press, 1971), p. 8.
32. Deena Weinstein, "Bureaucratic Opposition: Whistle-Blowing and Other Tactics," in Ron Westrum and Khalil Samaha (Eds.), *Complex Organizations: Growth, Struggle, and Change* (Englewood Cliffs, NJ: Prentice-Hall, 1984), p. 255.
33. Frank Vandiver, "Why Should My Conscience Bother Me?" in Robert Heilbroner, et al., *In the Name of Profit* (New York: Doubleday, 1972), pp. 3–31.
34. Dominic Gates, Steve Miletich, and Lewis Kamb, "Boeing Rejected 737 Max Safety Upgrades before Fatal Crashes, Whistleblower Says," *The Seattle Times* (October 3, 2019). Accessed from www.seattletimes.com/business/boeing-aerospace/boeing-whistleblowers-complaint-says-737-max-safety-upgrades-were-rejected-over-cost.
35. Alexandra Ma, "A Boeing Whistleblower Says He Tried to Raise Concerns about Sloppy 737 Max Production, but Was Ignored by the CEO, Board, FAA," *Business Insider* (December 10, 2019). Accessed from www.businessinsider.com/boeing-737-max-whistleblower-ed-pierson-ceo-faa-ntsb-ignored-2019-12.
36. Diane Vaughan, *The Challenger Launch Decision: Risky Technology, Culture, and Deviance at NASA* (Chicago and London: University of Chicago Press, 1996).
37. Douglas Martin, "Roger Boisjoly, 73, Dies; Warned of Shuttle Danger," *The New York Times* (February 3, 2012). Accessed from www.nytimes.com/2012/02/04/us/roger-boisjoly-73-dies-warned-of-shuttle-danger.html?_r=0.
38. Layton, *The Revolt of the Engineers*, passim.
39. Ralph Nader, Peter J. Petkas, and Kate Blackwell, *Whistle Blowing: The Report on the Conference on Professional Responsibility* (New York: Grossman, 1972), p. 52.

40. Mark Dowie, "Pinto Madness," September/October 1977. *Mother Jones.* www.motherjones.com/politics/1977/09/pinto-madness.
41. N. G. Leveson, "The Therac-25: 30 Years Later," *Computer* 50, 11, pp. 8–11, November 2017; and Sara Baase, "8.2 Case Study: The Therac-25," pp, 425–420, in *A Gift of Fire: Social, Legal, and Ethical Issues for Computing Technology*, 4th ed. (Upper Saddle River, NJ: Pearson, 2013).
42. B. D. Mittelstadt, P. Allo, M. Taddeo, S. Wachter, & L. Floridi, "The Ethics of Algorithms: Mapping the Debate," *Big Data & Society* 3, 2 (2016).
43. Gary Smith, "High-Tech Redlining: AI is Quietly Upgrading Institutional Racism," *Fast Company* (November 20, 2018). Accessed from www.fastcompany.com/90269688/high-tech-redlining-ai-is-quietly-upgrading-institutional-racism.
44. A. Tsamados, N. Aggarwal, J. Cowls, et al. "The Ethics of Algorithms: Key Problems and Solutions," *AI & Society* 37 (2022): 215–230.
45. Robert K. Merton, "Science and Technology in a Democratic Order," *Journal of Legal and Political Sociology* 1 (1942): 115–126. Reprinted frequently.
46. Edouard I. Kolchinsky, Ulrich Kutschera, Uwe Hossfeld, Georgy S. Levit, "Russia's New Lysenkoism," *Current Biology* 27, 19 (2017): R1042–R1047.
47. Philip Ball, *Serving the Reich: The Struggle for the Soul of Physics Under Hitler* (Chicago, IL: University of Chicago Press, 2014). Excerpt available at www.scientificamerican.com/article/how-2-pro-nazi-nobelists-attacked-einstein-s-jewish-science-excerpt1.
48. CDC, "The Untreated Syphilis Study at Tuskegee Timeline." Accessed from www.cdc.gov/tuskegee/timeline.htm.
49. See U.S. Health and Human Services at www.hhs.gov/ohrp/regulations-and-policy/belmont-report/index.html.
50. L. M. Beskow, "Lessons from HeLa Cells: The Ethics and Policy of Biospecimens," *Annual Review of Genomics and Human Genetics* 17 (2016): 395–417.
51. Emily Davies, (2021). "70 Years Ago, Henrietta Lacks's Cells Were Taken without Consent: Now, Her Family Wants Justice," *The Washington Post* (October 4, 2021).
52. *Moore v. Regents of the University of California*. 793 P.2d 479 at 490 (Cal. 1990) and Rebecca Skloot, "Taking the Least of You. The Tissue-Industrial Complex," *The New York Times Magazine* (April 16, 2006).
53. W. B. Carlson & R. Welker, "The Whammy Line As A Tool For Fostering Moral Imagination." Paper presented at 2001 Annual Conference, Albuquerque, NM (June 2001). Paper available at https://peer.asee.org/the-whammy-line-as-a-tool-for-fostering-moral-imagination.
54. This is a good place to introduce the "Online Ethics Center for Engineering and Science": https://onlineethics.org. It has numerous resources, case materials, collections, and tools, as well as local programming and webinars for examining ethics and values issues in science and engineering research and practice.
55. Meika Loe, *The Rise of Viagra: How the Little Blue Pill Changed Sex in America* (NYU Press, 2004).
56. Diane Vaughan, "The Dark Side of Organizations: Mistake, Misconduct, and Disaster," *Annual Review of Sociology* 25, 1 (1999): 271–305.

Chapter 24

Governing Technology

Prior chapters put forth the claim that organizations play a major role in determining the kind of technologies we have. To be sure, these organizations are not all-powerful. When business firms are involved, consumers can affect technological change by deciding whether to buy their products. As employees, we might be able to exert some influence through our acceptance or rejection of new workplace technologies or the direction that firms are taking as producers of new technology. But not all of the major decisions affecting the trajectory of technological change are made in the private sector. Government organizations influence technological choices to a large extent. In a democratic society, this should expand the power of citizens to get the technologies they want. But do they? In a technological society, what are the various kinds of technological politics? Does governmental involvement in the development of technology necessarily result in more beneficial technologies? Are democratic ideals realized in the operation of government policies and processes? These are some of the issues that will be addressed in the following survey of the government's role in shaping technological change.

Government Actions and the Shaping of Technology

The involvement of the federal government in technological change is as old as the Republic. The United States of America emerged during an era when technological change was beginning to accelerate, and the framers of the Constitution believed that the government should encourage technological advance through the granting of patents and copyrights to those who had produced new things.

Article I, Section 8, of the Constitution stipulates that Congress is to "promote the Progress of Science and Useful Arts by securing for limited Times to Authors and Inventors the Exclusive Rights to their respective Writings and Discoveries."

The government's granting of patent rights provides the legal framework for the monopolistic control of technological advances generated by individuals and private organizations. The government also supports other types of infrastructure that are essential to the creation, development, and use of modern technologies. Since the ultimate basis of technological advance is the growth of knowledge, governments at all levels do much to indirectly sponsor technological advance through their support of education. Federal aid to individual students has supported the education of large numbers of scientists, engineers, and technicians.

Moreover, in recent decades, institutions of higher learning have been major sources of technological advance through their research activities. These activities were strongly encouraged by the passage of P.L. 96-517, commonly known as the Bayh-Dole Act, in 1980. Under its provisions, universities receiving federal research funds that aid in the creation of innovative products and processes can obtain patents on them.[1] Universities are involved with more than 4,000 patent licensing agreements annually, accounting for about $2 billion in licensing fees. A few patents have been spectacularly successful from a financial standpoint; the gene-splicing technology developed by Herbert Boyer and Stanley Cohen brought $200 million in revenues to Stanford University and the University of California, San Francisco. But patenting has not produced financial windfalls for most institutions; most of the revenues have gone to a small number of institutions.[2] Moreover, critics have voiced concerns that the pursuit of these revenues has undermined some of the basic values of the university, such as the pursuit of knowledge for its own sake and the free transmission of knowledge. Michael Crow, President of the Arizona State University, made the point that the U.S. academic research support system serves as a de facto technology policy in an environment where negative ideas about interference in the market prevent coherent de jure technology policy.[3]

In addition to maintaining the patent system and supporting education and university research, government programs also provide the auxiliary goods and services that are essential for the use of particular technologies. For example, high-speed air travel requires more than just jet aircraft; equally important are government-supported airports, air traffic controllers, navigation aids, route allocations, weather forecasts, and safety inspections. To take another obvious example, the effective use of automobiles is dependent on a network of roads and highways, as well as driver training in schools, the enactment and enforcement of traffic laws and safety standards, and so on.

The case of the automobile brings up another reason for government involvement in technological matters. To repeat a theme briefly noted earlier in the book, every significant technological change brings with it an assortment of auxiliary consequences, some of them pernicious. The private automobile has generated many benefits, but it has also created a number of problems. Negative externalities, discussed earlier, are costs that are not directly taken into account when a person buys a product. In operating a car, the owner incurs a number of personal costs: gas, insurance, repairs, and monthly payments that seem to go on forever. At the same time, the operation of the car results in costs for society as a whole: increased pollution, congestion, and so on. Since these costs are not included in the market transaction, governmental action is necessary. The negative externality of automobile-generated air pollution is commonly addressed by regulations that require the installation of emission-control devices. Alternatively, the government could assess an emission tax, thereby making it financially advantageous for an owner to have a car equipped with these devices, as is the case in several countries in the European Union.

Some government policies and activities are not aimed directly at technological change, but they can nonetheless strongly influence its scope and direction. Tax policies can affect technological change by encouraging capital investments that often embody new technologies. One element of current tax policy is based on the idea (which is not universally held by economists) that capital investment can be stimulated by taxing the income derived from investments at a lower rate than other sources of income. As a result, individuals and firms supposedly are motivated to invest their money in new machines and other productive goods. Capital investment can also be encouraged by allowing the accelerated depreciation of capital equipment. This means that the cost of a machine or other piece of equipment with a useful life of, say, 20 years can be used to offset taxes during a much shorter period. As a result, a firm's immediate profits are increased. Since business executives tend to be primarily concerned with short-term performance, accelerated depreciation can stimulate capital investments that embody new technologies.

The government can also encourage the development and use of specific technologies by offering tax credits or deductions to those willing to invest in these technologies. For many years, the federal tax code provided an important benefit for oil producers by granting a depletion allowance that was used to offset corporate taxes. One result was the expansion of petroleum-based energy technologies. Whatever argument one might make about free markets, oil, gas, and coal were subsidized to the tune of about $20 billion in 2020.[4] A different approach was taken when the energy crises of the 1970s generated a strong interest in alternative sources of energy. One governmental response was the sponsorship of alternative energy technologies; during the crises the federal government provided $100 billion in tax credits to individuals and institutions investing in wind-powered electrical generators, the economic viability of which had been closely tied to government subsidies.[5]

Subsidies were also employed to stimulate the sales of electric automobiles. Until they began to be phased out in 2019, state and federal governments subsidized a deduction of $7,500 from the sticker price of an electric car. This was justified due to the environmental damage caused by conventional gasoline-fueled vehicles and the need to provide an alternative. But government programs do not always have straightforward consequences that benefit everyone equally. On average, purchasers of electric cars have incomes well above the national median, so the electric-car subsidy amounted to a government handout to the wealthy.[6] The changes in the automotive markets in California and France, which have pledged to only sell zero-emission vehicles by 2035 and 2040, respectively, will be interesting to follow.[7]

As we have seen, innovation does not always respect existing social and legal arrangements, and the courts may be called upon to resolve issues generated by technological change. To take one example, until fairly recently, biological parenthood was not a problematic concept; every child had a father and mother who jointly contributed to his or her conception. Since 1978, however, it has been possible to use in vitro fertilization (IVF) for this purpose.

IVF has led to a number of legal conundrums centering on the definition of parenthood. One such issue has arisen as hundreds of men have had their sperm frozen for future artificial insemination because they suffered from a terminal disease, were soldiers about to be deployed to a combat zone, or simply did it as a precautionary measure. Some of these men did in fact die prior to a successful insemination and birth, giving rise to this question: Are children that are the product of post-death insemination entitled to the Social Security benefits that are paid to children whose father has died? In Florida, one man's widow filed for benefits on behalf of twins that had been conceived after her husband's death. The agency denied her claim, only to have the decision overturned by an appeals court. But in 2012, the United States Supreme Court unanimously overturned the lower court's decision because Florida barred inheritance for posthumously conceived children.[8] But this was not the end of the matter, as the laws in individual states take different approaches to the inheritance rights of posthumously conceived children.[9] What is fair and proper in cases such as these?

Artificially assisted reproduction will, if anything, become more common in the near future; children can be born decades after the death of their biological father, given that frozen sperm can retain viability for up to 100 years. It is also possible for embryos to be implanted in a surrogate mother long after the death of the woman who was the source of the eggs that had been fertilized. Issues such as these will become even more prominent in the future as advances in reproductive technologies have the potential to alter the essential nature of parenthood and family. Judges and legislators will confront many issues and challenges regarding child custody, inheritances, and government benefits.

Paying for Science and Technology

In addition to dealing with the consequences of emerging technologies, the federal government is directly involved in the generation of technologies through its support of research and development (R&D). The scope of government involvement in R&D has grown massively in the postwar era.[10] In 1940, the federal government spent only $67 million on the direct support of science and technology, the equivalent of about $1.03 billion today.[11] In 2014, the federal government budgeted 127.2 billion for R&D.[12] Over the years, the space program has benefited from an impressive amount of federal funds. Since its creation in 1958, NASA has been a major recipient of federal R&D funds. By the time of the last lunar landing, in late 1972, the Apollo program had absorbed $19.4 billion, about $115 billion in today's dollars.[13] Today, it accounts for $33 billion for fiscal year 2023.[14] The cost of the construction of the International Space Station has made major claims on the science and technology budgets of several nations; from the beginning of construction in 1998 until 2019, the tab for the International Space Station came to $150 billion, of which the United States contributed $100 billion, making it likely the most expensive human project of all time.[15]

Government expenditures on military R&D have waxed and waned as different governments have come to power and as perceptions of military threats have oscillated. In the 1984 fiscal year, a massive peacetime military buildup resulted in military R&D expenditures of $31.8 billion, almost 70 percent of the total federal R&D budget.[16] The waning of the Cold War resulted in a significant reduction in the proportion of R&D funds allocated to defense. For fiscal year 2022, the expected Defense R&D budget is 36.7 percent of the total, at $62.4 billion.[17] Since 9/11, the federal government has greatly increased antiterrorism research and development. Support for technologies aimed at foiling terrorist attacks has increased significantly, as has R&D for code breaking. The extent of funding for the latter is classified, but it is known that the National Security Agency, the highly secretive organization that oversees this research, possessed the world's fastest computer in 2009 and is sponsoring the construction of an even faster one today. The raw data for these supercomputers will come from everything from personal emails and telephone records to bookstore purchases and travel itineraries, all of it intercepted and aggregated in a structure being built in Utah that will be five times the size of the U.S. Capitol.[18]

Although the military sector absorbs a large portion of federal R&D resources, the funds allocated for other purposes are still considerable. By supplying about half of the funds used for the nation's R&D efforts, the federal government plays a very large role in the sponsorship of scientific and technological advance. Consequently, the allocation of federal R&D funds can significantly influence the direction of technological advance. To take a notable example, from fiscal year 2003 to 2012, the federal government spent a total of nearly $40 billion on energy R&D. Of that sum, 25.4 percent went for fossil fuel R&D, and another 25.9 percent went to nuclear R&D, while R&D funds for energy efficiency and renewable energy accounted for 16.4 and 17.1 percent, respectively.[19] But as times have changed, the proportion going to renewable and clean energy has transformed. The proposed FY 2023 Federal appropriations for renewable energy quadrupled from FY 2022 to $3.2 billion and about half of the R&D budget, surpassing expenditures on nuclear energy entirely.[20]

But Is It Really Necessary?

Why has government involvement in technological development increased to such a marked degree in the postwar era? In part, it is due to the recognition that our prosperity is tied to our ability to advance technologically. Given the fierce competition within the international economy, the failure to maintain high technological levels can have severe consequences. Japan recognized this fact, and its Ministry of International Trade and Industry (MITI, now METI, the Ministry of Economy, Trade, and Industry) aggressively sponsored the development of selected high-technology industries such as semiconductors, genetic engineering, robots, and composite materials. Not only had it provided research funds, MITI also organized cooperative joint research efforts between firms in the same industry so that they could do collectively what they could not do individually.[21]

The government of the United States has been less explicit in its sponsorship of technological advance, but it still has played a major role. The development of computers and semiconductors during the early 1950s was extensively supported by federal funds for research and development. Equally important, government procurement of the resulting products created a market at a time when sales within the private sector did not seem promising.[22] In similar fashion, government procurement of aircraft engines and other components provided a strong stimulus for technological advances that spilled over into commercial aviation.[23] Military needs were of course the main focus of these efforts. Accordingly, it can be argued that during the postwar era the United States has had a de facto policy of supporting technological development, but unlike Japan and China more recently, much of it has been driven by military rather than commercial considerations.

Important though they are, military requirements are not the sole justification for government support of technological advance. Although the market system has served as midwife to an abundance of innovative technologies, it cannot be expected to stimulate every type of technological innovation. Certain kinds of technological advances have little appeal to individual entrepreneurs, for they cannot be easily sold to individual consumers. Moreover, once these technologies are in place, they presumably benefit everyone, whether they pay for them or not. This is the so-called free-rider problem. For example, some people may be ardent supporters of a cleaner environment, but if they were to individually pay for a power plant's smokestack scrubbers, everyone in the community would benefit equally, even though most of them made no financial contributions to the installation of the scrubbers. Under these circumstances, the government has to take the lead by mandating the development and installation of emission-control equipment or assessing a pollution tax that stimulates their installation. The costs would be borne by the operators of the power plant, who would likely pass them along to consumers in the form of higher utility bills. Alternatively, the government could subsidize the installation of scrubbers, and offset the subsidy by levying a tax.

Some activities produce benefits that are considerably greater than the returns that accrue to the individuals or firms that engage in them. In contrast to negative externalities, these can be labeled "positive externalities." In some cases, the potential social benefits may be great, but the private rate of return may be so small that no private party is willing to undertake them. This is frequently the case with basic research that has no direct payoff but provides a foundation for subsequent technological applications. Orphan drugs are similar: for patients with very rare diseases, it is difficult to motivate the private sector to provide treatments and the Orphan Drug Act of 1983 has stimulated action. Tax and patent protections, and basic research support from the government, help to incentivize solutions.[24]

Once the research has been done, the knowledge gained can be made available to everyone at little additional cost, thereby increasing its benefits to society as a whole. Under these circumstances, it

Figure 24.1 Pipetting and laboratory glassware: typical biotechnology tools. (Source: Shutterstock/Pressmaster)

makes sense for the government to sponsor basic research, paying for it through taxation. This is precisely what has happened: about half of the "pure" or "basic" scientific research conducted in the United States is financed by the federal government, while the remainder is funded by colleges, universities, and private industry.[25]

Government Institutions for the Guidance of Technology

A modern nation's economic and military security is closely tied to its ability to generate and absorb new technologies. At the same time, simply coping with all of the consequences of technological change can require the leadership and organization that only government can provide. For these reasons, many governments have established special departments and ministries charged with the promotion, control, and governance of technology. But this has not been the case in the United States. There is no cabinet-level Department of Technology, only a collection of congressional committees and government agencies that involve themselves in a plethora of separate issues and decisions. Some of these agencies reside within the executive branch as parts of cabinet-level departments (such as the Department of Energy's Nuclear Regulatory Commission), while others (such as NASA and the National Science Foundation) are quasi-independent agencies directly under the president.

The executive branch is best situated to bring some degree of coherence to technology policy, and in fact the president and his staff have on occasion played significant roles in determining the course of a particular technology by influencing legislation, designating agency heads, impounding funds, and using the other tools available to the president. The importance of having the president well advised on scientific and technological matters was recognized in 1957 with the formation of the Presidential Science Advisory Committee (PSAC) within the Eisenhower administration. PSAC was largely concerned with matters relating to defense and space, and when its members failed to support President Nixon's policies on missile defense and the construction of a supersonic airliner, PSAC was abolished in 1973 and some of its functions were transferred to the National Science Foundation.

A formal structure for providing scientific and technical advice for the president was reconstituted in 1976 with the formation of the Office of Science and Technology Policy (OSTP).[26] The director of OSTP also serves as a member of the National Science and Technology Council, which was established by a 1993 executive order of President Clinton. In addition to the OSTP director, the council includes the president, vice president, and several cabinet-level officials and high-level administrators. Among its responsibilities are providing the president and senior staff with technical advice, ensuring that policies of the executive branch are based on sound science, and coordinating the science and technology policy-making process within the executive branch of the OSTP.[27] The White House is also the domain of the chief technology officer, a position created by President Obama in 2009 and whose efforts have been directed toward making the government more effective and accessible through the use of current technology.[28]

Some presidents have had more interest in science and technology policies than others, but direct presidential involvement in the shaping of these policies has been infrequent in all administrations. On a few occasions, however, it has been highly significant. President

Reagan's Strategic Defense Initiative (the "Star Wars" ballistic missile defense system) represented a major redirection of American defense policy, yet it began as an idiosyncratic effort. The president publicly announced the program only five days after informing his science adviser, while the joint chiefs of staff, the secretary of state, and the secretary of defense got only two days' notice. The chief scientist for the Pentagon learned of the proposal only nine hours before the speech that presented the program to the nation.[29]

With presidential involvement in the setting of technology policy largely confined to a few major policy decisions, the day-to-day shaping of technological policy tends to be the business of Congress and various government agencies. Here, the disjointed nature of American technology policy is apparent. There is no unified approach to the budgeting of funds for research and development: over a dozen appropriation subcommittees in each house of Congress take a hand in determining the budgets of the agencies they oversee. These committees and subcommittees often have to address issues directly connected to technological matters. Congressional involvement with technological issues was recognized in 1972 by the creation of the Office of Technology Assessment (OTA), which had as its purpose the provision of information to Congress to help guide its decisions about technological policy, and to give Congress more leverage over the president in science and technology matters.[30] OTA investigated a wide selection of topics that ranged from an examination of the physiological basis of drug addiction to an evaluation of the Social Security Administration's plans to upgrade its computer facilities. In its early years, OTA was criticized for not taking on highly controversial topics, but it eventually became involved in contentious matters like the effectiveness of a proposed antimissile defense system. In the latter case, the publication of its report was delayed by a lengthy classification review, and in the end several key chapters were deleted due to the objections of the Department of Defense.[31] It is likely that the Department of Defense's opposition had more to do with the negative tone of the report than with concerns about its revelation of military secrets.

As the past example indicates, OTA sometimes found it difficult to achieve a balance between the goals of its overseers in the federal government and the objectivity and neutrality characteristic of scholarly inquiry. Given the contentiousness of many technological issues and the frequent difficulty of separating facts from values, OTA was in an inherently precarious situation. Being on the "wrong side" in some controversial issues undercut OTA's support in Congress, especially after Republicans gained control of both houses in 1994. Despite strong bipartisan agreement on the value of OTA outside Congress, the office was dismantled one year later. Its neutrality was part of its undoing. As far as the actual working of Congress is concerned, policies relating to technology emerge from the many committees and subcommittees of the two legislative branches. Legislation affecting the development and use of technology is forged in bodies as disparate in their concerns as the House Armed Services Subcommittee and the Senate Agriculture Committee. These committees and subcommittees often work in conjunction with federal agencies and departments, such as NASA or the Department of Defense, that administer a particular technological sector.

How Government Policies Shape Technological Change

The most direct way that Congress determines the course of technological development is through passing laws that either forbid the use of a particular technology or regulate the way it is used. Indeed, the very notion that the government has a legitimate right to regulate private industry originated with the need to address a problem presented by a new technology—the explosion of riverboat steam boilers in the first half of the nineteenth century.[32] The Franklin Institute in Philadelphia was given the task of investigating the construction and operation of boilers. Its report provided a wealth of information about boilers and their components, as well as a set of recommendations regarding their operation and maintenance, many of which were subsequently written into law by Congress. Today, government regulations cover a vast number of technological concerns, everything from the siting of nuclear power plants to the length of trucks.

Along with the passage of laws and regulations, the federal government affects the course of technological development through the budgetary process; that is, by deciding to pay for certain kinds of technologies and not others. In some important cases, government financial support can be crucial. The supersonic transport (SST) plane literally never got off the ground because in 1971, Congress voted not to appropriate any more funds for its development. Also, as we saw earlier, the George W. Bush administration severely restricted stem-cell research that used human embryos. Conversely, some technologies, such as civilian nuclear power, have received a large amount of financial and legal support from the federal government that has been crucial to their development and use.

Governmental decisions that affect technological development do not take place in a political vacuum. It often happens that technologies are sponsored because they have substantial outside support. Gaining government interest in addressing a technological issue is a crucial first step, and one that cannot be taken for granted. Here again, the piecemeal approach to governing technological change is evident. There is no clear and comprehensive approach to deciding what technological goals should (and should not) be pursued so that a governmental agenda can be set. Instead of being the product of calm deliberation, an issue is often put on the agenda because of some "focusing event"; an oil embargo forces a consideration of the energy technologies we use; the launch of Sputnik by the Soviet Union demonstrates the apparent need for an accelerated U.S. space program; a serious accident reveals material and organizational defects in the operation of nuclear power plants.

A sense of crisis and the need for decisive action also emerge when a social movement is directed at a particular technology. Some of these have taken on well-established technologies that had remained under the radar until they became the subject of media exposés, notably hydraulic fracturing, or "fracking," for the extraction of oil and gas. But large-scale protests are more likely to emerge when a new technological possibility lies just over the horizon. Technologies that have already been established, whatever their unfortunate consequences, simply do not generate the kind of concerns that new ones do.[33] Cars, along with giving us our prized freedom of mobility, also befoul our environment, encourage ugly sprawl, and in 2019, killed an estimated 38,800 people in the United States, the equivalent of a major commercial airline crash every two days. Yet few people consider bringing them under tighter governmental regulation or even abolishing them altogether.

Only a few technologies have generated much popular opposition. Such diverse technologies as the chlorination of water, the extensive use of medical X-rays, microwave ovens, and birth-control pills, are unevenly subject to governmental control.[34] In fact, it is not entirely clear why some technologies and not others become the focus of widespread concern. But it is certainly true that a technology that produces large-scale failures, such as a nuclear reactor accident, is more likely to evince opposition than one that produces a series of small-scale failures, even though in aggregate their consequences may be great, as has been the case with automobile accidents or pollution. And it is also the case that media exposure can greatly heighten opposition to a technology with evident problems.[35] Again, this sort of exposure is more likely when the failure is large and spectacular rather than diffuse and significant only in the aggregate.

Examples of widespread public debate over new technologies are infrequent; most innovations emerge with little public attention. But this does not mean that politics has been irrelevant. As noted earlier, many new technologies owe their development and diffusion to sponsorship by particular governmental agencies. Since there is no centralized approach to the governance of technology, many key decisions are made at the middle levels of government, that is, by government agencies and congressional committees. It is here that decisions are made to sponsor a particular technology and to create the conditions necessary for its success. Technology policy is the sum of many separate policy actions undertaken at this level. Under these circumstances, the ambitions of individual governmental agencies and their administrators can determine the course of technological development, for an agency has a strong interest in pushing technologies that are in accordance with its own goals and having them adopted as national policy.[36] This is often achieved with little fanfare. Most decisions about technological policy do not generate much political heat, and a government agency may successfully promote a particular technology with little or no public discussion or debate.

The importance of government sponsorship can be seen in the history of nuclear power in the United States. In the mid-1950s, the electrical utilities, the greatest potential users of nuclear power, had little reason to make use of the new technology. According to one utility company president, "We weren't anxious to get into nuclear power, and I don't think any other company in its right mind wanted to get into it either."[37] To overcome this reticence, the key government agency, the Atomic Energy Commission, engaged in a large-scale selling job. A demonstration plant was built at Shippingport,

Pennsylvania, in 1954, and technical assistance was offered to the industry, along with subsidized prices for nuclear fuel.[38] This is not to say that the Atomic Energy Commission's actions were illegitimate or that civilian nuclear power owes its existence solely to government support, but it does show how national policy can be shaped by the interests of a particular government agency.

Since some government agencies are more powerful than others, their programs forge ahead while other agencies' projects languish. As a result, potentially significant programs are not enacted, or if enacted they are chronically underfunded, while more dubious ones move ahead. A few agencies are more influential than others because of the greater technical abilities of their staff and their greater interest in science and technology. But in many cases, what really matters is who their clients are. Government agencies do not operate in splendid isolation; they are often closely tied to a particular set of clients. As we have seen with the Federal Communications Commission in an earlier chapter, even when the task of the agency is regulation, it may be "captured" by the industry that it is supposed to regulate. This happens because the agency needs the support of the industry it oversees in order to maintain its legitimacy and resources. An agency may even delegate to the industry some of the regulatory tasks that it is charged with performing. For example, much of the inspection work performed before the Federal Aviation Administration certifies the airworthiness of a new airplane is actually done by engineers and technicians in the employ of the plane's manufacturer. The deficiencies of this delegation became apparent only after two Boeing 737 MAX were involved in crashes that killed everyone onboard.[39]

If an industry has a strong interest in a particular technology, it may exert a great deal of influence, and Congress and the relevant government agencies will lend their support to that technology. In turn, members of Congress may find a particular technology to be politically attractive because it offers "pork barrel" benefits to their constituents. Some technologies, especially those in the military sector, have forged ahead because a project has contractors and subcontractors in a large number of congressional districts. Members of Congress quite naturally are likely to support programs that provide jobs and other economic benefits for their states and districts and are willing to support projects that benefit the constituents of other representatives in return for support of projects that benefit their own.

In sum, as with many other government policies, the direction of technological policy often results from the interaction of an "iron triangle" of congressional committees, government agencies, and the key groups whose interests are closely affected by governmental decisions. The decision to sponsor a new technology, along with the shape that it takes as it is developed, is often determined by these triangles. This means that the government usually does not sponsor the development of new technologies as part of a general effort to find solutions to public problems. Rather, a technology is sponsored because a coalition can be formed from the members of the triangle, who see to it that the selected technologies develop in accordance with their own needs and interests.

It also happens that once under way, the development of a new technology builds up momentum. The personnel and agencies charged with its development have a strong interest in its continuation, as do potential users. As a result, a technological policy gets locked in place and options are closed off. In the absence of outside pressures, projects are destined to move ahead. Even when a

Figure 24.2 Government processes shape technology directly and indirectly, through priority-setting, legislation, taxation, R&D support, and regulatory activities. (Source: Shutterstock/VideoFlow)

technology has manifest problems of considerable magnitude, the sponsors of the technology will tend to "throw good money after bad," a sunk cost problem, for once the commitment has been made and the program is underway, it often is far more difficult to terminate a program than to continue with it.

This is one of the greatest defects of our system of governing technology. Technological change is an uncertain process, with many unforeseeable consequences. Often efforts to forecast the precise consequences of a new technology so that policies can be constructed in advance to deal with them turn out to be a failure. The best that can be hoped for is that technologies can be stopped or modified if this should prove necessary. This is easier said than done. The effective governance of technology is faced with a fundamental paradox: change is easiest at the beginning, but this is when the need for it cannot be foreseen. As Ernest Fitzgerald, the whistleblower we briefly met in an earlier chapter, said of military projects, "There are only two phases to a major program. The first phase: 'It's too early to tell.' The second phase: 'It's too late to stop.'"[40]

In this way, technology appears to be out of control, as many technologies seem to be driven by their own momentum. But this is not really what is happening; seemingly out-of-control technologies move ahead because they suit the real interests of individuals and organizations. Were vested interests to be taken away, there would be far less reason to continue with the development or deployment of the technology. Technologies do not simply stand or fall on their own merits; their success or failure can often be attributed to the political support that they receive. And this political support will likely reflect the distribution of power both inside and outside the corridors of government. Under these circumstances, is there any hope that an unorganized citizenry can have any influence over the way technology develops?[41] It is to this, our last topic, that we will next turn.

The Democratic Control of Technology

The decisions and activities of experts, enterprise managers, and government officials have powerfully affected the kinds of technologies that have been developed, where they have been applied, and how their costs have been met. Most of us have been little more than the consumers of technologies that have been developed by people and organizations largely unknown to us. This process has produced spectacular successes and spectacular problems. Technology has extended our lives, lightened some of our burdens, increased our wealth, and kept us amused. At the same time, the advance of technology has left us with some very unpleasant byproducts. This is not the place to attempt to determine if the good has outweighed the bad, for so much depends on what we value. Which way of living is preferable, that of the !Kung San with its relaxed work patterns but few material possessions and limited control of the environment, or our own with its abundance of goods but frantic work schedules? In any event, for most of us the choice is moot; unless we decide to become nomadic hunter-gatherers or subsistence farmers, our lives will be lived in a world shaped by modern technology. At issue is not the acceptance or rejection of technology; what matters is how much individual influence we have over the choice of technologies that we use now and will use in the future.

As individuals, are we doomed to the passive acceptance of technological changes that have been decreed by others? Even if technological development has produced more good than bad it has proceeded in a fashion that has put it beyond our individual control. It seems as though we have struck a bargain whereby we consume the fruits of technological advance in return for delegating to others the power to determine the technologies that shape the basic contours of our lives—everything from what we eat, to how we work, to the way we are entertained.

Most people seem to believe that this is a fair bargain. As we have seen in the first chapter of this book, there is not a great deal of anguish about our reliance on modern technology. Then, too, it must be remembered that democratic participation has its costs. Above all, the expansion of the number of people engaging in technological decision making can disrupt the orderly processes favored by managers, engineers, and bureaucrats. This follows from the simple arithmetic of participation; the more parties there are to a decision, the harder it is to decide. The result is often a stalemate—"my vote cancels yours." Everybody ends up with veto power, but no one is able to act in a positive fashion. At best, the result is compromise, which may be a reasonable conclusion in some cases, such as the modification of nutritional standards; but in matters that require decisive action, such as stopping the spread of weapons of mass destruction or significantly mitigating climate change, a decision to compromise may be worthless or even dangerous.

Does this mean that there is no middle ground between an apathetic acceptance of technological change decreed by others and a paralyzing stalemate? Can democratic participation be reconciled with technological advance? Before these questions can be answered, it is obviously necessary to have some workable definition of democracy. A vast amount of thinking and writing has gone into this issue, and we cannot presume to break any new ground here. But most of us could agree that a democratic process is one in which the people as a whole are able to participate in making meaningful choices about the things that affect their lives. The capitalist economic system does contain some democratic elements, for consumers can choose to buy or not buy particular products, and in so doing they affect the course of technological change. There has been a "democratic" affirmation of gas-electric hybrid cars, but not of battery-powered ones. And certainly, consumer choice has dictated the shift to smaller cars whenever gasoline prices shoot up. But these are choices made after the fact. Consumers can choose only between existing alternatives; they cannot determine which alternatives will be made available to them. This is a bit like voting in a referendum; the electorate can decide an issue, but it might not determine which issues are put on the ballot in the first place. Here, it is clear that in the earliest phases of technological development, inclusivity on various dimensions improves technological outcomes and decision process.[42]

In similar fashion, democratic choice can be exercised by voting for or against candidates because of their stance on a key technological issue. There have been local elections that have hinged on a candidate's position on the fluoridation of water or on nuclear power, but these have been rare. Elections are usually fought over a variety of issues; technological policy has been but one and usually a minor one at that. And even where a technological issue is at the center of a campaign, the voters' voices are heard only after a technology is already in place, as with referenda on nuclear power. There is virtually no popular input into whether or not completely new technologies should be developed. In this sense, democratic, and market, controls are at best reactive.

Is it possible to have democratic control of technology that goes beyond these after-the-fact votes? In some ways, the advance of technology has increased the potential for democratic participation in government. Most importantly, the electronic media have heightened our awareness of political issues. It has often been remarked, for example, that the widespread protests against the Vietnam War were stimulated by the horrible realities of warfare appearing on television day after day. At the same time, however, exposure does not guarantee an enlightened response. As we have seen, the expansion of newspaper reading that was triggered by the invention of the steam-powered rotary press stimulated a great outpouring of jingoist sentiment and the fanning of imperialist ambitions in the United States. More recently, the rise of the internet, cable TV, and social media have exacerbated political divisions by allowing us to live in our own "echo chambers" where we only hear what we want to hear. Making matters worse is the mode of discourse often found in these media, uninformed, hyperbolic, and at times downright vile. Finally, many critics have decried the present-day debasement of the political process by slickly produced advertisements for candidates and ballot initiatives. It thus seems that advances in communication technologies have had mixed consequences for the democratic process. The electronic media in general and the internet in particular hold out the prospect of a more informed electorate, but the positive contributions of these media are often overwhelmed by the negative ones, just as what happened with the rise of the popular press generations earlier.

Advances in electronic media have also presented the possibility of direct democracy through a kind of electronic town hall. Instead of delegating decisions to elected officials, every citizen could use a website to vote on the pressing issues of the day. Such a system is certainly technologically feasible, but is it a practical possibility? Security questions aside, could voters be expected to be sufficiently well-informed to pass judgment on everything from levels of farm price supports to the wisdom of supplying arms to a revolutionary movement in some far-off land? And even if such a system were implemented, would it really result in a truly democratic order? It is obvious that people could not vote on every single issue, so some kind of rationing would have to be instituted. Some agency would have to determine what got on the "ballot" in the first place, and this would be a source of power in itself. The ability to set governmental agendas—the determination of which issues make it into the political arena—is a crucial political decision, and this system would do nothing to democratize this process. All voters could do would be to make choices according to a preplanned agenda.

Part of the technology and governance question is tied to near-universal questions about priority-setting. In the EU, countries have a longer trajectory of participatory research,[43] which entails end-users and concerned citizens having some input in to the research priorities of agencies, and sometimes specific problem-driven research. As we noted in an earlier chapter, Stoke's model of "Pasteur's Quadrant" helps us see the diversity of research enterprises: descriptive, basic, applied, and hybrid, which might make the way clear to a balanced portfolio of knowledge-seeking activities.[44] The precautionary principle is central to policy-making in Europe, although it is in tension with cost-benefit analyses that prioritize shorter-term evaluations of risks and rewards.[45] Labor laws are also much stronger in much of Europe than in the U.S., leading to effective policies around work boundaries, privacy, job displacement, and safety.[46] These factors lead to a maturity of participatory design, which involves end-users in problem formulation and technological assessment far earlier in the development process than simply being take-it-or-leave it proposition for consumers. To reiterate some earlier language: participatory design involves stakeholders as high-inclusion actors, rather than low-inclusion consumers.[47]

Given the limitations of direct democracy, we necessarily have to rely on some form of representative democracy. Unfortunately, the established institutions of representative democracy in the United States have not always guided technological development in ways that serve the public interest. The key political actors have been especially responsive to large, well-organized bodies, such as industries, labor unions, and other special interests. All too often the common good is lost to the special pleadings of these groups. Compounding the problem is the fact that the political process is fragmented, which allows government agencies and officials to cut deals with special interests while no one looks after the needs and concerns of the nation as a whole. Under these circumstances, government officials end up representing themselves and their powerful constituents rather than the people who voted them into office.

As we have noted previously, artifacts have politics, or, at least, they are often selected with particular political ends, and some forms of governance do seem to converge on technological systems. It seems to be the case that nuclear technologies require strong, centralized states to govern risks and prevent proliferation. And distributed energy systems, such as pervasive local solar, seem to both reflect and strengthen local governance processes.[48] Rather than considering these as matters of technological determinism, however, we need to think of these as rather open hypotheses, and reflecting sunk costs, path dependencies, and the durability of infrastructures which reflect priority setting and political processes. We are often entranced by the priority-setting of billionaires and entrepreneurs, generally without participatory discussion of the costs, benefits, and who will gain and lose, from new technological systems.

Does this mean that there is no hope that individual citizens can affect the course of technological change? Despite what has just been said, there is some room for optimism, and there are numerous occasions when democracy can assert itself in technological affairs. Democracy may work best when issues are close to home. To note two recent examples, concerns about their effects on local environments have motivated well-organized protests over fracking and the construction of an oil pipeline passing through the Midwest. In contrast, many pressing but seemingly far-off technological issues do not engage the interest of people distracted by their individual challenges of raising a family, meeting mortgage payments, or studying for exams. Burdened with mundane daily problems, it is all too easy for us to deny that there is an urgent need to do something about nuclear proliferation or climate change. The perception that immediate, local concerns are being affected by technological change can therefore energize greater levels of public perception than is the case when the issues seem too large and remote to concern us. There is considerable wisdom in the slogan that enjoins us to "think globally and act locally."

It must be admitted that local politics is uninteresting to many people, as the usually low levels of voter turnout for local elections attest. Moreover, participation at the grassroots level faces the same limits as other kinds of political participation; involvement can be time-consuming and stressful, and people would prefer to leave the tasks of governance to local officials. But at the same time, involvement in technological policies that have a direct impact on one's own community can reduce apathy and lethargy. And it might also be the case that expanding democratic participation and tying technological issues to local concerns would result in the emergence of technologies significantly different from those produced by our present system.[49] This is the argument of scholars ranging from British economist E. F. Schumacher, who gave us the 1973 book *Small is Beautiful*, about appropriate technologies and policies "as if people mattered," to recent work by Ruha Benjamin and her discussion of local, personal actions and engagements that can have rippling effects in enhancing local and global justice. As she learned both literally and metaphorically from the COVID-19 pandemic, very small things can have complex and extensive impacts on large and seemingly entrenched systems.[50]

Finally, while our governmental structures and processes have produced some technologies of very questionable value, we should also realize that no system of guiding technology will guarantee a utopia. No matter how advanced or benign our technologies, human life will always have its insecurities, disappointments, frustrations, and tragedies. It can even be argued that it is dangerous to seek perfection in both our technological and political systems. There is something inhuman about a society that cannot tolerate the fact that life does not always go according to plan. The invention and development of new technologies have been characterized as ongoing searches for the "best way," and elegantly satisfactory solutions to technological problems do emerge on a regular basis. But most of life is not like that. We will always have to reconcile ourselves to the fact that even the most advanced and responsive technologies will not solve many of our basic problems. As we have seen, the spectacular advances of many individual technologies should not blind us to the fact that some of the inherent difficulties of life are simply not amenable to technological solutions.

The Challenges of the Future

In considering the years ahead, it is well to recall the waggish statement that it is dangerous to make predictions, especially about the future. The course of technological development has produced many surprises, and it will continue to do so. Fifty years ago, few would have foreseen the emergence of smartphones with the power of the then-existing mainframes. Yet at about the same time, it was predicted that in the future electricity would be so inexpensive to produce that it would be "too cheap to meter."[51] Forecasting is a highly inexact art, and although it can be safely said that our lives will be strongly influenced by technological change, predictions beyond this general statement are fraught with hazards.[52]

Even so, the general contours of future technological developments can be discerned. Advances in transportation and electronic communication will bring the people of the world even closer together. The internet will be the basis for ever-increasing amounts of information, media presentations, and social interaction. Advances in nanotechnology, materials, 3D printing, and microprocessors will produce fundamental changes in the design and operation of many objects. The application of artificial intelligence could affect every aspect of work, play, and learning. And genetic engineering might allow nothing less than the redesign of human beings. Profound social changes will be necessary if we choose to make use of these technologies.

Technology is increasingly implicated in governance itself. For example, the spread of military technologies to civilian policing challenges basic civil liberties and potentially reduces police reputation and effectiveness.[53] Intensive surveillance is unequally applied, harming those in public housing through intensive regulation, yet protecting those in gated communities and private enclaves.[54] Similarly, algorithms which reproduce biases can be integrated into policing, surveillance, and policy. This algorithmic governance draws on the increased coordination of data sets and computer-based procedures for decision support.[55] Because decisions—whether in the workplace, traffic control, infrastructure development, or in terms of community resource allocation or police priorities—are made with intensive technological support, decisions have the illusion of objectivity and finality. Most of us are "low inclusion actors" in the construction, maintenance, and use of these data-intensive systems, even if we live with their consequences. The lack of transparency is a challenge for democratic accountability.

Of course, there is also the potential for the significant dislocations of what has been both praised and feared for nearly 30 years: "the jobless future."[56] It is marked by the end, at least in many sectors of the global workforce, of career-building long-term jobs. As we have seen elsewhere in the book, the "gig economy" which relies on subcontractors, consultants, and at-will employees is an example of technological changes which are convergent with new business and policy models. Enhanced technology facilitates this transient approach to jobs and work. But in other arenas, depth, expertise, and long-term commitments to work may still be essential. The impacts of AI and related robotics and automated systems will be felt differentially. Work that is easily replaced by automation is often repetitive and with a basic set of skills, requiring retraining for older workers and new forms of skill acquisition for younger ones. But what will people do to make a living? The U.S. "social safety net" of unemployment insurance is time-limited (as few as 12 weeks and at most 26).[57] Retirement and disability benefits are not structured to accommodate the problem of massive technological unemployment. There are some suggestions that universal basic income (UBI) may become increasingly viable, either through a "robot tax" or other taxation mechanisms.[58]

And yet, just because these things are possible does not mean that they should be done. We do not have to resign ourselves to the inexorable advance of technology while passively hoping that it produces more good than harm. Technology remains a human creation, and we have it in our power to determine its future course. Unfortunately, the power to make this determination is very unevenly distributed through society. A major reason for this unequal distribution of power is the unequal distribution of knowledge, and vice versa. This is especially unfortunate because a knowledgeable citizenry has always been essential to the functioning of democracy. In the words of Thomas Jefferson, "If we think [the people] not enlightened enough to exercise their control with a wholesome discretion, the remedy is not to take it from them, but to inform their discretion."[59] The expansion of citizens' knowledge is even more crucial than it was in the days of Jefferson when most adults worked as farmers, and humans had never traveled faster than the speed of a horse. The relentless advance of technology has produced a dizzying pace of change that at times leaves us gasping. Even so, it bears repeating that technology is our own creation and that its ultimate source is knowledge. But for technology to be truly beneficial, more than technical knowledge is required. Our challenge will be to develop and apply many different kinds of knowledge—ethical, philosophical, sociological, political, and economic—so that we can do a better job of defining our real needs and creating the technologies that serve them.

Figure 24.3 The cover art for *DIY Utopia: Cultural Imagination and the Remaking of the Possible*, edited by Amber Day, Rowman & Littlefield, 2016.

We hope that this book has given you some of the knowledge necessary to participate in shaping a future that will be strongly influenced by the technological decisions being made in the present. But acquiring knowledge is not enough; it is also necessary to apply that knowledge. And that will be up to you.

QUESTIONS FOR DISCUSSION

1. Do you think that the government should establish an agency similar to Japan's Ministry of External Trade and Industry as a way of stimulating technological advance? What obstacles would such an agency confront in the American political climate?
2. Do artifacts have politics? Are there essential political arrangements that are required for certain kinds of technology? Are

redistributions of political power a good reason to adopt or not adopt technologies?

3. Very few elected officials in America have professional training in science or technology. Does this inhibit the effective governance of science and technology? Do you think that the American government would be substantially different if most senators and representatives previously worked as scientists or engineers instead of as lawyers, as is the case today?
4. Can you think of any potentially important technologies that have languished because they have lacked political support? How could they gain this support?
5. What is your definition of democracy? On the whole, has technological advance increased or decreased the amount of democratic participation in modern society? In the text, it is argued that citizen involvement in local issues will aid in the democratic shaping of technology. But local concerns have at times blocked technologies such as wind turbines that may contribute to the general good. Is it possible to address local concerns when pursuing projects that may benefit the nation as a whole?
6. Which emerging technologies will have the greatest impact on life during the next decades of this century? Will they have any unfortunate consequences? Should the government restrict or prevent any of them? Do you intend to exert any influence over these decisions? How will you do it?
7. *The Ministry for the Future* by Kim Stanley Robinson is a climate fiction work that engages both technical and sociopolitical innovations, and economic systems that need to change to intervene in global climate change. Are there other works of science fiction or speculative fiction that engage your imagination for a better world? What role does imagination play in sociotechnical transformation?

Notes

1. University of California Office of Technology Transfer, "The Bayh-Dole Act: A Guide to the Law and Implementing Regulations" (April 24, 2009). Accessed from www.ucop.edu/ott/faculty/bayh.html.
2. Richard Pérez-Peñanov, "Patenting Their Discoveries Does Not Pay Off for Most Universities, a Study Says," *The New York Times* (November 21, 2013).

Accessed from www.nytimes.com/2013/11/21/education/patenting-their-discoveries-does-not-pay-off-for-most-universities-a-study-says.html?_r=1.
3. Michael M. Crow and Christopher Tucker, The American Research University System as America's de Facto Technology Policy," *Science and Public Policy* 28, 1 (2001): 2–10.
4. William Gittins, "Which Energy Source Is Most Subsidized in the US?" *As* (March 28, 2022). Accessed from https://en.as.com/en/2022/03/29/latest_news/1648510535_730791.html.
5. U.S. Energy Information Agency, "Renewable Energy Explained: Incentives," (November 22, 2019). Accessed from www.eia.gov/energyexplained/renewable-sources/incentives.php.
6. Isaac Orr, "As $7,500 Federal Subsidies Expire, Electric Vehicles Are About to Get More Expensive," *Center of the American Experiment* (September 30, 2020). Accessed from www.americanexperiment.org/2019/09/as-7500-federal-subsidies-expire-electric-vehicles-are-about-to-get-more-expensive.
7. California Air Resources Board, "California Moves to Accelerate to 100% New Zero-Emission Vehicle Sales by 2035," (August 25, 2022). Accessed from https://ww2.arb.ca.gov/news/california-moves-accelerate-100-new-zero-emission-vehicle-sales-2035; and Simon Jessop, "COP27: France, Spain Latest to Pledge Halt to Gasoline-Driven Vehicle Sales," *Reuters* (November 16, 2022). Accessed from www.reuters.com/business/cop/cop27-france-spain-latest-pledge-halt-gasoline-driven-vehicle-sales-2022-11-16.
8. 11-159 Astrue v. Capato (05/21/2012)—Supreme Court; and Bill Mears, "Justices Deny Benefits for Child Conceived after Death of a Parent," *CNN* (May 21, 2012). Accessed from www.cnn.com/2012/05/21/justice/scotus-posthumous-conception-ruling.
9. See Alea Roberts, "Where's My Share?: Inheritance Rights of Posthumous Children," ABA (June 13, 2019). Accessed from www.americanbar.org/groups/litigation/committees/jiop/articles/2019/inheritance-rights-posthumous-children.
10. Matt Hourihan, "A Primer on Recent Trends in Federal R&D Budgets," *American Association for the Advancement of Science* (January 15, 2015). Accessed from www.aaas.org/news/primer-recent-trends-federal-rd-budgets.
11. National Science Federation, "Science and Technology in Times of Transition: The 1940s and 1990s," Table 1-3.
12. Congressional Research Service. U.S. Research and Development Funding and Performance: Fact Sheet (January 24, 2020). Accessed from https://fas.org/sgp/crs/misc/R44307.pdf, which included $30.8 billion for the National Institutes of Health and $11.5 billion for NASA.
13. See more history of the Apollo at this archived page: https://web.archive.org/web/20231020190614/www.history.nasa.gov/apollo.html.
14. See: USAspending.gov, National Aeronautics and Space Administration at www.usaspending.gov/agency/national-aeronautics-and-space-administration?fy=2023.
15. Loren Grush, "NASA Is Opening the Space Station to Commercial Business and More Private Astronauts," *The Verge* (June 7, 2029). Accessed

from www.theverge.com/2019/6/7/18656280/nasa-space-station-private-astronauts-commercial-business.

16. Mary Acland-Hood, "Statistics on Military Research and Development Expenditure," in Stockholm International Peace Research Institute, *World Armaments and Disarmament* (London: Taylor and Francis, 1984), pp. 169–70.
17. See the following Congressional Budget Office Document: Congressional Research Service: Federal R&D FY 2022. https://docs.google.com/viewer?url=https%3A%2F%2Fsgp.fas.org%2Fcrs%2Fmisc%2FR46869.pdf.
18. James Bamford, "Black Box," *Wired* 20, 4 (April 2012).
19. Fred Sisine, "Renewable Energy R&D Funding: A Comparison with Funding for Nuclear Energy, Fossil Energy, and Energy Efficiency R&D," *Congressional Research Service* (March 7, 2012). Accessed from www.fas.org/sgp/crs/misc/RS22858.pdf. The remaining 15.1 percent went for electric systems R&D.
20. See: www.nrel.gov, as well as President Biden's FY 2023 budget for reducing energy costs," at www.whitehouse.gov/omb/briefing-room/2022/03/28/president-bidens-fy-2023-budget-reduces-energy-costs-combats-the-climate-crisis-and-advances-environmental-justice; and "DOE Applied Energy Budget: FY22 Outcomes and FY23 Request" at www.aip.org/fyi/2022/doe-applied-energy-budget-fy22-outcomes-and-fy23-request. It remains to be seen exactly how the budget allocations develop.
21. Daniel Okimoto, *Between MITI and the Market: Japanese Industrial Policy for High Technology* (Stanford, CA: Stanford University Press, 1989), pp. 55–111. For a much less laudatory analysis of MITI's role in fostering technological and economic advance, see Richard Katz, *Japan, the System That Soured: The Rise and Fall of the Japanese Economic Miracle* (Armonk, NY: M. E. Sharpe, 1998).
22. Richard C. Levin, "The Semiconductor Industry," and Barbara Goody Katz and Almarin Phillips, "The Computer Industry," both in Richard R. Nelson (Ed.), *Government and Technical Progress* (New York: Pergamon Press, 1982).
23. David C. Mowery and Nathan Rosenberg, "The Commercial Aircraft Industry," in ibid.
24. M. Braun, S. Farag-El-Massah, K. Xu, and Timothy R Coté, "Emergence of Orphan Drugs in the United States: A Quantitative Assessment of the First 25 Years," *Nature Reviews Drug Discovery* 9, (2010): 519–522; The cost of the drugs remains a problem for individuals and health care systems, see: Lucio Luzzatto, et al., "Outrageous Prices of Orphan Drugs: A Call for Collaboration," *The Lancet* 392 (2018): 791–794,
25. Jeffrey Mervis, "Data Check: U.S. Government Share of Basic Research Funding Falls below 50%," *Science* (March 9, 2017). Accessed from www.sciencemag.org/news/2017/03/data-check-us-government-share-basic-research-funding-falls-below-50.
26. Bruce L. R. Smith, *The Advisers: Scientists in the Policy Process* (Washington, DC: Brookings Institution Press, 1992), pp. 155–88.
27. Office of Science and Technology Policy." Accessed from www.whitehouse.gov/ostp.

28. The position was vacant under President Trump for two years, but as of Fall 2022, the position remains unfilled in the Biden administration.
29. Richard Barke, *Science, Technology, and Public Policy* (Washington, DC: CQ Press, 1986), p. 43.
30. W. Henry Lambright, *Governing Science and Technology* (New York: Oxford University Press, 1976), p. 24.
31. Laura van Dam and Robert Howard, "How John Gibbons Runs through Political Minefields: Life at the OTA," *Technology Review* 91, 7 (October 1988): 50.
32. John G. Burke, "Bursting Boilers and Federal Power," *Technology and Culture* 7, 1 (Winter 1966): 1–23.
33. Allan Mazur, *The Dynamics of Technical Controversy* (Washington, DC: Communications Press, 1981), p. 97.
34. Ibid., p. 90.
35. Ibid., p. 111.
36. Lambright, *Governing Science and Technology*, p. 202.
37. James W. Kuhn, *Scientific and Managerial Manpower in the Nuclear Industry* (New York: Columbia University Press, 1966), p. 115.
38. John F. Hogerton, "The Arrival of Nuclear Power," *Scientific American* 218, 2 (February 1968): 21–31.
39. Tommaso Sgobba, "B-737 MAX and the crash of the regulatory system," *Journal of Space Safety Engineering*, 6, 4 (2019): 299–303.
40. "A Tribute to Pentagon Whistleblower Ernie Fitzgerald," at www.pogo.org/analysis/2019/02/remembering-ernie-fitzgerald and as quoted in Robert Bell, *Impure Science: Fraud, Compromise and Political Influence in Scientific Research* (New York: John Wiley and Sons, 1992), p. 80.
41. David Collingridge, *The Social Control of Technology* (New York: St. Martin's Press, 1980), p. 138
42. Erik Larson, "New Research: Diversity + Inclusion = Better Decision Making At Work," *Forbes* (September 21, 2017). Accessed from www.forbes.com/sites/eriklarson/2017/09/21/new-research-diversity-inclusion-better-decision-making-at-work.
43. Colin Fletcher, "Issues for Participatory Research in Europe." *Community Development Journal* 23, 1 (1988): 40–46.
44. Donald Stokes, *Pasteur's Quadrant: Basic Science and Technological Innovation* (New York: Brooking's Institute, 2011).
45. Johanna Thoma, "Time for Caution," *Philosophy & Public Affairs* 50 (2022): 50–89.
46. See "Technology at the Workplace – A European Overview of Employment Law Issues in a Modern Working Environment," at www.cms-lawnow.com/ealerts/2017/05/technology-at-the-workplace.
47. Finn Kensing and Jeanette Blomberg, "Participatory Design: Issues and Concerns," *Computer Supported Cooperative Work (CSCW)* **7** (1998): 167–185.
48. Langdon Winner, "Do Artifacts Have Politics?" *Daedalus* 109, 1 (1980): 121–36.

49. Richard E. Sclove, *Democracy and Technology* (New York and London: Guildford Press, 1995), pp. 119–38.
50. Ruha Benjamin, *Viral Justice: How We Grow the World We Want* (Princeton University Press, 2022); and E. F. Schumacher, *Small is Beautiful: A Study of Economics as if People Mattered* (NY: HarperCollins, 1973).
51. This prediction was made in 1954 by Lewis L. Strauss, the Chairman of the Atomic Energy Commission. It is usually assumed that he was speaking of electricity generated by nuclear fission power, but in reality, he was speaking about the future in general while perhaps thinking about the potential of fusion energy. See Thomas Wellock, "'Too Cheap to Meter:' A History of the Phrase," *U.S. Nuclear Regulatory Commission* (June 3, 2016). Accessed from https://public-blog.nrc-gateway.gov/2016/06/03/too-cheap-to-meter-a-history-of-the-phrase.
52. Herb Brody, "Great Expectations: Why Technology Predictions Go Awry," *Technology Review* (July 1991).
53. Jonathan Mummolo, "Militarization Fails to Enhance Police Safety or Reduce Crime but May Harm Police Reputation," *PNAS* 115, 37 (2018): 9181–9186.
54. Roy Coleman, "Eyes Everywhere: The Global Growth of Camera Surveillance." *Surveillance & Society* 10, 1 (2012): 99–102; and Torin Monahan, "Chapter 11: The Image of the Smart City: Surveillance Protocols and Social Inequality," in *Handbook of Cultural Security* (Cheltenham, UK: Edward Elgar Publishing, 2018); and T. Monahan, "Electronic Fortification in Phoenix: Surveillance Technologies and Social Regulation in Residential Communitie," *Urban Affairs Review* 42, 2 (2006): 169–192.
55. Christian Katzenbach and Lena Ulbricht, "Algorithmic Governance," *Internet Policy Review* 8, 4 (2019).
56. Stanley Aronowitz and William DiFazio, "The Jobless Future, 2nd ed. (Minneapolis: University of Minnesota Press, 1995).
57. See: Center on Budget and Policy Priorities, "Policy Basics: How Many Weeks of Unemployment Compensation Are Available?" at www.cbpp.org/research/economy/how-many-weeks-of-unemployment-compensation-are-available.
58. Antonio Cabrales, Penélope Hernández, and Angel Sánchez, "Robots, Labor Markets, and Universal Basic Income," *Humanities and Social Sciences Communications* 7, 185 (2020); and a European perspective on "robot taxes" at www.euractiv.com/section/social-europe-jobs/interview/universal-basic-income-paid-by-a-robot-tax-is-a-pretty-but-bad-idea.
59. Thomas Jefferson, "Letter to William C. Jarvis," (1820). Accessed from http://tcfir.org/opinion/Thomas%20Jefferson%20on%20Educating%20the%20People.pdf.

Index

Note: Page numbers followed by italicized letters *f* or *t* indicate figures and tables.